Two-Dimensional Gel Electrophoresis of Proteins

METHODS AND APPLICATIONS

Two-Dimensional Gel Electrophoresis of Proteins

METHODS AND APPLICATIONS

EDITED BY

JULIO E. CELIS
Aarhus University
Department of Chemistry
Division of Biostructural Chemistry
Aarhus, Denmark

RODRIGO BRAVO
European Molecular Biology Laboratory
Heidelberg, Federal Republic of Germany

1984

ACADEMIC PRESS, INC.

(Harcourt Brace Jovanovich, Publishers)

Orlando San Diego New York London
Toronto Montreal Sydney Tokyo

ACADEMIC PRESS, INC.
Orlando, Florida 32887

United Kingdom Edition published by
ACADEMIC PRESS, INC. (LONDON) LTD.
24/28 Oval Road, London NW1 7DX

Library of Congress Cataloging in Publication Data

Main entry under title:

Two-dimensional gel electrophoresis of proteins.

Includes index.
1. Proteins--Analysis. 2. Electrophoresis, Polyacrylamide gel. I. Celis, J. E. II. Bravo, R. (Rodrigo)
QP551.T86 1983 574.19'245 83-5022
ISBN 0-12-164720-X

PRINTED IN THE UNITED STATES OF AMERICA

85 86 87 88 9 8 7 6 5 4 3 2

Contents

Contributors xi
Preface xv

PART I. Methods and Satellite Techniques

1. Two-Dimensional Gel Electrophoresis: A Guide for the Beginner
RODRIGO BRAVO

I. Introduction 4
II. Materials and Solutions 5
III. Experimental Procedures 7
IV. Hints and Comments 15
V. Concluding Remarks 34
References 34

2. The QUEST System for Computer-Analyzed Two-Dimensional Electrophoresis of Proteins
JAMES I. GARRELS, JOHN T. FARRAR, AND CARTER B. BURWELL IV

I. Introduction 38
II. Problems to Be Faced in Development of Computerized Protein Data Bases 39
III. Strategy for the QUEST System 40
IV. The QUEST System As Seen by the User 45
V. Program Descriptions 75
VI. Perspectives 88
References 90

3. Detection of Polypeptides in Two-Dimensional Gels Using Silver Staining
CARL R. MERRIL AND DAVID GOLDMAN

I. Introduction 93
II. Staining Procedures 95
III. Mechanisms of Silver Stains 104
IV. Potential and Real Problems in Silver Staining 107
References 108

4. Applicability of Color Silver Stain (GELCODE® System) to Protein Mapping with Two-Dimensional Gel Electrophoresis
D. W. SAMMONS, L. D. ADAMS, T. J. VIDMAR, C. A. HATFIELD, D. H. JONES, P. J. CHUBA, AND S. W. CROOKS

I. Introduction 112
II. Variables that Affect Sensitivity and Reproducibility of the Color Silver Stain (GELCODE System) 113
III. Application of Color Silver Staining (GELCODE System) to Two-Dimensional Mapping of Proteins 117
IV. Computerized Data Acquisition of Color-Stained Images: Test for Stoichiometric Staining 122
V. Concluding Remarks 125
References 125

5. Electrophoretic Transfer of Proteins from Two-Dimensional Gels to Sheets and Their Detection
JANEY SYMINGTON

I. Introduction 128
II. Electrophoretic Transfer of Proteins 131
III. Probing the Transfer 146
IV. Removal of Probes 153
V. Examples and Applications of Protein Transfer Analysis 155
VI. Concluding Remarks 162
References 163

6. Correlation between Mouse and Human Two-Dimensional Gel Patterns: Peptide Mapping of Proteins Extracted from Two-Dimensional Gels
STEPHEN J. FEY, RODRIGO BRAVO, PETER MOSE LARSEN, AND JULIO E. CELIS

I. Introduction 170
II. Methods 170
III. Correlation between the Mouse and Human Two-Dimensional Gel Patterns 177

IV. Concluding Remarks 184
References 187

PART II. Applications of Two-Dimensional Gel Electrophoresis

7. Clinical Applications of Two-Dimensional Gel Electrophoresis
RUSSELL P. TRACY AND DONALD S. YOUNG

I. Introduction 194
II. The Electrophoresis System 196
III. Serum and Plasma 211
IV. Urine 220
V. Blood Cells 226
VI. Solid Tissue 230
VII. Other Specimens 233
VIII. Conclusions 234
References 235

8. Two-Dimensional Gel Electrophoresis for Studies of Inborn Errors of Metabolism
DAVID GOLDMAN AND CARL R. MERRIL

I. Introduction 241
II. Detection of Protein Polymorphism 242
III. Protein Mutation by Two-Dimensional Gel Electrophoresis 248
IV. Characteristic Patterns of Polypeptide Modulation 251
V. Conclusion 254
References 255

9. Adapting Two-Dimensional Gel Electrophoresis to the Study of Human Germ-Line Mutation Rates
J. V. NEEL, B. B. ROSENBLUM, C. F. SING, M. M. SKOLNICK, S. M. HANASH, AND S. STERNBERG

I. Introduction 260
II. Some Basic Facts Concerning Mutation 260
III. Why Attempt to Use 2-D Gels to Study Mutation 262
IV. Some General Considerations in the Design of a Program to Detect Mutations 263
V. Selection of Most Appropriate Material for Study 265
VI. Technical Considerations 269

VII. A Statistical Analysis of Factors Influencing Spot Location 272
VIII. Ability of the System to Detect Genetic Variants 274
IX. Nomenclature 282
X. Gel Analysis 288
XI. Steps Necessary to Identify a Putative Mutant 292
XII. Data Management 296
XIII. The Selection of Populations for Study 299
XIV. Concluding Comments 302
References 303

10. Expression of Cellular Proteins in Normal and Transformed Human Cultured Cells and Tumors: Two-Dimensional Gel Electrophoresis as a Tool to Study Neoplastic Transformation and Cancer

JULIO E. CELIS, RODRIGO BRAVO, PETER MOSE LARSEN, STEPHEN J. FEY, JAIME BELLATIN, AND ARIANA CELIS

I. Introduction 308
II. Transformation-Sensitive Polypeptides in Cultured Epithelial and Fibroblast Human Cells: Work in This Laboratory 311
III. Transformation-Sensitive Polypeptides of Known Identity 323
IV. Variation in the Expression of Mutant Actins in Chemically Transformed Human Diploid Fibroblasts: Actin in General 334
V. Phosphoproteins 337
VI. Polypeptide Synthesis in Human Tumors and Normal Tissue 343
VII. Conclusions 347
References 353

11. Application of Two-Dimensional Polyacrylamide Gel Electrophoresis to Studies of Mistranslation in Animal and Bacterial Cells

JEFFREY W. POLLARD

I. Introduction 363
II. Detection of Error-Containing Proteins by Two-Dimensional Polyacrylamide Gel Electrophoresis 364
III. Quantification of the Error Frequency of Protein Synthesis 373
IV. Assessment of the Error Frequencies of Protein Synthesis under Varying Physiological Conditions 379
V. Other Uses of the Two-Dimensional Gel Assay 388
VI. Two Out of Three Reading 390
References 392

12. Application of Two-Dimensional Gel Electrophoresis in Studies of Gene Expression during Early Plant Development
Z. R. SUNG

I. Introduction 397
II. Gene Expression during Early Embryogenesis 398
III. Somatic Embryogensis 399
IV. Gene Expression during Somatic Embryogenesis 399
V. Coordinate Regulation of Sets of Biochemical Phenotypes 403
VI. Gene Expression in Temperature-Sensitive Cell Lines Impaired in Embryogeny 405
VII. Gene Expression during Late Embryogeny 406
VIII. Gene Expression in Cotton Cotyledons 406
IX. Gene Expression in Germinating Seeds 407
X. Conclusion 411
References 412

Part III. Protein Catalogs

13. The Protein Catalog of *Escherichia coli*
FREDERICK C. NEIDHARDT AND TERESA A. PHILLIPS

I. Introduction 417
II. Identification of Proteins 418
III. Current *E. coli* Catalog 426
IV. Applications in *E. coli* Research 439
References 442

14. Catalog of HeLa Cell Proteins
RODRIGO BRAVO AND JULIO E. CELIS

I. Introduction 445
II. Two-Dimensional Gel Electrophoretic Separation of HeLa Cell Proteins 446
III. Some Characteristics of HeLa Cell Proteins 447
IV. Transformation-Sensitive Polypeptides 474
V. Conclusions 474
References 475

Index 477

Contributors

Numbers in parentheses indicate the pages on which the authors' contributions begin.

L. D. Adams (111), Diabetes and Atherosclerosis Research, The Upjohn Company, Kalamazoo, Michigan 49001

Jaime Bellatin[1] (307), Division of Biostructural Chemistry, Department of Chemistry, Aarhus University, DK-8000 Aarhus C, Denmark

Rodrigo Bravo[2] (3, 169, 307, 445), Division of Biostructural Chemistry, Department of Chemistry, Aarhus University, DK-8000 Aarhus C, Denmark

Carter B. Burwell IV (37), Cold Spring Harbor Laboratory, Cold Spring Harbor, New York 11724

Ariana Celis (307), Division of Biostructural Chemistry, Department of Chemistry, Aarhus University, DK-8000 Aarhus C, Denmark

Julio E. Celis (169, 307, 445), Division of Biostructural Chemistry, Department of Chemistry, Aarhus University, DK-8000 Aarhus C, Denmark

P. J. Chuba[3] (111), Diabetes and Atherosclerosis Research, The Upjohn Company, Kalamazoo, Michigan, and Kalamazoo College, Kalamazoo, Michigan 49001

S. W. Crooks (111), Diabetes and Atherosclerosis Research, The Upjohn Company, Kalamazoo, Michigan, and Kalamazoo College, Kalamazoo, Michigan 49001

John T. Farrar (37), Cold Spring Harbor Laboratory, Cold Spring Harbor, New York 11724

[1]Present address: MRC Laboratory of Molecular Biology, Cambridge, United Kingdom.

[2]Present Address: European Molecular Biology Laboratory, Heidelberg 6900, Federal Republic of Germany.

[3]Present address: Department of Immunology/Microbiology, Wayne State University, Detroit, Michigan 48202.

Stephen J. Fey (169, 307), Division of Biostructural Chemistry, Department of Chemistry, Aarhus University, DK-8000 Aarhus C, Denmark

James I. Garrels (37), Cold Spring Harbor Laboratory, Cold Spring Harbor, New York 11724

David Goldman (93, 241), Laboratory of General and Comparative Biochemistry and Laboratory of Clinical Science, National Institute of Mental Health, Bethesda, Maryland 20205

S. M. Hanash (259), University of Michigan Medical School, Ann Arbor, Michigan 48109

C. A. Hatfield (111), Western Michigan University, Kalamazoo, Michigan, and Hypersensitivity Disease Research, The Upjohn Company, Kalamazoo, Michigan 49001

D. H. Jones (111), Diabetes and Atherosclerosis Research, The Upjohn Company, Kalamazoo, Michigan, and Kalamazoo College, Kalamazoo, Michigan 49001

Peter Mose Larsen (169, 307), Division of Biostructural Chemistry, Department of Chemistry, Aarhus University, DK-8000 Aarhus C, Denmark

Carl R. Merril (93, 241), Laboratory of General and Comparative Biochemistry and Laboratory of Clinical Science, National Institute of Mental Health, Bethesda, Maryland 20205

J. V. Neel (259), Department of Human Genetics, University of Michigan, Ann Arbor, Michigan 48109

Frederick C. Neidhardt (417), Department of Microbiology and Immunology, University of Michigan, Ann Arbor, Michigan 48109

Teresa A. Phillips (417), Department of Microbiology and Immunology, University of Michigan, Ann Arbor, Michigan 48109

Jeffrey W. Pollard (363), MRC Human Genetic Diseases Research Group, Department of Biochemistry, Queen Elizabeth College, University of London, London W8 7AH, England

B. B. Rosenblum (259), University of Michigan Medical School, Ann Arbor, Michigan 48109

D. W. Sammons (111), Diabetes and Atherosclerosis Research, The Upjohn Company, Kalamazoo, Michigan 49001

C. F. Sing (259), Department of Human Genetics, University of Michigan, Ann Arbor, Michigan 48109

M. M. Skolnick (259), University of Michigan Medical School, Ann Arbor, Michigan 48109

S. Sternberg (259), University of Michigan Medical School, Ann Arbor, Michigan 48109

Z. R. Sung (397), Department of Plant Pathology and Department of Genetics, University of California, Berkeley, California 94720

Janey Symington (127), Institute for Molecular Virology, St. Louis University Medical Center, St. Louis, Missouri 63110

Russell P. Tracy[4] (193), Department of Laboratory Medicine, Mayo Clinic, Rochester, Minnesota 55905

T. J. Vidmar (111), Biostatistics, The Upjohn Company, Kalamazoo, Michigan 49001

Donald S. Young (193), Department of Laboratory Medicine, Mayo Clinic, Rochester, Minnesota 55905

[4]Present address: Department of Pathology, Strong Memorial Hospital, University of Rochester Medical Center, Rochester, New York 14642.

Preface

Eight years have elapsed since O'Farrell first introduced two-dimensional gel electrophoresis as a high-resolution technique for the separation of thousands of proteins from a given cell type. Since then there have been many technological advances that have led to improvements in the methods of detecting, quantifying, comparing, characterizing, and storing information contained in the gels. The technique, however, has not been fully standardized, and various laboratories use slightly different gel running conditions and sample preparation procedures. In most fields in which this technique is being applied there is no consensus as to how gels should be presented (e.g., with respect to the direction of the pH gradient). Despite these minor drawbacks it is timely to publish a volume on two-dimensional gel electrophoresis of proteins, especially because no comprehensive compilation of these techniques or data is available. This book attempts to present an overview of current procedures and to review a few areas of research in which this technique is currently being applied.

The book has been divided into three sections: General Methodology, Applications, and Protein Catalogs. The first chapter in Section I describes the methodology of two-dimensional gel electrophoresis. The following chapters describe computerized two-dimensional gel electrophoresis, silver staining, immunoblotting, and one- and two-dimensional peptide mapping. In most cases, a step-by-step guide to the techniques is given so that procedures may be easily repeated. Section II is dedicated mainly to applications of two-dimensional gel electrophoresis. Because of space limitations, it has not been possible to review all fields in which this technique is currently being applied. Rather we have chosen to reflect current research interests by paying special attention to expanding subjects. These include applications in clinical and cancer research, human genetics, protein biosynthesis, and gene expression in plants. Section III presents current protein catalogs of *Escherichia coli* and human HeLa cells. A catalog of mouse fibroblast proteins is also presented in Section I.

The book is suitable for young researchers as well as for senior scientists working with a wide variety of problems in molecular and cell biology, basic biochemistry, genetics, and clinical research.

We would like to thank the authors for their readiness to write the chapters and for their punctuality in providing manuscripts. Our gratitude is also due to the staff of Academic Press for their aid and cooperation in the completion of this work.

J. E. Celis
R. Bravo

PART I

Methods and Satellite Techniques

CHAPTER 1

Two-Dimensional Gel Electrophoresis: A Guide for the Beginner

RODRIGO BRAVO

European Molecular Biology Laboratory (EMBL)
Heidelberg 6900
Federal Republic of Germany

I. Introduction 4
II. Materials and Solutions 5
A. Materials 5
B. Stock Solutions and Buffers for the First Dimension 5
C. Stock Solutions and Buffers for the Second Dimension 6
III. Experimental Procedures 7
A. Labeling of Cells with [^{35}S]Methionine 7
B. Sample Preparation 7
C. Determination of Trichloroacetic Acid (TCA)-Precipitable Radioactivity in Samples 8
D. Procedures for the First Dimension 8
E. Procedures for the Second Dimension 11
F. Fluorography 13
G. Quantitation of Spots 15
IV. Hints and Comments 15
A. Effect of Storage or Heating on the Sample 15
B. Loading Capacity 17
C. A Simple Device for Casting the First Dimension Gels 17
D. Equilibration Time of the First Dimension 18
E. Simplifying the Assembly and Casting of the Slab Gels 18
F. Fluorography 19
G. Sensitivity of Detection 21
H. Percentage of the Initially Applied Radioactivity Recovered in the Second Dimension 22
I. Comparison of Complex Protein Mixtures 24
J. The Use of Coordinates and Markers 24
K. A Few Common Problems in Two-Dimensional Gels 28
V. Concluding Remarks 34
References 34

TWO-DIMENSIONAL GEL
ELECTROPHORESIS
OF PROTEINS

ISBN 0-12-164720-X

I. Introduction

The two-dimensional gel electrophoresis system that gives the best resolution for separating a complex mixture of proteins combines the techniques of isoelectrofocusing (IEF) in the presence of urea and a neutral detergent in the first dimension and slab gel electrophoresis under denaturing conditions using sodium dodecyl sulfate (SDS) in the second dimension (Klose, 1975; O'Farrell, 1975; Scheele, 1975; Iborra and Buhler, 1976). The separation makes use of two independent protein characteristics: one is the charge, which is reflected by the isoelectric point (pI), and the other is the molecular weight, which determines the mobility of the SDS–protein complexes in polyacrylamide gels (Weber and Osborn, 1969).

O'Farrell (1975) first demonstrated the great potential of such a technique when using isotopically labeled proteins by resolving more than a thousand polypeptides and detecting components as minor as 0.001% or less of the total cellular protein. This method has been extensively applied in resolving proteins from both prokaryotic and eukaryotic organisms and can be used for the separation of many types of cellular proteins [for references, see special issue of *Clinical Chemistry* (Vol. 28, No. 4, Part II, pp. 737–1092, 1982) and other chapters in this volume]. Unfortunately, basic proteins are not well separated in the IEF/SDS system, because in general they enter the IEF gel poorly. Even when more basic ampholytes are included, the extension of the pH gradient is very small, because in the presence of urea the basic region of the gradient is unstable. Furthermore, the few basic proteins that enter the IEF gel under these conditions always produce streaks. However, it is possible to overcome these problems using a nonequilibrium pH gradient electrophoresis (NEPHGE) in the first dimension as described by O'Farrell *et al.* (1977). The main differences between NEPHGE and IEF are that in the former the samples are applied to in the acidic side of the gel and the voltage × time product is smaller than it is in IEF. Under such conditions the pH gradient does not reach full equilibrium. As a result the proteins are not totally focused at their isoelectric point as they are in IEF gels. Nevertheless, most proteins in NEPHGE gels are separated according to their charge differences.

In general, special, not easily attainable apparatus is needed to improve the resolution of the technique, making an initial start difficult for the beginner (Garrels and Gibson, 1976; Garrels, 1979; N. G. Anderson and N. L. Anderson, 1978; N. L. Anderson and N. G. Anderson, 1978). Nevertheless, it is the author's experience that many studies can be done using the standard equipment found in any laboratory. However, when the highest resolution and fully quantitative analysis are required, then special apparatus is essential.

It is the purpose of this article to describe the IEF–SDS and NEPHGE–SDS techniques using a simple setup and to give the beginner enough basic knowledge for the critical evaluation of the results.

II. Materials and Solutions

A. Materials

Acrylamide, agarose, bisacrylamide, *N,N,N′,N′*-tetramethylethylenediamine (TEMED), and ammonium persulfate (APS) were obtained from Bio-Rad. Urea "ultrapure" was obtained from Schwartzman, and Tris-base, Tris-hydrochloride, 2,5-diphenyloxazole (PPO), β-mercaptoethanol, and glycine were obtained from Sigma. Ampholytes were obtained from LKB and Serva. Sodium dodecyl sulfate (SDS) and dithiothreitol (DTT) were from Serva. DNase I and RNase A were from Worthington. Nonionic detergent P40 (NP-40) was obtained from Fluka. Dimethyl sulfoxide (DMSO) was from Merck. [^{35}S]Methionine (SJ.204) was from Amersham. These reagents are routinely used in this laboratory with good results.

B. Stock Solutions and Buffers for the First Dimension

These solutions are similar to those described previously by O'Farrell (1975).

Lysis buffer: 9.8 *M* urea
2% (w/v) NP-40
2% Ampholytes pH 7–9
100 m*M* DTT
keep at −80°C

Overlay buffer: 8 *M* urea
1% Ampholytes pH 7–9
5% (w/v) NP-40
5% β-mercaptoethanol
keep at −80°C

Equilibration buffer: 0.06 *M* Tris-HCl pH 6.8
2% SDS

5% β-mercaptoethanol or 100 m*M* DTT
10% glycerol
keep at room temperature

Acrylamide solution: 28.38% acrylamide
1.62% bisacrylamide
filter and keep at 4°C for not more than 2 weeks

NP-40 solution: 10% (w/v) NP-40 in H_2O
keep at room temperature

H_3PO_4 solution: 1 *M*, keep at room temperature

NaOH solution: 1 *M*, keep at 4°C, not more than 2 weeks

10× nuclease solution: 1 mg/ml DNase I
500 μg/ml RNase A
0.5 *M* Tris, pH 7
50 m*M* $MgCl_2$
keep at −80°C

Agarose solution: 1% agarose
0.002% bromphenol blue
in equilibration buffer
keep at 4°C

C. Stock Solutions and Buffers for the Second Dimension

Solution A (resolution gel): 30% (w/v) acrylamide
0.15% (w/v) bisacrylamide
filter and keep at 4°C for not more than 2 weeks

Solution B pH 8.8 (resolution buffer) for 1000 ml: 200 ml 1.5 *M* Tris-HCl
800 ml 1.5 *M* Tris-base
keep at 4°C

Solution C pH 6.8 (stacking buffer) for 1000 ml: 160 ml 1.5 *M* Tris-base
titrate to pH 6.8 with 6 N HCl
keep at 4°C

Solution D (stacking gel): 10% (w/v) acrylamide
0.5% (w/v) bisacrylamide
filter and keep at 4°C for not more than 2 weeks

5× running buffer (1000 ml):	30.3 g Tris-base
	144 g glycine
	make up to 0.1% in SDS when diluting
	keep at room temperature

III. Experimental Procedures

A. Labeling of Cells with [^{35}S]Methionine

Cells can be grown in 0.25-ml flat-bottomed microtiter plates or in 35-mm dishes containing, respectively, 0.25 or 1.5 ml of Dulbecco's modified Eagle's medium (DMEM). To label cells grown in microtiter plates, the normal medium is replaced by 0.1 ml of homemade DMEM containing 1 mg/liter of cold methionine, 10% dialyzed fetal calf serum and 100 μCi of [^{35}S]methionine (Bravo and Celis, 1980a,b; Celis and Bravo, 1981; Bravo *et al.*, 1982). For labeling cells grown in 35-mm dishes, the medium is replaced by 1 ml of radioactive medium containing 500 μCi of [^{35}S]methionine.

If short term labeling is preferred (up to 3 h), DMEM without methionine is used because it increases significantly the radioactivity incorporated into proteins. This procedure is not recommended for long term labeling.

B. Sample Preparation

The two-dimensional gel system is sensitive to single charge changes that result in one protein giving several spots in the gel. Therefore, drastic conditions that cause chemical modifications of the proteins during sample preparation must be avoided.

When cells are grown and labeled in 0.25 ml microtiter wells the sample can be prepared with no special treatment. The labeling medium is aspirated carefully, and the cells are immediately lysed by adding 20–40 μl of lysis buffer. Samples are stored at −70°C as described by Bravo *et al.* (1982). Nucleic acids present in the sample in small amounts do not interfere with the first dimension.

If cells are grown and labeled in larger quantities, then treatment of the sample with DNase and RNase is necessary. The following procedure (previously described by Garrels, 1979) should be done in the cold to avoid protein degradation. Cells scraped in 20 m*M* Tris, 2 m*M* $CaCl_2$, at pH 8.8, and dispersed through a narrow gauge needle are first treated with 0.3% final concentration of SDS to solubilize the proteins. Then nuclease solution containing DNase I and RNase A described in Section II,B is added, and the sample is left in the cold until the solution is no longer viscous. Following freeze-drying the sample is resuspended in lysis buffer and stored at −70°C. This procedure is also adequate for samples to be developed by silver staining (see chapters by Merril and Goldman and by Sammons).

C. Determination of Trichloroacetic Acid (TCA)-Precipitable Radioactivity in Samples

A reliable method for the determination of the radioactivity of the samples applied to the gels is needed for quantitative studies in order to describe the percentage of radioactive label in a given protein compared to the total number of counts applied to the gel.

An appropriate dilution of the radioactive sample containing 0.1 mg/ml of bovine serum albumin (BSA) is precipitated with 5 ml of cold 10% TCA, mixed vigorously, and left on ice for 30 min. Millipore filters (HA 0.45 μm) are previously soaked in 10% TCA containing 1% Casamino acids for about 10 min. The sample is then filtered and washed three to four times with cold 10% TCA. The filter is dried and counted in a scintillation counter. All assays are done in duplicate.

D. Procedures for the First Dimension

1. PREPARATION OF TUBES

It is important that the gel is firmly attached to the walls of the tube during electrophoresis; therefore the tubes must be carefully cleaned before being used. The glass tubes should be thoroughly washed in Decon, rinsed with water, then boiled in 0.1 *M* HCl for 30 min, rinsed with double-distilled water, and allowed to dry. Gel tubes should have an inside diameter of 1.2 to 3 mm and an appropriate length that may vary from 12 to 20 cm. The tubes should all be of the same length and so should the gels, because these factors affect the reproducibility of the runs.

The bottom of the tubes is sealed with three layers of Parafilm to avoid leakage and to prevent the entry of air through the slightly porous Parafilm, which could affect polymerization. The tubes are marked to indicate the desired length of the gel (normally 1–2 cm from the top) and are placed vertically in a rack.

2. PREPARATION AND RUNNING OF THE FIRST DIMENSION

The procedure is essentially that described by O'Farrell (1975) and O'Farrell *et al.* (1977) with minor modifications (Bravo *et al.*, 1982).

To prepare 10 ml of gel mix for the first dimension add the following to a 125-ml side arm flask:

	IEF	NEPHGE
Urea (g)	5.5	5.5
Acrylamide solution (ml)	1.3	1.3
10% NP-40 (ml)	2.0	2.0
Distilled H_2O (ml)	1.7	2.0
Ampholyte pH 5–7 (ml)	0.6	—
Ampholyte pH 3–10 (ml)	0.12	—
Ampholyte pH 7–9 (ml)	—	0.25
Ampholyte pH 8–9.5 (ml)	—	0.25

The exact ampholyte composition can be varied according to the result desired.

Once all the reagents have been added, the flask is swirled until the urea is completely dissolved. This can be speeded up by placing the flask in a water bath at 37°C. The solution is degassed carefully by connecting the flask to a water pump. Then 7 μl of TEMED and 10 μl of 10% ammonium persulfate are added to the IEF solution. Twice as much TEMED and ammonium persulfate is necessary for the NEPHGE gels because the presence of basic ampholytes affects polymerization. The solution is then immediately loaded into the tubes using a long, narrow gauge blunt end needle. To avoid trapping air bubbles, the tip of the needle is inserted to the bottom of the tube and withdrawn slowly as the acrylamide mixture runs in. The tubes are filled up to a specified mark so that all gels are of the same length. The gel mixture is overlayed with 10 μl of H_2O and allowed to polymerize for at least 2 h. The Parafilm is then carefully removed by cutting rather than pulling so as to avoid damaging the bottom of the gels. They are loaded into a standard gel electrophoresis chamber, the lower reservoir being filled with 10 m*M* H_3PO_4 (IEF) or 20 m*M* NaOH (NEPHGE). Air bubbles are removed from the bottom of the gel by gently tapping the tube. The solution above the gels is removed by aspiration. An automatic device for casting several first dimension gels simultaneously has been described by N. G. Anderson and N. L. Anderson (1978) and Garrels (1982) (see also Section IV).

a. Isoelectrofocusing (IEF) Gels

The top of the gel is overlayed with 10 μl of lysis buffer, 10 μl of overlay buffer and enough 20 m*M* NaOH to fill the tubes. The upper chamber is then filled carefully with 20 m*M* NaOH, previously degassed, so as not to disturb the layers of solutions over the gel. The positive electrode is connected to the lower reservoir and the negative electrode to the upper reservoir. Gels (13-cm long) are prerun as follows:

200 V × 15 min
300 V × 30 min
400 V × 60 min or more

During the prerun, at each voltage the current reading should constantly decrease. If it starts increasing, the electrode connections and the solutions in the reservoirs should be checked.

After the prerun the NaOH from the upper chamber is removed and discarded. The liquid above the gels is aspirated, and the top of the gels is washed twice with 50 μl of distilled H_2O. At this point samples can be loaded.

The samples are conveniently loaded with a Gilson Pipetman or equivalent dispenser. The use of disposable tips prevents contamination between samples, a problem that is especially important in the use of highly radioactive material. Alternatively the sample can be introduced through the different layers that cover the gel with the aid of a Hamilton syringe. After loading, 10 μl of overlay buffer are carefully added, the tubes are filled with 20 m*M* NaOH, the upper chamber is replenished, and the gels are run for 18 h at 400 V.

For total cell extracts, the optimal sample volume should be less than 20 μl containing a maximum of 10 μg of protein and 0.5–1 × 10^6 cpm. However, if necessary, as much as 100 μg of protein can be applied but loss of resolution will result.

To remove the gels from the tubes, water can be injected around the gels or air pressure can be applied. Care should be taken not to break the gels. Gels are collected in screwcap tubes or Petri dishes containing 10 ml of equilibration buffer. They are allowed to equilibrate for 10–20 min and are then frozen on dry ice or in a −80°C freezer.

b. Nonequilibrium (NEPHGE) Gels

These gels are not prerun. The samples are applied and covered with 10 μl of overlay buffer, and the tubes are completed with 10 m*M* H_3PO_4. The upper chamber is filled with 10 m*M* H_3PO_4 and has the positive electrode connected to it. The lower reservoir is filled with 20 m*M* NaOH and is connected to the negative electrode. The gels are run at 400 V for 4.5 h. The current should decrease constantly during the running. The gels are removed and equilibrated as described for IEF gels. For a detailed description of these gels see O'Farrell *et al.* (1977).

E. Procedures for the Second Dimension

1. PREPARATION OF SLAB GELS

The slab gel techniques already in use are readily adapted to the two-dimensional system. Essentially the notched plate system described by Studier (1973) is used.

The gel is formed between two uniform glass plates, each about 3 mm thick. One plate is a rectangular glass of 17 × 18 cm and the other plate is the same size with a notch approximately 2-cm deep and 14-cm wide (the construction of a special leveled glass plate has been described by O'Farrell, 1975). Spacers are made of perspex or polyvinylchloride (PVC) 1–1.5-mm thick, 1-cm wide and 20-cm long. The spacers must be of uniform thickness to ensure an even gel throughout the slab. The glass plates should be cleaned carefully with detergent followed by several rinses in water and a final wash with ethanol to eliminate any remaining grease that could affect polymerization. The two glass plates and three spacers are mounted together, one spacer along the bottom and the other two running down the vertical sides of the plates, taking care that the vertical spacers are in close contact with the one at the bottom. The assembled plates are held together with strong clips which are positioned to press over the spacers. Several ways of sealing the assembled plates can be used. One of them is to slightly grease the spacers with vaseline before the assembly, taking care not to grease the plates. A second commonly used method is to pipet (with a Pasteur pipette) melted 2% agarose solution around the outside of the assembled glass plates and against the spacers to produce a seal.

The discontinuous SDS gel system described by Laemmli (1970) is used in the second dimension. Uniform gels of any acrylamide concentration as well as gradient gels can be used. The exact composition of the separation gel depends on the sample to be analyzed.

The system which the author normally uses is as follows. For other systems currently in use see the special issue of *Clinical Chemistry* (1982). To make 100 ml of resolution gel consult the following tabulation:

Strength (%)	Sol A (ml)	10% SDS (ml)	Sol B (ml)	H_2O (ml)	TEMED (μl)	10% APS (ml)
7.5	25	1	25	48.5	20	0.5
9	30	1	25	43.5	20	0.5
12	40	1	25	33.5	20	0.5
15	50	1	25	23.5	20	0.5
18	60	1	25	13.5	20	0.5

The solution must be degassed before adding TEMED and freshly made APS. Once these components are added the solution is mixed carefully and poured into the assembled vertical plates up to 2.5 cm from the top of the notched plate. The resolution gel is overlaid carefully with H_2O or with water-saturated isobutanol. Gels are left to polymerize for at least 1 h. The overlay is then removed and the gel is rinsed and dried with blotting paper before pouring the stacking gel solution. For analysis of total cellular extracts the best separating gel is 15%. For 20 ml of stacking gel use the following ingredients:

Strength (%)	Sol D (ml)	10% SDS (ml)	Sol C (ml)	H_2O (ml)	TEMED (μl)	10% APS (ml)
3	6	0.2	10	4	8	0.16
5	10	0.2	10	—	8	0.16

Normally the concentration of the stacking gel is 5%. This solution must also be degassed before adding TEMED and APS. Then it should be poured as far as the top of the notched plate and overlaid as described for the resolution gel.

2. APPLICATION OF THE FIRST DIMENSION ON TOP OF THE SECOND DIMENSION AND RUNNING OF THE SLAB GEL

Before the first dimension gel is applied on top of the second dimension, the stacking gel should be rinsed and dried with blotting paper. All the clamps are removed together with the bottom spacer and the plates are laid flat to facilitate the application of the first dimension gel. The latter is thawed, removed from the equilibrating solution, and carefully laid on top of the plates avoiding any stretching. The melted 1% agarose solution (2–3 ml) is used to keep the first dimension gel in place. It can be layered on with a Pasteur pipette from one end of the plate between the stacking and the first dimension gel. Any trapped air bubbles should be eliminated to avoid artifacts in the SDS gel. The gel is completely covered with the agarose solution and left to set for 3 to 5 min before the slab gel is clamped to the electrophoresis chamber. Running buffer is added to the upper and lower tanks. Air bubbles from the bottom of the gel must be removed to allow good contact with the buffer. This can be done by using a syringe with a bent needle. Electrodes are connected and the gels are run overnight at constant current (9–11 mA). At the end of the run the plates are removed from the apparatus and the gel is processed as desired.

A typical example of the results obtained when using IEF–SDS and NEPHGE–SDS systems for resolving [^{35}S]methionine-labeled polypeptides of a total cellular lysate is shown in Fig. 1.

Other methods for the assembly, casting, and running of many gels simultaneously have been described by N. L. Anderson and N. G. Anderson (1978), Garrels (1979), and by Jones *et al.* (1980).

F. Fluorography

Gels can be treated for fluorography immediately after electrophoresis without prior fixation (Bonner and Laskey, 1974). The gel is soaked in about 10 times its volume of DMSO for 20 min with constant gentle shaking, followed by a second 20 min immersion in fresh DMSO. Gels are transferred to five volumes of a solution of 20% (w/v) PPO in DMSO and shaken for 2 h, after which they are immersed in H_2O for 15 min followed by a second 15 min washing. Gels are dried under heat and vacuum for 45 to 60 min. Excess drying should be avoided. Dried gels are exposed to an appropriate X-ray film such as Kodak XAR or XS at −40°C or lower. The films can be presensitized as described by Laskey

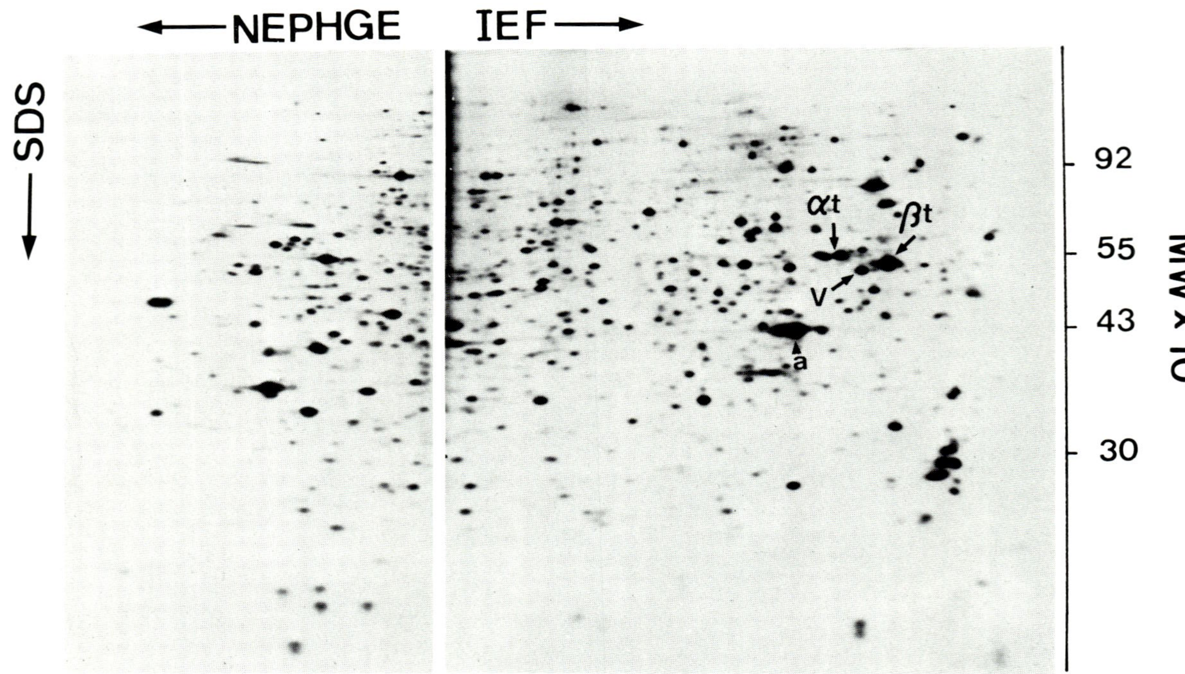

Figure 1. Two-dimensional gel maps of acidic (IEF) and basic (NEPHGE) polypeptides of asynchronous HeLa cells. Both fluorograms have been slightly overexposed in order to detect minor components. In IEF the pH ranges from 7.5 (left) to 4.5 (right). In NEPHGE the pH ranges from 7.5 (right) to 9.5 (left). A total of 1169 polypeptides (855 acidic and 314 basic) have been detected. a, actin; αt, α-tubulin; βt, β-tubulin; v, vimentin.

and Mills (1975). Inadequate removal of DMSO produces artifactual blackening of the film.

G. Quantitation of Spots

When a limited number of spots needs to be quantitated, a very reliable method is direct counting of the gel as described by Bravo *et al.* (1982). The selected spots are cut out of the dried gel, put directly into scintillation fluid, and counted for 5 to 10 min. If the gel has been previously treated for fluorography, the counting efficiency for ^{35}S labeled proteins is as high as 50%.

To check that the spots have been cut out properly, the gel is reexposed and the film is compared with the original. The main limitation of this procedure is that it allows quantitation only of those proteins that contain more than 0.01% of the total radioactive label in a sample.

When a detailed quantitation of the whole gel is needed, the use of a scanner together with the aid of a computer is required. Several computer programs have been described that enable the accurate quantitation and analysis of the proteins resolved by two-dimensional electrophoresis (Bossinger *et al.*, 1979; Garrels, 1979; Vo *et al.*, 1981; Lemkin and Lipkin, 1981; Anderson *et al.*, 1981; see also the special issue of *Clinical Chemistry*, 1982, and chapter by Garrels *et al.*).

IV. Hints and Comments

A. Effect of Storage or Heating on the Sample

Samples can be kept at −70°C for several months in lysis buffer without any detectable effect on protein separation, even when they have been frozen and thawed a few times (see Fig. 2b). Heating of the samples to 80 to 100°C for 1 min in lysis buffer is enough to produce an enormous effect on the dimension as shown in Fig. 2c. This is mainly due to carbamylation of the proteins by isocyanate formed by decomposition of urea, with a subsequent change in the protein's pI. It is clear from this figure that a series of artifactual spots with different pI values are produced from one original polypeptide. Under no circumstances should the sample be heated over 37°C when in lysis buffer.

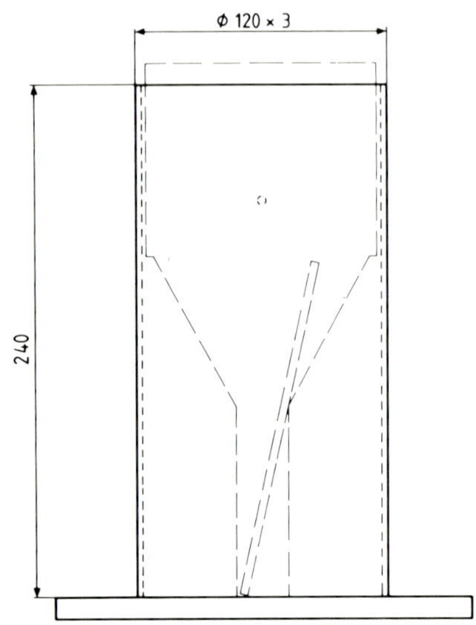

Figure 3. Simple device for casting several first-dimension gels. The inner cylinder, which has an attached funnel, is tightly fitted to a plastic vial containing the gel solution. The glass tubes are carefully introduced into the vial and the whole system is placed in the outer cylinder, which has been previously filled with water up to the desired height. At first the water enters slowly through the small hole of the funnel, overlaying the solution without mixing. Later the water enters through the bigger hole of the cylinder.

D. Equilibration Time of the First Dimension

During this step the protein can diffuse and a certain amount is lost: how much depends on the length of time of incubation. For example, if an equilibrium time of 2 h is used, as much as 50% of the total counts in the gel can be lost. Because this step is needed to avoid streaking in the SDS gel, it cannot be eliminated without altering the running conditions of the second dimension. Therefore, it is recommended to equilibrate the gel for a period not longer than 30 min.

E. Simplifying the Assembly and Casting of the Slab Gels

A simple device for casting four gels at a time is shown in Fig. 4. The lateral spacers have been glued to the notched plate and cover the whole length of the glass. The bottom spacer has been eliminated. Vaseline or

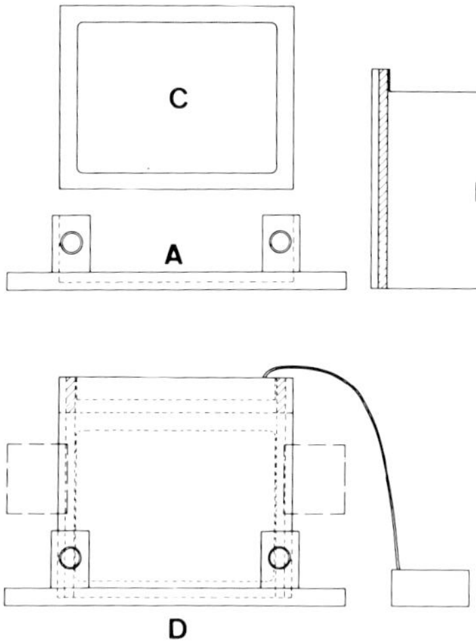

Figure 4. Device for casting four slab gels. In the top are shown the grooved acrylic base with lateral supports (A), notched glass plate with glued spacers (B), and a PVC frame (C). The assembled device is shown in (D). PVC frames are placed between each set of glass plates, clamped together and tightened with the screw on the lateral supports.

agarose is used for sealing the sides of the plates. Four plates are clamped together, separated by a frame of PVC (1.5-mm-thick), and placed in the device as shown in Fig. 4. The bottom of each plate is sealed with a small volume of acrylamide solution of the same strength as the separating gel. After the plug has polymerized, the separating gel is poured simultaneously into the four plates with a peristaltic pump.

F. Fluorography

The use of fluorography is recommended when working with a sample of low radioactivity or one available in very limited quantities. The enormous increase in the sensitivity of detection when using this technique combined with exposures at low temperatures was well demonstrated by Laskey and Mills (1975), see also Fig. 5.

In the original procedure a solution of 20% (w/v) PPO–DMSO was used to obtain the maximum sensitivity for ^{3}H-labeled proteins. Garrels (1979) has shown that for ^{35}S-labeled proteins a 10% (w/v) PPO–DMSO solution can also be used and produces full sensitivity. In order to avoid water contamination of the PPO–DMSO solution, replacement of the DMSO used in the washes is recommended after every second use. The

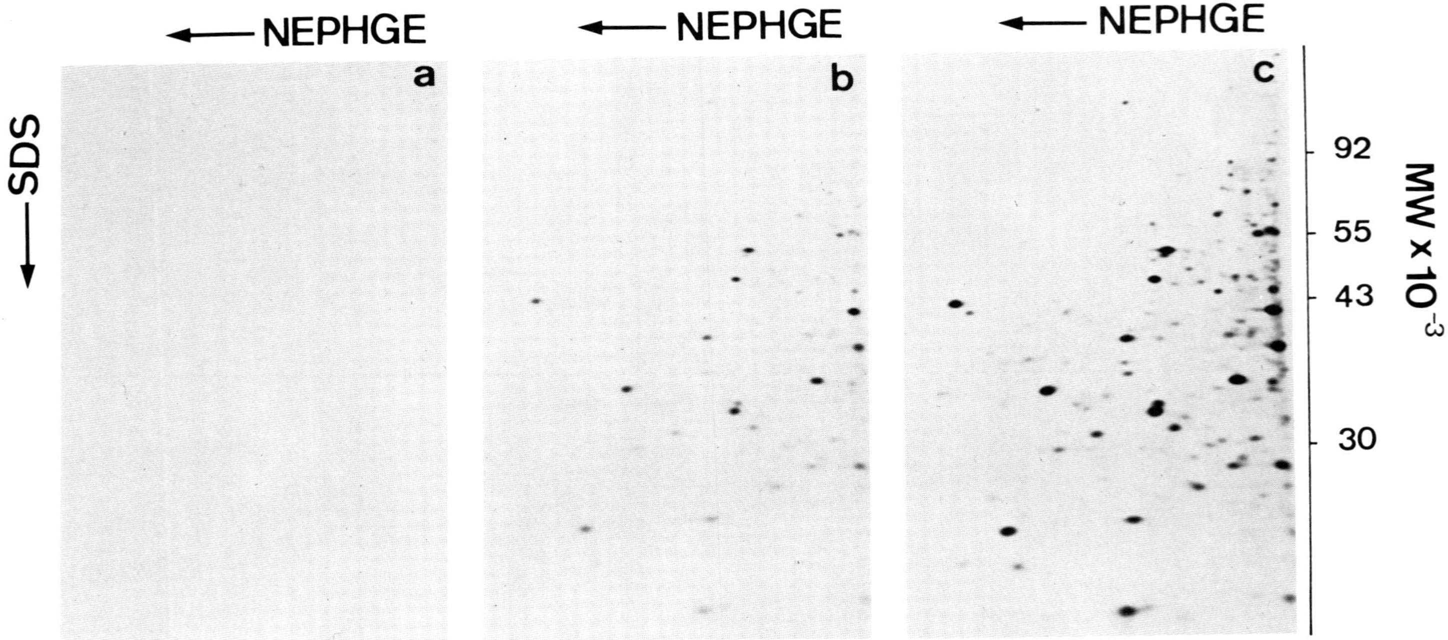

Figure 5. Effect of fluorography and low temperature on the sensitivity of detection: (a) gel without fluorographic treatment exposed at room temperature; (b) gel treated for fluorography exposed at room temperature; (c) gel exposed at −80°C after fluorography. In all cases identical numbers of counts were applied and the gels were exposed for the same period of time.

PPO–DMSO solution thus does not need to be replaced every time it is used. Care must be taken to keep the concentration of PPO constant by adding solid to the solution. If the PPO starts to precipitate due to the presence of water, then the solution must be replaced.

PPO can be recovered by precipitation as follows: 3 volumes of 10% ethanol are added and the mixture allowed to stand for 15–20 min. The precipitate is then collected by filtration and washed with 15 to 20 volumes of water before drying at room temperature.

G. Sensitivity of Detection

The two-dimensional gel electrophoresis system is sensitive enough to detect minor protein species that contain between 0.0001 and 0.00001% of the total radioactivity applied (O'Farrell, 1975; Garrels, 1979; Duncan and McConkey, 1982). Because the average rare mRNAs constitute 0.001–0.005% of the total RNA in eukaryotic cells, it should be possible to detect the proteins encoded by these mRNAs if they are translated in proportion to their abundance. This means that, in principle, the ability of the technique to reveal trace components is not limited by the sensitivity of the detection method.

The main problem in detecting minor proteins is the great difference in abundance of the various polypeptides that constitute an eukaryotic cell. Major proteins may account for as much as 3 –5% of the total counts while less abundant proteins may account for 0.0001% or less. Thus if minor proteins are detected using fluorography, the major components will be overexposed by a factor of 10^4–10^5 times, producing a spot with an area at least 15–20 times larger than its minimal size (fluorography spreading). This results in a considerable loss in the resolving power of the technique.

In complex biochemical phenomena such as growth regulation, cellular transformation, or hormone response, the changes that occur are usually quantitative rather than qualitative (see chapter by Celis *et al.*). Therefore, these studies will depend heavily on the resolution and sensitivity of the system. A reliable quantitative method such as scanning of the fluorographs and computer processing of the data is also required. In these situations, when maximum sensitivity and resolution are needed, larger gels in both dimensions and running under conditions to obtain spots of a minimal size are strongly recommended (see chapter by Garrels *et al.*). An example of such type of separation is shown in Fig. 6. One of the greatest advantages of this system is the possibility of expos-

Figure 6. Increased resolution in two-dimensional gels. Typical separation of acidic proteins (IEF) when larger gels in both dimensions are used. (IEF, 1.2 × 200 mm; SDS gel, 200 × 200 mm). The IEF and SDS gels have the same composition as described in Section III.

ing the fluorograms for longer times without diminishing seriously the resolution of the gels due to fluorographic spreading.

H. Percentage of the Initially Applied Radioactivity Recovered in the Second Dimension

During each of the different steps of the two-dimensional technique, a certain amount of the initial protein applied is lost. A schematic representation of the recovery after each step is shown in Fig. 7. As can be seen from the figure, when this system is used (as described in Section III) for the analysis of total cellular lysates neither IEF–SDS nor NEPHGE–SDS alone is sufficient, so they must be used together. According to the values presented in the figure, if proteins are analyzed in IEF gels, approximately 30–35% of the total radioactivity applied does not enter the gel. For nonequilibrium gels, 70–75% of the total labeled

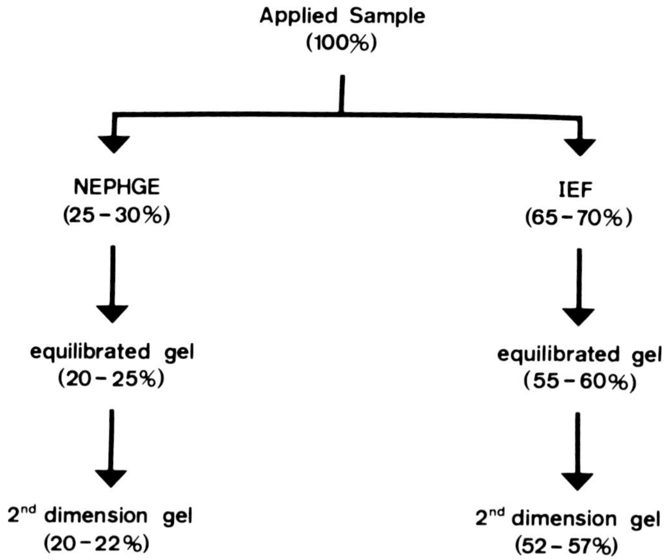

Figure 7. Percentage of recovery of the applied sample when using NEPHGE–SDS or IEF–SDS systems. The values correspond to the percentages of radioactivity present in relation to the total applied. To determine the radioactivity that entered the first dimension, the gels were cut in 0.5-cm slices, each of which was incubated in 30% H_2O_2–2% NH_4OH at 37°C overnight. Samples were counted in 10 ml of Biofluor. The amount of radioactivity lost during the equilibration step was calculated by counting the incubation media. The percentage of radioactivity present in the slab gel was determined by direct counting of the gel. After fluorography and drying, the gel was cut into 2-cm^2 pieces, each of which was put into scintillation fluid and counted for 10 min. In all cases the values are the average of at least four independent determinations. The counts were corrected according to the counting efficiency calculated for each determination. The sample analyzed consisted of asynchronous HeLa cells labeled for 20 h with [^{35}S]methionine. Labeling of the cells, sample preparation, and running of the gels were performed as described in Section III.

protein is left out of the gel. From that point on, in both cases the losses in each of the following steps are proportionally very similar. In the IEF–SDS system the amount of radioactivity present in the slab gel consists of 52–57% of the total label applied. For the NEPHGE–SDS system this consists of 20–22% of the total. Because only 5–7% of the polypeptide components of a labeled cell are common to both systems, if both systems are used for the separation of radioactive polypeptides of a total cellular lysate, more than 95% of the components are analyzed, assuming that the losses of protein in the steps following the first dimension are not selective.

I. Comparison of Complex Protein Mixtures

The comparison of two-dimensional protein patterns from different sources usually involves a search for changes in a few polypeptides among 1000 or more. In many cases the differences can be clearly detected by visual inspection when several gel patterns are compared side by side. However, there are cases in which the differences between the protein maps are unclear due to small variations in the running conditions that make comparisons very difficult. To settle these uncertainties it is useful to run a mixture of the two samples on the same gel and compare the results with the separate runs of the individual samples in order to identify the origin of the spots. As shown in Fig. 8, using this approach it is possible to recognize the origins of many proteins and therefore to use them as a "reference constellation" when looking for other differences. Nevertheless, there are circumstances in which this is not sufficient, and a more convincing method, such as double-label autoradiography, is needed. In this case the protein samples to be compared are labeled with different isotopes and coelectrophoresed as described by McConkey (1979) and Choo *et al.* (1980). When using this technique, careful attention must be given to the amount of radioactive label of each type applied to the gel. They should be adjusted so that the intensities of the spots with both isotopes are approximately equal under identical periods of exposure. This will ensure that weaker spots are detectable and that major proteins will not be too overexposed.

Another convenient technique that can be used with two-dimensional gels is the partial peptide mapping of proteins. It is very useful when searching for homologous proteins from diverse sources or when looking for differences between polypeptides of unknown identity (see Fey *et al.*).

J. The Use of Coordinates and Markers

One of the major problems of communication between different laboratories using the two-dimensional system is the lack of a precise means of referring to the exact location of a given polypeptide resolved by this technique. Furthermore, each laboratory uses its own two-dimensional system.

Considering that the two-dimensional system resolves proteins based on their charge (first-dimension) and molecular weight (second-dimension) properties, the most direct way to localize a polypeptide in the gel

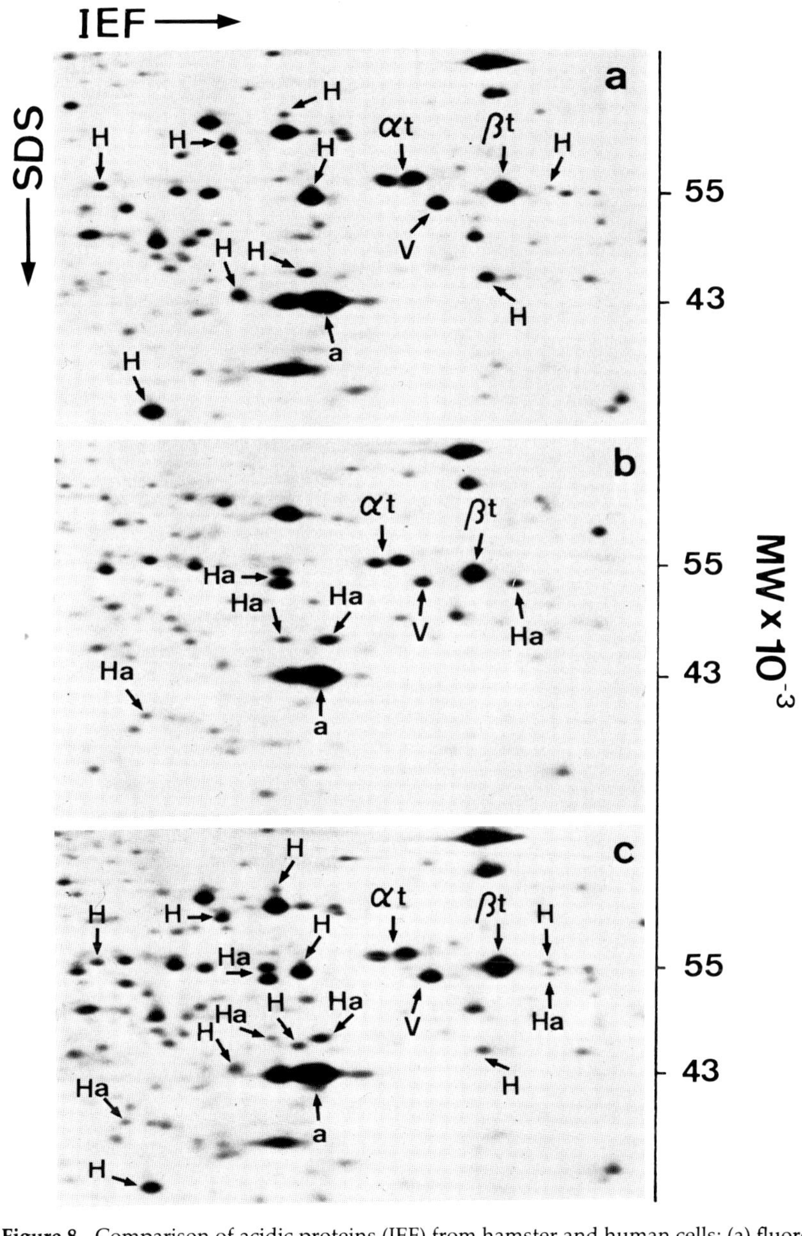

Figure 8. Comparison of acidic proteins (IEF) from hamster and human cells: (a) fluorograph of HeLa cell polypeptides; (b) fluorograph of Chinese hamster ovary (CHO) cell polypeptides; (c) coelectrophoresis of HeLa and CHO polypeptides. H, human-specific polypeptides; Ha, hamster-specific polypeptides; a, actin; αt, α-tubulin; βt, β-tubulin; v, vimentin. Only a fraction of the fluorograms is shown.

pattern would be the use of pI and MW as coordinates. There are good reasons not to use the pH as the *x* axis. The presence of a high concentration of urea (9 *M*), which causes the reduction of hydrogen ion activity (Bull *et al.*, 1964), makes the pH measurement very inaccurate. Urea also has a direct effect on the pH electrode, as described by Ui (1971). Furthermore, isoelectric points determined in urea are altered due to the denaturation of the protein. Added to this, there is great discrepancy among the pI values of some major cellular polypeptides reported in different publications (Peterson and McConkey, 1976; Strand and August, 1978; Krystal *et al.*, 1978; Milcarek and Zahn, 1978; Bravo and Knowland, 1979; Garrels, 1979). The problem is, then, what to use as reference points on the *x* axis. For samples from whole cells one of the most reliable and simple ways is to use certain identified polypeptides of the sample as internal markers (e.g., actin, tubulin, tropomyosin, vimentin, and lactic dehydrogenase) and to refer to the *x* axis position of a protein in relation to one of these selected standards, as described by Bravo *et al.* (1981a), Fey *et al.* (1981), Celis and Bravo (1981) and Bravo and Celis (1982). For other types of samples the reference problem can be solved by adding internal standards such as a mixture of carefully chosen and readily available purified proteins of known pI values, as described by Comings and Peters (1981) or carbamylated proteins, as shown by Anderson and Hickman (1979).

The use of MW as the *y* axis is recommended, but the apparent MW of the proteins determined from the gel must be considered with caution because many polypeptides present anomalous behavior when the concentration of bisacrylamide in the gel is changed. Figure 9 shows examples of proteins whose apparent molecular weight changes by several thousands when the concentration of bisacrylamide is increased. For example, polypeptides 1 and 2 have a very similar molecular weight to α-tubulin (55 K) when a low bisacrylamide concentration is used (Fig. 9a); however, when the concentration is increased (Fig. 9b), the difference in the apparent MW between the two polypeptides is nearly 5K. Other polypeptides indicated in the figure also show drastic differences in their apparent molecular weights when run under these circumstances. If the percentage of acrylamide is varied while the acrylamide/bisacrylamide ratio is kept constant, the apparent molecular weight of the polypeptides shown in the previous example remains unchanged.

Considering the problems mentioned above, when comparing results from different laboratories it is recommended to exchange the samples in question in order to analyse them in the system routinely used in each laboratory. Whenever possible immunological (immunoblotting or immunoprecipitation) or peptide mapping techniques should be employed

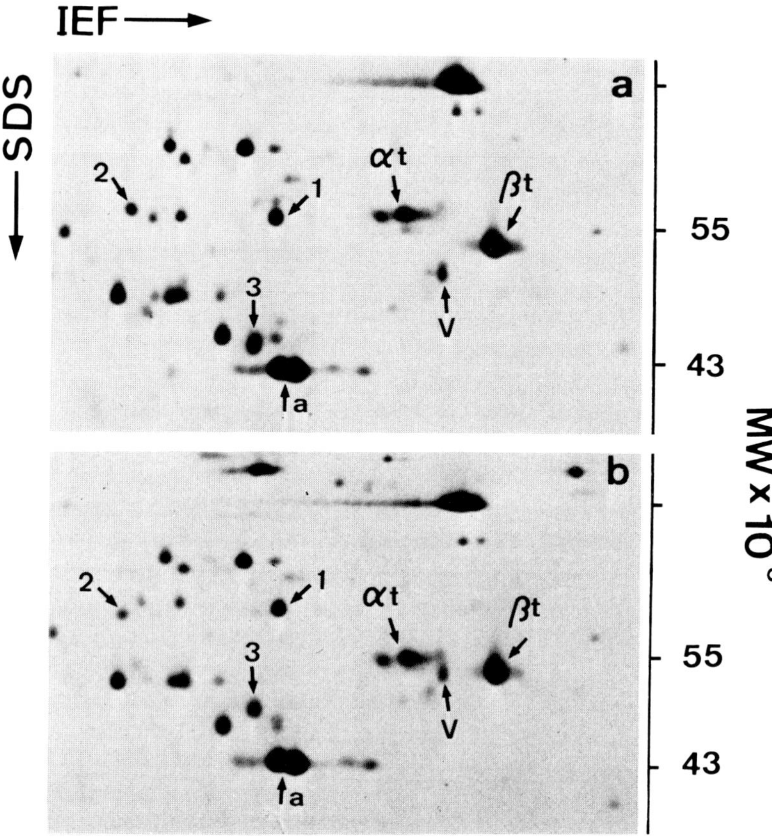

Figure 9. Effect of cross-linking on the apparent molecular weight of some proteins. (a) The ratio of acrylamide to bisacrylamide in the second dimension was 1 : 200. (b) The ratio of acrylamide to bisacrylamide used in the second dimension was 1 : 30. The size of the slab gels was identical. Only a small fraction of the gel is shown. a, actin; αt, α-tubulin; βt, β-tubulin; v, vimentin.

for further confirmation of the results (see chapters by Symington and by Fey *et al.*).

A two-dimensional gel protein map, in which all cellular polypeptides are numbered, can be used as a catalogue to which the precise location of a given polypeptide can be referred. This has proved to be of great value in the study of the response of a given cell type at the level of its polypeptide synthesis under several physiological conditions, or when analyzing the components of the different cellular structures or organelles, as described by Anderson (1981), Bravo and Celis (1980a), Bravo *et al.* (1981b,c, 1982), Comings (1982), Celis and Bravo (1981), Giometti and

Anderson (1981), and Willard and Anderson (1981) (see also chapter by Neidhart and Phillips).

K. *A Few Common Problems in Two-Dimensional Gels*

A difficult problem for the beginner is to decide when a technique is working properly, and if not, to determine how to make it work. Some examples of typical artifacts that, in the author's experience, appear frequently have been selected and are discussed below.

1. FIRST-DIMENSION GELS

Figures 10 and 11 show different degrees of resolution obtained when [^{35}S]methionine labeled polypeptides of a total cellular lysate are analyzed under the conditions described in Section III. Clearly, the resolution can vary enormously. In Fig. 11b the great majority of the proteins are focused and the number of polypeptides detected is high. This quality of separation can be reproducibly obtained. Interestingly, the gel patterns shown in Figs. 10a,b are also very reproducible. Therefore, reproducibility does not necessarily mean quality. The problem is how to obtain the highest resolution or to determine the causes of unsatisfactory gel patterns. If the resolution is poor, as in Fig. 11a, the first thing to do is to replace old solutions, especially the NaOH and acrylamide stocks. In many cases this is enough to improve the gel resolution. For more severe problems, such as those presented in Figure 10, there are several alternatives. First, it is recommended to increase the concentration of ampholytes up to 4% and then to vary the ratio of the ampholyte mixture (pH 3.5–10 and pH 5–7). It is also highly recommended to test and mix different batches and brands of ampholytes. This last procedure often gives very good results.

There are some proteins that are difficult to focus properly, either because they are heavily phosphorylated or because of some other intrinsic property. A typical example of this anomalous behaviour is shown by α-tubulin, (Fig. 12). The relative position of this protein and of the number of isospecies resolved varies with different batches of ampholytes, even though the pH range in the isoelectrofocusing gels is kept constant. Interestingly, proteins migrating close to α-tubulin, including β-tubulin, do not exhibit this behaviour. This specific example is given to demonstrate that the appearance of only one α-tubulin in a two-

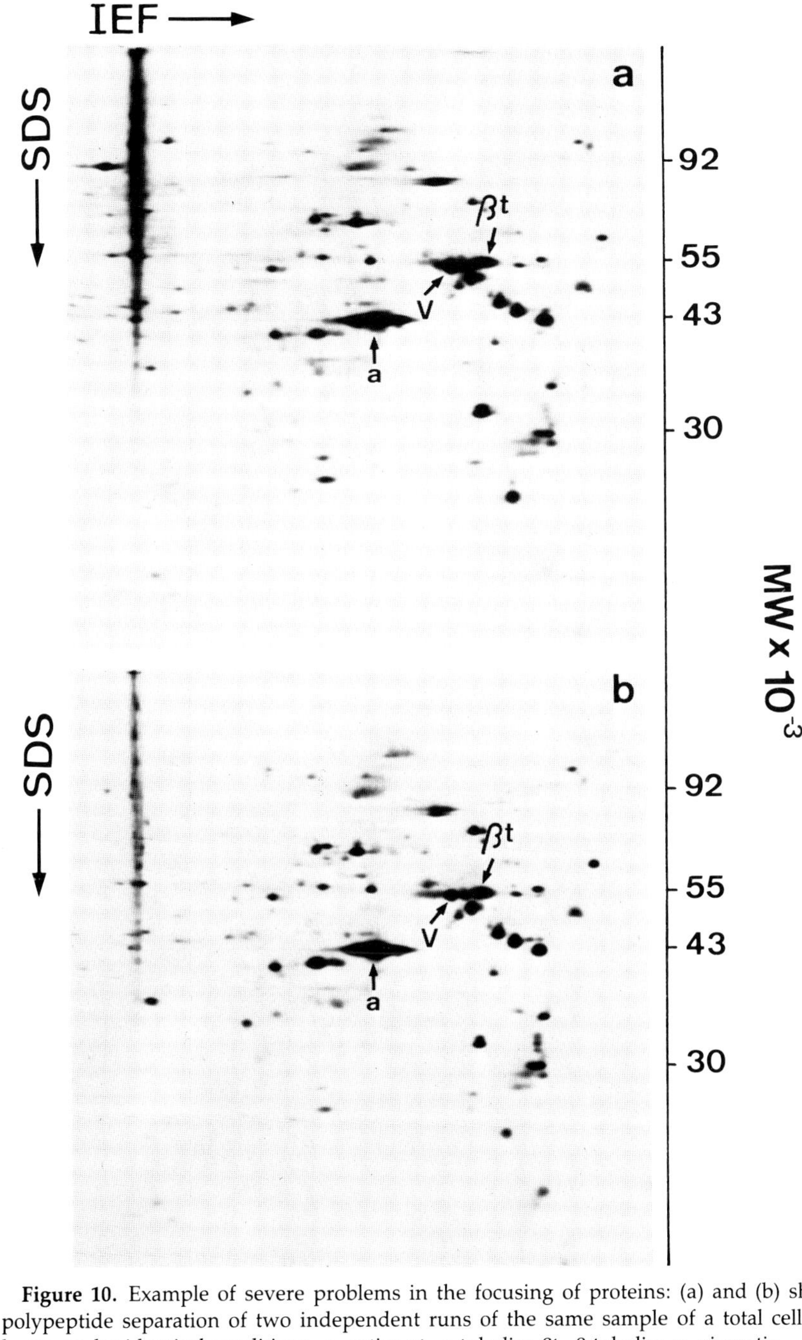

Figure 10. Example of severe problems in the focusing of proteins: (a) and (b) show polypeptide separation of two independent runs of the same sample of a total cellular lysate under identical conditions. a, actin; αt, α-tubulin; βt, β-tubulin; v, vimentin.

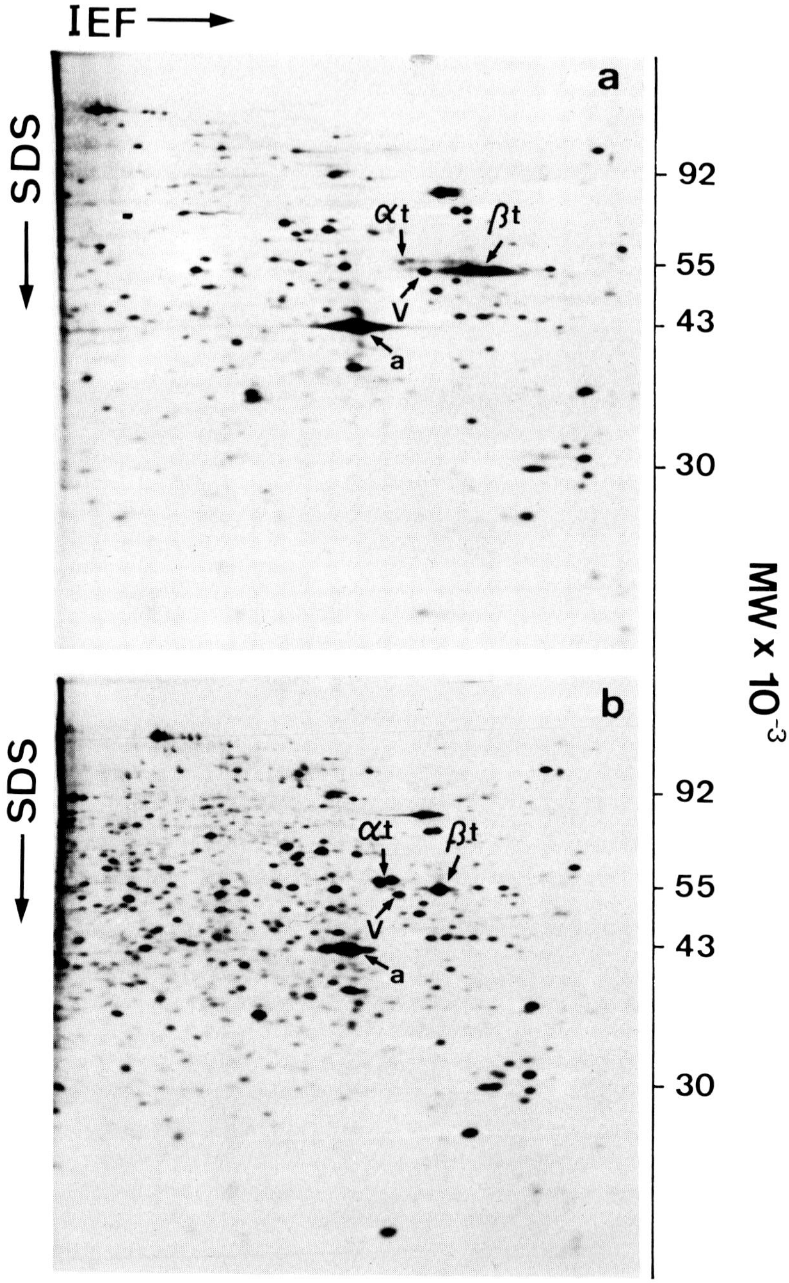

Figure 11. Typical examples of poorly and well-resolved acidic proteins (IEF) from a total cellular lysate: (a) fluorogram of poorly resolved proteins; (b) gel pattern of well-resolved proteins. In this case the resolution of the separation was improved by mixing ampholytes of different brands (see text). The same sample was analyzed in both cases.

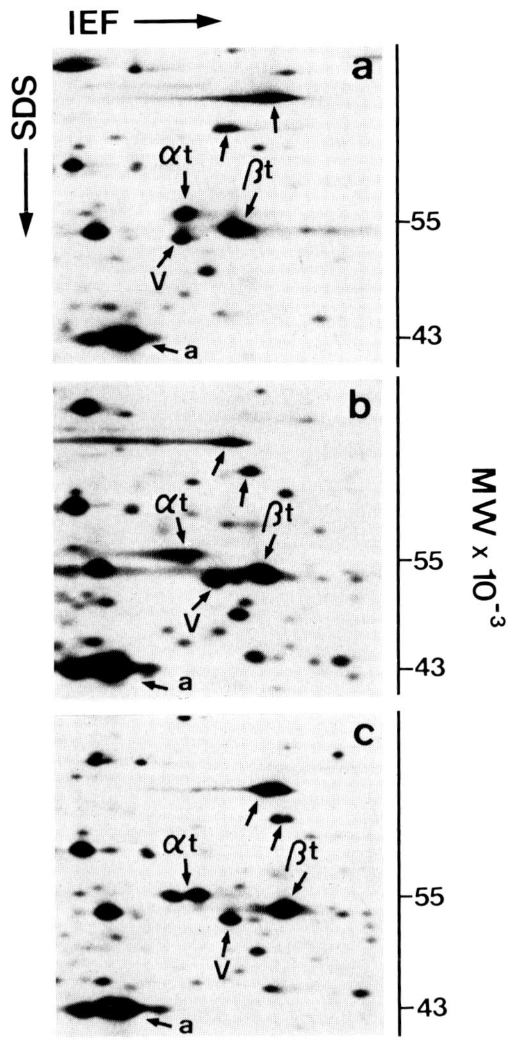

Figure 12. Erratic behavior of some proteins in the isoelectric focusing gel. In all three cases (a, b, and c) the samples were run under identical conditions, except that the ampholyte batches used were different. The behavior of the polypeptides indicated by arrows is very reproducible for each of the examples shown. Only a small area of the second-dimension gel is shown. a, actin; αt, α-tubulin; βt, β-tubulin; v, vimentin.

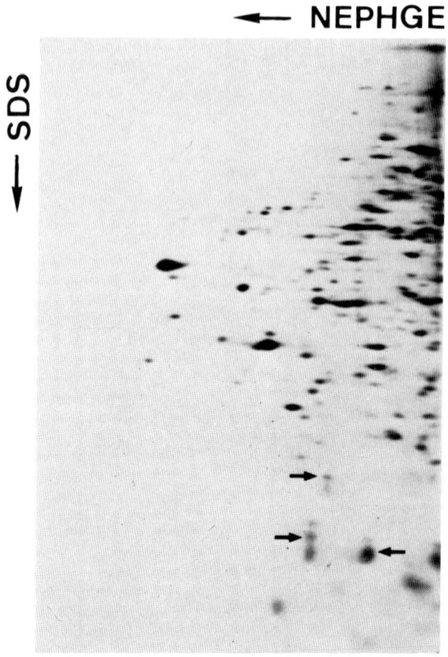

Figure 13. Smearing of low MW proteins in the second dimension. Polypeptides that show smear in the SDS gel are indicated with an arrow.

dimensional gel does not necessarily mean that the specific cell or tissue under study expresses only one of the α-tubulin isospecies.

2. SECOND-DIMENSION GELS

Figure 13 illustrates an artifact involving low molecular weight proteins in an SDS gel. As shown in this figure, in the lower part of the gel the spots present a smear, which in many cases produces a double spot as indicated by the arrows. In general, low molecular weight proteins show this anomalous behavior when the concentration of SDS is too high. This can be caused by an excess of agarose solution (2% SDS) used to fix the first-dimension gel.

Another problem that occurs as a result of mistakes in adjusting the pH of the resolution buffer (solution B) is illustrated in Fig. 14b. From this figure it is clear that the relative mobility of the proteins has been altered with respect to a normal pattern. This can be easily detected because the running time of the tracking dye (bromphenol blue present

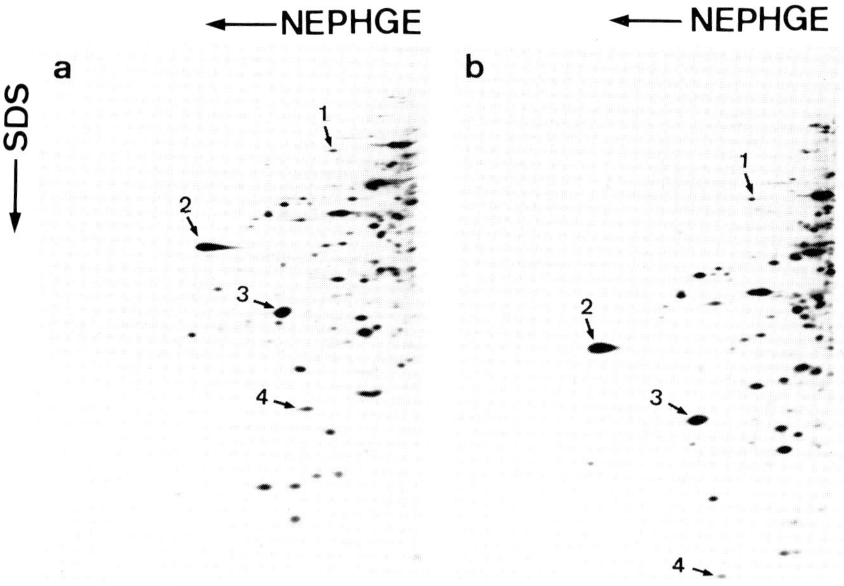

Figure 14. Changes in the relative mobility of proteins in the second dimension: (a) basic proteins of a total cellular lysate presenting a normal migration; (b) altered mobility of proteins due to wrongly adjusted pH of the resolution buffer. Corresponding proteins are indicated by numbers.

in the agarose solution) is longer. The pH problem develops when the stock resolution buffer is prepared from a concentrated solution of Tris-base by titration. To avoid this problem it is recommended to prepare the resolution buffer as described in Section II,B.

V. Concluding Remarks

The technique of two-dimensional gel electrophoresis allows the direct detection of a large number of cellular proteins, many of which have never been previously characterized. Although two-dimensional gel electrophoresis cannot be used to directly determine the functions of the proteins, it gives us the overall magnitude of a biological response in terms of the number and amplitude of the changes in the different cellular polypeptides.

Two-dimensional gel electrophoresis, one of the most sensitive analytic techniques in molecular biology, in conjunction with the recombinant DNA and hybridoma technologies will provide unique opportunities for the analysis and understanding of the function of many cell molecules in the near future.

Acknowledgments

The assistance of Ines Benner in the preparation of the manuscript is gratefully acknowledged.

References

Anderson, N. G., and Anderson, N. L. (1978). Analytical techniques for cell fractions. XXI. Two dimensional analysis of serum and tissue proteins: Multiple isoelectric focusing. *Anal. Biochem.* **85,** 331–340.

Anderson, N. L. (1981). Identification of mitochondrial proteins and some of their precursors in two-dimensional electrophoretic maps of human cells. *Proc. Natl. Acad. Sci. USA* **78,** 2407–2411.

Anderson, N. L., and Anderson, N. G. (1978). Analytical techniques for cell fractions. XXII. Two dimensional analysis of serum and tissue proteins: multiple gradient-slab gel electrophoresis. *Anal. Biochem.* **85,** 341–354.

Anderson, N. L., and Hickman, B. J. (1979). Analytical techniques for cell fractions. XXIV. Isoelectric point standards for two dimensional electrophoresis. *Anal. Biochem.* **93,** 312–320.

Anderson, N. L., Taylor, J., Scandora, A. E., Coulter, B. P., and Anderson, N. G. (1981). The TYCHO system for computer analysis of two-dimensional gel electrophoresis patterns. *Clin. Chem.* **27,** 1807–1820.

Bonner, W. M., and Laskey, R. A. (1974). A film detection method for tritium-labelled proteins and nucleic acids in polyacrylamide gels. *Eur. J. Biochem.* **46,** 83–88.

Bossinger, J., Miller, M. J., Vo, K. P., Geiduschek, E. P., and Xuong, N. H. (1979). Quantitative analysis of two-dimensional electrophoresis. *J. Biol. Chem.* **254,** 7986–7998.

Bravo, R., and Celis, J. E. (1980a). A search for differential polypeptide synthesis throughout the cell cycle of HeLa cells. *J. Cell Biol.* **84,** 795–802.

Bravo, R., and Celis, J. E. (1980b). Gene expression in normal and virally transformed mouse 3T3B and hamster BHK21 cells. *Exp. Cell Res.* **127,** 249–260.

Bravo, R., and Celis, J. E. (1982). Up-dated catalogue of HeLa cell proteins: Percentages and characteristics of the major cell polypeptides labeled with a mixture of 16 [^{14}C]-labeled amino acids. *Clin. Chem.* **28,** 766–781.

Bravo, R., and Knowland, J. S. (1979). Classes of proteins synthesized in oocytes, eggs, embryos, and differentiated tissues of Xenopus laevis. *Differentiation* **13,** 101–108.

Bravo, R., Bellatin, J., and Celis, J. E. (1981a). [^{35}S]-methionine labeled polypeptides from HeLa cells. Coordinates and percentages of some polypeptides. *Cell Biol. Int. Rep.* **5,** 93–96.

Bravo, R., Fey, S. J., Small, J. V., Mose-Larsen, P., and Celis, J. E. (1981b). Coexistence of three major isoactins in a single Sarcoma 180 cell. *Cell* **25,** 195–202.

Bravo, R., Celis, A., Moses, D., and Celis, J. E. (1981c). Distribution of HeLa cell polypeptides in cytoplasts and karyoplasts. *Cell Biol. Int. Rep.* **5,** 479–489.

Bravo, R., Small, J. V., Fey, S. J., Mose-Larsen, P., and Celis, J. E. (1982). Architecture and polypeptide composition of HeLa cell cytoskeletons. Modification of cytoarchitectural proteins during mitosis. *J. Mol. Biol.* **159,** 121–143.

Bull, H. B., Breese, K., Ferguson, G. L., and Sevenson, C. A. (1964). The pH of urea gels. *Arch. Biochem. Biophys.* **104,** 297–304.

Celis, J. E., and Bravo, R. (1981). Cataloguing human and mouse proteins. *Trends Biochem. Sci.* **5,** 281–283.

Choo, K. H., Cotton, R. G. H., and Danks, D. M. (1980). Double-labelling and high precision comparison of complex protein pattern on two-dimensional polyacrylamide gels. *Anal. Biochem.* **103,** 33–38.

Comings, D. E. (1982). Two-dimensional gel electrophoresis of human brain proteins. I. Technique and nomenclature of proteins. *Clin. Chem.* **28,** 782–789.

Comings, D. E., and Peters, K. E. (1981). Two-dimensional gel electrophoresis of nuclear particles. *In* "The Cell Nucleus" (H. Busch, ed.), Vol. 9, pp. 89–118. Academic Press, New York.

Duncan, R., and McConkey (1982). How many proteins are there in a typical mammalian cell? *Clin. Chem.* **28,** 749–755.

Fey, S. J., Bravo, R., Mose-Larsen, P., Bellatin, J., and Celis, J. E. (1981). [^{35}S]-Methionine labelled polypeptides from secondary mouse kidney fibroblasts: Coordinates and one-dimensional peptide maps of some major polypeptides. *Cell Biol. Int. Rep.* **5,** 491–500.

Garrels, J. I. (1979). Two-dimensional gel electrophoresis and computer analysis of proteins synthesized by clonal cell lines. *J. Biol. Chem.* **254,** 7961–7977.

Garrels, J. I. (1983). Quantitative two-dimensional gel electrophoresis of proteins. *Methods Enzymol.* **100,** 411–423.

Garrels, J. I., and Gibson, W. (1976). Identification and characterization of multiple forms of actin. *Cell* **9,** 793–805.

Giometti, C. S., and Anderson, N. L. (1981). A variant of human nonmuscle tropomyosin found in fibroblasts by using two-dimensional electrophoresis. *J. Biol. Chem.* **256,** 11840–11846.

Iborra, G., and Buhler, J. M. (1976). Protein subunit mapping. A sensitive high resolution method. *Anal. Biochem.* **74,** 503–511.

Jones, M. I., Massingham, W. E., and Spragg, S. P. (1980). Electrophoresis tank for running multiple slab gels. *Anal. Biochem.* **106,** 446–449.

Klose, J. (1975). Protein mapping by combined isoelectric focusing and electrophoresis in mouse tissues. A novel approach to testing for induced point mutations in mammals. *Humangenetik* **26,** 231–243.

Krystal, G., Ratner, J. B., and Hamkalo, B. A. (1978). Partial purification and characterization of the intercellular bridge from cultured mouse cells. *Proc. Natl. Acad. Sci. USA* **75,** 4977–4981.

Laemmli, U. K. (1970). Cleavage of structural proteins during the assembly of the head of bacteriophage T4. *Nature* (*London*) **227,** 680–685.

Laskey, R. A., and Mills, A. D. (1975). Quantitative film detection of ^{3}H and ^{14}C in polyacrylamide gels by fluorography. *Eur. J. Biochem.* **56,** 335–341.

Lemkin, P. F., and Lipkin, L. E. (1981). GELLAB: A computer system for 2D gel electrophoresis analysis. I. Segmentation of spots and system preliminaries. *Comput. Biomed. Res.* **14,** 272–297.
McConkey, E. H. (1979). Double-label autoradiography for comparison of complex protein mixtures after gel electrophoresis. *Anal. Biochem.* **96,** 39–44.
Milcarek, C., and Zahn, K. (1978). The synthesis of ninety proteins including actin throughout the HeLa cell cycle. *J. Cell Biol.* **79,** 833–838.
O'Farrell, P. H. (1975). High resolution two-dimensional electrophoresis of proteins. *J. Biol. Chem.* **250,** 4007–4021.
O'Farrell, P. Z., Goodman, H. M., and O'Farrell, P. H. (1977). High resolution two-dimensional electrophoresis of basic as well as acidic proteins. *Cell* **12,** 1133–1142.
Peterson, L. J., and McConkey, E. H. (1976). Non-histone chromosomal proteins from HeLa cells. *J. Biol. Chem.* **251,** 548–554.
Scheele, G. A. (1975). Two-dimensional gel analysis of soluble proteins. Characterization of guinea pig exocrine pancreatic proteins. *J. Biol. Chem.* **250,** 5375–5385.
Strand, M., and August, J. T. (1978). Polypeptide maps of cells infected with Murine Type C leukemia or sarcoma oncovirus. *Cell* **13,** 399–408.
Studier, F. W. (1973). Analysis of bacteriophage T7 early RNA's and proteins on slab gels. *J. Mol. Biol.* **79,** 237.
Ui, N. (1971). Isoelectric points and conformation of proteins I. Effect of urea on the behaviour of some proteins in isoelectric focusing. *Biochim. Biophys. Acta* **229,** 567–581.
Vo, K. P., Miller, M. J., Geiduschek, E. P., Nielsen, C., Olson, A., and Xuong, N. H. (1981). Computer analysis of two-dimensional gels. *Anal. Biochem.* **112,** 258–271.
Weber, K., and Osborn, M. (1969). The reliability of molecular weight determinations by dodecyl sulfate–polyacrylamide gels. *J. Biol. Chem.* **244,** 4406–4412.
Willard, K. E., and Anderson, N. G. T. (1981). Two-dimensional analysis of human lymphocyte proteins. II. An assay for lymphocyte effectors. *Clin. Chem.* **27,** 1327–1334.

CHAPTER 2

The QUEST System for Computer-Analyzed Two-Dimensional Electrophoresis of Proteins

JAMES I. GARRELS
JOHN T. FARRAR
CARTER B. BURWELL IV
Cold Spring Harbor Laboratory
Cold Spring Harbor, New York

I. Introduction 38
II. Problems To Be Faced in Development of Computerized Protein Data Bases 39
III. Strategy for the QUEST System 40
IV. The QUEST System As Seen by the User 45
A. The QST Data Management System 46
B. The QUEST-2D Gel Laboratory 49
C. Image Analysis Facility 50
D. Scanning and Spot Quantitation 58
E. Matching Procedures 60
F. Data Reduction to GELSPOTS and SMPSPOTS 65
G. The IPMANAGER 67
H. Analysis of Reduced Data 67
V. Program Descriptions 75
A. Programs for Spot Detection and Quantitation 78
B. The Automatic Matching Program 80
C. Programs to Create and Update Standards 82
D. Assignment of Species Spot Numbers 83
E. Programs for Final Data Reduction 84
F. Processing Times and Storage Requirements 86
G. Future Program Enhancements 87
VI. Perspectives 88
References 90

TWO-DIMENSIONAL GEL
ELECTROPHORESIS
OF PROTEINS

ISBN 0-12-164720-X

I. Introduction

When high-resolution, two-dimensional gel electrophoresis was introduced by O'Farrell in 1975, it generated great excitement and fascination. The unprecedented level of resolution suggested that even very minor protein components of the cell could be detected, and if computers could be used to quantitate and match the complex patterns, it seemed possible to compare the amounts of 1000 or more proteins in different types of cells and tissues. Furthermore, if the technique could be sufficiently standardized, the data from many experiments done in many laboratories could be accumulated in protein data bases. In the clinical laboratory, a data base of human protein information might be used to reveal proteins that were abnormal in size, charge, or amount in diseased tissues (see also chapter by Tracy and Young). In the basic research laboratory, protein data bases might be used to correlate information from many experiments, revealing the patterns of induction, repression, and modification for each detectable protein of normal, mutant, and transformed cells.

The concept of large protein data bases at first seemed quite futuristic. It was not yet known how many of the spots on the gels were artifactual, how reproducible the gel patterns would remain over time, or how to program computers to quantitate and match the complicated spot patterns. However, as two-dimensional gel electrophoresis became an established laboratory technique, as the methodology was improved (Garrels and Gibson, 1976; O'Farrell *et al.*, 1977; N. G. Anderson and N. L. Anderson, 1978; N. L. Anderson and N. G. Anderson, 1978; Garrels, 1979a) and as computers became more common in biological laboratories, the general attitude concerning protein data bases turned from skepticism to expectation. Now, nearly eight years after the initial goals were set, the first data bases are starting to emerge. Software development work has been underway in several laboratories, and various approaches to spot detection and pattern matching have been explored. Reports from most of these groups are contained in the April 1982 issue of *Clinical Chemistry* (Vol. 28, No. 4, Part II, pp. 737–1092), which is entirely devoted to two-dimensional gel electrophoresis. In several laboratories, software systems aimed at data-base generation have been developed and are now being used to generate limited data bases (Garrels, 1979a; Lipkin and Lemkin, 1980; Anderson *et al.*, 1981; Vo *et al.*, 1981; Skolnick *et al.*, 1982).

In this chapter, we first present the special problems to be faced in using two-dimensional gel electrophoresis for data-base development and describe the strategies adopted in our laboratory to deal with these problems. We then present a description of our complete system, the QUEST system, for computer-analyzed, two-dimensional gel electrophoresis. The system encompasses a specialized gel laboratory, optimized procedures for two-dimensional gel electrophoresis, and a large software system for image analysis and data management. The emphasis in this article is on showing, from the user's point of view, how data from two-dimensional gels can be analyzed, compared, and reduced to permanent data-base files and how useful information can be extracted from these files. The equipment and methods used in our electrophoresis laboratory have been presented elsewhere (Garrels, 1979a, 1983) and will not be described in detail here.

An analysis of the rat skeletal muscle cell line L6 has been chosen to demonstrate the use of the QUEST system. It is not within the scope of this article to present a full study of this differentiating cell line. A preliminary study of this cell line by two-dimensional gel electrophoresis has already been published (Garrels, 1979b) and a more complete study including many more samples and controls will be published elsewhere.

II. Problems To Be Faced in Development of Computerized Protein Data Bases

The first problems to be faced by any user of two-dimensional gel electrophoresis are the practical ones of the technique itself. Many manual procedures requiring operator skills are still involved, and high-quality reagents are needed throughout the process. Insufficient attention to detail and lack of quality control at any step can lead to diminished resolution and to changes in the spot patterns (see also chapter by R. Bravo). Because most proteins are recognized only by their position within the surrounding pattern, a change that affects the relative positions of the proteins is a serious problem. Even in laboratories that have developed dedicated facilities and exercise a high level of quality control, undefined changes in the reagents or electrophoresis conditions occasionally occur, leading to variations in quality or reproducibility.

Second, any laboratory that succeeds in the routine production of high-quality gels faces a formidable data analysis problem. It is extremely difficult to analyze even one gel by manual methods, and the processing of many two-dimensional gel images requires dedicated computing and scanning facilities. The computers and scanning hardware needed for two-dimensional gel analysis are expensive but readily

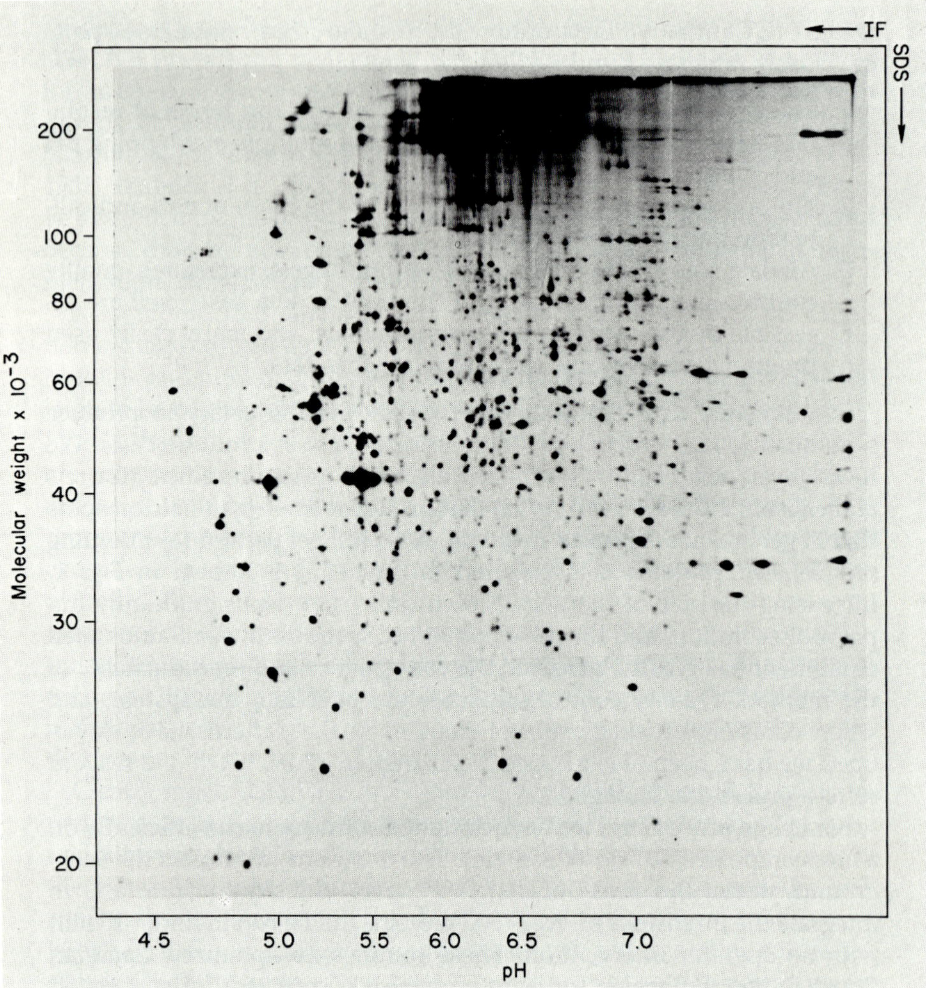

Figure 1. Separation of radiolabeled proteins by a single two-dimensional gel. Proteins of cultured muscle cells were radiolabeled with [^{35}S]methionine, solubilized, resolved on a two-dimensional gel, and detected by exposure to x-ray film. In this gel, a broad range pH gradient was used for isoelectric focusing in the first dimension (IF), and a 10% slab gel was used for SDS-electrophoresis in the second dimension (SDS). This type of gel resolves both the most acidic and the most basic proteins that can be detected by isoelectric focusing.

much more resolution than any one gel. Figure 2 shows how the proteins of rat myoblasts can be spread out into four overlapping two-dimensional gels, using two narrow isoelectric focusing pH ranges and two second-dimension gel acrylamide concentrations (compare to Fig.

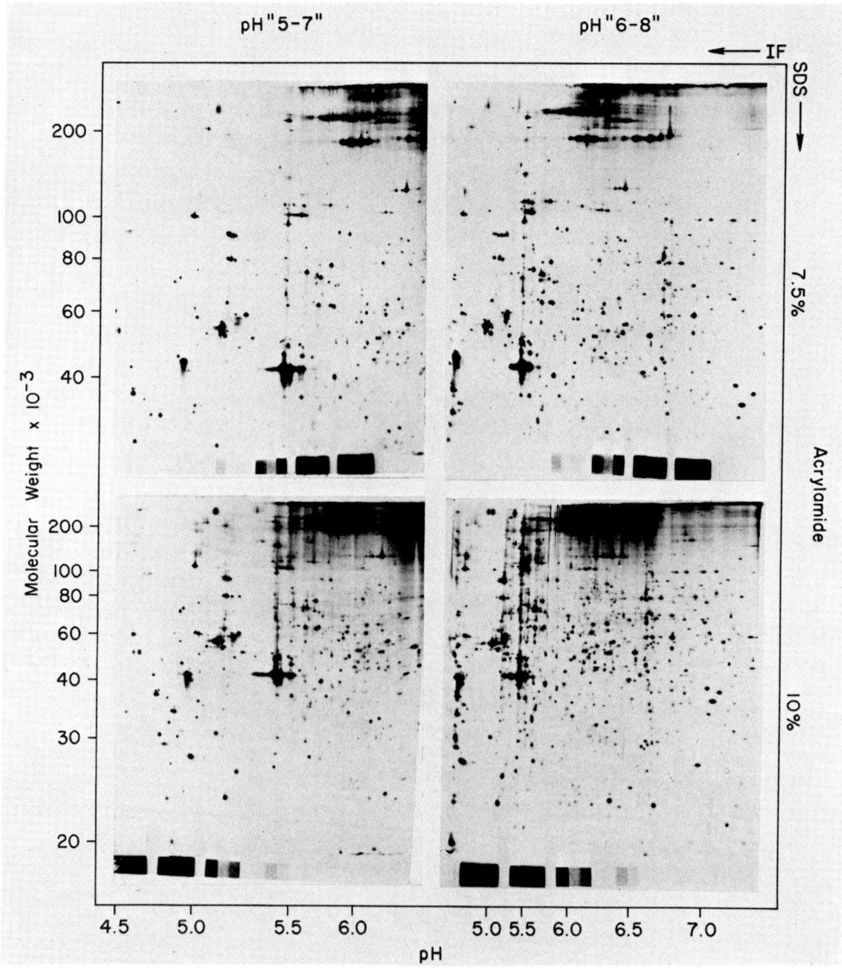

Figure 2. Separation of radiolabeled proteins on multiple two-dimensional gels. Proteins from the same cell culture as analyzed in Fig. 1 were loaded onto four two-dimensional gels, differing in pH range and slab gel acrylamide concentration as indicated. Each gel provides optimal resolution for proteins within a particular range of charge and size, and the combination of gels provides much more resolution than the single gel shown in Fig. 1. The segmented strips below each gel are calibration gels used to determine the relationship between film density and radioactivity per unit area in the gel.

1). In each type of gel, the proteins are resolved with a different pattern of surrounding spots. Particularly, changes in the second-dimension gel composition affect each protein differently (some will migrate faster, some slower, with respect to their neighbors), so that proteins that overlap on one type of gel probably do not overlap on another type of gel. This allows many more of the spots to be reliably quantitated, and it provides extra information to be used in the identification of each protein before it is logged in the data-base files. When studying diverse cell types, we can require that a protein be detected in the expected position on two or more gels. For each protein in the data-base files, its position on each type of gel will in general be known. If a spot is found at the position expected for a known protein on a gel containing 10% acrylamide, but not at the expected position on a gel containing 12.5% acrylamide, then we must conclude that this protein is not present and that a different protein overlaps its position on the 10% gel.

As a corollary to our use of multiple gels, we must use a data base that is not tied to any particular standard spot pattern. We have standard spot patterns for each of the gel types we run, and the system supports as many different standards as the users might desire. The programs include facilities for matching the standards to one another so that every protein logged in the data base is known in most or all of the separate standards. If new technology or improved techniques cause a change in the positions of the spots on the gels, the new and old patterns can be correlated, and all data can then be expressed in the new coordinate system.

Our strategy for combining the data from multiple gels and multiple exposures of each gel is best explained at this time by a simple overview of the data reduction process. As diagrammed in Fig. 3 we begin with a radiolabeled sample that is electrophoresed on multiple gels (for example, three different pH ranges). Each of the three gels is exposed to film three times to detect the full range of intensities without film saturation. (In practice, three or more gels are run of most samples, and up to four exposures may be required.) The role of the data processing system is to reduce this large amount of data into a single file called SMPSPOTS (SSP), representing the best quantitative description of each protein in the original sample. In the first stage of data reduction the spots are extracted from the film images, integrated, and stored in a file called FILMSPOTS (FSP). The FSP files are matched to standard filmspot (STD) files to yield MATCHLIST (MTL) files. In the second stage of data reduction (the MERGE process), the FSP and MTL files are used to create one GELSPOTS (GSP) file for each gel. Finally, in the RECONCILE process, the data from each of the gels is combined into a single SMPSPOTS (SSP) file, which contains the best quantitative value for each protein, expressed as a fraction of the total radioactivity incorporated into all proteins of the sample.

One might expect that the SMPSPOTS files would be the most permanent and fundamental files in the system, but this is not the case. Each

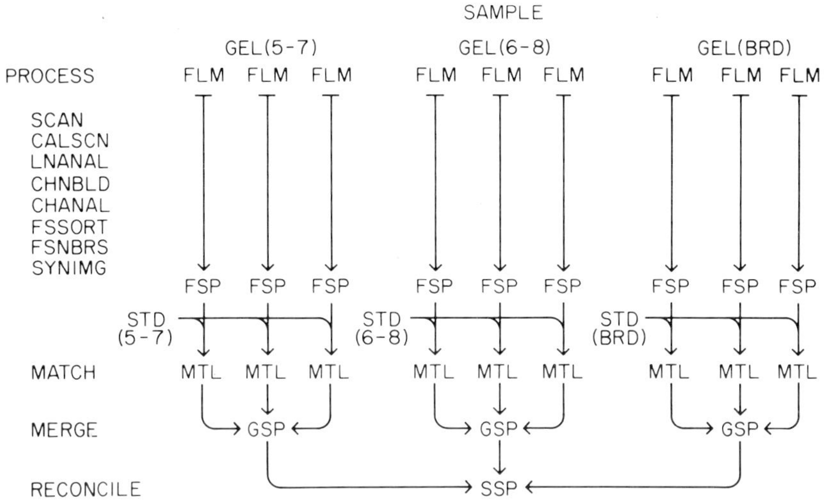

Figure 3. Flow chart for data reduction. Each sample for electrophoretic analysis is applied to one or more gels, and each gel is exposed to film for several different intervals. The role of the computer analysis system is to reduce the data from these films into a single SMPSPOTS (SSP) file. The SSP file has one entry for each protein of the sample, and each entry records the best determination of protein intensity, expressed as a fraction of the total radioactivity in the sample.

SMPSPOTS file is just a computational derivative of the FILMSPOTS files, the STD files, and the MATCHLIST files. It is these files that are most important. After the initial data analysis and generation of a SMPSPOTS file, new exposures might become available, another gel might be run, or one of the matches might be reinterpreted in a complex area. Meanwhile spots may have been deleted from, or new spots added to, the STD files. These changes do not lead to complications; if an automatic sequence of programs is run, all films for a given sample are automatically rematched (based on the latest set of landmarks and edit codes), all GELSPOTS files are regenerated, and a new SMPSPOTS file is created. This strategy allows the data base to be dynamic and constantly improving.

IV. The QUEST System As Seen by the User

The operation of the QUEST system will be explained in this section from the point of view of the user. The users include the scientists who produce samples for analysis and the technical staff who carry out most of the electrophoresis and data processing procedures. Because all aspects of the QUEST system are computerized, the users must frequently enter, access, and manipulate computerized data records. The scientist must describe his experiment, identify his samples, and specify the types of gels and exposures he desires. He needs to be able to follow the progress of his data through the system knowing when the exposures are complete, when the scans and matches are done, and when the final SMPSPOTS files are available. Finally the scientist examines the data, checks the matches and standards if necessary, and runs the data analysis programs to compare the proteins in each sample to other samples from his experiments and to other samples in the QUEST data base. The technical staff needs to keep complete records on all the samples, gels, exposures, scans, and matches that are completed or requested. In the gel laboratory samples must be scheduled for electrophoresis, gels must be scheduled for exposure to film, and quality control information concerning all operations must be recorded. In the image analysis center data reduction programs must be run on a routine basis and numerous data files must be managed.

These and other data management functions are handled by a data management system, called QST, which allows entry and access of data records and the initiation of computational processes without any specialized computer knowledge. Forms displayed on the screens of intelligent terminals provide the interface between the users and the computer. Because these forms are such a central part of the entire QUEST system, their use will be described first, and many of the forms will be shown in the following sections to facilitate the description of the QUEST system.

A. The QST Data Management System

Users of the QUEST system must become familiar with many of the forms that can be displayed on the terminal screens, but they do not need to know the names of any programs, data files, or commands to the operating system of the host computer. The forms have simple names such as SAMPLE, GELDAT, SCAN, or MATCH (see examples in Fig. 5), and they contain self-explanatory information. Most of the area of a form consists of protected fields that cannot be written into by the user; only the white (unprotected) areas are for data entry. All of the unprotected fields on the form constitute one data record, and any number of data records can be created for display on each form.

The terminals used for the QST system (Lear–Siegler ADM-42) allow up to eight different screen formats to be loaded into the eight "pages" of memory within the terminal. Over 190 forms have been created and can be loaded from the computer into the terminal using the menu form, which is always loaded onto the first page (see Fig. 4). The commands to load forms and particular data records are entered using the function keys of the terminal. After a brief introduction to the forms and function keys, a user can display or modify any data records in the QST system. One of the keys displays a blank form for creation of a new data record, another key causes the data record to be saved, and other keys are used to display particular data records. A search function key can be used to find a record containing a particular sequence of characters, and a related forms key is provided for rapid access to additional information. As an example of the related forms function, if the cursor is placed in the sample number field of GELDAT and the related forms key is pressed, the SAMPLE form and the appropriate data record will be displayed. Finally, one of the function keys is used to access a HELP record providing more information on the use of each particular form.

Aside from entry and retrieval of data, the next most important use of the forms is to run programs. Many of the forms are connected to specific programs that are called to run each time a record is saved from the given form. These programs, called QST ancillary or QSTACP programs, read and write from the data records stored on disk, and do calculations that might be as simple as determining the weights of reagents to be used in making a solution (see the ELECTRODE form in Fig. 5B for composition of the electrode buffer), or they may initiate more complex calculations involving data from many forms, such as processing a request for gels and exposures using the MAKREQUES form (see Fig. 5A), or they may initiate image processing (IPP) programs (see below). Providing input to a program by filling in the fields of a self-explanatory form is much easier than the more conventional method of responding to a series of questions, one at a time, from the computer. The fields that are seldom changed can be preentered into a default record that is automatically displayed prior to the creation of each new data record. Another advantage of using the forms to control programs is that the input entries are permanently stored in the data record, and output information is written by the program directly into the same data record. A rerun of the program is as simple as resaving the data record with any of the input fields modified or not as desired.

The QST data management system allows the user to concentrate on the data as displayed on the forms, and it gives him a great deal of power to manipulate data without ever knowing the names of the data

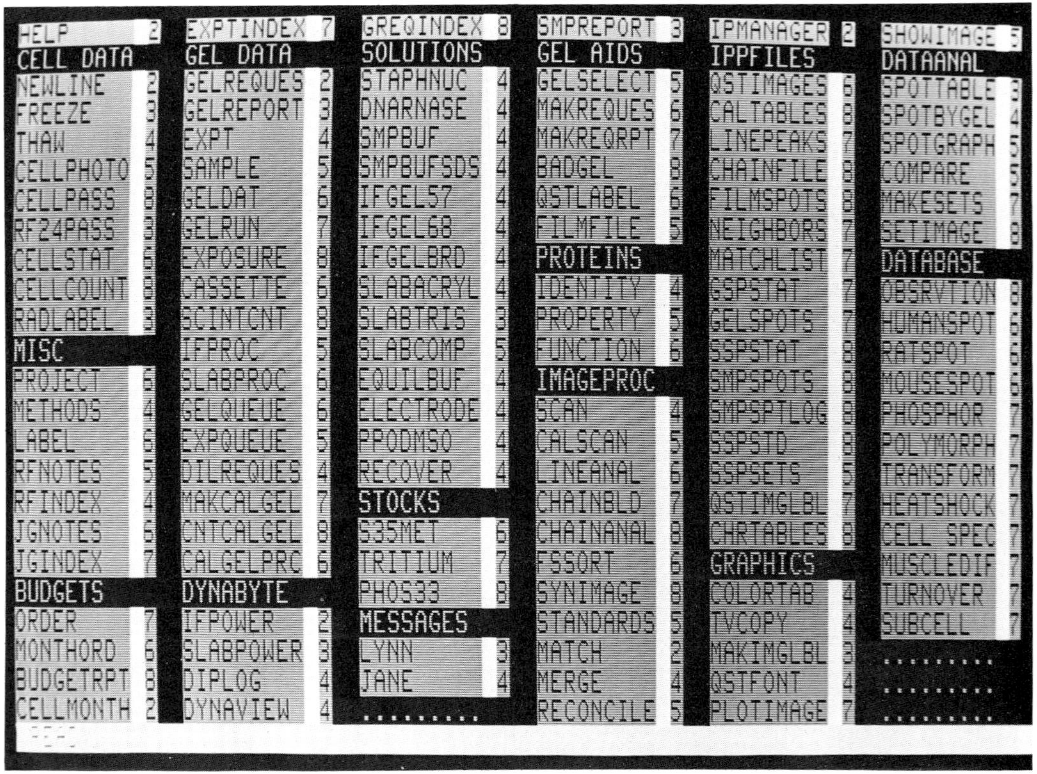

Figure 4. Menu page of the QST terminals. This display from the terminal screen shows the selection of forms available to users of the QUEST system. Each name in gray is a form name; names in black are headers dividing the menu into logical groups. Using the cursor (the black square presently at IPMANAGER) and an appropriate function key, the user can select and load any form into terminal memory. The selection of forms and memory page numbers (the number after each name) can be configured independently for each QST terminal.

files or the programs. In the remainder of this section, we show many of the forms and explain what processes are controlled by each, but we will not in general discuss the programs or data files by name. Further information on each of the major programs will be presented in the Program Description section (see below).

B. The QUEST-2D Gel Laboratory

The gel laboratory contains specially designed equipment to facilitate the routine production of 20 large (20 × 20 cm) two-dimensional gels each day. This equipment includes casting apparatus for isoelectric focusing and slab gels, special electrophoresis chambers, a gel dipping system for fluorographic processing, large gel dryers, and special film exposure holders. Much of the gel processing is monitored and controlled from a microcomputer (Dynabyte Basic Controller), which can also communicate with the main computer. The procedures of gel electrophoresis have not been substantially changed since an earlier publication (Garrels, 1979a); an up-to-date description of the methods and some of the equipment is provided by Garrels (1983).

All aspects of two-dimensional gel electrophoresis are carried out in our laboratory by skilled technicians. Scientists prepare their samples and enter into the computer a description of their experiments and their samples. On the EXPT form (Fig. 5A), a name for the experiment and for each sample is entered, as well as general labeling and sample preparation procedures. The user gives the sample volumes and the computer assigns the sample numbers. After the trichloroacetic acid (TCA)-precipitable radioactivity in the sample has been assayed (Garrels, 1983), the scientist can request his gels and exposures using the self-explanatory MAKREQUES form (Fig. 5A). To obtain uniform exposure densities, exposures are requested by exposure value (EV). Because the exposure value is proportional to total grains on the film, exposures of the same EV should have the same overall level of darkness. Given the EV, the computer will calculate the necessary exposure time based on the number of counts loaded on the gel, the decay since the gel was run, the nature of the isotope, and the type of film used.

Each gel and exposure request is written by the computer onto a GELREQUES and a GELREPORT form (Fig. 5A). Here the user can see each individual request and can determine exactly which gels have been run and which exposures are complete. The GELREPORT form gives the projected dates (in small letters) for exposures that are not yet complete. Each night the computer reviews all GELREQUES and GELREPORT records to update its internal gel and exposure queues, and it automatically creates the EXPOSURE records (Fig. 5B) scheduling the gels for exposure to film in efficient groups. A GELRUN record (Fig. 5B) is created from the list of requested gels and guides the preparation and loading of samples onto the first-dimension gels. A CASSETTE form (Fig. 5B) keeps a list of all of the cassettes (light-tight boxes loaded with gels being exposed to film), and it indicates each day which cassettes should be unpacked for film development. Other forms, such as ELECTRODE (Fig. 5B), specify formulas for electrophoresis solutions and provide a place for recording quality control information.

The master record for each sample is maintained in the SAMPLE form (Fig. 5C), and a master record for each gel is maintained in the GELDAT form (Fig. 5C). These records are of central importance to operation of the gel laboratory and for the data processing operations. Nothing need be entered into these forms directly by the user; they are maintained and updated completely by the programs of the QST data management system using input information from other forms such as EXPT, MAKREQUES, and SCINTCNT (Fig. 5D).

C. *Image Analysis Facility*

The equipment of film scanning and image analysis in our laboratory consists of a PDP-11/60 minicomputer, an Optronics P-1000 film scanner with a 10-in image capacity, and a Grinnell GMR-27 color raster display system. The computer has 256 kilobytes of main memory and two large disk drives (254 megabytes and 88 megabytes capacity). The scanner reads film absorbance at intervals of 50, 100, or 200 microns, and for each

---

Figures 5A–G. Some important forms used in the QUEST system. The individual forms ▶ are largely self-explanatory, and the use of each is described in the text. The top line of each form contains the form name, the form version number, the number of the data record currently on display, and the total number of data records created for this form. The data record number should be read as "experiment number" for the EXPT form, as "sample number" for the SAMPLE form, as "gel number" for the GELDAT form, etc. The second line holds the date and time the record was created and the last date and time the record was modified. The data in the top two lines are automatically entered by the QST program. Users can type into the white areas of the forms; the black or gray areas are protected and cannot be written over by users. The bottom line (line 25) is a message line and is not part of the form.

DATA RECORD 267 OF 450 TOTAL RECORDS
08-AUG-81 TIM:03:08:39 MOD BY: DAT:16-AUG-81 TIM:02:47:40
L6 time course
Cell type: L6
Labeled 1 35mm dish of L6-04-04-05 plated 04-AUG. Cells are about half confluent and not fusing. Well distributed but there are a few old tubes from previous plating.
Changed medium to DME-met + 7.5% NCS, washed once, final vol is 0.5 ml. Added 1 vial 35S met batch 26 for 2 hr.
L6 blasts, day 3.5, prefusion
Labeled second dish, same protocol, 2 hr. Not fusing
L6 blasts, day 4.5, prefusion
Labeled third dish, same protocol, 2 hr, cells fusing rapidly
L6 in fusion, day 5.5
Labeled fourth dish, same protocol, 2 hr, midfusion
L6 in fusion, day 6.5
Labeled fifth dish, same protocol, 2 hr, mid-late fusion
L6 in fusion, day 7.5
MAKREQUES V. 1 DATA RECORD 214 OF 511 TOTAL RECORDS
DAT:11-AUG-81 TIM:05:33:08 MOD BY: DAT:16-DEC-82 TIM:10:26:20
SAMPLES BY EXPERIMENT #
Run samples from these expt records:
EXCLUDE THESE SAMPLES
INCLUDE THESE SAMPLES
1349 1350 1351
GEL TYPES
SAMPLE VOLUME
MODE OF DETECTION
Fluorography
Autoradiography
Stain
If dilution is not specified, dilution and volume will be chosen for a load of 600000 dpm/gel relative to request date.
EXPOSURES
(Same format as GELREQUES)
GEL REQUEST NAME
All films will go into a file of this name.
L6 time course, first 3 samples
SYSTEM MESSAGES
GELREQUEST MADE. RESAVES WILL NOT BE PROCESSED EXCEPT TO ADD EXPOSURES
READY
GELREQUES V. 1 DATA RECORD 678 OF 1343 TOTAL RECORDS
ENTERED BY:SYS DAT:11-AUG-81 TIM:05:35:00 MOD BY: DAT:01-DEC-82 TIM:07:51:10
L6 time course, first 3 samples
EXP 1 EXP 2 EXP 3 EXP 4 EXP 5 EXP 6 EXP 7 EXP 8
GELREPORT V. 3 DATA RECORD 678 OF 1343 TOTAL RECORDS
ENTERED BY:SYS DAT:11-AUG-81 TIM:05:36:00 MOD BY:SYS DAT:16-DEC-82 TIM:12:58:00
L6 time course, first 3 samples
READY

Figure 5A

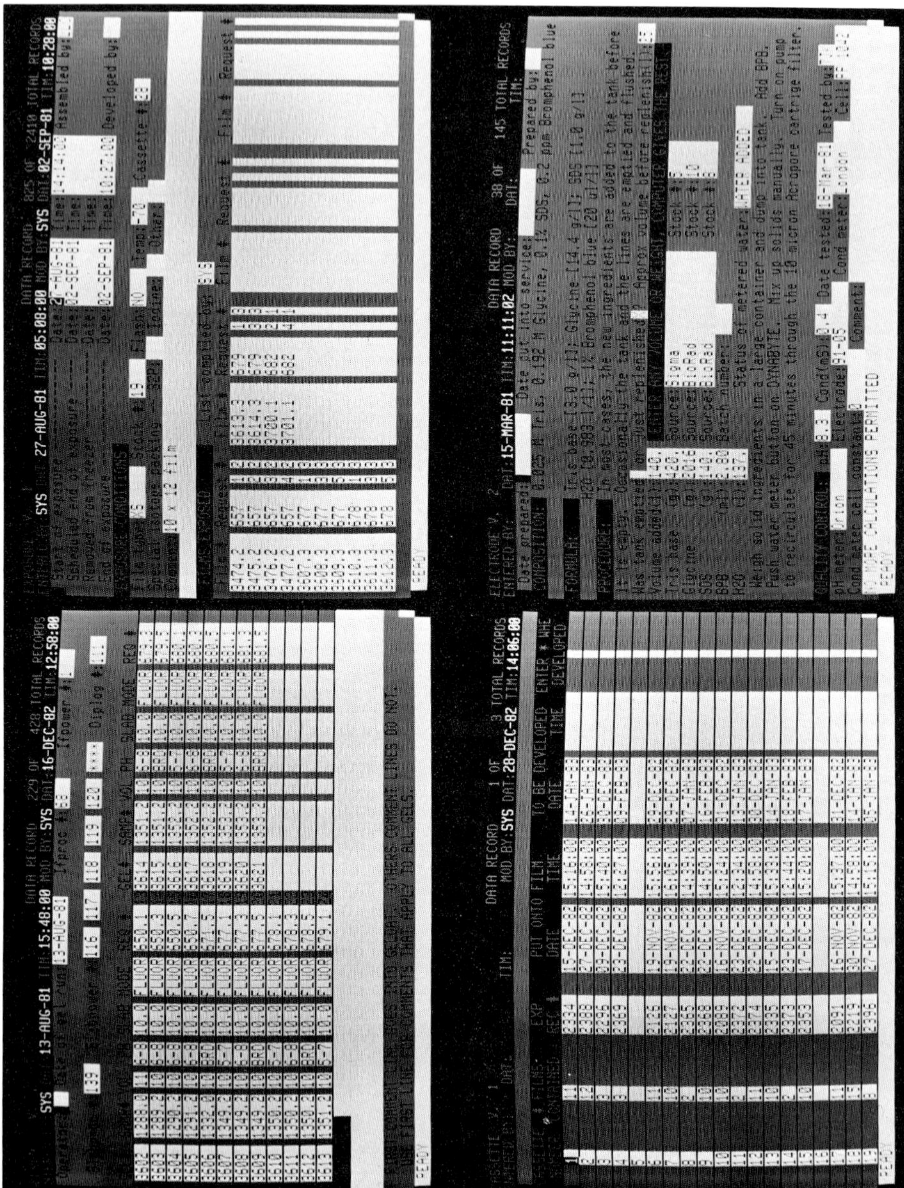

Figure 5B

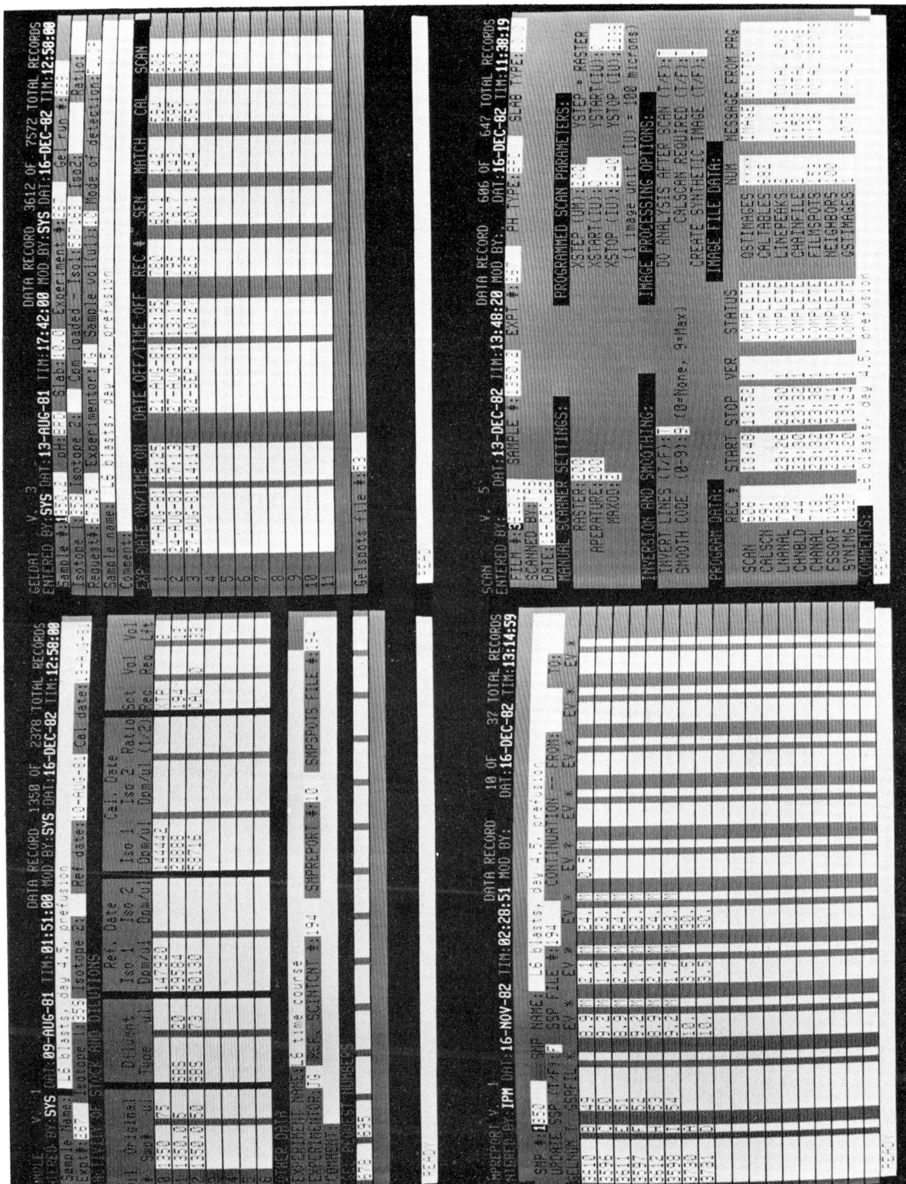
DATA RECORD 1358 OF 2378 TOTAL RECORDS
STOCK AND DILUTIONS
SMPREPORT V. 1
DATA RECORD 10 OF 37 TOTAL RECORDS
GELDAT V. 3
DATA RECORD 3612 OF 7572 TOTAL RECORDS
SCAN V. 5
DATA RECORD 606 OF 647 TOTAL RECORDS
MANUAL SCANNER SETTINGS:
PROGRAMMED SCAN PARAMETERS:
INVERSION AND SMOOTHING:
IMAGE PROCESSING OPTIONS:
PROGRAM DATA:
IMAGE FILE DATA:
READY

Figure 5C

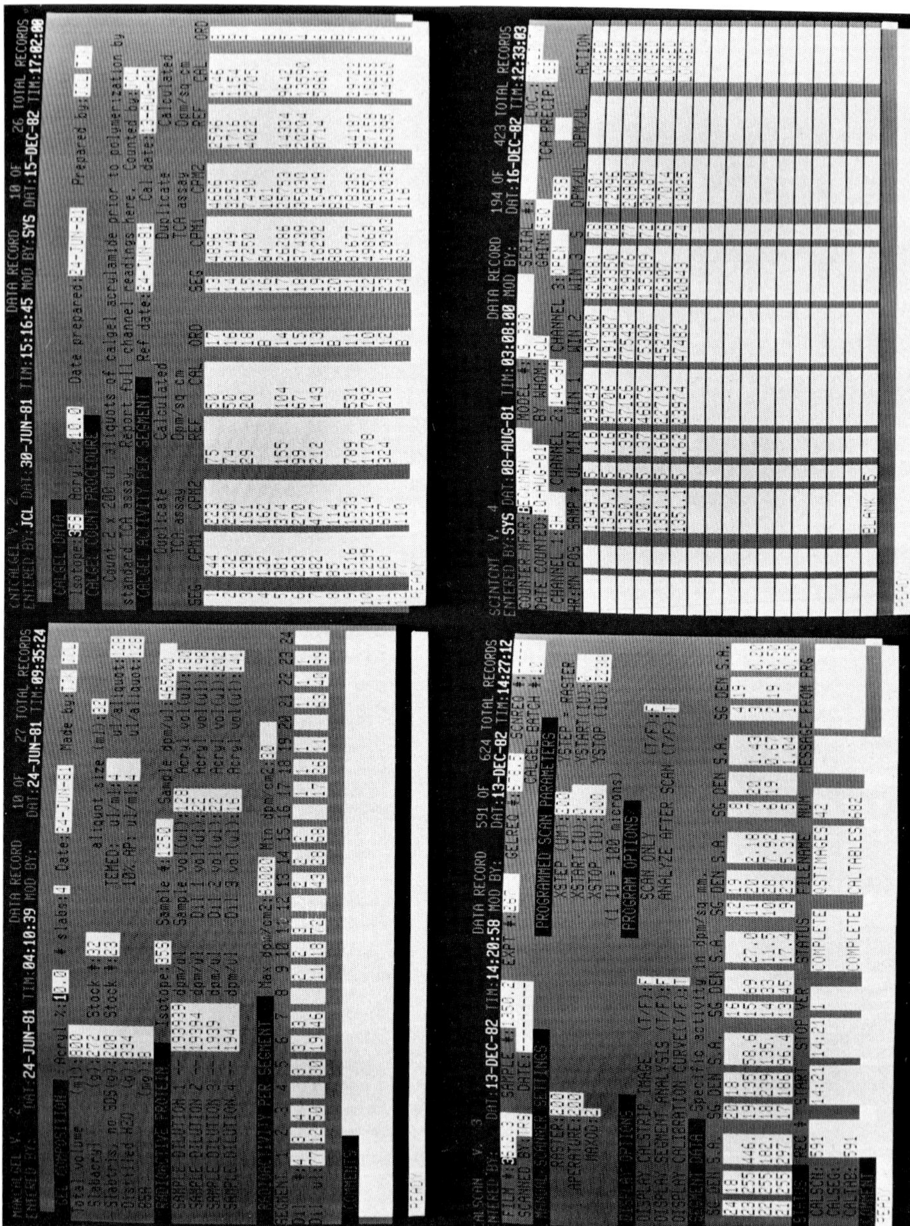

Figure 5D

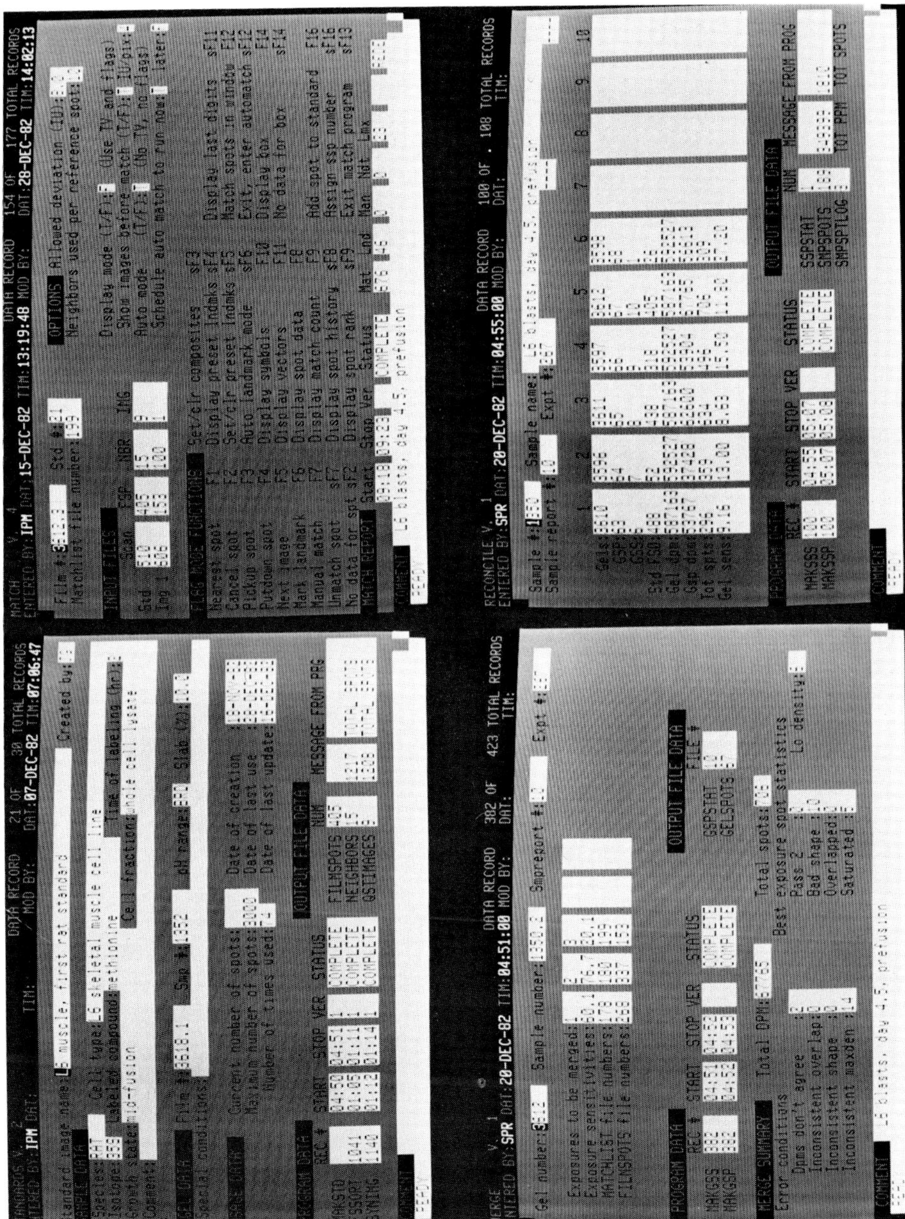
STANDARDS V. 2
DATA RECORD 21 OF 30 TOTAL RECORDS
SAMPLE DATA
GEL DATA
USAGE DATA
PROGRAM DATA
OUTPUT FILE DATA
COMMENT
MATCH V. 4
DATA RECORD 154 OF 177 TOTAL RECORDS
INPUT FILES
OPTIONS
FLAG MODE FUNCTIONS
MATCH REPORT
MERGE V. 1
DATA RECORD 382 OF 423 TOTAL RECORDS
PROGRAM DATA
MERGE SUMMARY
OUTPUT FILE DATA
COMMENT
RECONCILE V. 1
DATA RECORD 100 OF 108 TOTAL RECORDS
PROGRAM DATA
OUTPUT FILE DATA
COMMENT

Figure 5E

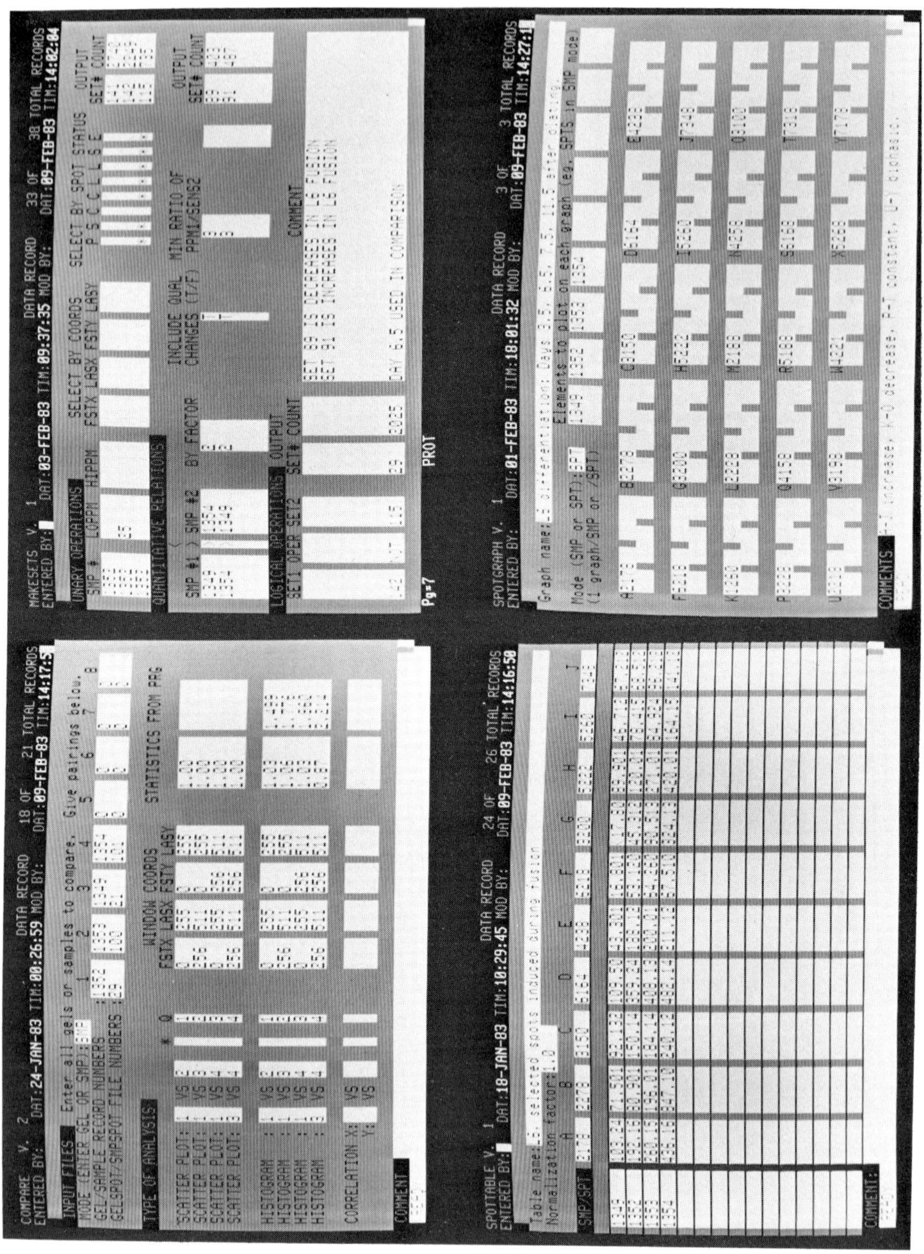

Figure 5F

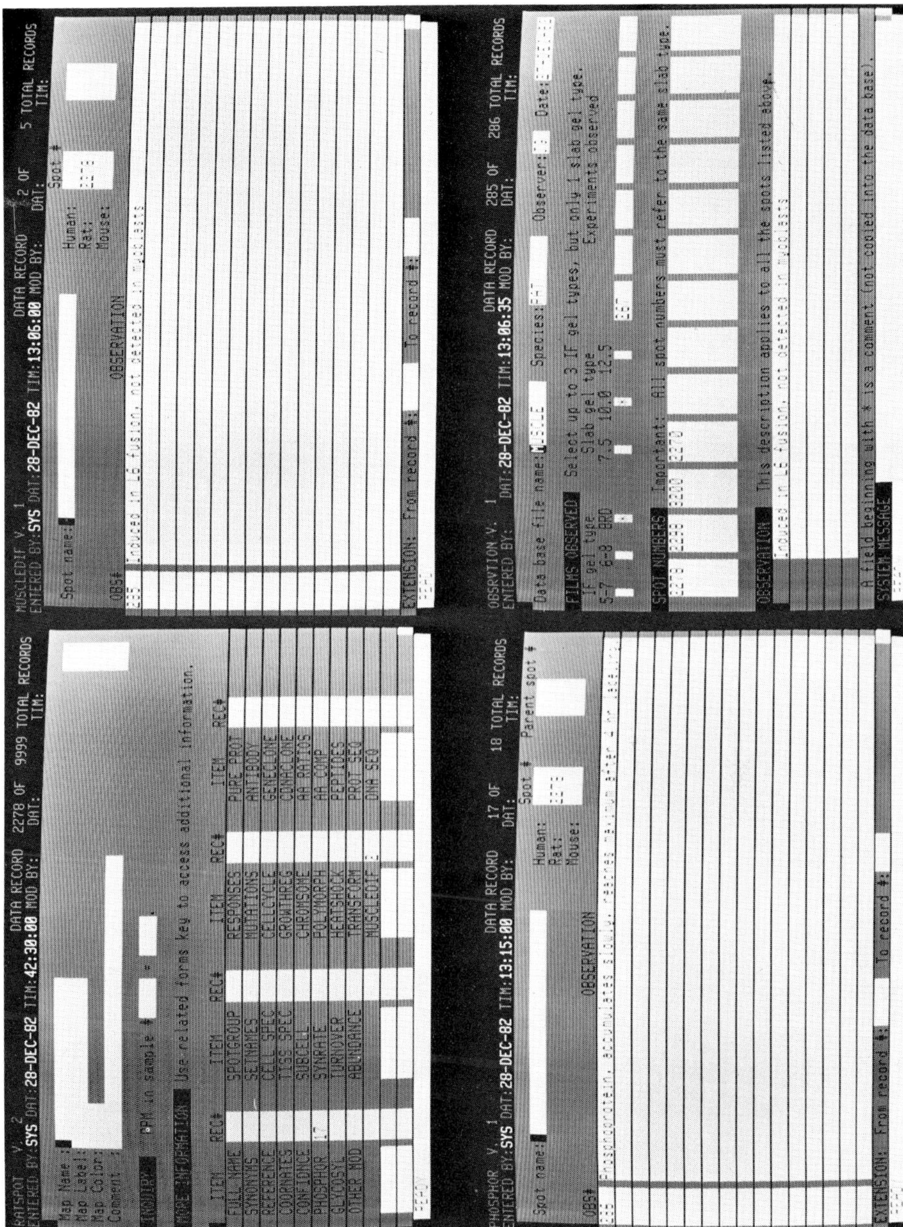
KATSPOT V. 2
ENTERED BY:SYS DAT:28-DEC-82 TIM:42:30:00 MOD BY:
DATA RECORD 2278 OF 9999 TOTAL RECORDS
DAT:
TIM:
Map Name:
Map Label:
Map Color:
Comment:
INQUIRY
PM in sample #
MORE INFORMATION
Use related forms key to access additional information.
ITEM REC# ITEM REC# ITEM REC# ITEM REC#
MUSCLEDIF V. 1
ENTERED BY:SYS DAT:28-DEC-82 TIM:13:06:00 MOD BY:
DATA RECORD 2 OF 5 TOTAL RECORDS
Spot name:
Human:
Rat:
Mouse:
Spot #
OBSERVATION
OBS#
Induced in L6 fusion, not detected in myoblasts
EXTENSION: From record #: To record #:
PHOSPHOR V. 1
ENTERED BY:SYS DAT:28-DEC-82 TIM:13:15:00 MOD BY:
DATA RECORD 17 OF 18 TOTAL RECORDS
Spot name:
Human:
Rat:
Mouse:
Spot #
Parent spot #
OBSERVATION
OBS#
EXTENSION: From record #: To record #:
OBSRVTION V. 1
ENTERED BY: DAT:28-DEC-82 TIM:13:06:35 MOD BY:
DATA RECORD 285 OF 286 TOTAL RECORDS
Data base file name: MUSCLE
Species:
Observer:
Date:
FILMS OBSERVED
Select up to 3 IF gel types, but only 1 slab gel type.
IF gel type
Slab gel type
Experiments observed
5-7 6-8 BBD
7.5 10.0 12.5
SPOT NUMBERS
Important: All spot numbers must refer to the same slab gel type.
OBSERVATION
This description applies to all the spots listed above.
Induced in L6 fusion, not detected in myoblasts
A field beginning with * is a comment (not copied into the data base).
SYSTEM MESSAGE

Figure 5G

reading it reports to the computer an 8-bit number scaled for maximum output at a film density of 2 OD units. Images are stored on disk and can be displayed directly on the color monitor. The Grinnell system has resolution of 512 × 512 pixels (picture elements) with 8 bits of information for each pixel. A color lookup table converts each 8-bit density reading to a selected color for display. A cursor is maneuvered across the screen by a joystick providing the necessary interactive information needed for image editing and matching.

D. Scanning and Spot Quantitation

In the QUEST system only films (autoradiograms or fluorograms) are processed. These are trimmed to a size of 22 × 22 cm and mounted on the Optronics rotating drum scanner. Under computer control the film is scanned in 200 × 200 μm increments, using a light beam which passes through a 200 μm aperture. A typical scan requires about 3 min and produces over 1 million bytes of image data. The image data is smoothed as it is scanned (by averaging adjacent elements within the scan line) and stored in image files on the disk.

The scan program is initiated from the SCAN form (Fig. 5C). This form requires as input only the film number. The image size and increments are pre-entered as default values, and the output file number is assigned automatically. The running time and the output file number are permanently recorded in the program log area at the bottom of the form. The SCAN form is multifunctional, not only controlling the scanning program, but also directing the flow of the series of image analysis programs that follow the scan.

A calibration scan is required before image analysis can begin. Calibration gels are prepared as described previously (Garrels, 1979a, 1983) by mixing radioactive protein into a multilayered gel, which is subsequently cut into thin strips. The preparation of these gels is assisted by the MAKCALGEL form (Fig. 5D), and the determinations of radioactivity in each segment of the calibration gels are logged into the CNTCALGEL (Fig. 5D) form. A calibration gel is processed with each two-dimensional gel through fluorographic processing, drying, exposure to film, and film development. The image of the calibration gel is scanned under the direction of the CALSCAN (Fig. 5D) form.

To read the calibration curve the image of the calibration gel is displayed on the color monitor as a series of colored segments. The operator reads the best representative density of each segment of the calibration gel by selecting the color level that best matches each segment of the gel image. Each color level corresponds to a density, which is entered into the CALSCAN form. The computer determines the batch number of each calibration gel, which is encoded into the order of the segments, looks up the specific activity of each segment on the date the gel was run (using the appropriate CNTCALGEL data record), and then plots the calibration curve relating density and dpm/mm^2. A calibration table, consisting of a specific activity for every density level from 0 to 255, is automatically read from the curve and recorded in a data file called CALTABLES. The CALTABLES file number is recorded on the CALSCAN form and on the SCAN form for the same film. Once the image file (from SCAN) and the CALTABLES file (from CALSCAN) are ready, then image analysis can begin.

The SCAN form directs the subsequent image analysis. Image analysis can proceed directly after completion of the SCAN and CALSCAN, or it can be delayed and run overnight using the IPMANAGER form (not shown). The steps in image analysis are line analysis, chain building, chain analysis, sorting of the filmspots, and creation of a synthetic image. The first three steps lead to the creation of the FILMSPOTS file, which contains detailed information on the intensity, location, and quality of each detected spot. The sorting program creates a smaller sorted version of the FILMSPOTS file, called NEIGHBORS. After the sort, the ten nearest neighbors to each filmspot are compiled and recorded in the NEIGHBORS file. Finally, a synthetic image is created from the magnitude and shape parameters of the detected spots. The synthetic image looks like the original image, except that streaks and noise are missing, and the spots have idealized Gaussian shapes. An example of an original image and its corresponding synthetic image, as seen on the color display monitor, are presented in Plate 1.

The sequence of programs controlled from the SCAN form is run automatically with no further input needed from the operator. The QST ancillary programs find the location of each input file, find space on the disk for each new output file, and record this information in the various file management data records, leaving the operator completely unburdened with file management. The original scanned image, a QSTIMAGES file, will be deleted after processing is complete. The intermediate files (LINEPEAKS and CHAINFILE) are of no concern to the operator, and they will be automatically deleted as space is needed. The most important output file is FILMSPOTS, which is permanent. The NEIGHBORS file and the synthetic image, another QSTIMAGES file, may be deleted if space is required, but each can be regenerated in a few minutes from data in the FILMSPOTS file. The FILMSPOTS, NEIGH-

BORS, and synthetic image (QSTIMAGES) files provide input to the next stage of image analysis, which is the matching of the spots from the film image to a standard spot pattern.

E. Matching Procedures

1. FILMSPOTS FILES AND STANDARDS

The creation of the FILMSPOTS file, abbreviated FSP, completes the data reduction for one film exposure of one gel. The scanned image has been converted to a series of data records that describe each spot in terms of total dpm, total integrated density, maximum density, shape and fit parameters, coordinates, quality, and status codes. These data may be sufficient for users interested only in the spots from a particular film, but in general the FILMSPOTS data will be further reduced by the MERGE procedure to yield the GELSPOTS files and by the RECONCILE

Plate 1. Computerized spot detection. These images, displayed on the monitor of the computer graphics system, show the accuracy and sensitivity of spot detection. The image on the left is a portion of the scanned image of the gel shown in Fig. 1. The film image has been smoothed but is otherwise unprocessed. The image on the right is a synthetic image constructed from the list of detected spots, their intensities, half-widths, and coordinates. Note that the background streaks and haze have been subtracted and that most overlapped spots are properly resolved.

Plate 2. Matching a film image to a standard image. The split-screen image, taken from the color monitor, is a typical view of an interactive matching session. The image on the right is the standard image representing all spots so far detected in muscle cells on this type of gel, and the image on the left is part of the synthetic image shown in Plate 1. The spots marked with a plus sign are landmarks entered by the operator. The lettered spots are matches made by the automatic matching program based on the landmarks given. Further details of matching and editing are given in the text.

Plate 3. Protein sets. The MAKESETS form was used to select from the data base files those proteins that are increased or decreased during myogenesis. The images shown in the left three panels are part of a synthetic image for day 3.5 myoblasts, and the images shown in the right 3 panels are part of a synthetic image for day 11.5 myotubes. The members of each selected set are highlighted in yellow. The sets displayed are (middle left) proteins repressed twofold or more during myogenesis, (middle right) proteins induced twofold or more during myogenesis, (lower left) proteins that change less than twofold during myogenesis, and (lower right) proteins that are transiently repressed during fusion (lower at Day 6.5 than at Days 3.5 and 11.5). These sets can be compared with the gel photographs in Fig. 6, and graphic profiles of many of these proteins are presented in Fig. 9.

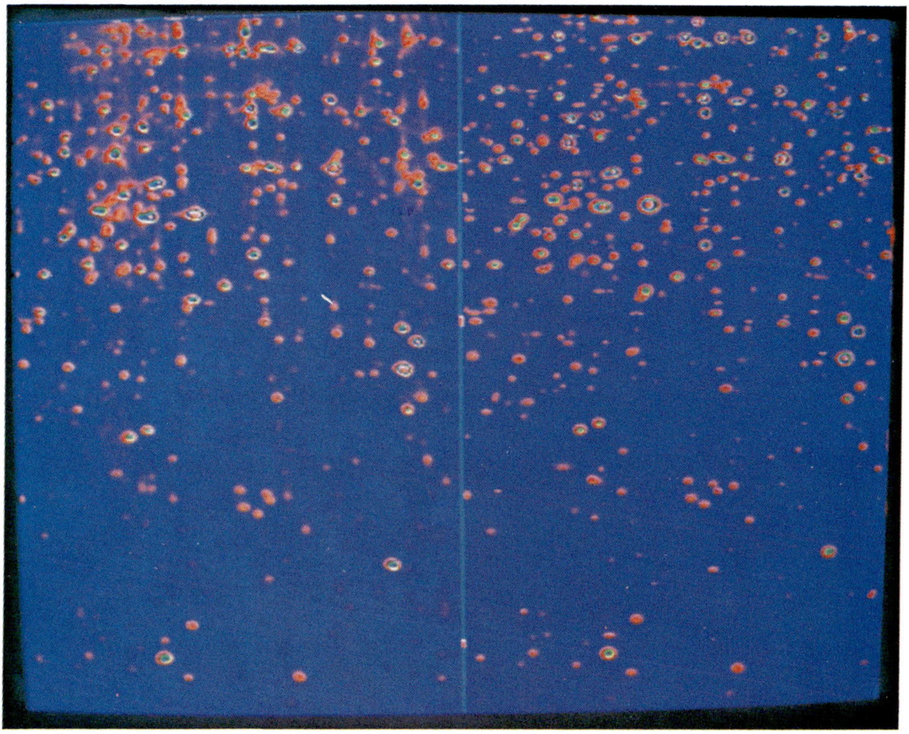

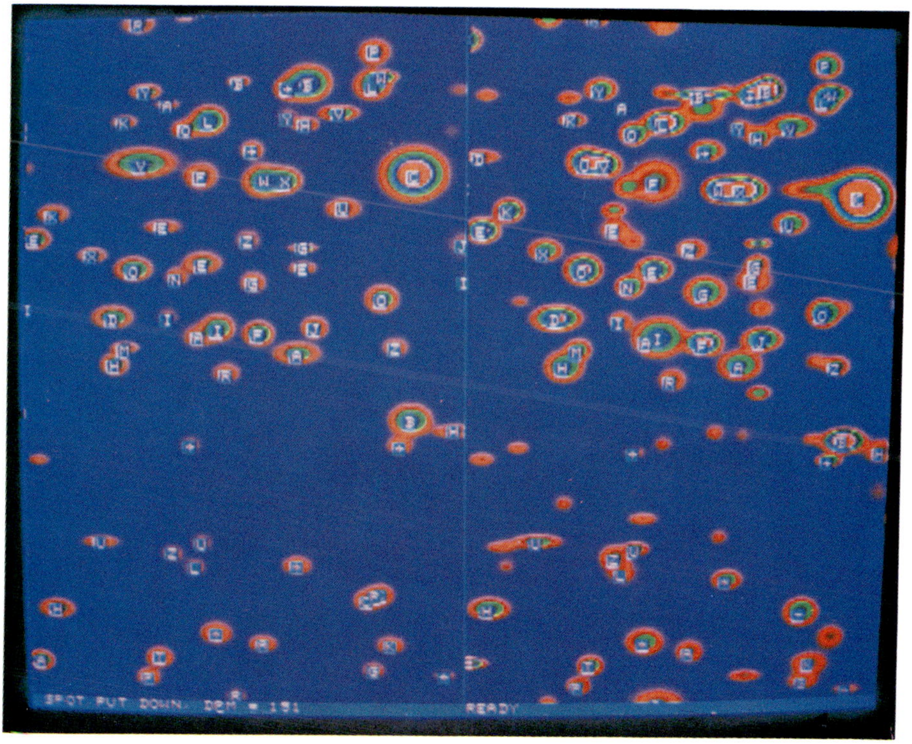
SPOT PUT DOWN.
READY

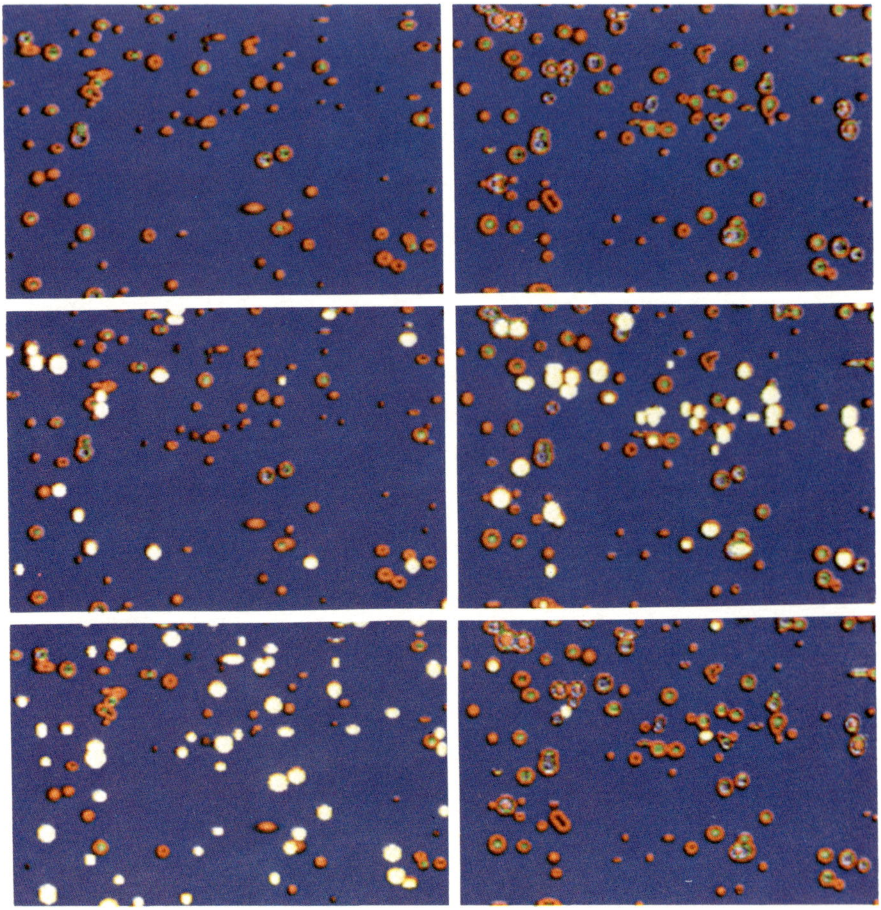

procedure to yield the final SMPSPOTS file (see Fig. 3). These procedures, which combine data from multiple films and multiple gels, can be performed only after the spots on each film have been identified by matching to the spots of standard images.

The standard images are synthetic images created from a standard FILMSPOTS file, abbreviated STD. The STD file, an associated NEIGHBORS file, the standard image file, and a corresponding record displayed on the STANDARDS form (Fig. 5E) are collectively termed a standard. To create a new standard, some descriptive information is entered into a new STANDARDS data record, and a film to be used as the initial standard pattern is specified. The FILMSPOTS file for the specified film is then automatically copied and modified (as described in the Program Description section) to become the new STD file. Next, the associated NEIGHBORS file and standard synthetic image (one of the QSTIMAGES files) are generated under direction of the STANDARDS form. The new standard image is carefully edited to remove artifactual spots due to streaks and noise (using the editing features of the interactive matching program), and as the standard is used for matching to other films, spots not previously included in the standard will be added. After each matching session in which standard spots have been added or deleted, the STANDARDS form directs the recompilation of neighbor relationships and the creation of a new standard synthetic image.

Different standards are required for each gel type (ampholyte range and acrylamide concentration) and for each experimental species to be studied. Separate standards may also be required for particular cell types or subcellular fractions if a new sample does not have enough spots in common with an existing standard for efficient matching. The standards provide a fixed set of spot numbers (STD record numbers) and coordinates that become associated with the corresponding FILMSPOTS records during matching. After each film representing a given gel has been matched to a common standard, the MERGE process can begin. The final stage of data reduction, the RECONCILE process, can begin only after the standards have been matched to one another. Both the matching of FILMSPOTS to standards (FSP to STD) and the matching of standards to each other (STD to STD) are processes initiated and directed using the MATCH form (Fig. 5E).

2. INTERACTIVE MATCHING

The MATCH form (Fig. 5E) requires the entry of a film number and a standard number (or of two standard numbers) and the selection of either display or automatic mode. In the display mode, the images are shown side by side on the color TV monitor, with the standard on the right. An example of the display for interactive matching is shown in Plate 2. The function keys of the QST terminal are redefined during interactive matching to control the matching, editing, and information functions of the MATCH program. Single keystrokes on a second keyboard are used for image control functions such as panning and zooming through each of the images. A cursor controlled by a joystick is used to select spots on the screen for matching and editing.

Prior to automatic matching, the images are displayed for entry of landmarks. The landmarks are well-defined spots selected by the operator as reliable matches to be used for initial registration by the automatic matching program. Those spots that do not run in a reproducible position between the standard and the film image can be marked as "manual matches," indicated by an asterisk (*), and those spots of the standard image that are not properly represented on the film image either because of a defect in the gel or because of severe overlapping with other spots can be marked "no data," indicated by a red square on the standard image. Both the "manual matches" and the "no data" spots are ignored by the automatic matching program.

Some other important editing functions are "cancel spot," "pickup" and "putdown," and "add spot to standard." Spots of either image can be cancelled (i.e., permanently deleted) from the respective file. Multiple spots that arise due to streaks or improper splitting can be combined using the "pickup" and "putdown" keys. All spots "picked up" are cancelled (marked "C"), but their intensities are summed to become a new spot at the location chosen for the "putdown," indicated by a "P." An example of two combined spots is seen in the lower right of the images in Plate 2. Finally, spots of the film image that are not presently members of the standard can be assigned coordinates and entered into the standard FSP file by a single keystroke. These spots are indicated by the symbol "A," and an example can be seen near the top of the images in Plate 2.

3. AUTOMATIC MATCHING

If the automatic mode is selected on the MATCH form, no TV display is presented, and an automatic matching program proceeds to match the two spot patterns based on landmark matches previously entered from the interactive mode. The automatic matching process is optimized for speed and therefore includes no display or editing functions. It runs to completion in about 5 min; summary information is logged at the bottom of the MATCH form. The details of the automatic matching algorithm are presented in the Program Description section. The output of the automatic matching program can be displayed and checked by re-saving the original MATCH data record in display mode (see Plate 2).

4. STRATEGY AND PROCEDURES FOR MATCHING A FILMSPOTS FILE TO A STANDARD

Output from the interactive and automatic matching programs goes into the permanent output file, called MATCHLIST. The MATCHLIST file contains one record for each record in the STD file, that is, something is recorded in the output file for each spot defined in the standard. Each standard spot will be scored as either a landmark match, a manual match, an automatic match, unmatched, or "no data." Unmatched spots are presumed to be undetectable in the film analyzed, and their dpm value is presumed to be less than the level of sensitivity of the film analyzed. "No data" spots could be absent or present at any dpm value; the film simply provides no data.

The fundamental strategy of matching is to start with a standard pattern related to the spot pattern being analyzed. The spots of the standard pattern are "known" and can be expected to appear in subsequent images. By searching for a match to each of the standard spots, we can match most of the valid spots in the FILMSPOTS file without having to be concerned about the extraneous filmspots that represent only streaks and noise. After the automatic match, the operator can easily recognize which of the unmatched spots are real, and these can be easily added to the standard.

The matching procedure requires alternating use of the interactive and automatic matching facilities. The interactive, or display, mode must be entered first to make the initial landmark matches. Usually the first interactive session is done quickly, marking between 10 and 30 obvious landmark matches in scattered regions of the image. An automatic match is run next, followed by reloading the files in interactive mode. The results of the automatic match can be displayed using paired alphabetical symbols, as shown in Plate 2, or by the display of a field of vectors representing the coordinate differences between the matched spots in the two images (not shown). Any bad matches can be detected as a disturbance in the smooth field of displacement vectors. The operator can quickly find regions where the match is incomplete due to insufficient landmarks (in distorted areas or areas with large distances between the spots) and areas of incorrect matching (usually regions where many of the filmspots represent streaks). The marking of additional landmarks in these regions causes more spots to be explicitly matched and increases the probability that the surrounding spots will be matched correctly by the automatic matching program. As the matched images are examined, the operator can indicate which unmatched spots of the film image represent new spots that should be added to the standard. The program then assigns standard coordinates to the new spots (based on matched spots in the surrounding area) and enters a new standard spot record in the STD file. After editing, the automatic match is run once more and then is not usually edited further until after a SMPSPOTS file has been generated. At that time, the match can be reentered to examine problem spots or to verify spots of special experimental interest (see the MRKSET function below).

Since the FSP, STD, and MATCHLIST files are preserved indefinitely, any MATCH data record can be resaved in either display mode or automatic mode at any future time. The display mode redisplays the split-screen image (as seen in Plate 2) with symbols to indicate the currently matched spots and the latest editing status, and it allows further editing. The automatic mode initiates a new automatic match based on the currently defined landmark matches and edit codes in the MATCHLIST file and based on the spots currently defined in the STD file. If the match has been reedited, or if the standard has been modified, a rerun of the automatic match brings the MATCHLIST file up to date. If no reediting has been done and if no changes have been made to the standard, then a rerun of the automatic match program simply reproduces the previous MATCHLIST file.

5. MATCHING ONE STANDARD TO ANOTHER

If two standard numbers are specified into the MATCH form, then two STD files will be accessed and two standard images will be displayed on the TV monitor. (The program can distinguish a standard number from a film number because the former is an integer and the latter is a decimal number.) The matching of one standard to another is done only for the purpose of assigning the same species spot number (SSP number, explained in the next section) to the spots that represent the same protein. No automatic match is allowed because the patterns from different types of gels differ in unpredictable ways. The operator, however, can usually determine the location of the corresponding spots in each gel based on similarities of migration in one dimension and similar intensities. As the operator marks the corresponding spots, using the manual matching facilities of the interactive match program, the

SSP number is transferred from one STD file (the one displayed on the right) to the other, and the manual match is recorded in a MATCHLIST file.

F. Data Reduction to GELSPOTS and SMPSPOTS

The final stages of data reduction can be controlled from the SMPREPORT form (Fig. 5C). This form provides a summary of each gel and each film that is available for the specified sample. For each film the actual exposure value is reported, followed by a status symbol to indicate if the film has been scanned (S) or matched (M). An asterisk (*) entered after each GELSPOTS (GSP) file number will request a new merger of the films for that gel, and a request to update SSP will cause all the gel data to be re-reconciled, resulting in a new SMPSPOTS file. A complete sample update (remerging all exposures, followed by a new reconcile) can also be requested by entering the sample number into the IPMANAGER form (see later).

The merger of FILMSPOTS data into GELSPOTS is directed by the MERGE form (Fig. 5E), which can be entered directly or indirectly through a merge request in SMPREPORT. Only the gel number need be specified; all other fields are entered by the system. Up to 5 different exposures can be merged at once, yielding the two output files GSPSTAT and GELSPOTS. Each of these files contains one data record for each spot in the standard (STD) file. The GSPSTAT file is a large, temporary file which holds much of the FILMSPOTS data for each of the 5 possible exposures. The merging process involves the selection of the exposure that best represents each spot (and the second best exposure for each spot), followed by writing the best dpm value into the GELSPOTS file. These files provide input to the next stage of data reduction, the RECONCILE process.

The last stage of data reduction, the RECONCILE process (see form in Fig. 5E), entails several new concepts. Whereas GELSPOTS (and GSPSTAT) files use the same record numbers as their respective standard (STD) file, the SMPSPOTS records must be numbered according to a different system. Furthermore, the spot intensity data, which is expressed simply as disintegrations per minute (dpm) in the FILMSPOTS and GELSPOTS files, must be converted to a different system. The solution to the latter problem is handled simply by dividing the dpm value for each protein by the dpm loaded onto the respective gel. This causes each protein to be quantitated as a fraction of the total radioactive protein, and in this form, data from several different gel types can be averaged. Therefore, in the SMPSPOTS files, each entry represents the best (usually averaged) determination of radioactive incorporation into one protein, expressed as a fraction of the total incorporation into all proteins. For convenience these values will usually be expressed as parts per million (PPM).

To solve the numbering problem we have established a four-digit numbering system to encompass all the proteins we expect to detect from the various samples analyzed within a given species. These numbers are called RATSPOT (RSP) numbers, MOUSESPOT (MSP) numbers, HUMANSPOT (HSP) numbers, or generically, SPECIESSPOT (SSP) numbers. Each detected spot in each of our standards is assigned one of these numbers. The purpose of matching standards to each other is to be sure that spots that represent the same protein will have the same SSP number. The primary record of SSP assignments is recorded within the STD records, and newly added spots are automatically assigned new SSP numbers each time a standard is updated (by resaving the STANDARDS data record). The way the numbers are distributed across the standard patterns and the method for assignment of new numbers is discussed in the Program Description section.

The RECONCILE process operates by working backwards from the SMPSPOTS file. For each sample, a SMPSPOTS file of 9999 data records will be created, one for each possible species spot in the SSP numbering system. To reconcile the data, the program looks for each possible species spot in each of the available GELSPOT files and then "reconciles" any ambiguities or inconsistencies that might be present. If any gelspot is severely overlapped, has a bad shape, is faint or is saturated, or was flagged as inconsistent during the merge process (possibly from a bad match), that data point will be discarded in favor of data from gels in which this situation does not occur. If a spot is known to be closely overlapped by another spot in one of the standards (i.e., a composite in the standard image), data from that gel alone cannot unambiguously identify the spot. If data from another gel are available, then by looking for each spot in its expected place in the second gel pattern, the ambiguity can be resolved. If a given species spot is undetected on any of the gels where its location is known, then the species spot must be considered undetected. The decision-making processes of MERGE and RECONCILE (see Section V.) are fixed in the programs and are not subject to user modification, assuring that these processes will be standardized and repeatable. If experience suggests better strategies, the programs can easily be modified. The program version number used for each analysis is recorded on the RECONCILE form for each analysis.

Users insure the integrity of the SMPSPOTS data by matching the film images to the standard images and by matching the standard images to one another. In matching one standard to another, it is difficult at first to unambiguously match all of the minor spots because standard patterns representing different gel types are not superimposable. This causes some of the proteins to be known by different SSP numbers in different standards. This is not a serious problem; it simply prevents taking full advantage of the RECONCILE process for better spot identification and better quantitation. As the data accumulate, enough information will be available in the data-base files to match most of the minor proteins of the standards according to similar patterns of regulation and tissue specificity.

G. The IPMANAGER

Those programs that are computationally intensive and do not require human interaction can be queued to run at later times using the IPMANAGER facility of the QUEST system. Such programs include the spot detection and integration programs run from the SCAN form, the automatic matches run from the MATCH form, the updating of standard neighbors and image files controlled from the STANDARDS form, and the updating of SMPSPOTS files controlled from the SAMPLE form. To queue any of these programs for later execution, the operator types the name of the controlling form and the record number into the IPMANAGER form (not shown). Whenever the computer is not being used for interactive work, these programs are automatically run.

H. Analysis of Reduced Data

The collection of SMPSPOTS files, each representing the best data available for 2000 or more proteins from each sample analyzed, constitutes a growing and powerful data base. Several programs have been written to compare and correlate data from these files, and more will be written in the future as new experiments suggest new forms of analysis.

Our examples of reduced data are taken from the analysis of the differentiating rat skeletal muscle cell line, L6. Shown in Fig. 6 are portions of the pH 3.5–10, 10% acrylamide gels for four stages of differentiation (see figure legend for details). In this analysis, seven different gel types were analyzed for each sample (all three pH ranges for 10% and

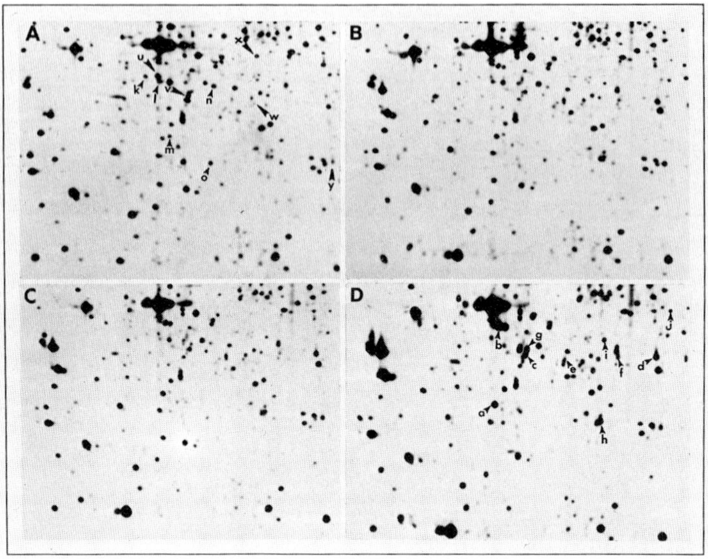

Figure 6. Four stages of differentiation in L6 muscle cells. Shown are portions of two-dimensional gels (broad pH range, 10% second dimension as in Fig. 1) for L6 muscle cells at the following stages of differentiation: (A) myoblasts, 3.5 days after plating (contains a few contaminating myotubes); (B) mid-fusion, 6.5 days after plating; (C) mid-fusion, 7.5 days after plating; and (D) well-differentiated, 11.5 days after plating. The area shown includes the actins (overexposed near the top), the skeletal muscle tropomyosins (major induced spots on the left), and one of the skeletal muscle myosin light chains (large induced spot at lower left). Indicated are proteins that increase during myogenesis (letters a–j), proteins that decrease during myogenesis (letters k–o), and proteins that change in an irregular manner (letters u–y). The lettered spots correspond to those shown in the spot sets of Plate 3 and in the spot graphs of Fig. 9.

12.5% slab gels, and a pH 6–8, 7.5% gel), and for most gels three exposures were analyzed. A standard filmspots (STD) file has been created for each of these gel types, and the standard images have been matched to one another.

1. MEASURES OF DATA QUALITY

The user who is performing quantitative comparisons of the data in the SMPSPOTS files will be concerned about the quality of each data entry. How confident can the user be that a quantitative change in a particular protein is real? How can he or she know which proteins were detected as sharp and clearly defined spots, and which were detected only as faint, overlapped, or poorly shaped spots? How can the user automatically remove the poorly resolved spots from his or her comparisons? These questions have been addressed in several ways. As the user obtains the PPM value of any spot using the SPOTTABLE form (Fig. 5F), he or she immediately can know on how many gels the spot was detected, and the standard deviation of its intensity on those gels. The numbers reported in the SPOTTABLE have five digits. Only three are significant; the fourth is the standard deviation (divided by ten and rounded up), and the fifth is the number of gels from which the spot was quantitated.

If a user wants to know what was the actual PPM of the spot from each gel, he can examine the SPOTBYGEL form (not shown). This form will tell the actual PPM obtained from each gel, and in one of its modes, it will report the spot status codes to indicate which spots are overlapped, composite, saturated, have bad shape, or have an error flag due to an inconsistency detected during the MERGE or RECONCILE processes. Using the MAKESETS form (explained below), it is possible to select those spots that are less than ideal based on any of the status codes. If desired, the selected spots can be excluded from further consideration by the data analysis programs.

2. SAMPLE COMPARISONS

To determine the relatedness of any two samples, we use the form COMPARE (Fig. 5F). On this form, we specify the samples that are to be compared, either as histograms of spot ratios, as scatter plots, or as correlation plots. The histograms and scatter plots help to reveal how many proteins differ in intensity between two samples, and they can give us a quantitative measure of relatedness. For comparison of muscle cultures at various stages of differentiation, the samples prepared from L6 muscle cultures were resolved on 7 different types of gel (three pH ranges for 10 and 12.5% gels, plus a pH 6–8, 7.5% gel), and each gel was exposed to film three or four times. A standard was generated for each gel, and the films were carefully matched to these standards. The standards were also matched to one another so that the position of most proteins is known to the computer for each gel type. The matched data were processed using the MERGE and RECONCILE procedures as described earlier, and a single SMPSPOTS file was created for each of four muscle cell samples. These database files, each containing entries on at least 1800 spots, were used to create the histograms and scatter plots shown in Figs. 7 and 8, and for the spot graphs shown in Fig. 9.

In each histogram, the ratios of protein intensities are plotted on a logarithmic scale. All spots detected in both samples, excluding spots

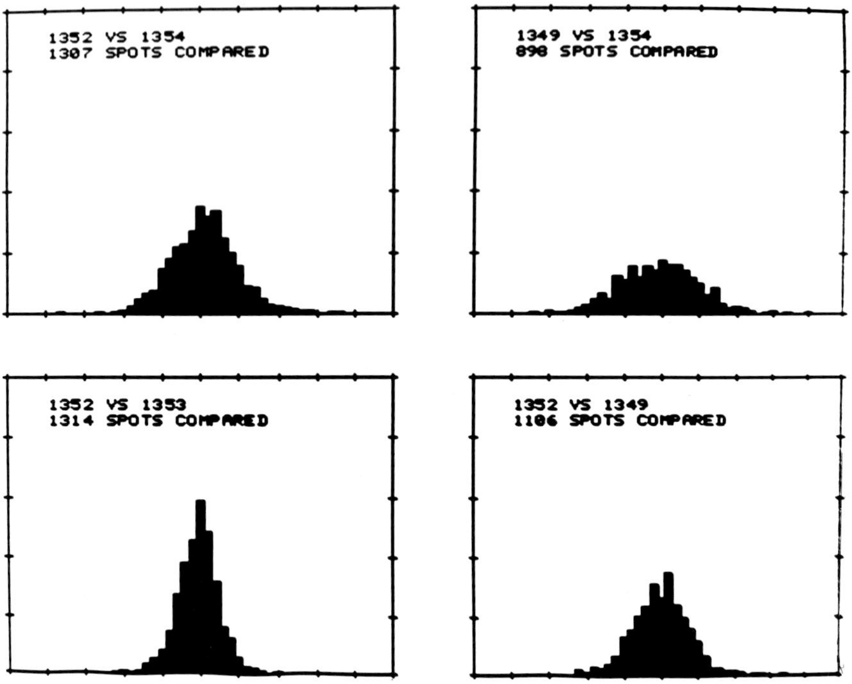

Figure 7. Comparison of differentiating L6 cells by histograms of protein intensity ratios. Proteins of the four samples, as shown in Fig. 6, were quantified, matched to standards, and compared pairwise. For each comparison, intensity ratios were computed for proteins detected in both samples, and the logarithm (base 2) was used to construct a histogram. The midpoint on the x-axis represents a ratio of 1, and each division represents a two-fold higher (to the right) or lower (to the left) ratio. The y-axis represents the relative number of proteins in each bar of the histogram. The standard deviation of these histograms can be used as a measure of relatedness for the samples compared. The samples compared and the protein intensity ratios corresponding to one standard deviation are (lower left) Day 6.5 versus Day 7.5, 1.53; (lower right) Day 6.5 versus Day 3.5, 1.81; (upper left) Day 6.5 versus Day 11.5, 1.92; (upper right) Day 3.5 versus Day 11.5, 2.22. In a control experiment (not shown) using duplicate fibroblast samples, the ratio corresponding to one standard of the histogram was 1.35.

flagged as composites, severely overlapped, or having bad shapes, were plotted. The standard deviation of each histogram is a measure of sample relatedness, and these values are reported for each comparison in the legend to Fig. 7. In a control experiment, 1023 proteins were compared from duplicate samples of a rat fibroblast line, and the relatedness value (ratio corresponding to one standard deviation of the histogram) was 1.35, which is somewhat smaller than the relatedness value (1.53) for Day 6.5 versus Day 7.5 fusing myotubes, and is much smaller

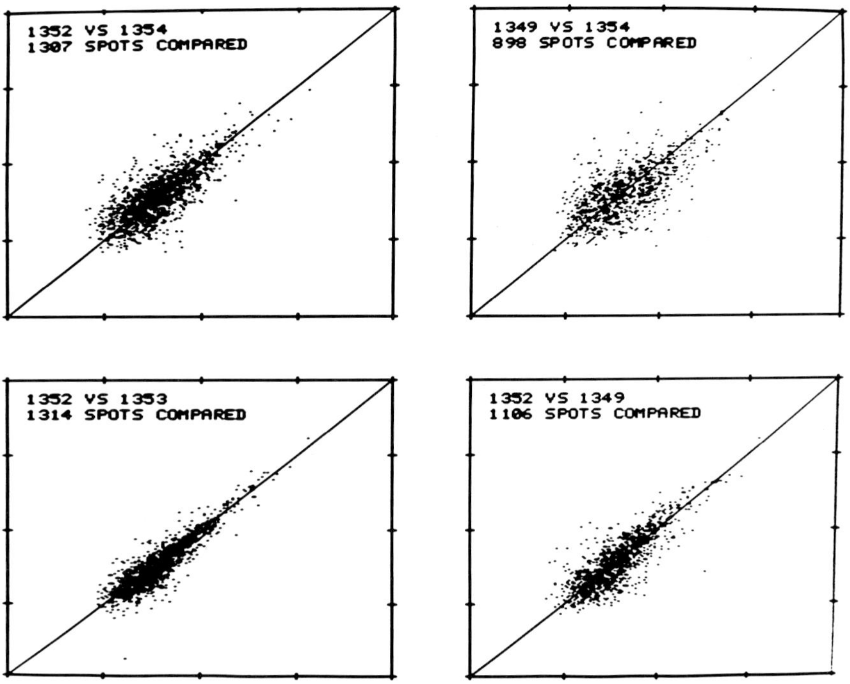

Figure 8. Comparison of differentiating L6 cells by scatter plots. Each scatter plot, like each histogram of Fig. 7, is a comparison of the proteins in two samples. Each protein detected in both samples is plotted as a point according to its intensity in the first sample (x-axis) and its intensity in the second sample (y-axis). Each axis is a logarithmic scale; the origin represents 1 PPM and each division represents a 10-fold higher intensity. The samples compared in each panel are the same as in Fig. 7.

than the relatedness value (2.22) for Day 3.5 myoblasts versus Day 11.5 myotubes.

The scatter plots (Fig. 8) present the same data in a different format. For the scatter plots, each protein is plotted as a point according to its intensity in the two samples (represented by logarithmic scales for both the x and y axes). The scatter plots reveal how much the proteins of the two samples differ as a function of intensity.

A correlation plot (not shown) is a method for comparison of two biological response patterns. For each protein of the correlated samples, increases and decreases in its relative PPM value are plotted on the x-axis for the first response (an experimental treatment versus a control), and its increases or decreases in relative PPM for a second response (another experimental treatment versus another control) are plotted on the y-axis. Any correlation between the two responses is quickly re-

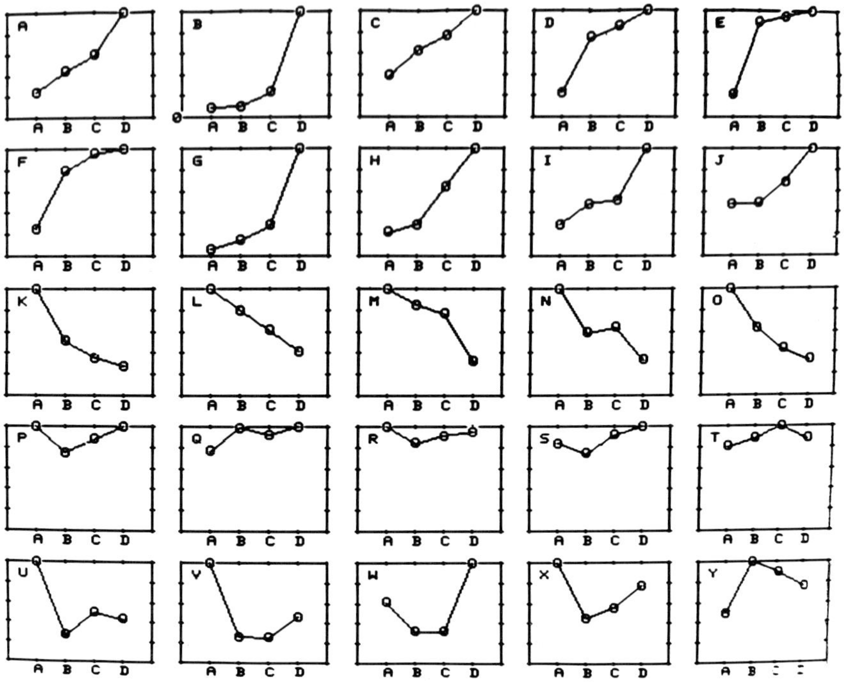

Figure 9. Comparison of differentiating L6 cells by spot graphs. Some of the data from the L6 data base files has been plotted on the TV monitor using the SPOTGRAPH form. Each graph presents data for one of the proteins seen in the portion of the gels shown in Fig. 6. In each graph, the protein intensities (fraction of total incorporation over a 2-h labeling period) are plotted for the four stages of myogenesis. The vertical scales have been normalized so that the data on each graph are plotted with the highest data point at full scale. The graphs in the top two rows show some of the different patterns of protein increase during L6 myogenesis, the third row shows proteins that decrease, the fourth row shows proteins that do not change substantially, and the fifth row shows proteins with unusual patterns of regulation. The letter of each graph refers to one of the lettered spots in Fig. 6.

vealed and can be expressed statistically as a correlation coefficient. Examples of correlation plots have been presented in previous publications (Garrels, 1979a; Garrels and Schubert, 1979).

3. SPOT SETS

To select the particular proteins that respond to changes of experimental manipulations and to correlate data from larger numbers of samples, we can generate spot sets using the form MAKESETS (Fig. 5F). Three basic types of sets can be created: sets based on "unary" relationships (involving intrinsic spot properties and selected from a single input file), sets based on quantitative relationships between two SMP-SPOTS files, and sets based on logical relationships between two other sets. Unary sets can be selected based on a specified range of spot intensities, spot coordinates (not yet implemented), and on the basis of spot status codes (as described previously). Sets based on quantitative relationships can be selected according to the amount of increase or decrease between two samples. Qualitative differences (spots present in one sample but not in the other) can be included in these sets by indicating the amount by which a spot must exceed the level of detection to be included in the set. Logical sets are created from other sets by the logical operations AND, OR, and NOT.

The sets constructed by combinations of unary, quantitative, and logical operations are used for data analysis in three ways. The members of a set can be listed, using the LISTSETS form (not shown), they can be marked graphically on any standard image, using the program MRKSETS, and they can be converted into an image, using the form SETIMAGES (not shown). The MRKSETS program is especially useful for marking selected spots for special attention during interactive matching. Problem spots, spots with large intensity changes in a given experiment, and spots of special scientific interest can be explicitly marked and rechecked during an interactive matching session to be sure that each has been properly resolved and properly matched. The use of the SETIMAGES form to produce synthetic images from spot sets is useful for graphic displays and plotted maps, especially when one image is overlaid on another. The set images for the proteins induced and repressed during differentiation in the L6 muscle cell line are shown in Plate 3. The images shown correspond to a portion of the gels shown in Fig. 6.

4. TABULAR AND GRAPHIC DISPLAYS

Two important forms, SPOTTABLE and SPOTGRAPH (Fig. 5F), allow either tabular or graphic presentation of the intensities of particular spots in particular samples. SPOTTABLE is the basic tool for retrieval of quantitative information for specific spots in selected samples. As mentioned above, the data displayed by SPOTTABLE contain information on the standard deviation and number of gels used in the quantitation of each spot. SPOTGRAPH facilitates the plotting of up to 25 graphs of SMPSPOTS data on one color TV screen. In SPT mode, the spots of interest can be plotted across the experimental series of samples (which may represent a time course, a dosage series, etc). Alternatively, one graph can be plotted for each sample (SMP mode) and the relative amounts of a selected group of spots can be displayed in each sample. The plotted graphs presented in Fig. 9 (generated from the SPOTGRAPH record shown in Fig. 5F) show the changes of some of the L6 muscle cell proteins (as marked in Fig. 6) during differentiation in culture.

5. THE TEXTUAL DATA BASE

Another aspect of the data base involves a further use of the QST data management system. As spots become identified with known proteins, and as other attributes are attached to the spots, such as tissue specificity, hormonal responses, phosphorylation, etc, we would like a centralized system for recording all these observations. To this end we have created the forms RATSPOT (Fig. 5G), MOUSEPOT, and HUMANSPOT. Each of these forms has 9999 allocated data records, one for each spot in the SSP numbering system. Each record provides a place for the name and basic properties of the spot, and it provides pointers into a large variety of other possible types of records. Using the related forms key, it is possible to move directly from a particular spot record to other data records that display information concerning phosphorylation, tissue specificity, polymorphisms, and much more. The entries to this system are made using the OBSRVTION form (Fig. 5G). Here the user enters all the applicable spot numbers along with the experiments and gel types being observed, followed by a written observation. When the record is saved, the observation is written into the next available record of the appropriate textual data-base form (e.g., PHOSPHOR or MUSCLEDIF as shown in Fig. 5G), and pointers to this record are inserted in all of the specified spot data records. The entry into the OBSRVTION form is itself permanent and is cross-referenced from the data-base form, so that if the observation is later questioned, the name of the observer and gels he was observing will be known.

6. THE PLOTTED SPOT MAPS

Storage of a large number of observations by spot number requires some convenient way to get to know the spot numbers and their positions within the gel patterns. We use the form PLOTIMAGE (not shown) to create a plotted contour map of any standard image at any magnification. Usually the images are magnified 6–8 times to give a wall-sized map on which the spots and their labels are clearly distinguished. The plot program writes onto each image a separate label file (made using the form MAKIMGLBL, not shown). The label file can contain any of the following for each spot: the SSP number, the name of the spot if known, and a separate label denoting some characteristic property of the spot. These wall charts allow all users to see the standard maps and the spot labels, and to look up the SSP numbers if necessary. A portion of the L6 map is shown in Fig. 10.

Since the plotted maps look very much like the gel images, it is possible to visually correlate the spots on a map with the spots on a gel exposure. If a user has a new gel with an interesting change in an unknown (to him) spot, he can get its number from the map and then use the data-base facilities of the computer to obtain its intensity in other samples and to read whatever has been previously recorded for this spot. The printed maps are a good way to transfer information between laboratories, and eventually annotated maps similar to these may be as commonplace on laboratory walls as are the maps of intermediary metabolism.

V. Program Descriptions

The software system for a dedicated two-dimensional gel electrophoresis and image analysis facility should provide a large variety of computational, data management, and image processing services, as described in the previous section. These services should be accessible at multiple terminals for use by scientists and technicians with no special computer training. To be practical in a scientific laboratory, such a system must be easy to enhance and modify without a large programming staff, and it should be able to run on a minicomputer.

In designing the QUEST system, we decided to give highest priority to ease of use. This led to the development of formatted screens to handle virtually every user need, including data entry, data access, and control of programs. Programs were written to cross-reference and index the user-entered data as much as possible, and other programs were written to manage the allocation and deletion of files needed for the image processing programs. As a result, data files such as GELDAT and SAMPLE are compiled automatically by the system, and programs such as SCAN or CALSCAN can be initiated merely by entry of the film number and the initials of the operator.

The second priority was ease of program development and modification. This required use of standardized subroutines for many shared functions, such as data-base access and intertask communications, and it required use of standardized conventions of program layout and variable naming. The result of this effort has been a system of over 200

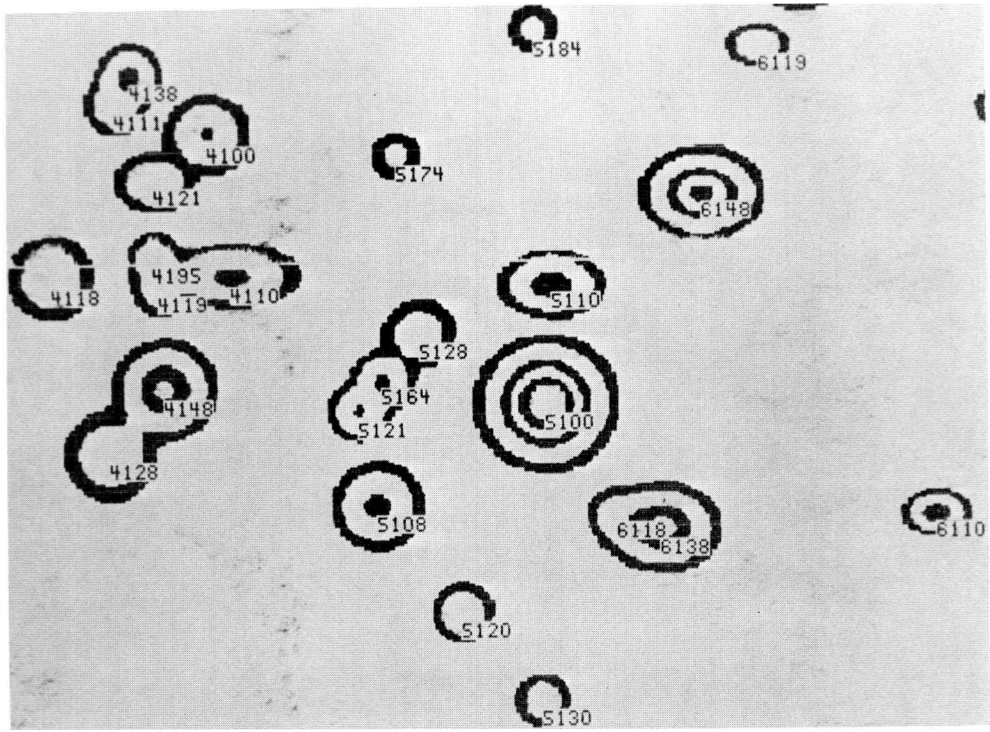

Figure 10. A portion of the L6 standard map. The map is constructed from a synthetic image of the standard FILMSPOTS file. The contour lines represent levels of intensity just as the colors represent intensity levels on the TV display. The RATSPOT numbers have been assigned automatically by the computer, with the first digit increasing from left to right and the second digit increasing from bottom to top. The area of the map shown corresponds to the lower right portion of the gels shown in Fig. 6.

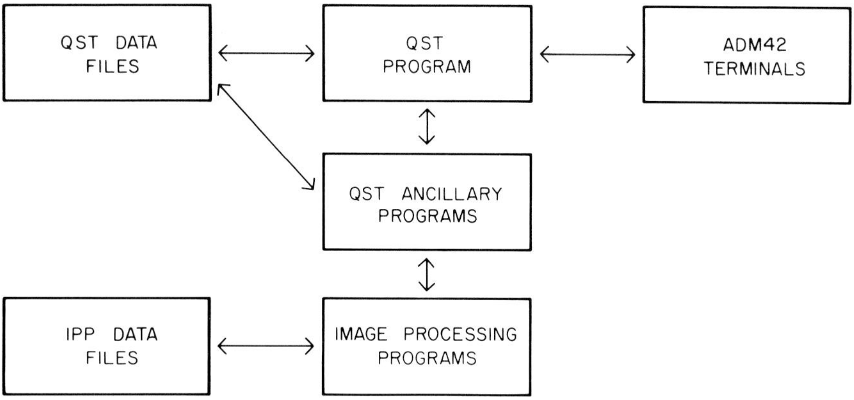

Figure 11. Relationships between the QUEST programs and the data files. The QUEST system is an integrated system of intercommunicating programs and data files. The QST program manages data flow to and from the QST terminals. The QST ancillary programs process formatted data (QST data) and in some cases communicate with the image processing programs that are specialized for processing large amounts of numerical data (IPP data files).

programs, each written in a standardized format and consistent style and each able to access data from the data base or pass data to any other program in the system.

We decided not to place emphasis on conservation of disk space or on speed of program execution if these considerations led to compromises in the previously mentioned objectives. It seemed likely that as our processing and data storage requirements outgrew our initial system, faster processors and larger storage devices would come into the affordable range. This assumption has proven justified.

The QUEST system has been implemented on a PDP-11/60 computer using the RSX-11M operating system (version 3.1), and all programs are written in Fortran (Fortran 4-Plus). Some specialized calls to the operating system (nonstandard Fortran calls) are necessary, but these are found in only a few specialized subroutines, which could be modified to adapt the software to another host computer.

There are three major types of programs and two main types of data files in the QUEST system. The programs communicate with one another and with the data files as diagrammed in Fig. 11. Each ADM-42 terminal is controlled by a separate copy of a program called QST. This program handles the reading and writing of forms and data records between the QST DATA files on disk storage and the ADM-42 terminal. The QST DATA files store formatted (character) data exactly as it is received from the terminal. A master table loaded into each QST program specifies which forms can be accessed from each terminal, the "page" of the terminal on which each will be displayed, and the name of an ancillary program (if any) to be called to run each time a new or modified data record is read from the terminal. The ancillary program (QSTACP) takes input from the record just saved, and it can read and write to any QST DATA record in the QST data management system. When the QSTACP program returns an acknowledgment to the QST program, the current data record is redisplayed on the terminal. Many of the QSTACP programs do calculations and data management for the gel laboratory, but a subset called the QSTIPP programs are used to provide input to the image processing (IPP) programs. A QSTIPP program reads input parameters from the form, finds the input data files and space for the output data, and sends this information to the IPP program through a shared memory area (QCOM). The IPP programs can be smaller and faster because they do not have the input overhead. They access only IPP data files (large, unformatted data files for images, filmspots, etc), but they cannot access the formatted QST DATA records.

File management and image processing are logically distinct functions which are conveniently separated into the QSTIPP and IPP programs. File locations and other header information is accessible to the user or to QSTIPP programs on forms entitled QSTIMAGES, FILMSPOTS, SMPSPOTS, etc. The QSTIPP programs normally handle all file allocation, purging, and deleting of unneeded IPP files (e.g., unneeded image files), but the user can perform the same functions manually if desired, simply by displaying the file headers on a form and changing their status. The IPP programs have specialized buffers and input/output (I/O) routines optimized for rapid access of the IPP DATA files. Data can be moved in either direction between disk and memory while data processing continues. Most IPP programs are limited by processor speed and not by I/O delays.

A. *Programs for Spot Detection and Quantitation*

The SCAN program controls the Optronics P-1000 scanner, receiving data from one scan line at a time. As each line is received it is inverted and smoothed before being stored in the QSTIMAGES file. Inversion is carried out for operator convenience, otherwise all users would have to be trained to mount films upside down. Smoothing removes some of the background noise due to film grain. For speed, smoothing is done using only adjacent points within each scan line. The degree of smoothing (selectable from the SCAN form) is determined by weighting the average of each pixel with its immediate neighbors in the scan line.

Images are reduced to spots by the three programs: LNANAL (line analysis), CHNBLD (chain building), and CHANAL (chain analysis). These programs have been described in detail previously (Garrels, 1979a), and only an outline of their function is given here. The line analysis program analyzes each line of the image to find the peaks of density. These peaks are integrated [after subtraction of local background and conversion of density to disintegrations per minute per unit area (dpm)], and the parameters that describe the intensity, shape, and position of each line peak are recorded in the LINEPEAKS file. After line analysis the CHNBLD program rearranges the line peak data to build "chains." A chain of line peaks is the series of line peaks from adjacent scan lines that represent an entire spot (or several overlapping spots).

Line analysis has been enhanced with regard to integration of line peaks for faint spots. Very faint spots are usually underrepresented because only a small area near the center (the "tip of the iceberg") is detected above background. After longer exposure such spots have a much greater area and consequently a higher value for dpm. Examination of many well-developed spots showed that total line peak dpm is always at least eight times higher than the dpm of the single highest pixel in the line peak, but many faint spots have ratios much smaller than this. Based on this information, we have built a correction into the line analysis program to increase any line peak dpm values that would otherwise be less than eight times the dpm of the highest pixel.

The chain analysis program is the largest of the three spot detection programs and requires the most computer time. It analyzes each chain of line peaks by fitting Gaussian profiles over the chain and resolving it into one or more spots. It uses line peak intensities and positions to determine how the chains should be split. Spots that overlap horizontally or diagonally can be easily resolved and accurately fitted. If a chain is composed of more than two spots (which often occurs in crowded areas), an optimizing routine is used for the best fitting of multiple Gaussians. This program has been much enhanced over the version described previously (Garrels, 1979a), but the principles remain the same.

Output from the chain analysis program goes into the FILMSPOTS files. The FILMSPOTS record is large (128 bytes) because it is used to record many parameters describing each spot and to provide for work space for use during matching. The FILMSPOTS records are not sorted, and their positions within the file are never changed. A smaller file called NEIGHBORS contains some of the FILMSPOTS data (dpm, position, status codes) and the list of ten nearest neighbors to each spot. The

NEIGHBORS file is sorted by x coordinates, and each points to its corresponding FILMSPOTS record. For some purposes, the data in NEIGHBORS is sufficient; for other purposes, the NEIGHBORS file provides an index to the FILMSPOTS file.

B. The Automatic Matching Program

An automatic matching (AUTMAT) program has been developed that is fast and highly accurate for comparison of related spot patterns. It takes as input a FILMSPOTS (FSP) file, a standard filmspot (STD) file, a NEIGHBORS file for each of the above, and a MATCHLIST file containing the initial landmark information. Based on the landmarks previously supplied by the operator, the surrounding spots of the images will be matched and logged into the MATCHLIST file. Each spot of the standard image, except those matched manually and those marked "no data," are candidates for automatic matching. There is a record in the MATCHLIST file for each standard spot; those that cannot be matched are simply logged as unmatched and are presumed to have an intensity below the level of sensitivity of the film analyzed.

Two algorithms are fundamental to the AUTMAT program, and they will be described separately. One is the procedure, encoded in the MATSPT subroutine, for taking a given standard spot and, based on coordinate information from previously matched spots, finding a match in the FILMSPOTS file. The second algorithm is the procedure for determining the order in which the standard spots should be matched. We describe the latter procedure first.

The first standard spots selected for matching are those nearest to the landmarks. At first the landmarks provide the only available coordinate information for matching the two images. As we match the spots near the landmarks, we can use the new match information to progressively match spots farther and farther from the landmarks. Four passes are made over the image using a different spot selection strategy each time. On the first pass the matching begins at the landmarks and propagates outward simultaneously from all landmarks. The spots first selected for matching are those standard spots that are neighbors to a landmark. If any were successfully matched then their neighbors are in turn selected for matching and so on for many generations, until finally on one generation no new matches are made. At each generation all neighbors of the new matches are rematched, including spots matched on previous generations. This means that the matches made in the early generations are constantly being rematched as more and more of their neighbors become matched. Sometimes a spot matched incorrectly in one of the early generations will be unmatched or matched to a different spot during one of the later generations.

The second pass is like the first, except that it is less selective. The first pass considers only high-quality standard spots that do not have overlaps or bad shape codes. The second pass does not exclude spots on the basis of quality. On the second pass spot selection begins again with the neighbors of the landmarks and proceeds outward as before. All spots from the first pass will be rematched and some additional spots will be matched, but this time the coordinate information from all the first pass matches can be used for improved accuracy. This strategy assures that the most reliable matches are made first and that subsequent matches can be made from the best possible coordinate information.

The third pass matches all spots that were not selected for matching on one of the first two passes because they were not reached by one of the trees propagating from a landmark. There is no propagation by neighbors; these spots are simply submitted to the matching subroutine to see if a match can be found on the basis of all previous matching data.

On the fourth and final pass all spots are once more submitted to the matching subroutine to be rematched based on the information provided by all previous matches. Those spots closest to the landmarks are matched first, then those of the next generation, etc. Unlike the first two passes this pass does not match all the neighbors of each new match: it simply matches each spot once. The matches made on the fourth pass are logged in the MATCHLIST file.

The actual matching is done by the MATSPT subroutine. To find the FILMSPOT that best matches a selected standard spot, MATSPT must determine the relationship between the two coordinate systems. There is no attempt at any stage to calculate a global mapping function to relate all spots on the two images; instead, a local mapping function is calculated each time a spot is to be matched. The local mapping function is based on the coordinates of previously matched spots in the local area, and as more and more of the spots become matched the local mapping functions get better and better.

To find a match for a selected standard spot, the MATSPT program will determine how many of the neighbors of the standard spot have been matched. If fewer than four have been matched it will then determine how many of the neighbors of the neighbors (extended neighbors) have been matched. Then for all available matches among these spots it makes a list of the standard coordinates (X0,Y0) and the FILMSPOT coordinates (X1,Y1). If this list contains three or more spots, then a linear regression is applied to the data to determine the straight line that best fits the X0 versus X1 data and another line to fit the Y0 versus Y1 data. The slopes of these lines measure the local stretch (or compression) between the two images. If fewer than three matches are available, then only a relative displacement between the coordinate systems can be calculated. Once the displacements and slopes have been calculated, then a set of coordinates can be calculated, predicting where the selected standard spot should appear in the film image. If one or more spots fall within a predefined radius of this point, then the closest spot is considered to be the match. This match will in turn be used to calculate subsequent matches, and each match will itself be redetermined several times, as more of the surrounding spots become matched, before the match process is complete.

By matching each standard spot multiple times and by recalculating the local mapping function each time new information is available, an optimal mapping of the entire image can be achieved. Frequently spots that were matched incorrectly in the early stages of matching will be matched correctly at a later time when more of the surrounding spots have been matched. This matching program does not "get lost" if one or a few spots are matched incorrectly because the matching propagates outward from all landmarks at once and because so much redundancy and rechecking is built into the program. In areas with good landmarks the automatic matching is more reliable than human matching. The computer can simultaneously consider all surrounding matches and the local distortion factors whereas the human operator can match the spots only by sliding one film across the other. Near the edges and in regions of low spot density, the MATSPT subroutine occasionally makes mistakes or fails to match spots. Human editing is needed after automatic matching to add further landmarks in poorly-matched regions (due to low density or severe distortion) and to check complex regions where many standard spots overlap. The automatic matching can only be as reliable as the standard used, and it is the duty of the operator to ensure that only well-defined and accurately positioned spots are included in the standard.

C. Programs to Create and Update Standards

The standards are central to the matching and final data reduction processes. Each standard is described by a data record displayed on the STANDARDS form and by three IPP DATA files: a standard filmspots (STD) file, a NEIGHBORS file, and a synthetic image (QSTIMAGES) file. A new standard is created by the MAKSTD program directed from the STANDARDS form. The FILMSPOTS file derived from the specified film is copied into a new standard filmspots (STD) file that has space allocated for 3000 standard spots. As it is copied the FILMSPOTS records are sorted and distributed throughout the new file, leaving many null (unused) records for definition of new spots that might later be added to the standard. Some of the fields of the standard filmspots records are redefined and used to log some information on standard spot history and usage. Recorded in each standard filmspots record is the original film number (or film from which the spot was added), the total number of times the standard spot has been matched to a FILMSPOT, the total dpm of all matched FILMSPOTS, the total x deviation (distance between expected and actual position of the matched spot) and total y deviation for all matches, the maximum dpm of a matched FILMSPOT, and the film number for the matched spot with the highest dpm. These fields allow later assessment of the usage of each standard spot. If a standard spot is added but never used for matching, it might be considered an artifact and later removed. If the initial coordinates of the standard spot are slightly misaligned, the examination of average x and y deviations will reveal this fact and allow the coordinates to be later readjusted. The dpm value used to represent each standard FILMSPOT in construction of the synthetic image used for matching can be changed, if desired, from that of the original film to a value representing the average of many gels.

Other fields of the standard FILMSPOTS record allow definition of a spot quality. Some spots may be marked as composites if it is known that the spot represents two or more unresolved spots. Others may be marked as permanent landmarks, indicating that they are prominent, nonoverlapped spots that are reliable landmarks in virtually all gels. The shape and overlap information from the original film is maintained for the standard spot, and space is reserved for computation of overlap between standard spots even though they may arise from different gels. Finally, the standard FILMSPOTS file contains a field for the species spot number.

D. *Assignment of Species Spot Numbers*

The species spot (SSP) number, as described in Section IV,F, is the master number for each distinct protein detected within the species. There is one record in each SMPSPOTS file for each possible SSP number (1–9999). Conversion of data from the numbering system of each of the various standards to the single SSP numbering system is the role of the RECONCILE process, as outlined in Section IV,F. As the RECONCILE process gathers information about each species spot from the various gels that have been analyzed, it needs to know where each species spot runs on each standard. A table relating SSP numbers to the FILMSPOTS numbers in each standard is required. This table is a large disk file called SSPSTD. The rows of the table represent SSP numbers and the columns represent the standards (from any species) that have been created by MAKSTD. Each entry into the table is a standard filmspot (STD) record number. Special columns at the right of the table provide summary information to indicate which SSP numbers have been used by any standard spot within the species.

SSP numbers are assigned to standard spots by a program called STDTAB that is run automatically each time a standard is created or updated. The STDTAB program first updates the column for the current standard and the summary column of the SSPSTD table. It then looks for spots of the current standard that do not yet have assigned SSP numbers, and it chooses new numbers for these spots. The new SSP number must be a number that has not been used for any spot in any standard. (Spots representing the same protein on different standards will have different SSP numbers until the respective standards have been matched to one another.)

Assignment of new SSP numbers is not random; the numbers are chosen so that the first digit indicates position from left to right in the standard image and the second digit is chosen to indicate position from bottom to top. When each standard is first created, the STDTAB program divides the standard image into ten columns, each containing an equal number of spots, and each column is divided in turn into twenty boxes, each with an equal number of spots. The column coordinates and the box coordinates within each column are permanently recorded at the end of the standard column in the SSPSTD table. Each time a new SSP number is to be assigned, the STDTAB program first finds the box in which the spot lies, and then assigns an unused SSP number appropriate to that box. This distribution of the SSP numbers is done only for the convenience of the users; none of the programs require any special distribution of the SSP numbers. As seen on the spot maps (Fig. 10), the number distributions help the user to find any spot on the map easily.

E. Programs for Final Data Reduction

Most proteins of any sample analyzed by the QUEST system will be detected on more than one gel and on more than one film exposure of each gel. After each film has been reduced to a FILMSPOTS file, and after the spots in each FILMSPOTS file have been matched to a standard, we are ready for final data reduction. The MERGE process produces a

GELSPOTS file representing the best dpm value for each spot of each gel, and the RECONCILE process combines the gel data into a single SMPSPOTS file. No user intervention is required during the MERGE or RECONCILE processes. The rules for deciding which data to keep, which to discard, and which to average together are built into the programs. Both users and programmers should, however, be aware of the decision-making processes involved in final data reduction.

The MERGE process involves two programs, MAKGSS and MAKGSP. The former gathers together the relevant data into a temporary file called GSPSTAT, and the second program analyzes this data to produce a GELSPOTS file. The GSPSTAT file contains the following types of information for each spot in up to five exposures of a gel: integrated dpm (both by fitting and by addition of line peaks), integrated optical density, maximum density, degree of overlap with neighboring spots on the left and right, a shape code, and a quality code. The shape code indicates whether the line peak intensities in the spot rise and fall across the width of the spot (as for a normal spot) or whether the line peak intensities rise only or fall only (which might indicate either noise or a shoulder split from the side of another spot). The quality code indicates whether the spot was recognized during initial Gaussian fitting to the "chain" during chain analysis or whether the spot was apparent only after subtraction of the more prominent spots from the chain.

To choose the best dpm value for each spot, the MAKGSP program examines the spot data from each exposure and discards problematic spots if better data is available from another exposure. It first discards spots based on the quality code, then on the basis of shape, then on the basis of overlap, and finally on the basis of density, discarding saturated spots and spots of low density. The number of the exposure that best represents the spot is then logged into the GSPSTAT file, and if two exposures represent the spot adequately, the second best exposure is also recorded. Finally a consistency check is made and an error will be flagged if:

1. The dpm from two adequate exposures does not agree.
2. The spot has a maximum density too low for accurate quantitation.
3. The spot has a "rising only" shape in one exposure, but a "falling only" shape in another exposure.
4. The spot is overlapped on different sides in different exposures.

The latter two conditions often indicate inconsistent matching and the error codes assist in finding these spots when the matches are re-edited.

The RECONCILE procedure is carried out by the two programs MAKSSS and MAKSSP. The MAKSSS program gathers information for each of the 9999 possible species spots. Using the SSPSTD table it looks into the STD files and the GSPSTAT files to obtain all information pertaining to each species spot from up to 10 gels. This information is logged in a temporary file called SSPSTAT. These parameters logged in SSPSTAT include the same information as those used in the MERGE process, but in addition flags are included to indicate those spots that are a composite of two or more unresolved spots in either the standard or in the film image. The second program, MAKSSP, examines the spot data logged in SSPSTAT to determine the best quantitative value (in terms of parts per million of the applied sample) for each species spot. Data from poorly resolved or problematic spots are discarded if better data are available from another gel. First to be eliminated, if possible, are the composite spots, followed by spots with a merging error, spots with bad shape, and finally spots with overlap. The spot data from the remaining gels are averaged and those that fall more than two standard deviations from the mean are discarded. The best PPM value is logged into SMPSPOTS, the standard deviation is encoded into the fourth digit of the PPM value, and the number of gels averaged is placed in the fifth digit of the PPM value. In addition, a status byte composed of bits to indicate saturation, overlap, merge errors, and bad shape is constructed and logged in an accessory data file called SMPSPTLOG. The latter file is utilized by the SPOTBYGEL form, which was described in the previous section.

F. Processing Times and Storage Requirements

Processing times for spot detection and integration are reasonable (20 min or less for an image with 1000 spots), especially since most of the scans done each day are queued in the IPMANAGER for overnight processing. The automatic matching program is also fast (5 min or less). The limitations of the present system come in terms of processor memory and overall throughput. A 16-bit minicomputer has been adequate for development and testing of the system, but we now have many programs and many users, putting a strain on the computer memory space for running multiple programs concurrently. Fortunately, a variety of 32-bit minicomputers have been introduced while our system was under development, and these are now the hardware of choice for most laboratory applications such as ours. These newer computers allow more programs to run simultaneously and allow the use of very large programs. Entire images can be loaded into memory, eliminating the programming complications and processing delays entailed by frequently reading portions of the image from disk.

Overall throughput is limited by the time needed for interactive graphics. The present system supports only one user at a time and only a portion of one image can be held in the display memory at one time. Much of the operator's time is spent waiting for the display memory to be reloaded from the computer. Fortunately, much more powerful graphics equipment has been introduced while our system was under development and an upgrade to this equipment should not be difficult. The system of choice now should allow two or more full (1000 × 1000 byte) images to be loaded into display memory at once and should permit real-time panning and zooming of the display, controlled by a joystick or trackball. Overlay planes can be used to show (or to suppress) the editing marks without interfering with the image memories. For faster indication of spot positions, the films to be matched could be mounted on a digitizer tablet (lighted from the back) and landmark spots could be indicated directly on the films with a stylus.

Disk storage has become limiting, but better equipment and falling costs promise to alleviate the problem. We prefer to store FILMSPOTS data for as many films as possible so that we can always go back to check and re-edit our matches. Ideally about 1000 FILMSPOTS files would be on line. Each contains on average about 1000 records of 128 bytes for a total required storage of 128 megabytes. We also need to store at least as many images as can be processed in one day, which is presently in the range of 20 or 30, and each image requires about 1.5 megabytes of storage. Finally, a large amount of storage is needed for our gel laboratory data and for our textual data base (records containing written observations in several categories for thousands of RATSPOT, MOUSESPOT, and HUMANSPOT proteins). We presently have about 60 megabytes devoted to such use. Therefore, a system to support routine gel analysis and development of a sizable data base and should include at least two large (over 200 megabyte) disk drives.

G. Future Program Enhancements

Other limitations of the QUEST system will be overcome with further program enhancements. A two-dimensional Gaussian fitting routine would be useful for some clusters of overlapped spots. The present routine splits horizontal overlaps well but often does not properly handle spots with vertical overlap. If the overlapped spots come from the same chain, they will be flagged in the FILMSPOTS file as overlapped, and data from other gels where the spots are not overlapped will be preferentially used. However, spots with vertical or diagonal overlap are often split into different chains and do not have any reported overlap. A program to recognize spot clusters based on neighbor relationships could be developed. The coordinates of these spot clusters could be passed to an optimizing two-dimensional fitting routine, and the improved Gaussian parameters could be logged into the FILMSPOTS file. Additionally, a two-dimensional reintegration of selected spots could be implemented as one of the MATCH editing functions.

As an extension of our analysis of multiple exposures, we would like to develop a routine that can analyze double-labeled gels by time decay. Having two calibration curves, one for each isotope, and knowing the total amount of each isotope loaded onto the gel, it should be possible to analyze exposures done several months apart to accurately determine the ratio of a long- and a short-lived isotope (such as ^{3}H and ^{35}S) at every point on the gel. With this additional information the identification and accurate matching of minor proteins from different cell types should be much more reliable.

A totally automatic program for segmentation and analysis of calibration gels should be written. The computer can find the segments of equal density in the calibration gel image, average these densities, find their order, and then plot them versus the specific activity of the radioactive protein incorporated into each segment. The comparison of calibration gels processed on the same day and exposed for the same length of time would provide a redundancy check for the accuracy of the calibrations.

Further data analysis programs of many types will be written as needs arise. For example, programs are needed to plot the distribution of proteins as a function of molecular weight, pH, and intensity. A program to calculate the number of spots versus intensity should statistically compensate for the fraction of the gels that is covered by spots at each exposure level and therefore unavailable for detection of additional minor spots. Given the standardized form of the data-base files, it should not be difficult to write a variety of higher-level programs to correlate and manipulate this data.

VI. Perspectives

The large amount of information available for two-dimensional gels was almost overwhelming when the O'Farrell system was introduced. Even today, it remains difficult to extract the full information content from each two-dimensional gel. Computers have been programmed to process two-dimensional gel images, removing background noise and integrating the spot intensities. However, as with many forms of information, the data devoid of context have little meaning. The detection and quantitation of minor, unnamed proteins has little value unless these proteins can be detected repeatedly in different experimental situations. The body of accumulated data about each protein, which we call a data base, provides the context needed to bring out the full information content of each two-dimensional gel.

A classification of protein data bases by levels of complexity can help to give perspective to present and future data-base development. The levels of data-base complexity are separated by technological barriers largely involving the reproducibility of two-dimensional gel electrophoresis. The simplest (Level One) data bases can be created from a series of gels run as part of one experiment in one laboratory, using a single type of cell. This data base serves for the analysis of one experiment, but each new experiment requires a new data base. A Level Two data base can be constructed when the temporal barrier is overcome, allowing data from experiments done at different points in time to be compared. Level Three data bases contain data from more than one type of cell (within the same species) derived from experiments done over time in one laboratory. This type of data base requires more than just reproducibility; it requires high resolution and some form of accessory information, such as separation on multiple gel types or the use of several amino acid labels, in order to match the patterns representing diverse cell types reliably. Finally, a Level Four data base can be constructed only when the electrophoresis methods are standardized and automated sufficiently to allow detailed exchange of two-dimensional gel data between laboratories. A Level Four data base could eventually include a very large body of data about the proteins of a species, based on experiments carried out in many laboratories with many cell types over a long period.

The experience gained with Level Three data bases will be important for the eventual construction of Level Four data bases. The QUEST system is designed to support a large Level Three data base, and Level Three data bases will probably result from several of the other software systems described elsewhere (Lipkin and Lemkin, 1980; Anderson *et al.*, 1981; Vo *et al.*, 1981; Skolnick *et al.*, 1982). Each laboratory involved in data-base development must explore systems of protein nomenclature, methods of data management, and methods of user interaction. Furthermore, each laboratory with a Level Three data base will necessarily have a strong interest in maintaining and improving the reproducibility of the gels, and this experience should accelerate the move toward Level Four technology.

For the immediate future, it seems likely that computer-analyzed two-dimensional electrophoresis will become an important tool, spreading to many laboratories as less expensive equipment and better software becomes available. Many laboratories will use it only to quantitate proteins of special interest, and some laboratories will use it to develop Level One and Level Two data bases. To the extent that high resolution and reproducibility can be maintained, some laboratories will begin to develop sizable Level Three data bases. The establishment and use of such data bases are fundamental to the systematic study of the proteins involved in complex biological events such as differentiation, growth control, and transformation. Through this work, many new proteins will be named and characterized, and even before Level Four data bases are possible, much of the information in the lower level data bases can be exchanged and compiled (at least for several hundred major proteins). With the experience gained from the use of these systems and with improved technologies, it should eventually become possible to contemplate a universal numbering system and a master catalog of the names and properties for many of the human proteins and for the proteins of many other living organisms.

Acknowledgments

We thank Dr. James D. Watson for constant support during the development of the QUEST facility. We thank Dr. Robert Franza for his numerous helpful suggestions and for careful review of the manuscript. For technical assistance, we thank Patricia Smith, Thomas Kelly, Jessica Leibold, Lynn Cascio, and Jane Emanuele, and for programming assistance, we thank Steven Norris. We are thankful to Brian Hall and Lissa Evans for proofreading, and to Phil Renna and Mike Ockler for photographic assistance. This work was supported by grants to J.I.G. from the National Science Foundation (PCM-7809215 and PCM-7922769), from the National Institutes of Health (5-RO1-GM26298), from the Cystic Fibrosis Foundation, and from the Muscular Dystrophy Association.

References

Anderson, N. G., and Anderson, N. L. (1978). Analytical techniques for cell fractions. XXI. Two-dimensional analysis of serum and tissue proteins: multiple isoelectric focusing. *Anal. Biochem.* **85,** 331–340.

Anderson, N. L., and Anderson, N. G. (1978). Analytical techniques for cell fractions. XXII. Two-dimensional analysis of serum and tissue proteins: multiple gradient-slab electrophoresis. *Anal. Biochem.* **85,** 341–354.

Anderson, N. L., Taylor, J., Scandora, A. E., Coulter, B. P., and Anderson, N. G. (1981). The TYCHO system for computer analysis of two-dimensional gel electrophoresis patterns. *Clin. Chem.* **27,** 1807–1820.

Garrels, J. I. (1979a). Two-dimensional gel electrophoresis and computer analysis of proteins synthesized by clonal cell lines. *J. Biol. Chem.* **254,** 7961–7977.

Garrels, J. I. (1979b). Changes in protein synthesis during myogenesis in a clonal cell line. *Dev. Biol.* **73,** 134–152.

Garrels, J. I. (1983). Quantitative two-dimensional gel electrophoresis of proteins. *Methods Enzymol.* **100,** 411–423.

Garrels, J. I., and Gibson, W. (1976). Identification and characterization of multiple forms of actin. *Cell* **9,** 793–805.

Garrels, J. I., and Schubert, D. (1979). Modulation of protein synthesis by nerve growth factor. *J. Biol. Chem.* **254,** 7978–7985.

Lipkin, L. E., and Lemkin, P. F. (1980). Data-base techniques for multiple two-dimensional polyacrylamide gel electrophoresis analyses. *Clin. Chem.* **26,** 1403–1412.

O'Farrell, P. H. (1975). High resolution two-dimensional electrophoresis of proteins. *J. Biol. Chem.* **250,** 4007–4021.

O'Farrell, P. Z., Goodman, H. M., and O'Farrell, P. H. (1977). High resolution two dimensional electrophoresis of basic as well as acidic proteins. Cell **12,** 1133–1142.

Skolnick, M. M., Sternberg, S. R., and Neel, J. V. (1982). Computer programs for adapting two-dimensional gels to the study of mutation. *Clin. Chem.* **28,** 969–978.

Vo, K.-P., Miller, M. J., Geiduschek, E. P., Nielsen, C., Olson, A., and Xuong, N.-H. (1981). Computer analysis of two-dimensional gels. *Anal. Biochem.* **112,** 258–271.

CHAPTER 3

Detection of Polypeptides in Two-Dimensional Gels Using Silver Staining

CARL R. MERRIL
DAVID GOLDMAN
Laboratory of General and Comparative Biochemistry and Laboratory of Clinical Science
National Institute of Mental Health
Bethesda, Maryland

I. Introduction. 93
II. Staining Procedures. 95
A. A Negative Image Silver Stain. 95
B. A Positive Image Silver Stain. 96
C. A Modified Positive Image Silver Stain. 98
D. Recycling for Increased Sensitivity. 99
E. Other Image Intensification and Destaining Procedures. . . . 99
F. Sensitivity, Quantitation, and Protein Detection. 100
G. Silver Quenching of Radioactively Labeled Proteins. 103
H. Colored Protein Images with Silver Stains. 104
III. Mechanisms of Silver Stains. 104
IV. Potential and Real Problems in Silver Staining. 107
References. 108

I. Introduction

Separation techniques, such as two-dimensional gel electrophoresis, are capable of resolving thousands of proteins in complex mixtures. However, the most commonly used nonradioactive detection method, Coomassie Blue staining, lacks the sensitivity to detect proteins present in trace amonts. Concentration of samples is often counterproductive, resulting in distortions of gel patterns by the overloading of more abun-

TWO-DIMENSIONAL GEL
ELECTROPHORESIS
OF PROTEINS

ISBN 0-12-164720-X

dant proteins. Use of autoradiography with radioactively labeled proteins permits the detection of trace proteins (10^{-4} to 10^{-5}% of the total protein in a sample) as demonstrated by O'Farrell (1975). However, many samples cannot generally be studied by radioactive methods for the following reasons:

1. Experiments utilizing whole animals often require prohibitively large amounts of expensive radioactive precursors.
2. Radioactive labeling at the level generally necessary for detection of trace proteins in human clinical studies is unacceptable for ethical reasons.
3. *In vitro* labeling often alters the physical characteristics of the proteins to be studied.
4. Radiolabeling and detection is usually more expensive and time consuming than most staining procedures.

The inability to detect trace proteins limited the applications of two-dimensional gel electrophoresis in many biological and medical problems. Recognition of this limitation led to an extensive search for new protein stains in this laboratory. Heavy metal stains (uranium, lead, and mercury), fluorescent stains [fluorescamine and 2-methoxy-2-4-diphenyl-3-(2*H*)-furanone (MDPF), Hoffman La Roche, New Jersey], and chemiluminescent stains were tried. Sensitivities of these stains was less than or at best equivalent to Coomassie Blue. It was not until we applied a modified histological silver stain that Merril *et al.* (1979) and Switzer *et al.* (1979) obtained a stain which was 100-fold more sensitive than Coomassie Blue. The use of a histological silver stain was described by Merril *et al.* (1979) and Switzer *et al.* (1979). A histological silver stain had been used in agarose gels to stain cerebrospinal fluid proteins (Kerenyi and Gallyas, 1972, 1973; Verheeka, 1975), but it did not achieve widespread acceptance, which may be due in part to an inability to eliminate staining artifacts and Verheeka's (1975) report of quantitative irreproducibility. Our attempts to adapt this histological stain were abandoned because of our difficulties in using it with polyacrylamide gels and the expense of the required reagents. However, Hubbell *et al.* (1979) and Lischwe *et al.* (1979) demonstrated histological silver stains that are highly specific for certain proteins such as nucleolar proteins, and Gambetti *et al.* (1981) adapted a silver stain to stain neurofilament polypeptides in polyacrylamide gels.

General use of histological silver stains as adapted by Merril *et al.* (1979), Switzer *et al.* (1979), Kerenyi and Gallyas (1972, 1973), and Verheeka (1975) to detect proteins separated on polyacrylamide and agarose gels respectively was hindered by the fact that many of these stains took hours to perform, used considerable quantities of silver or other expensive reagents, and gave variable results. Attempts to simplify the staining procedures were hampered by the general lack of knowledge concerning the chemical basis of the selective reduction of ionic to metallic silver in most histological stains. In some cases the reduction may be catalyzed by reducing agents in the tissues, whereas in other cases differential silver staining appears to be based on physical effects in the tissues as described in the review by Thompson and Hunt (1966).

The highly developed field of photographic chemistry, from which many histological staining techniques were borrowed, offered the means to develop new highly sensitive silver stains. In photography photons initiate reduction of silver ions on the surface of silver halide crystals. This photoreduced metallic silver on the light-exposed crystals acts as a catalyst resulting in differential reduction of silver halide crystals during chemical image development. Chemical development in photography usually relies on the use of organic reducing agents in alkaline solutions. Given the extensive efforts that have been made in developing and understanding photographic chemistry, an attempt was made to utilize this knowledge in developing new silver stains for polyacrylamide gels. Efforts were made to minimize the use of expensive reagents and time while simplifying the procedures and maximizing the sensitivity (Merril *et al.*, 1981a,b, 1982). During this development effort with photochemical techniques, it became apparent that both negative and positive image stains could be produced. The positive image stains have proven to be the most sensitive and the negative image stains have been of interest for the insight they have provided on stain mechanisms.

II. Staining Procedures

A. A Negative Image Silver Stain

First fix the gel in 50% methanol–12% acetic acid (v/v) and then wash with deionized water. Soak the washed gel in 0.2 *M* silver nitrate solution in the dark for 20 min and briefly rinse with deionized water. The gel is then soaked in photographic developer (such as Kodak D76 in a 1:5 dilution) for 10 min followed by a 5 min rinse with photographic fixer and three 20-min water washes. All of the steps prior to the final water washes should be performed in total darkness.

The negative image obtained shows clear regions in the portions of the gel containing protein and background regions are brownish-gray (Fig. 1). The production of such a negative image indicates that the protein has affected the reducibility of the silver in the region of the gel occupied by the protein. This image may be reversed by exposure to light during development or by the assistance of chemical reversal procedures as shown by Merril *et al.* (1981a,b, 1982). This negative silver stain is less sensitive than the positive silver stains by about a factor of ten.

B. A Positive Image Silver Stain

As positive image stains are generally easier to analyze than negative image stains, a highly sensitive and simple reversal silver stain was developed by Merril *et al.* (1981a,b, 1982). In this stain, image reversal is facilitated by the use of potassium dichromate. To stain a gel, the following steps are performed (with gentle agitation):

Step 1. *Gel fixation* in 20% w/v trichloroacetic acid (TCA) or 50% methanol–12% acetic acid (v/v) for 1 h. Polyacrylamide gels thinner than 0.5 mm and agarose gels should be fixed in TCA to prevent loss of protein. Thicker gels require a longer fixation time.

Step 2. *Secondary fixation* with 10% glutaraldehyde for 30 min. The use of glutaraldehyde as secondary fixative increases the sensitivity of the silver stain (D. Goldman and C. R. Merril, unpublished).

Step 3. *Wash* gel twice with 10% ethanol and 5% acetic acid (v/v) [or 10% methanol and 5% acetic acid (v/v)]. Each wash should be for 15 min with a minimum of 10 gel volumes. This step allows the gel to swell to normal size and assists in washing out contaminating buffers and ions from the electrophoresis which might otherwise reduce the sensitivity of the stain. Gels thicker than 1 mm require additional washes. The purity of the methanol may affect the sensitivity of the stain as noted by Merril *et al.* (1982).

Step 4. *Potassium dichromate* (3.4 m*M*) solution containing 3.2 m*M* nitric acid is used to soak the gel for 15 min. Addition of too much nitric acid reduces the sensitivity of the stain, as does a prolonged period (longer than 15 min) in this solution. The dichromate facilitates image reversal.

Step 5. *Silver nitrate* (12 m*M*) solution is used to soak the gel for 20

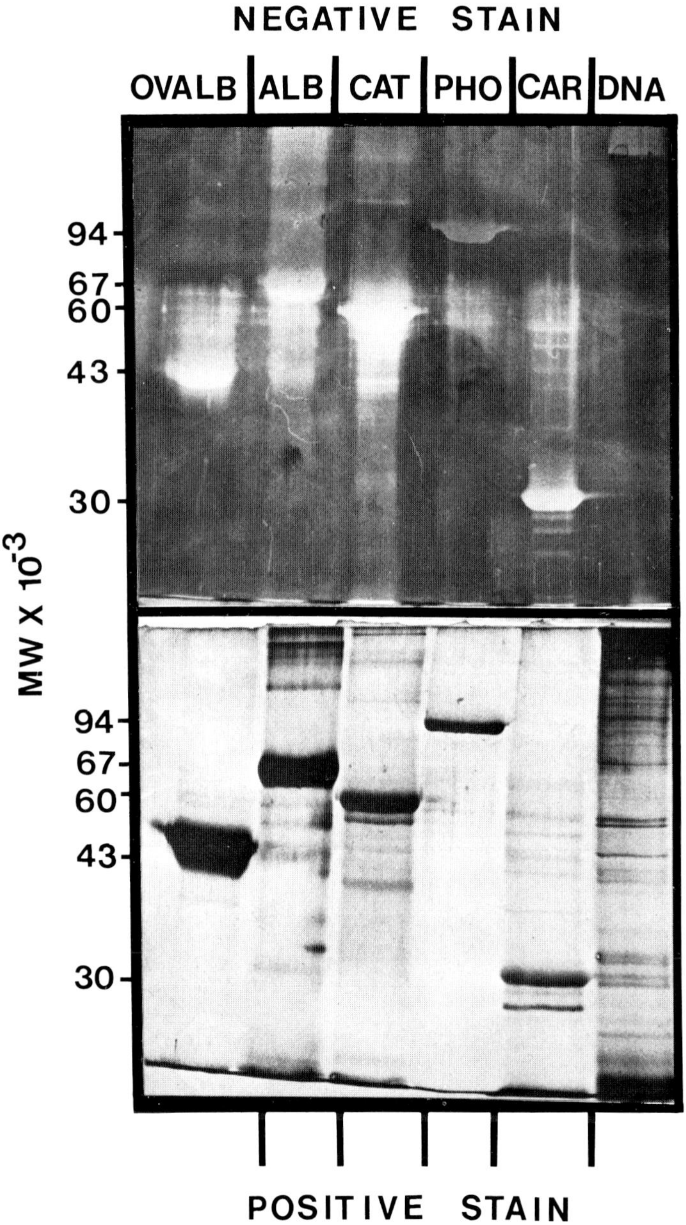

Figure 1. Purified proteins (ovalbumin, albumin, catalase, phosphorylase B, carbonic anhydrase) and DNA fragments stained with both negative and positive image silver stains. The negative stain must be performed in total darkness or in very-subdued light to prevent photoreversal and production of a positive stain effect.

min. The silver nitrate may be made in advance and kept as a 10-fold concentrated stock solution in a brown or amber bottle.

Step 6. *Develop the gel image* with 5 gel volumes of a solution containing 0.28 *M* sodium carbonate and 0.5 ml formaldehyde (37% commercial formaldehyde) per liter. This solution is added rapidly and then discarded. An additional 10 volumes of this solution is then added and the image is allowed to develop with gentle agitation. In this step a precipitate of silver salts is rapidly formed; to prevent this precipitate from adsorbing to the surface of the gel the sodium carbonate–formaldehyde solution should be changed at least twice. The sodium carbonate is used to make the gel alkaline so that the formaldehyde can reduce the ionic silver to the metallic form. The formic acid formed from the formaldehyde during the reduction is buffered by the sodium carbonate. For maximum sensitivity image development is allowed to continue until a yellowish background appears. This usually takes 15 to 20 min with a 1-mm-thick gel and longer with thicker gels. When the image is sufficiently developed the process is stopped.

Step 7. *Stop* image development by placing the gel in 3% (v/v) acetic acid for 5 min. The gel should be washed at least twice in water prior to storage as acetic acid will etch the silver image, eliminating faint bands or spots. The gels may be soaked in 3% glycerol 5 min and then dried between dialysis membranes under a vacuum at 80 to 82°C for 3 h. This results in a transparency that is relatively permanent and easy to store. If the gel has been properly washed the image becomes more intense with drying. For autoradiography or fluorography gels are dried onto Whatman 3 MM filter paper.

C. A Modified Positive Image Silver Stain

It is desirable for some applications to combine the dichromate and silver nitrate in a single solution, particularly for 7% or lower concentration polyacrylamide gels and for some agarose gel applications. Steps 1 and 2 are performed as described above (fixing agarose is more efficient with TCA). In Step 3, the gel is placed in a solution that is 19.5 m*M* silver nitrate, 1.34 m*M* potassium dichromate, and 13.5 m*M* sulfuric acid for 20 min. Maximal sensitivity may be enhanced with some gel preparations if the gel is placed on a light box and illuminated during the last 5 min of Step 3. The gel image is then developed normally as described in Steps 5 and 6 above (omitting Step 4).

D. *Recycling for Increased Sensitivity*

At the point in Step 6 when the background begins to yellow, spot intensity development begins to plateau. Addition of fresh sodium carbonate–formaldehyde solution will not greatly enhance spot densities, probably because most of the nonreduced silver ions have diffused out of the gel. If the silver nitrate is replaced, additional spot density can be achieved as was demonstrated by Merril *et al.* (1982) (Table I). To achieve this additional sensitivity stop development in the first cycle after performing Step 5, just as the yellowish background is beginning to appear. Soak the gel in 3% acetic acid (v/v). Rinse the gel with an additional 3% acetic acid solution and recycle through Steps 4–6 of the positive image stain procedure. Dichromate is not needed during recycling. Recycling can be repeated again and again to further intensify minor spots; however, background darkening may become a problem.

E. *Other Image Intensification and Destaining Procedures*

Because the gel silver image is similar to a photographic image, intensification and destaining procedures may be adapted from photographic formulas. These formulas are reviewed and described by Wall *et al.* (1975). The recycling procedure described above intensifies the image by

TABLE I

Effects of Recycling on Density

Cycle	Protein spot density (mean)[a]	Background density (mean)
1	10.2	0.4
2	20.2	0.6
3	25.6	1.1

[a] Mean protein spot density was determined by measuring the densities of 20 *E. coli* proteins separated by two-dimensional electrophoresis. Densities were calculated using a computer program which finds the modal density in the vicinity of each protein spot, subtracts this background from the average density and multiplies this density by the spot area.

the addition of silver; other metals such as copper, mercury, chromium, and uranium may also be used. Some intensification procedures increase image densities proportionally, whereas others, such as those employing copper iodide and mercuric chloride, are superproportional. They increase the image contrast by intensifying optically dense areas more rapidly. Reagents such as uranium, mercuric iodide, and chromium are subproportional in that they selectively intensify optically less dense regions.

Modification of chemical concentrations and procedure timing are usually necessary because most polyacrylamide gels are thicker and have different diffusion properties than photographic emulsions. To destain silver images, we have found the following method most useful:

Step 1. Dissolve 37 g of sodium chloride and 37 g of cupric sulfate in 850 ml of deionized water. Add concentrated ammonium hydroxide until all of the precipitate is dissolved and a deep blue solution is achieved. This solution is then brought to a final volume of 1 liter with deionized water.

Step 2. Sodium thiosulfate, 436 g, is dissolved in 1 liter of deionized water. The solutions in Steps 1 and 2 may be stored in glass bottles.

Step 3. Just prior to use, equal volumes of solutions from Steps 1 and 2 are mixed and used directly if total destaining is required. They may be diluted (1 : 10 or 1 : 100) with water if light silver deposits are to be removed or the image lightened.

It is advisable to photograph the gels prior to and during destaining to preserve the image, as it is difficult to stop destaining at a precise point. Gels may be restained after destaining by repeating Steps 2–6 of the positive image stain. Washing should be quite thorough to remove any trace of the destaining reagents prior to restaining; otherwise the gel will become opaque in Step 4.

F. Sensitivity, Quantitation, and Protein Detection

Proteins have been detected at concentrations as low as 0.02 ng/mm^2 on 1-mm-thick gels by Merril *et al.* (1982). The positive image silver stain procedure described above has been shown with eight purified proteins to be linear over a 50-fold range in concentration from 0.02 ng/mm^2 to 2 ng/mm^2 (Fig. 2 and Table II). Above 2 ng/mm^2, the stain becomes nonlinear as spot densities reach saturation as shown by Merril *et al.* (1982). This nonlinearity problem may be alleviated and the dynamic range of

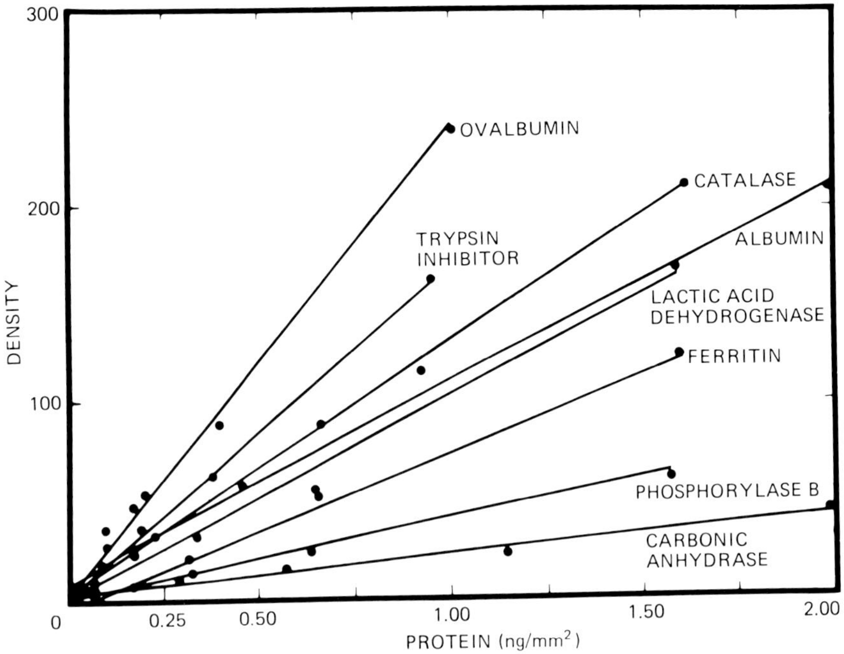

Figure 2. Silver stain density versus protein concentration for purified proteins (Pharmacia) electrophoresed in 10% polyacrylamide gels (0.88 mm thick) at 20 mA/gel. The gel images were developed for 9 min in sodium carbonate–formaldehyde solution. Band areas and densities were measured by computerized microdensitometry.

TABLE II

Linearity of Silver Staining

Proteins[a]	Total protein range (ng)	ng protein/mm^2	Number of bands measured	Slope[b]	*Y* intercept[b]	Correlation coefficient[b]
Albumin	5.4–270	0.05–2.27	5	101.7	9.0	0.995
Ovalbumin	7.2–360	0.02–1.01	5	231.5	2.7	0.998
Trypsin inhibitor	8.0–200	0.04–0.96	5	165.8	2.4	0.996
Ferritin	25.0–250	0.06–1.60	4	83.3	12.4	0.990
Carbonic anhydrase	4.0–200	0.06–2.85	5	26.0	0.8	0.983
Phosphorylase B	6.4–640	0.06–6.24	5	45.8	3.4	0.998
Catalase	3.6–180	0.03–1.59	5	127.0	8.4	0.990
Lactate dehydrogenase	9.6–240	0.06–1.57	5	107.5	4.4	0.990

[a] Proteins (Pharmacia purified molecular weight markers) were separated in a 10% polyacrylamide gel and densities and band areas determined with computerized microdensitometry.

[b] Slopes, Y intercepts and correlation coefficients were determined by linear regression analysis.

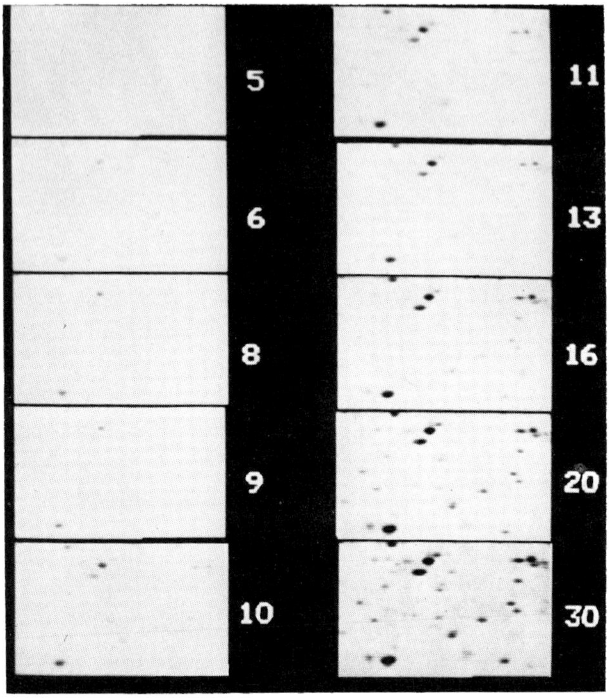

Figure 3. Development of an image of *E. coli* proteins by a silver stain. The image was photographed at time points ranging from 5 to 30 min during development in sodium carbonate–formaldehyde solution. By photographing or recording the image to obtain image development time versus silver density, it is possible to greatly expand the dynamic range of the silver stain. As an example, the protein observed in the bottom left region of the gel at 6 min through 13 min is no longer available for accurate densitometry after 16 min due to saturation problems.

the stain may be extended by recording and analyzing the image during development (Fig. 3). It should be noted that the relationship between density and concentration of protein, or the slope of the staining reaction, is characteristic for each protein (Fig. 2). For example, Merril *et al.* (1982) demonstrated that ovalbumin has a slope nine times that of carbonic anhydrase. Quantitative use of the silver stain is possible if constitutive or marker proteins are present on each gel so that densities can be normalized. Care must be taken to work within the linear range of the stain. Most ^{14}C-labeled proteins detectable by autoradiography can be detected by silver strain. In crude ^{14}C-labeled lysates of *E. coli,* one spot out of two hundred spots analyzed on an O'Farrell two-dimensional gel electrophoretic system (as visualized by autoradiography) did not stain with a single cycle of the reversal silver stain described above. However, this spot was shown to stain by Merril *et al.* (1982) when the recycling technique was employed. It should be noted that occasionally proteins may be visualized by silver staining that are not visible in radiolabeled cell lysates, depending on cellular growth and labeling conditions as shown by Bravo and Celis (1982; see also chapter by Bravo and Celis).

G. *Silver Quenching of Radioactively Labeled Proteins*

If gels stained with silver are used for autoradiography or fluorography, some quenching may be observed (Table III) as shown by Van Keuren *et al.* (1981). The histological silver stain as demonstrated by Merril *et al.* (1979), and Switzer *et al.* (1979) almost completely quenches the detection of ^{3}H-labeled compounds. This quenching makes the histological silver stain incompatible with fluorographic detection of ^{3}H-labeled proteins. However, the positive image photochemical silver stain described here causes less quenching with ^{3}H-labeled proteins and quenching with ^{14}C-labeled proteins is barely perceptible as was shown by Van Keuren *et al.* (1981). One can best use the photochemical silver stain for ^{3}H-labeled proteins by photographing the stained gel and then destaining it prior to fluorography. Spot densities on the X-ray film will then be quenched by approximately 50%. Quenching is most likely due

TABLE III

Effect of Silver Staining on Detection of Labeled Proteins

Stain	Fluorography of ^{3}H-labeled proteins	Autoradiography of ^{14}C-labeled proteins
Unstained	100[a]	100[a]
Coomassie Blue	85 ± 3	93 ± 4
Histological silver stain	N.D.[b]	53 ± 3
Photochemical silver stain	N.D.	98 ± 6
Destained histological silver stain	N.D.	48 ± 7
Destained photochemical silver stain	43 ± 9	81 ± 10

[a] Mean percent densities ± S.E.M. of 13 fibroblast proteins separated on two-dimensional gels on unstained gels.

[b] N.D., not detectable due to severe quenching.

to the effects of residual silver in the gel. Other chemicals are not as likely to participate as quenching agents because both the histochemical and photochemical stains share in common only the reagents silver nitrate and formaldehyde. Removal of most of the visible silver from the gel by photographic reducer does not fully restore detection of labeled proteins. This is probably due to the residual silver remaining in the gel following treatment with photographic reducer. This residual silver can be demonstrated on the dried gels that have been cleared of their silver image. Upon drying a faint silver image of major proteins appears. Furthermore, silver can be detected in cleared gels by electron beam X-ray spectral analysis.

H. Colored Protein Images with Silver Stains

Most silver stains produce some colored bands or spots. Such coloration was also observed in silver-based photographic processes. Color production has been shown to be dependent on three variables: silver grain size, the refractive index of the gel or emulsion, and the distribution of silver grains in the gel. In general smaller grains (less than 0.2 μm in diameter) transmit reddish or yellow-red light. Larger grains (above 0.3 μm) give bluish colors and still larger grains produce black images. The first observation of colored spots in silver stained polyacrylamide gels was reported in a study of human cerebrospinal fluid proteins by Goldman *et al.* (1980). This study utilized a histochemical silver stain. Lipoproteins stained blue while glycoproteins appeared as yellowish, brown, or red spots as demonstrated by Goldman *et al.* (1980). By modifying silver stain procedures color effects can be enhanced as shown by Sammons *et al.* (1981); however, saturation and negative staining effects are usually accentuated by such modifications making quantitation more difficult (see also chapter by Sammonds).

III. Mechanisms of Silver Stains

In all of the stains, histological and photochemical, production of the image involves the reduction of ionic silver to metallic silver. This reduction is differentially affected by proteins (or other polymers such as DNA, RNA, or lipopolysaccharides) as demonstrated by Sommerville and Wang (1981), Tsai and Frasch (1982), and Goldman and Merril (1982). To produce an image of the polymer pattern in a gel, treating the gel with silver nitrate followed by a reducing agent (as described in the "negative image stain") is sufficient. This result strongly suggests that the proteins or polymers in the gel are altering the local physical conditions for the silver ions, and that it is this physical effect which causes the differential staining.

One model that might describe this effect involves differential exclusion of silver ions in the region of the gel occupied by charged polymers. However, experiments with radioactive silver do not indicate this to be the case. Silver at a concentration similar to that used in these silver stains displays a relatively uniform distribution, as measured by autoradiography with a radioactive silver tracer (silver, ^{110M}Ag) in the gel (Fig. 4b). Furthermore, the ability to effect image reversal, as in the positive image photochemical silver stain, also argues against the exclusion of silver ions from the regions of gel occupied by the proteins. If they were excluded, then this region would not develop an image, which it does in the positive stain.

Another model invokes physical effects that could be due to charges on the protein molecules. These charges might affect the silver ions and their reactivity directly or might affect them indirectly by altering the adjacent water structure. This idea is supported by the observation that the addition of ions that can interfere with charge effects can also eliminate silver staining. For example, addition of 0.5 *M* ammonium nitrate to the silver nitrate solution prevents silver staining both in gels and in many histological silver stains (ammonium nitrate was chosen because it does not precipitate silver and it does not significantly alter the pH of the silver nitrate solution).

Although many silver stains that are generally useful for detecting proteins and other biopolymers separated on gels appear to depend on physical effects, it is apparent that some silver stains also depend on direct chemical reactions. One type of silver reaction is demonstrated by the use of a $10^{-9}M$ $AgNO_3$ solution (Fig. 4a); this binding has been shown to be reversible. The histologically derived silver stains probably owe some of their specificity to chemical reactions, and it has been suggested by Gambetti *et al.* (1981) that this specificity may provide a step toward clarifying the mechanism of silver staining.

Investigation of the silver staining of proteins and other biopolymers separated on gels should give additional insights into understanding the chemical basis of many of the currently used histological stains. In some histological studies the reduction of silver appears to be catalyzed by tissue containing significant amounts of reducing agents, such as ascor-

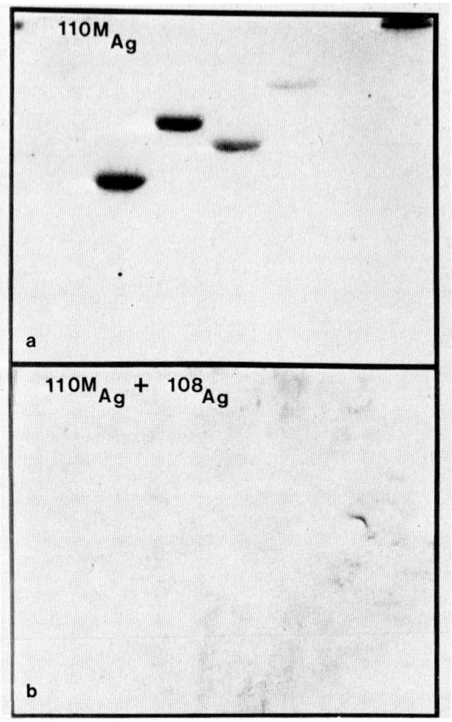

Figure 4. Autoradiography by radioactive silver (^{110M}Ag) of a gel containing five purified proteins separated by electrophoresis in one dimension. The gel was first soaked in a 10^{-9} *M* $AgNO_3$ solution containing ^{110M}Ag (Amersham, Y20186, 116Ci/M)(a). At this concentration there is sufficient specific binding to visualize the position of the purified proteins. When additional nonradioactive silver (^{108}Ag) in the form of silver nitrate was added to the solution to bring the concentration of silver nitrate to 0.1 *M* (a concentration similar to that used in silver staining), the distribution of radioactive silver is fairly uniform(b). Autoradiography was performed by placing the gel in a plastic bag and then placing individually wrapped X-ray film over this preparation for overnight exposures.

bic acid (Thompson and Hunt, 1966). In other cases aldehydes have been shown to play a general role. In some tissues neither aldehyde-creating nor aldehyde-blocking reagents appreciably affect silver staining (reviewed by Thompson and Hunt, 1966). Thompson and Hunt (1966) have also reviewed examples of tissue silver staining that may be based on physical interface phenomena, such as silver staining of tissue reticulum and collagen. In these studies electron microscopic observations suggest that staining differences depend on whether fibrils embedded in a glycoprotein matrix are in a random array or are organized into bundles.

Understanding the mechanisms underlying reversal of a negative image on a gel to a positive image will require considerable effort. Many of the chemicals that are used in photochemical processes involved in image reversal are oxidizing agents such as acid dichromate, potassium ferricyanide, ammonium persulfate, and acid potassium permanganate. These oxidizing agents are believed to function by destroying the latent surface image prior to development of an internal image as suggested by Hillson (1977). These same chemicals have been used to facilitate the development of a positive image in a variety of gel silver stains. In some of the histological silver stains ammonia and ammoniacal silver compounds may facilitate the image reversal process. In this regard it has been reported by Mees (1952) that photographic emulsions prepared with ammonia are much more sensitive to image reversal effects. Interestingly, dithiothritol, a reducing agent that has not been used in photographic image reversal processes, has been used by Morrissey (1981) in place of the acid dichromate step of the positive image silver stain.

IV. Potential and Real Problems in Silver Staining

Although current silver stains for protein detection are highly sensitive and fairly easy to perform, they are not without problems. The major loss of sensitivity in silver staining is usually due to inadequate water purity. Deionized water with a conductivity of less than 1 μMHO is required in the preparation of all reagents, including wash and fixing solutions. The second major cause of loss of sensitivity is due to inadequate fixation of proteins prior to staining. Gels thinner than 0.5 mm require fixation in 20% trichloracetic acid, as do agarose gels. Gels thicker than 1.0 mm require additional wash steps prior to staining. Thick gels also benefit from the addition of deionized water washes between Step 3, (potassium dichromate) and Step 4, (treatment with silver nitrate). Occasionally surface artifacts will mar a gel. These can be minimized by rapidly changing the sodium carbonate–formaldehyde during initial image development, Step 5, and by handling gels carefully. Pressure, finger prints, and surface drying all cause surface artifacts. Occasionally, two horizontal lines will be observed at 60,000 and 67,000 daltons. The cause of these lines is not known. Some investigators have implicated mercaptoethanol. They have suggested that the lines can be eliminated by reducing the amount employed.

References

Bravo, R., and Celis, J. E. (1982). Up-dated catalogue of Hela cell proteins: percentages and characteristics of the major cell polypeptides labeled with a mixture of 16 ^{14}C-labeled amino acids. *Clin. Chem.* **28**, 766–781.

Gambetti, P., Autilio-Gambetti, L., and Papasozomenos, S. G. H. (1981). Bodian's silver method stains neurofilamont polypeptides. *Science (Washington, D.C.) 213,* 1521–1522.

Goldman, D., and Merril, C. R. (1982). Silver staining of DNA in polyacrylamide gels, linearity and effect of fragment size. *Electrophoresis* **3**, 24–26.

Goldman, D., Merril, C. R., and Ebert, M. H. (1980). Two-dimensional gel electrophoresis of cerebrospinal fluid proteins. *Clin. Chem.* **26**, 1317–1322.

Hillson, P. J. (1977). Disposition of the latent image as determined by variation of exposure and development techniques. *In* "The Theory of the Photographic Process" (T. H. James, ed.), 4th ed., p. 174. Macmillan, New York.

Hubbell, H. R., Lawrence, I. R., and Hsu, T. C. (1979). Identification of a silver binding protein associated with the cytological silver staining of actively transcribing nucleolar regions. *Cell Biol. Rep.* **3**, 615–622.

Kerenyi, L., and Gallyas F. (1972). A highly sensitive methods for demonstrating proteins in electrophoretic, immunoelectrophoretic and immunodiffusion preparations. *Clin. Chim. Acta* **38**, 465–467.

Kerenyi, L., and Gallyas, F. (1973). Über probleme der quantitativen auswertung der mit physikalischer entwicklung versilberten agar elektrophoretogramme. *Clin. Chim. Acta* **47**, 425–436.

Lischwe, M. D., Smetana, K., Olsom, M. O. J., and Busch, H. (1979). Proteins C23 and B23 are major nucleolar silver staining proteins. *Life Sci.* **25**, 701–708.

Mees, C. E. K. (1952). Solarization and the Herschel, Sabattier and Albert effects. *In* "The Theory of the Photographic Process" (T. H. James, ed.), 1st ed., p. 276. Macmillan, New York.

Merril, C. R., Switzer, R. C., and Van Keuren, M. L. (1979). Trace polypeptides in cellular extracts and human body fluids detected by two-dimensional electrophoresis and highly sensitive silver stain. *Proc. Natl. Acad. Sci. USA* **76**, 4335–4339.

Merril, C. R., Dunau, M. L., and Goldman, D. (1981a). A rapid sensitive silver stain for polypeptides in polyacrylamide gels. *Anal. Biochem.* **110**, 201–207.

Merril, C. R., Goldman, D., Sedman, S. A., and Ebert, H. M. (1981b). Ultrasensitive stain for proteins in polyacrylamide gels shows regional variation in cerebrospinal fluid proteins. *Science (Washington, D.C.)* **211**, 1437–1438.

Merril, C. R., Goldman, D., and Van Keuren, M. L. (1982). Simplified silver protein detection and image enhancement methods in polyacrylamide gels. *Electrophoresis* **3**, 17–23.

Morrissey, J. H. (1981). Silver stain for proteins in polyacrylamide gels: a modified procedure with enhanced uniform sensitivity. *Anal. Biochem.* **117**, 307–310.

O'Farrell, P. H. (1975). High resolution two-dimensional electrophoresis of proteins. *J. Biol. Chem.* **250**, 4007–4021.

Sammons, W., Adams, L. D., and Nishizawa, E. E. (1981). Ultrasensitive silver-based color staining of polypeptides in polyacrylamide gels. *Electrophoresis* **2**, 135–141.

Sommerville, L. L., and Wang, K. (1981). The ultrasensitive silver "protein" stain also detects nanograms of nucleic acids. *Biochem. Biophys. Res. Commun.* **102**, 53–58.

Switzer, R. C., Merril, C. R., and Shifrin, S. (1979). A highly sensitive silver stain for detecting proteins and peptides in polyacrylamide gels. *Anal. Biochem.* **98**, 231–237.

Thompson, S. W., and Hunt, R. S. (1966). Gridley's reticulum stain. *In* "Selected Histochemical and Histopathological Methods," pp. 798–802. Thomas, Springfield, Illinois.
Tsai, C. M., and Frasch, C. E. (1982). A sensitive silver stain for detecting lipopolysaccharides in polyacrylamide gels. *Anal. Biochem.* **119**, 115–119.
Van Keuren, M. L., Goldman, D., and Merril, C. R. (1981). Detection of radioactively labeled proteins is quenched by silver staining methods: quenching is minimal for ^{14}C and partially reversible for ^{3}H with a photochemical stain. *Anal. Biochem.* **116**, 248–255.
Verheeka, P. (1975). Agar gel electrophoresis of unconcentrated cerebrospinal fluid. *J. Neurol.* **209**, 59–63.
Wall, E. J., Jordan, F. I., and Caroll, J. S. (1975). Intensification (Chap. 6) and Reduction (Chap. 7). *In* "Photographic Facts and Formulas," pp. 168–189. Prentice-Hall, Englewood Cliffs, New Jersey.

CHAPTER 4

Applicability of Color Silver Stain (GELCODE® System) to Protein Mapping with Two-Dimensional Gel Electrophoresis

D. W. SAMMONS
L. D. ADAMS
Diabetes and Atherosclerosis Research
The Upjohn Company
Kalamazoo, Michigan

T. J. VIDMAR
Biostatistics,
The Upjohn Company
Kalamazoo, Michigan

C. A. HATFIELD
Western Michigan University
Kalamazoo, Michigan
and Hypersensitivity Diseases Research
The Upjohn Company
Kalamazoo, Michigan

D. H. JONES
P. J. CHUBA
S. W. CROOKS
Diabetes and Atherosclerosis Research
The Upjohn Company
Kalamazoo, Michigan
and Kalamazoo College
Kalamazoo, Michigan

I. Introduction. 112
II. Variables that Affect Sensitivity and Reproducibility of the Color Silver Stain (GELCODE System). 113
A. Thickness of Gel. 114
B. Time and Temperature in Reducing Solution. 115

TWO-DIMENSIONAL GEL
ELECTROPHORESIS
OF PROTEINS

ISBN 0-12-164720-X

C. Fixation Alcohol and Time of Fixation. 116
D. Fixation Acid. 116
E. Reagents. 116
III. Application of Color Silver Staining (GELCODE System) to Two-Dimensional Mapping of Proteins. 117
A. Distinguishing Overlapping Spots. 117
B. Characterizing Proteins in Posttranslationally Modified Charge Trains. 118
C. Tracking Proteins in Urine. 119
D. Modulation of Proteins by the Drug Tunicamycin. 120
E. Use of Color to Identify Proteins in Different Species: Application to Gene Mapping of Human Chromosomes. . . . 120
F. Use of Animal Models to Find Marker Proteins for Disease. . 121
G. Subcellular Fractionation. 122
IV. Computerized Data Acquisition of Color-Stained Images: Test for Stoichiometric Staining. 122
A. BioImage™ System. 122
B. Stoichiometry of GELCODE Stain: Integrated Intensity versus Concentration. 123
V. Concluding Remarks. 125
References. 125

I. Introduction

There are approximately 3000 known genetic diseases, some of which are known to be caused by a single defective protein. In others the genetic origin is polygenic. The disease may be heterogeneously expressed and/or modulated by environmental factors such as stress and diet. Other diseases, including those associated with micro- and maccrovascular defects, may be secondarily acquired from metabolic disturbances created by prolonged chronic diseases and aging. In each of these disease processes qualitative and/or quantitative changes in protein gene products are expressed during progressive stages, thereby masking the primary genetic lesion. Thus in order to identify and distinguish the few defective proteins responsible for a particular disease, discriminating techniques must be developed and applied.

Two-dimensional gel electrophoresis (2-DGE), if developed to its ultimate potential, has a greater possibility of finding many of the defective proteins responsible for diseases initiated by genetic, metabolic, and toxic disturbances than any other currently available technology. The strategy of 2-DGE technology involves the following steps:

1. A tissue sample is collected and prepared.
2. Proteins in the sample are separated from each other in a suitable gel matrix.
3. The separated proteins are usually detected with a chemical stain or autoradiography.
4. Physical data about proteins are usually acquired from the gel by scanning.
5. The compiled information is analyzed and comparisons with other separated samples are made to identify the significant differences in protein patterns between different gels.
6. Relevant literature and clinical data bases are searched and associations between defective protein and the disease in question are made.

This chapter will deal with the third step of 2-DGE technology: detection of proteins by a chemical stain procedure. The method (GELCODE® system, Upjohn Diagnostics) is a silver stain designed to produce colored protein spots in the gel matrix.

In order for color silver staining to be applicable to protein mapping with 2-DGE and to be useful in our quest for understanding the molecular basis of disease processes certain criteria must be met.

1. The variables of the method which affect stain sensitivity and reproducibility must be understood and controlled.
2. The method should have unique applications that are otherwise impossible with other conventional stains including the mostly monochromatic silver stains.
3. The color-stained 2-D gel pattern should be amenable to computerized scanning and data acquisition.

II. Variables that Affect Sensitivity and Reproducibility of the Color Silver Stain (GELCODE System)

The procedural steps are listed in Table I and are basically the same as described in the original method (Adams and Sammons, 1981; Sammons *et al.*, 1981) except for the following changes:

1. The fixation is done in 5% acetic acid.
2. The duration of fixation is overnight.
3. The washing steps are done in H_2O for four changes of 1 h each.

TABLE I

Color Silver Staining Procedure for 1.5 mm Gel[a,b]

Steps	Solutions	Duration of agitation
Fix	50% EtOH + 5% acetic acid	Overnight (16 h)
Wash	H_2O	1 h
↓	H_2O	1 h
↓	H_2O	1 h
↓	H_2O	1 h
Equilibrate gel	$AgNO_3$ (1.9 g/liter)	1 h
↓		
Rinse	H_2O	10–20 sec
↓		
Reduce Silver	HCHO (7.5 ml/liter) in 0.75 N NaOH	10 min
↓		
Enhance color	Na_2CO_3 (7.5 g/liter)	1 h
↓	Na_2CO_3 (7.5 g/liter)	1 h
↓	Na_2CO_3 (7.5 g/liter)	1 h

[a] The silver reduction and color enhancement steps are performed with reagent solutions equilibrated to 25°C.

[b] After the color stained images stabilize (approximately 4 h after last Na_2CO_3 step) the wet gel can be scanned and the image stored on magnetic tape, or the gel may be stored in an airtight bag with a small amount of 0.75% Na_2CO_3.

4. The sodium borohydride is omitted from the reduction step.
5. An extra change of Na_2CO_3 is made during the color enhancing step.

The gels can be scanned on the BioImage system (see Section IV) 4 h after the last Na_2CO_3 change. Alternatively, they may be stored in airtight, heavy duty, plastic food storage bags (obtained from local food stores) with a small amount of 0.75% Na_2CO_3. If the stored gel remains moist with the Na_2CO_3 then they may be stored indefinitely without significant loss of stain intensity.

A. *Thickness of Gel*

The internal anatomy of a gel stained with the GELCODE system is illustrated in Plate 1. Within the silver stained gel, indicated by two arrows in Plate 1A, are two dark layers of silver that are deposited equidistant from each gel surface and are separated by a clear zone in the center of the gel. The thickness and/or density of these two layers contributes to overall background ranging from dark brown in 3.0-mm gels to light orange in 0.75-mm gels. The line thickness is the same for 0.75 and 1.50 mm, but is three times thicker in 3.0-mm gels. The arrow in Plate 1B indicates a protein spot through which the internal silver layer extends. The protein zone is stained brown in this instance but may be red and blue as well. Protein spots appear yellow when the internal layer is prevented from forming throughout the protein zone; this is due to the intrinsic nature of some proteins. The internal silver layer has not been reported with other silver stain methods (Switzer *et al.*, 1979; Allen, 1980; Oakley *et al.*, 1980; Merril *et al.*, 1981, 1982; Poehling and Neuhoff, 1981; Wray *et al.*, 1981). *Negative* protein spots have been commonly observed with several modifications of the ammoniacal silver based stains as well as in the monochromatic method of Merril *et al.* (1982; see also chapter by Merril and Goldman). It is reported by Merril *et al.* (1982) that negative spots cannot be quantitated. The GELCODE stain, because of its amber internal silver layer, allows detection of the negative (yellow) protein spots as well as their accurate quantitation (see Figs. 1 and 2).

B. *Time and Temperature in Reducing Solution*

Time in the reducing solution also affects the background and sensitivity. Under standard staining conditions described in Table 1 the internal silver layer of the 0.75-mm gel is lighter in density than that of a 1.5-mm-thick gel. The sensitivity is correspondingly reduced and the color brightness is diminished when the gel thickness is reduced (data not shown). To test whether loss of sensitivity can be compensated, the time the gel is in the reducing solution was varied. As seen in Plate 1 the intensity of the internal silver layer in 0.75-mm gels increases with decreasing times of 5 min (Plate 1C), 3 min (Plate 1D) , and 1 min (Plate 1E). When 0.75-mm gels are developed in the reducing solution for 3 min the sensitivity and colors are comparable to 1.5-mm gels developed for 10 min (data not shown). For 0.2-mm gels the time in the reducing solution should be 15 sec (data not shown).

Increasing reagent temperature by 5°C above the recommendation in Table I may significantly reduce desirable background and sensitivity of the procedure (data not shown). Thus control of time and temperature variables of the stain reaction is essential for optimal sensitivity and reproducibility.

C. *Fixation Alcohol and Time of Fixation*

As illustrated in Plate 2, both alcohol reagent and time of fixation are also critical variables affecting color reproducibility and sensitivity. Comparison of Plates 2A and 2B shows the resulting poor color when methanol is substituted for ethanol in the fixation step. The length of fixation time was overnight (16 h) and the concentration was 50% alcohol. Thus if the color is desirable to an investigator then ethanol must be used during the fixation step. The length of time in the fixation also needs to be regulated. As shown in Plates 2C and 2D, protein loss during the washing and staining steps are excessive if inadequate fixation times are used. For example, in Plates 2C and 2D, the fixation times were 1 and 4 h, respectively. The 4-h fixation preserves more protein than 1 h but not nearly as much as the 16-h time shown in Plate 2B.

D. *Fixation Acid*

In addition to alcohol the fixation acid can also affect color reproducibility. In Plate 3 duplicate 2-D gels of a plasma sample were separately treated. Plate 3A was stained as described in Table I and Plate 3B was similarly treated except that an equivalent amount of nitric acid was substituted for the acetic acid. Comparisons of protein colors reveal some interesting observations. When nitric acid is substituted for acetic acid in the fixation only selected proteins stain a different color.

Specifically, haptoglobin is a protein which is the same color whether acetic acid or nitric acid is used. In Plate 3A the fibrinogen α-chain and a protein within the haptoglobin train (indicated by an arrow) stain yellow. If, however, nitric is substituted for acetic acid, fibrinogen α-chain stains red and protein migrating within the haptoglobin train stains brown. All three spots of Apo A-1 lipoprotein (HDL) appear as rust-colored spots in acetic acid and brown in nitric. Therefore these data support the hypothesis that the colors observed in 2-D gels after GELCODE silver staining reflect inherent physical and chemical properties of individual proteins (Sammons *et al.*, 1981).

E. *Reagents*

Dark backgrounds and streaking can affect sensitivity and color reproducibility. Inappropriate concentrations of stain reagents, improper mixing of stain reagents, impure reagents, and inadequately cleaned electrophoresis tanks can lead to a dark background. The use of high purity reagents and careful techniques are essential elements in avoiding excessive background and streaking.

Two kinds of streaking unrelated to protein may be seen. The first type of streaking is indicated in Plate 4 with arrows labeled 1 and 2. This streaking has been previously reported in the literature (Guevara *et al.*, 1982) and probably relates to excessive amounts (usually 5–10%) of β-mercaptoethanol in either the sample and/or equilibrated first-dimension gel. We found that 2% β-mercaptoethanol completes the protein disulfide reduction and reduces the extent of streaking. Another trick that may reduce the β-mercaptoethanol-related streaking is to filter the stock mercaptoethanol through a 0.2-μm filter to remove precipitates that are often present in the stock bottles of β-mercaptoethanol. Also, while loading the sample and handling the gel one should wear gloves to prevent transfer of human skin protein contaminants into the gel. The second type of streaking, indicated by arrows 3, 4, and 5 on Plate 4, originates from particulates in reagents or on glassware and equipment and will be referred to as point streaking. To eliminate vertical point streaking, one should (1) carefully clean glass surfaces and equipment and (2) filter stock solutions of SDS, buffers, acrylamide, and urea through 0.2-μm filters to remove insoluble particulates. Failure to remove the particulates prior to gel preparation will result in their entrapment within the polyacrylamide gel. During the electrophoresis they will streak from their original position within the gel and may extend from a few millimeters to several centimeters in length (see arrows 3 and 4 in Plate 4). Both types of streaking can be observed in other silver stain methods, although not as extensively because they are less sensitive methods (Ochs *et al.*, 1981). Therefore the best solution to the problem is to eliminate the contaminants, not the high sensitivity of this silver stain method.

III. Application of Color Silver Staining (GELCODE System) to Two-Dimensional Mapping of Proteins

A. Distinguishing Overlapping Spots

The first appreciated advantage of differentially colored silver-stained 2-D protein patterns was to distinguish overlapping spots (Adams and

Sammons, 1981; Sammons *et al.*, 1981). Plate 5 is a comparison of proteins observed in two different Chinese hamster erythrocyte lysates. In Plate 5A a blue protein from an animal is indicated by arrow no. 2 and a brown overlapping protein is indicated by arrow no. 5. In Plate 5B the blue no. 2 protein from a different animal is missing, thereby completely exposing the entire brown protein no. 5. If one had used a monochromatic detection method such as silver, an organic dye, or radiolabel, the proteins no. 2 and no. 5 would have been mistakenly observed to be the same protein. A mistaken interpretation that might have resulted from comparing 5A to 5B after monochromatic staining would be that there is less protein at the indistinguishable positions 5 and 2 in the erythrocytes of Plates 5B than in 5A. In fact, the blue protein no. 2 seen in Plate 5A is shifted to the right to position 3 and protein no. 5 is similar in both animals. Thus colored spots in 2-D gel electrophoretograms aid in the interpretation of protein patterns when two proteins have similar x,y coordinate positions.

B. *Characterizing Proteins in Posttranslationally Modified Charge Trains*

Charge train heterogeneities in 2-D gel patterns are now well recognized and are due to posttranslational modifications (see also chapter by Bravo). The protein entities of a charge train may vary tremendously in spot morphology, concentration, and x,y coordinate position. In Plate 6 the human plasma proteins were treated for 6 h with and without neuraminidase. The haptoglobin charge train shifted position and the protein is almost entirely in two spots. The α_1-antitrypsin subunits move from their original positions but the relative amounts of the second and third (numbered from left to right) spots change so that after neuraminidase treatment most of the third spot has shifted into the second position. In the C_3 activator charge train all five subunits shift equally and relative proportions of each are retained. Three distinct G_4 charge trains have previously been identified with color silver stain (Sammons *et al.*, 1981). Two are blue and one is orange. However, after 6 h of neuroaminidase treatment, four spots are seen instead of three charge trains. Removal of sialic acid from the orange charge train does not affect color and it is distinguishable from the blue G_4 proteins even though it is no longer a charge train. Other proteins, such as fibrinogen α chain, are unaffected by neuraminidase treatment suggesting the absence or inaccessibility of sialic acid residues. Thus sialic acid residues, although contributing to charge trains of some proteins, do not contribute in a major way to the color of the silver-stained proteins. Therefore using color silver stain we can identify modified proteins after neuraminidase treatment and distinguish them from closely related members of the same protein family even though their spot morphology and/or x,y coordinate position is significantly shifted.

An additional parameter such as color is particularly useful in preventing a spot of similar morphology and position from being mistaken for a member of a given charge train. This is illustrated by the arrow in Plate 6 pointing out the haptoglobin train. The indicated yellow protein as well as the orange one immediately to its left is not affected by neuraminidase treatment of plasma, thereby further suggesting that the yellow and orange proteins are not part of the haptoglobin charge train family. Using immunoprecipitation prior to 2-DGE, Anderson and Anderson (1977) also showed that the yellow and orange proteins were not part of the charge train family.

C. *Tracking Proteins in Urine*

Proteinuria is a classical indication of disease states and numerous attempts have been made to characterize urinary proteins by classical electrophoresis in order to make a diagnostic correlation of the protein pattern and disease. With two-dimensional gel electrophoresis it has become possible to examine the urinary protein composition (Anderson *et al.*, 1979; see also chapter by Tracy and Young). At least two protein to disease correlations have been made for disease and one for heavy metal poisoning (Roto, 1978; Frearson *et al.*, 1981; Tracy *et al.*, 1982). It has been assumed that the presence of nonplasma proteins in the urine reflects tissue leakage from target organs of disease and/or toxic insults. Given 50,000 possible proteins it indeed is a challenge to identify the many potential proteins that may be shed into the urine. Color is particularly helpful in this regard. Plate 7 illustrates an example: a red protein (see arrows) is first observed in urine of juveniles (Plate 7B), increased in retired breeders, and not evident in very old rats. To determine if the protein was derived from the kidney the organ was subfractionated into cortex, medulla, and glomeruli. A red protein of identical molecular weight (17,000) and charge (about pI 5) was observed predominately in the cortex fraction (Plate 7A). Lesser amounts of the red protein was seen in the medulla, but it was not observed in 2-D gel patterns from the glomeruli or plasma (data not shown). As illustrated by the arrows in Plate 7A, the kidney cortex protein is a charge train of four red spots and was found to disappear as the animal ages. As the red protein charge train in the urine (Plate 7D) increases, the amount of red protein in the kidney cortex (Plate 7C) decreases until the 17,000 dalton red protein is completely absent from both (Plates 7E and 7F). Thus color readily aids in assigning the origin of red 17,000 dalton urinary protein to the kidney cortex.

D. Modulation of Proteins by the Drug Tunicamycin

Drug toxicity is of paramount importance to the drug industry, experimental medicine, the basic research alike. Plate 8 illustrates the utility of color in following alterations in posttranslationally modified proteins. For example, tunicamycin is known for its ability to block glycosylation of glycoproteins (Takatsuki *et al.*, 1971; Lehle and Tanner, 1976). In this experiment human heptoma cells were grown in tissue culture with the drug for varying times. The cells were harvested, 2-DGE was performed, and the 2-D gel was stained with GELCODE. Comparison of untreated (Plate 8A) and drug treated (Plate 8B, 1 day; Plate 8C, 4 days) samples on 2-D gels revealed a group of yellow proteins whose steady state level decreases with time (arrow 1) of drug exposure. Correspondingly other proteins appear with time (arrow 2); the appearing proteins have the same x coordinate positions (charge), but have a decreased y coordinate (molecular weight). Because the appearing protein is the same color as the disappearing protein, this suggests that the appearing protein is newly synthesized without formation of dolichal bound α-acetylglucosamine derivatives (Takatsuki *et al.*, 1971). Thus it follows that the disappearing protein with its normal carbohydrate moieties is being degraded in its normal fashion but is not being replaced since tunicamycin blocked the glycosylation pathway. Therefore this experiment suggests that color silver stain and 2-DGE can be uniquely combined to gain new insights into the mechanism of drug action by identifying the exact protein gene products affected by the drug being studied.

E. Use of Color to Identify Proteins in Different Species: Application to Gene Mapping of Human Chromosomes

In Plate 9, a comparison of protein gene products on 2-D gels of human erythrocyte lysate (Plate 9A) to Chinese hamster (Plate 9B) eryth-

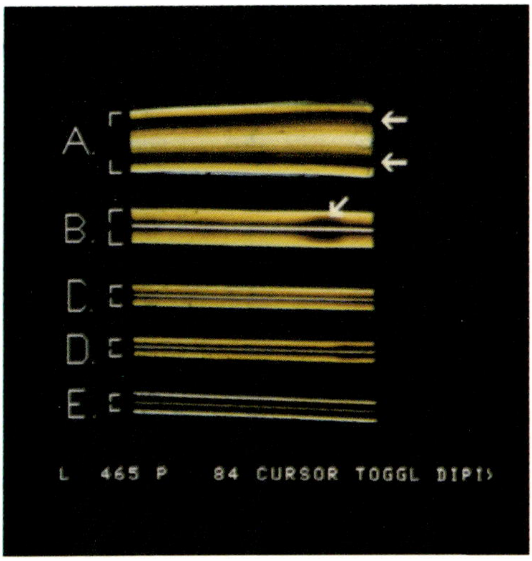

Plate 1. The effect of gel thickness on background. The stain procedure was done exactly as described in Table I except that the time in reducing solution was varied as follows: (A) 3 mm gel, no change; (B) 1.5 mm gel, no change; (C) 0.75 mm gel, time = 5 min; (D) 0.75 mm gel, time = 3 min; (E) 0.75 mm gel, time = 1 min. The cross-sections in this figure were scanned with the BioImage system and photographed from the monitor.

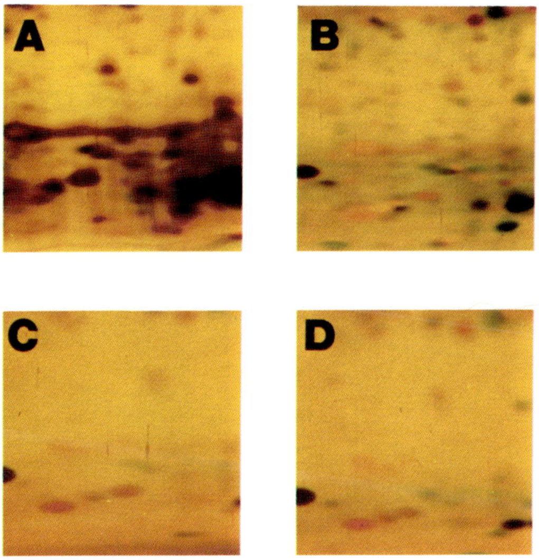

Plate 2. The effects of varying fixation alcohol and time on protein fixation and protein spot color. Replicate gels of the same liver extract were stained as described in Table I. However, in some the fixation reagents and times differed as follows: (A) Methanol was substituted for ethanol, no change in time; (B) No change in reagents or time; (C) No change in reagents, 1 h fixation instead of 16; (D) No change in reagents, 4 h fixation instead of 16.

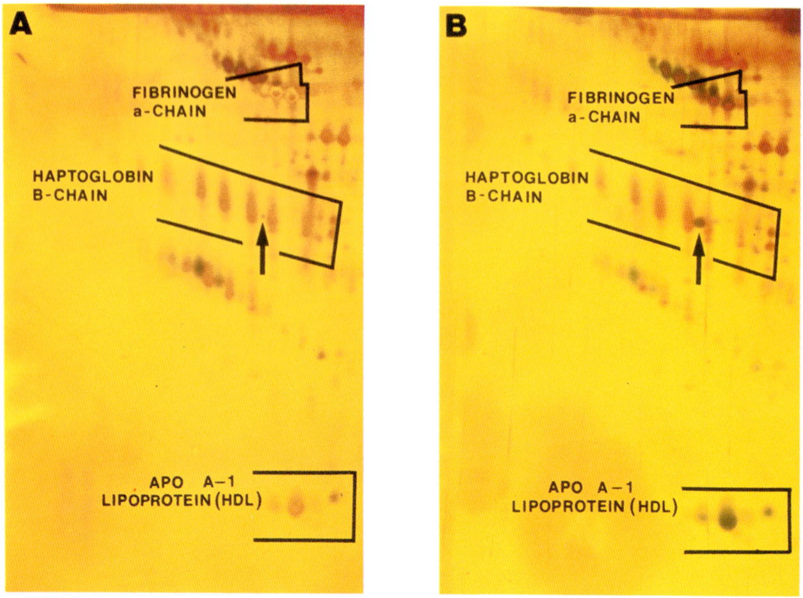

Plate 3. The effect of fixation acid on protein spot color. (A) 1*M* acetic acid + 50% ethanol; (B) 1*M* HNO_3 + 50% ethanol.

Plate 4. Streaking artifacts unrelated to protein. The 2-D gel is a blank to which protein was not added in the first dimension. Arrows 1 and 2 indicate streaking due to β-mercaptoethanol. Arrows 3, 4, and 5 indicate streaking due to particulates in other reagents.

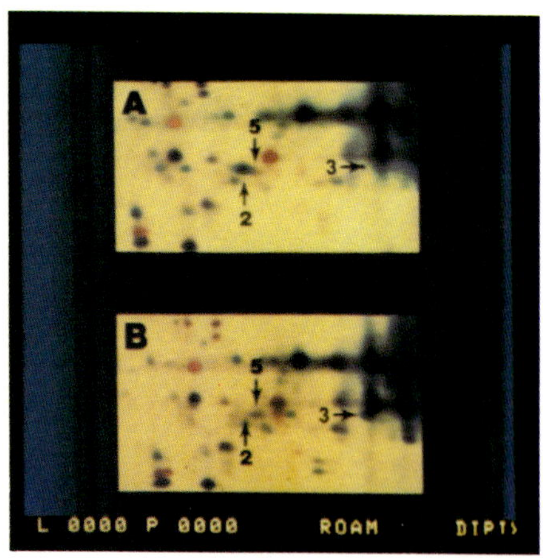

Plate 5. The discrimination of overlapping spots with color silver stain. (A) A region of a hamster erythrocyte lysate 2-D gel protein pattern; (B) A region of a different hamster erythrocyte lysate 2-D gel protein pattern.

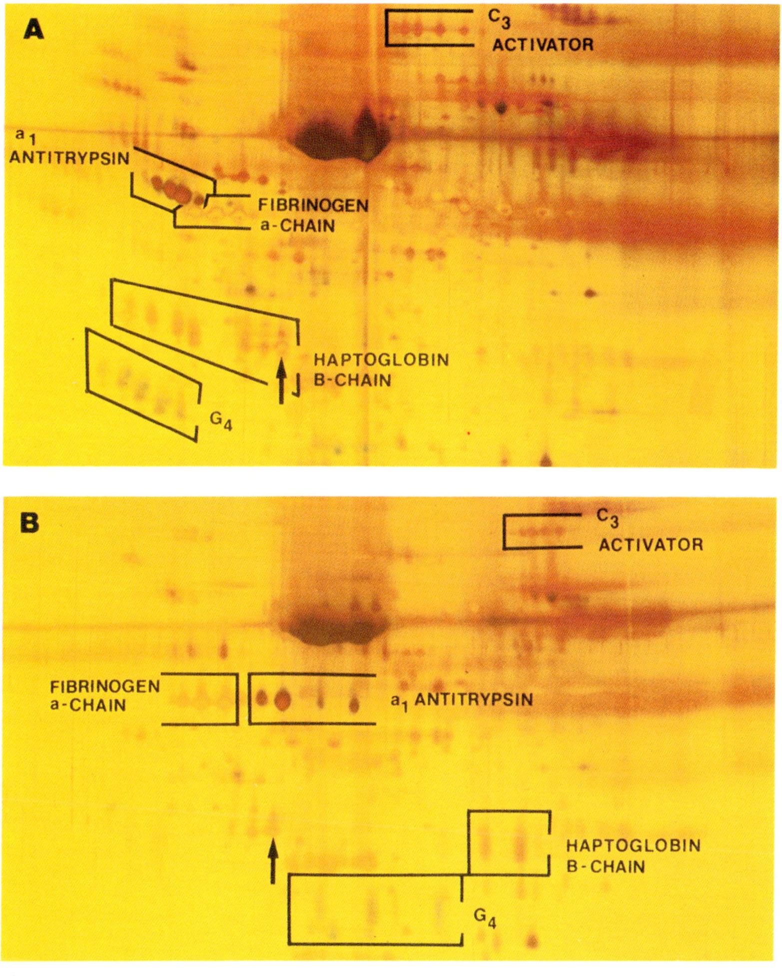

Plate 6. The use of color to characterize posttranslationally modified proteins. Human plasma was treated with (A) and without (B) neuraminidase prior to two-dimensional gel electrophoresis. The sample was incubated for 6 h with a standard concentration of neuraminidase.

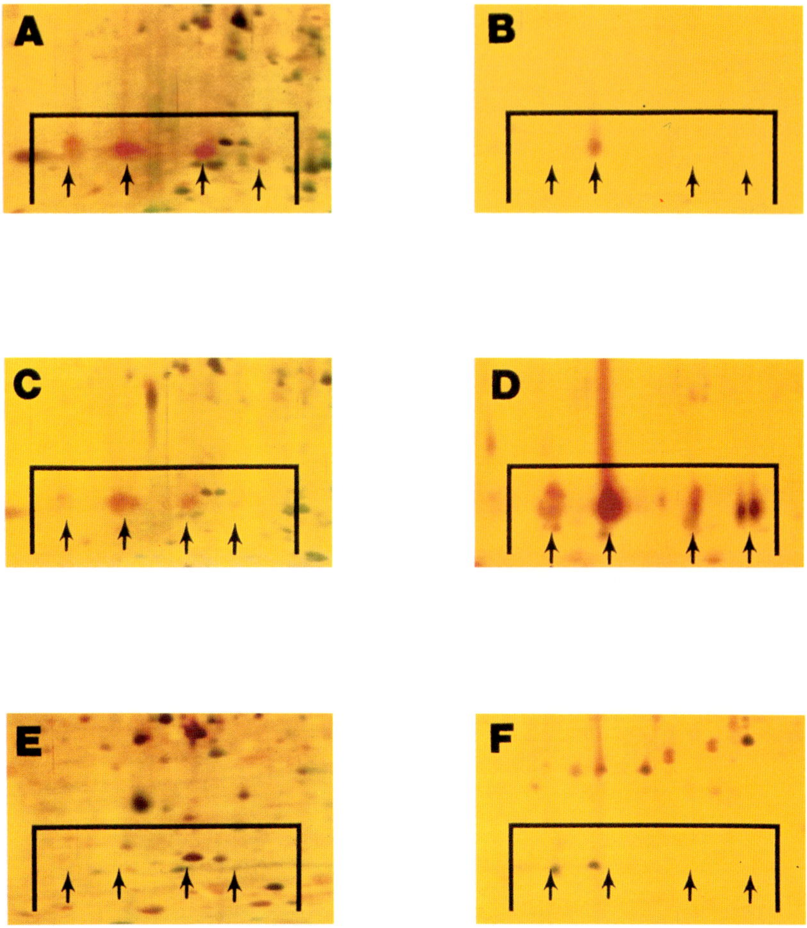

Plate 7. The tracking of urinary proteins to the kidney cortex. Urine was collected from juvenile (64 days/305 g), retired breeder (221 days/587 g), and aged (795 days/656 g) rats. The kidney was isolated and the cortex was punched out with a cork borer. Standard morphological landmarks were used to reduce contamination by the medulla. Each figure is representative of the appearance of samples from several rats in each group. (A), (C), and (E) show cortex protein patterns of juvenile, retired breeder, and aged rats, respectively; (B), (D), and (F) show representative urinary protein patterns seen in juvenile, retired breeder, and aged rats, respectively. The region of the gels displayed corresponds to an approximate molecular weight of 17,000 and an isoelectric point of about 5.

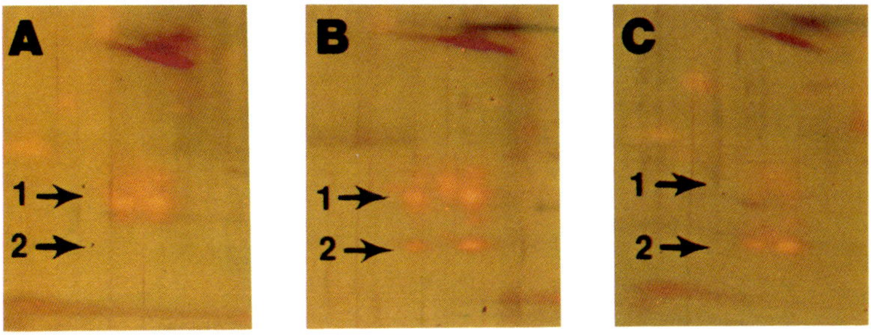

Plate 8. The modification of cellular proteins by drug treatment. Human hepatoma cells were cultured with and without 0.5 μg/ml of tunicamycin. (A) shows a region of the protein pattern derived from cells without drug treatment; (B) shows a region of the protein pattern derived from cells treated with the drug for 1 day prior to harvest; (C) shows the same region as A and B but is derived from cells treated for 5 days.

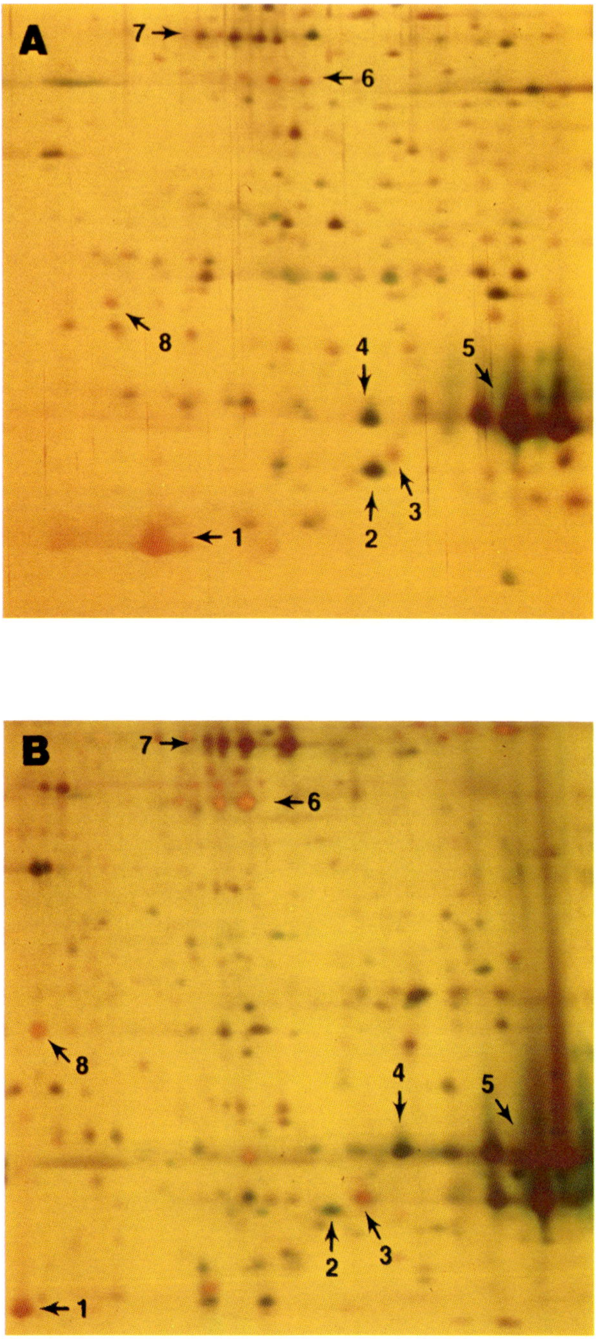

Plate 9. The comparison of 2-D protein patterns of human and Chinese hamster erythrocyte lysates. (A) A partial human erythrocyte lysate 2-D protein pattern; (B) A partial Chinese hamster erythrocyte lysate 2-D protein pattern.

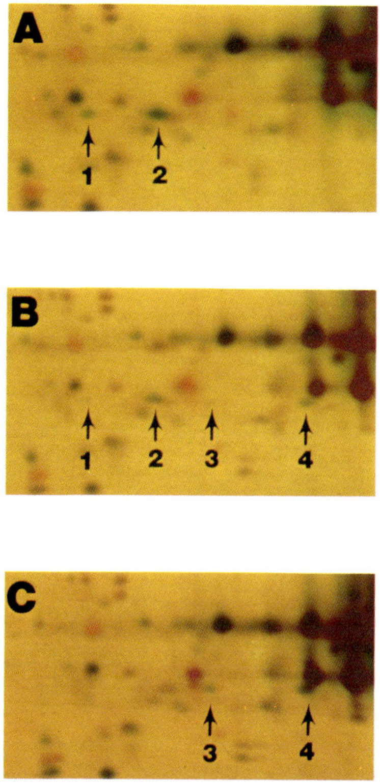

Plate 10. The use of two-dimensional gel electrophoresis to search for markers of human disease in animal disease models. (A), (B), and (C) are the same regions of 2-D protein patterns from a representative nondiabetic hamster, the F-1 heterozygous offspring of a cross between nondiabetic and diabetic hamsters, and a diabetic hamster, respectively. Positions 1 and 2 indicate nondiabetic proteins, and positions 3 and 4 indicate diabetic proteins.

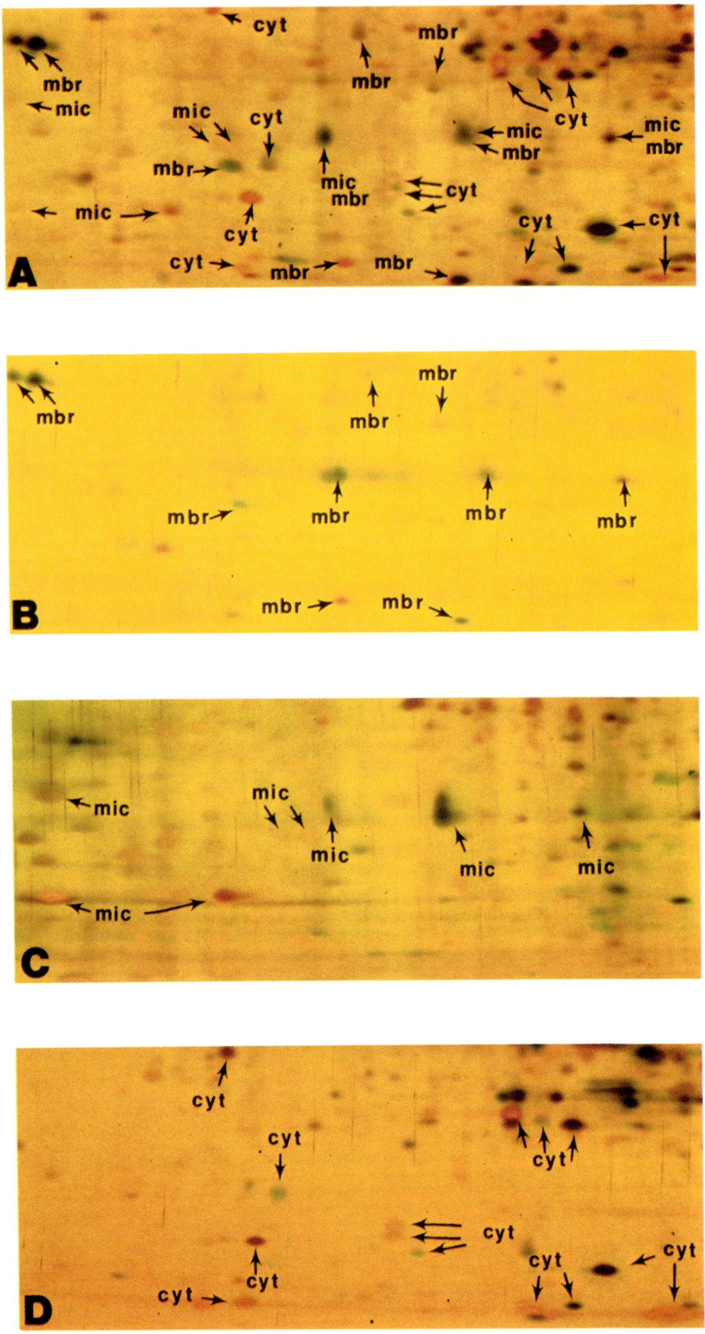

Plate 11. Subcellular fractionation improves 2-D protein pattern resolution and protein characterization. Hamster liver was subfractionated into membrane, microsomes, and cytoplasm. (A) shows the unfractionated sample; (B) the membrane fraction; (C) the microsome fraction; and (D) the cytoplasm. Subcellular-specific proteins are labeled mbr, membrane; mic, microsome; and cyt, cytoplasm.

rocyte lysate is made. Selected protein gene products in each species are numbered 1–8 and illustrate the conservation of a spot color characteristic from hamster to human in many proteins. Even though in some cases the isoelectric point is different, the similar color of a spot allows certain identification of the same protein across species (see spots 1, 4; see also chapter by Fey *et al*). Thus by using the color as a third parameter it is possible to use purified animal protein to find an analogous human protein in a 2-D map of human tissues even though the x or y coordinate position may be different.

Close comparison of Plate 9A (human) and Plate 9B (Chinese hamster) red blood cell lysates suggests that 2-DGE and color silver staining can be successfully used with a great degree of accuracy to map genes to their chromosome in somatic animal–human cell hybrids. Devine *et al.* (1982) used 2-DGE and color silver staining of protein gene products to map their genes to human chromosomes 21 and X. In these studies the similarity of color aided in the identification of the expressed human protein gene products from the mouse background.

F. *Use of Animal Models to Find Marker Proteins for Disease*

The Chinese hamster has been used as an animal models for human diabetes (for a review see Gerritsen, 1982). Using 2-D gel electrophoresis we have identified a protein that apparently segregated with the diabetic syndrome over years of breeding (Sammons *et al.*, 1981). Recently we have demonstrated two protein gene products that show a significant association in genetic studies with glycosuria in the diabetic subline X (Sammons *et al.*, 1983). Plate 10 illustrates the diabetes associated protein gene products. There are two blue protein gene products at positions 1 and 2 in the nondiabetic sublines (Plate 10A) and blue protein gene products 3 and 4 in the diabetic subline (Plate 10B). In the diabetic subline (Plate 10B) there is a brown spot similar to that seen in Plate 5 (no. 5); however, the blue protein at position no. 2 is absent. The protein gene product is shown to have a genetic basis through a cross of nondiabetic to diabetic animals because the heterozygote (Plate 10C) has half of the expected protein in each position (quantitation data is not shown). Because the protein is a diabetic marker in the animal model a corresponding protein might be found in human diabetics. In fact, the protein labeled no. 2 in the human red cell lysate (Plate 9A) is almost certainly the same as in the Chinese hamster (Plate 10A); it is blue and is located just to the left and below the red spot, which is also identically positioned in 2-D gels of both species.

G. Subcellular Fractionation

In Plate 11A, a portion of a 2-D protein pattern of an unfractionated liver sample is shown. Unfortunately there are many more protein spots than can be entirely resolved from each other. Thus to simplify the patterns the liver sample was fractionated into microsome, membrane, and cytoplasmic fractions. The silver-stained membrane, microsome, and cytoplasm subcellular fraction 2-D patterns are illustrated in Plates 11B, 11C, and 11D, respectively. The comparison of the constellation of spots between unfractionated and fractionated patterns allows documentation of microsomal-specific, membrane-specific, and cytoplasmic-specific proteins (see also chapter by Bravo and Celis). Some proteins seem to be shared by both microsome and membrane while others are specific. Color of the protein spots confirm and characterize the proteins even if the 2-D gels are imperfectly matched in their x,y coordinate dimensions. The use of protein color to measure cross-contamination of proteins in subcellular fractions may be more definitive than the enzyme assays previously used for this purpose. Finally, subcellular fractionation is shown to be an effective way to simplify the 2-D patterns and, in some cases, to resolve overlapping spots sharing identical x,y coordinate positions. Additionally, subcellular fractionation allows enrichment for minor organelle-specific proteins that would be obscured by more abundant proteins from other cellular structures in whole cell extracts.

IV. Computerized Data Acquisition of Color-Stained Images: Test for Stoichiometric Staining

A. BioImage™ System

A data acquisition system for 2-DGE was configured by and purchased from the BioImage Corporation, Ann Arbor, Michigan. The system components are a wet gel flatbed scanner, minicomputer and disc, color cathode ray tube, tape drive, printer, and camera. The color-stained wet gels are scanned through red, green, and blue filters and the digitized image is stored on the disc or on magnetic tape in a 1024 × 1024 array with a resolution of 185 × 185 microns per digital value. The BioImage software analysis of a protein spot is initiated by displaying in full color the gel on the system monitor. Spot boundary, area, color parameters from each of the three filters (channels), and quantification data (integrated intensity) are automatically determined for each protein spot by positioning the display cursor within the spot boundary.

B. *Stoichiometry of GELCODE Stain: Integrated Intensity versus Concentration*

In Figs. 1A and 1B the log integrated intensity is plotted versus the log concentration for each of the four colored spots in multiple gels for each

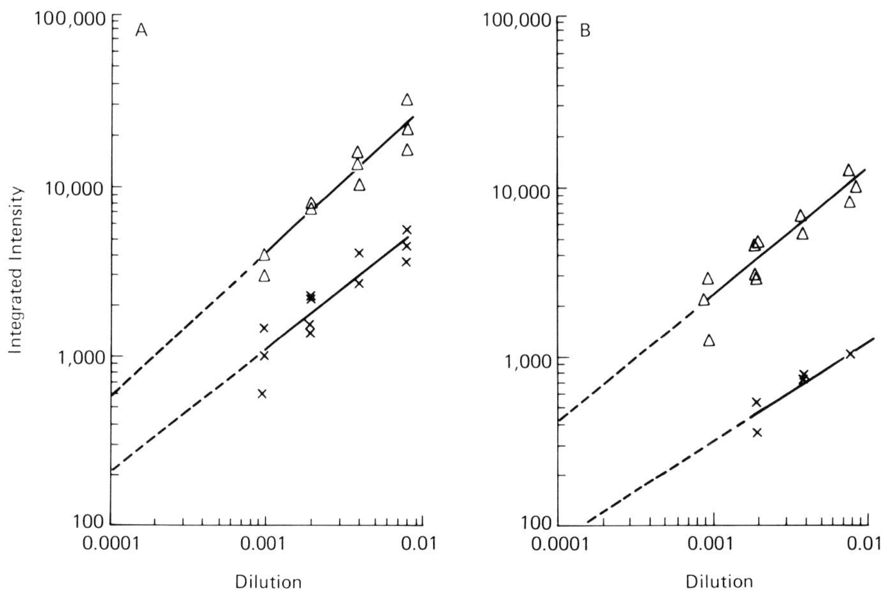

Figure 1. The linear relationship between integrated intensity and concentration for blue, brown, red, and yellow spots. A protein sample was serially diluted 1/16 through 1/1024 and 2-D gels were run for each concentration. The integrated intensity was automatically determined with BioImage software after the spot of interest was interactively designated in each 2-D gel image. The linear range of the data was determined with a standard statistical package for linear regression analysis. (A) △, blue spot, $R^2 = 0.95$ with a slope of 0.86; x, red spot, $R^2 = 0.86$ with a slope of 0.77. (B) △, brown spot, $R^2 = 0.87$ with a slope of 0.76; x, yellow spot, $R^2 = 0.91$ with a slope of 0.545.

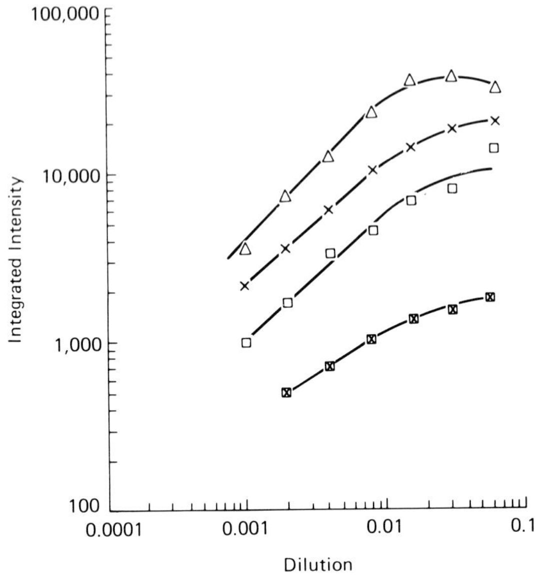

Figure 2. The stain saturation curve for integrated intensity and protein dilution for blue, brown, red, and yellow spots. The linear range data of Fig. 1 was combined with the nonlinear range and plotted as described in Fig. 12. x, brown spot; △, blue spot; □, red spot; ⊠, yellow spot are data points of the same protein on multiple gels.

dilution of protein. The data indicates a stoichiometric relationship between staining and protein concentration. The data for yellow spots show a diminished linearity range relative to blue, brown, and red proteins. Figure 2 illustrates a departure from linearity of the silver stain at different protein dilutions for blue, brown, red, and yellow protein. The integrated intensity value reaches a different plateau for each colored spot in the following order: blue > brown > red > yellow. Often in the nonlinear region the protein is a different color than in the linear part and therefore integrated intensity values in the nonlinear regions are not stoichiometrically related to protein concentration. The digital values for yellow protein spots are higher than background, and integrated intensity values for yellow spots less than 500 are the limit of detection. Digital values for blue, brown, and red protein spots are less than background and these proteins can be reliably quantified down to integrated intensity values of 300 (data not shown). Thus extrapolation of the linear range of the blue, brown, and red spots to their lower limits of detection reflects an extrapolated value predicted from the linear portion of the stain curve (dotted line in Fig. 1). If one accepts the linear extrapolation, then the linear integrated intensity range of blue, brown, red, and yellow protein spots is 300–40,000; 300–15,000; 300–6,000; and 500–900, respectively. This translates into an approximate linear concentration range for blue, brown, red, and yellow proteins of 100-, 85-, 78-, and 4-fold, respectively. These values for blue, brown, and red spots are approximately twice the linearity range reported for other silver stain methods. This is the first time negative or yellow spots have been reliably quantitated. Because there is an exact stoichiometric relationship between protein concentration and integrated intensity of a given GELCODE-stained spot these measurements of the integrated intensity with the BioImage system allow an accurate calculation of protein concentration from the general linear equation: log integrated intensity = intercept + slope (log concentration). Preliminary data suggest that all proteins with similar spectral properties (color) will have identical slopes (data not shown). Thus it is anticipated that a few calibration curves characteristic of each color class of protein spots as determined by multispectral digital image analysis (Vincent *et al.*, 1981) will allow eventual computerized determination of the absolute amounts of protein in each spot within a 2-D gel electrophoretogram.

V. Concluding Remarks

In summary, the color stain sensitivity and reproducibility, the unique application of color silver staining to 2-D mapping of proteins, and the stoichiometric relationship between integrated intensity of different colored protein spots and protein concentration makes the GELCODE system an ideal method for detecting protein patterns in 2-D gel electrophoretograms.

References

Adams, L. D., and Sammons, D. W. (1981). A unique silver staining procedure for color characterization of polypeptides. *In* "Electrophoresis '81" (Allen, R. C. and A. Arnaud, eds.), pp. 155–167. de Gruyter, Berlin.

Allen, R. C. (1980). Rapid isoelectric focusing and detection of nanogram amounts of proteins from body tissues and fluids. *Electrophoresis* **1**, 32–37.

Anderson, N. L., and Anderson, N. G. (1977). High resolution two-dimensional electrophoresis of human plasma proteins. *Proc. Natl. Acad. Sci. USA* **74**, 5421–5425.

Anderson, N. G., Anderson, N. L., and Tollaksen, S. L. (1979). Proteins of human urine. I. Concentration and analysis by two-dimensional electrophoresis. *Clin. Chem.* **25**, 1199–1210.

Devine, E., Brown, T., Krueger, L., Jenkins, E., Dutkowski, R., and Sammons, D. (1982). Assignment of proteins to human chromosome 21 and X using two-dimensional gel electrophoresis. *Am. J. Hum. Genet.* **134**, 170A.

Frearson, N., Taylor, R. D., and Perry, S. V. (1981). A protein in urine associated with muscle disease and muscle damage. *Br. Med. J.* **282**, 2002–2003.

Gerritsen, G. C. (1982). The Chinese hamster as a model for the study of diabetes mellitus. Proceedings of a task force on animals appropriate for studying diabetes mellitus and its complications. *Diabetes* **31**(*Suppl. 1*), 14–23.

Guevara, J., Johnson, D. A., Ramagali, L. S., Martin, B. A., Capetillo, S., and Rodriguez, L. V. (1982). Quantitative aspects of silver deposition in proteins resolved in complex polyacrylamide gels. *Electrophoresis* **3**, 197–205.

Lehle, L., and Tanner, W., (1976). The specific site of tunicamycin inhibition in the formation of dolichol bound *N*-acetylglucosamine derivatives. *FEBS Lett.* **71**, 167–170.

Merril, C. R., Goldman, D., Sedman, S. A., and Ebert, M. H. (1981). Ultrasensitive stain for proteins in polyacrylamide gels shows regional variation in cerebrospinal fluid proteins. *Science (Washington, D.C.)* **211**, 1437–1438.

Merril, C. R., Goldman, D., and VanKeuren, M. L. (1982). Simplified silver protein detection and image enhancement methods in polyacrylamide gels. *Electrophoresis* **3**, 17–23.

Oakley, B. R., Kirsch, D. R., and Morris, R. N. (1980). A simplified ultrasensitive silver stain for detecting proteins in polyacrylamide gels. *Anal. Biochem.* **105**, 361–363.

Ochs, D. C., McConkey, E. H., and Sammons, D. W. (1981). Silver stains for proteins in polyacrylamide gels: a comparison of six methods. *Electrophoresis* **2**, 304–307.

Poehling, H. M., and Neuhoff, V. (1981). Visualization of proteins with a silver "stain:" a critical analysis. *Electrophoresis* **2**, 141–147.

Roto, P. (1978). Urinary beta 2-microglobulin related to chronic low level exposure to metallic cadmium dust in Finnish cadmium processing workers. *Scand. J. Work Environ. Health* **4**, 131–136.

Sammons, D. W., Adams, L. D., Tollaksen, S. L., and Anderson, N. G. (1980). A genetic marker in spontaneously diabetic chinese hamsters. *Diabetes* **29**(*Suppl. 2*), Abstr. no. 146.

Sammons, D. W., Adams, L. D., and Nishizawa, E. E. (1981). Ultrasensitive silver-based color staining of polypeptides in polyacrylamide gels. *Electrophoresis* **2**, 135–141.

Sammons, D. W., Adams, L. D., Connell, M. A., and Gerritsen, G. C. (1983). A genetic marker for diabetes in Chinese hamster erythrocytes. *Diabetes* **32**(Suppl. 1), Abstr. no. 179.

Switzer, R. C., Merril, C. R., and Shifrin, S. A. (1979). A highly sensitive silver stain for detecting proteins and peptides in polyacrylamide gels. *Anal. Biochem.* **98**, 231–237.

Takatsuki, A., Arima, K., and Tamura, G. (1971). Tunicamycin, a new antibiotic. Isolation and characterization of tunicamycin. *J. Antibiot.* **24**, 215–223.

Tracy, R. P., Currie, R. M., Kyle, R. A., and Young, D. S. (1982). Two-dimensional gel electrophoresis of serum specimens from patients with monoclonal gammopathies. *Clin. Chem. (Special Issue, Part II)*, pp. 900–908.

Vincent, R. K., Hartman, J., Barrett, A. S., and Sammons, D. W. (1981). Multispectral digital image analysis of color two-dimensional color electrophoretograms. In "Electrophoresis '81" (Allen, R. C. and Arnaud, P., eds.), pp. 371–381, de Gruyter, Berlin.

Wray, W., Boulikas, T., Wray, V. P., Hancock, R. (1981). Silver staining of proteins in polyacrylamide gels. *Anal. Biochem.* **118**, 197–203.

CHAPTER 5

Electrophoretic Transfer of Proteins from Two-Dimensional Gels to Sheets and Their Detection

JANEY SYMINGTON
Institute for Molecular Virology
St. Louis University School of Medicine
St. Louis, Missouri

I.	Introduction	128
	A. General Description of the Protein Transfer Analysis Procedure	128
	B. Development of the Protein Transfer Analysis Procedure	129
II.	Electrophoretic Transfer of Proteins	131
	A. Principle and Relevant Factors for Optimizing Transfer	131
	B. Preparation of Two-Dimensional Polyacrylamide Gels for Transfer	132
	C. Transfer Sheets and Apparatus	134
	D. Transfer to Nitrocellulose Membrane	139
	E. Transfer to Diazo Paper	142
	F. Transfer to Positively Charged Nylon Membrane	145
III.	Probing the Transfer	146
	A. Principle and Relevant Factors for Optimizing Detection	146
	B. Immunoprobing: Chromogenic System for Visualization	149
	C. Immunoprobing: Radiolabeled Components for Visualization	150
	D. Alternative Probes	153
IV.	Removal of Probes	153
	A. Removal of Probes from Nitrocellulose Membrane	153
	B. Removal of Probes from Diazo Paper	154
	C. Removal of Probes from Positively Charged Nylon Membrane	154
V.	Examples and Applications of Protein Transfer Analysis	155
	A. Examples of the Detection of Specific Proteins on Transfers from Two-Dimensional Gels	155
	B. Applications and Potential of the Method	158
VI.	Concluding Remarks	162
	References	163

TWO-DIMENSIONAL GEL
ELECTROPHORESIS
OF PROTEINS

ISBN 0-12-164720-X

I. Introduction

In order to understand the mechanisms of virus replication, cell transformation and many other biological processes, one must identify specific proteins found in complex mixtures. Such identification has become important in the diagnosis of disease and is becoming of increasing industrial importance in the development of biosynthetic products. Much progress has occurred in the characterization of viral, cellular, and body fluid proteins. We can now combine a new technique, protein transfer analysis (PTA), with two-dimensional polyacrylamide gel electrophoresis (2-D PAGE) in order to identify and further characterize particular proteins. This system affords exquisite specificity, definition, and sensitivity.

A. General Description of the Protein Transfer Analysis Procedure

In the PTA procedure transfers are made from gels to sheets of nitrocellulose membrane, diazo paper, or positively charged nylon membrane. The transfer process maintains the resolution present in the gel and facilitates detection of specific proteins or classes of proteins.

Detection systems include treatment of the transfers with antibodies, lectins, and other probes. The specific proteins are visualized by use of radiolabeled second antibody or iodinated *Staphylococcus aureus* protein A (^{125}I-labeled PA). This can be used as a general reagent because it binds with high affinity to many immunoglobulins (Kessler, 1975; Goding, 1978; Langone, 1978). One may also visualize proteins by using antibodies conjugated with enzymes that will react with a chromogenic substrate. Radiolabeled probes such as ^{45}Ca, ^{125}I-labeled protein, [^{32}P]DNA, ^{125}I-labeled RNA (Bowen *et al.*, 1980) may help to elucidate the function of specific transferred proteins. Protein transfer analysis is so important that it has been called a third *specificity* dimension for the analysis of proteins (Anderson *et al.*, 1982b). For example, particular DNA-binding domains of fibronectin have been studied by PTA using limited proteolysis and [^{3}H]DNA as a probe on the transfers (Hoch, 1982).

Characterization of antibodies in serum and monoclonal preparations (Shen *et al.*, 1980; Nielsen *et al.*, 1982) has become an increasingly important use of protein transfers. Transfers are also useful in the examination of mixed antigen populations; they help identify that antigen toward which antibodies present in a particular preparation are directed (Tsang *et al.*, 1983). Protein transfer analysis has the advantageous ability to detect nonprecipitating complexes. Reiser and Wardale (1981) have combined PTA with two-dimensional partial proteolysis mapping to localize the antigenic determinants of peptides of SV40 proteins. This approach was also used by Yurchenco *et al.* (1982) to study spectrin. In the future further important data about protein structure will be obtained by the use of PTA. It will be invaluable for securing information on molecular relationships, precursors, and processing. In performing such analyses antibodies to small synthetic peptides made to a specific region of a genome will be useful. Protein transfer analysis will also become important in characterizing the products of *in vitro* synthesis.

Protein transfer analysis does not require *in vivo* labeling. This characteristic is a great asset because it enables the investigator to study more readily tissue and fluid samples from living donors as well as autopsy material. Symington *et al.* (1981) have shown that a protein present in a concentration of 1 in 100,000 in whole cell extracts is easily detectable by PTA. Reiser and Wardale (1981) have detected as little as 25 pg upon direct application of purified antigen to sheets.

B. Development of the Protein Transfer Analysis Procedure

Immunoprecipitation is a powerful tool for the identification of particular proteins present in cell lysates or other mixtures. However, it suffers from the problem of coprecipitation of unrelated proteins along with the protein being studied. This problem is solved by separation of protein mixtures on sodium dodecyl sulfate (SDS)-polyacrylamide gels before treatment with antiserum. Gels can be incubated with antibodies or lectins and the complexes formed on or near the surface detected by color formation, if enzyme conjugates are used, or by autoradiography (Burridge, 1976, 1978; Olden and Yamada, 1977; Adair *et al.*, 1978). However, these procedures are time consuming as many washes are required to obtain an acceptably low background.

Using a procedure similar to Southern blotting, Renart *et al.* (1979) were able to transfer proteins from polyacrylamide–agarose composite gels to diazobenzyloxymethyl (DBM) paper. They could detect specific proteins more rapidly and efficiently on these blots than in gels. The covalently bound proteins reacted with antibodies and resulting complexes were visualized with ^{125}I-labeled PA. The transfer of protein from gels to sheets is often referred to as "Western blotting". The gels used by Renart and coworkers were cross-linked by reagents that were cleaved by periodate or alkali before transfer. However, some proteins are altered by these treatments and transfer efficiency is low.

In a related approach Erlich *et al.* (1979a,b) identified specific proteins by use of $F(ab)_2$ fragments or lectins coupled covalently to DBM paper. The sheets were applied to one- or two-dimensional gels and proteins that reacted with the coupled affinity ligand were selectively bound as they moved by capillary action from the gel to the paper. They were then visualized by autoradiography if labeled or, if unlabeled, by a "sandwich" assay using antiserum and ^{125}I-labeled PA. Antigen-coupled DBM paper could also be used in this manner for the detection of specific immunoglobulins separated in isoelectric focusing gels.

Bowen and co-workers (1980) used a modified blotting method in which proteins were transferred by diffusion from SDS–urea–polyacrylamide gels to nitrocellulose. Reasonably good replicas were obtained. However, even after two days transfer was not complete. Towbin *et al.* (1979) introduced the use of an electric current to assist in the transfer of protein from polyacrylamide gels to nitrocellulose. Transfer of proteins was more rapid and more nearly quantitative. The electrophoretically transferred proteins that were immobilized on nitrocellulose could still be recognized by specific antibodies. They visualized the protein–antibody complexes by an indirect method, using fluorescein- or peroxidase-conjugated or ^{125}I-labeled second antibody.

Electrophoretic transfer of proteins from polyacrylamide gels to DBM paper was employed by Stellwag and Dahlberg (1980) and by Bittner and co-workers (1980). They found that most proteins transferred well and one-dimensional (1-D) gel patterns showed resolution similar to the original gel. An advantage is the covalent coupling of transferred proteins.

Symington *et al.* (1981) separated proteins from whole KB cell extracts by 2-D PAGE (O'Farrell, 1975) and transferred these electrophoretically to DBM paper. The high resolution characteristic of this type of gel was maintained. They utilized antisera and ^{125}I-labeled PA followed by autoradiography to detect specific viral proteins and found that more immunologically reactive forms could be discriminated than on 1-D gel transfers. Transfer sheets carrying covalently bound proteins could easily be "erased" (antibody and ^{125}I-labeled PA removed) and reprobed many times.

Reiser and Wardale (1981) used diazophenyl thioether (DPT) paper for PTA of 2-D gels. It is easier to prepare than DBM paper and is equally amenable to the use of immunoautoradiography for detection of the covalently bound proteins.

Recent modifications have been made to the electrophoretic transfer technique for transfer of proteins from 2-D gels containing SDS in the second dimension to nitrocellulose. Burnette (1981) achieved greater transfer by extending the time of electrophoresis (12 h or longer). He used ^{125}I-labeled PA for detection of the complexes on nitrocellulose. Anderson *et al.* (1982b) used the same buffer as Towbin and co-workers (1979) but omitted the methanol to facilitate elution. They got a good transfer of proteins in 4 h with only the most abundant proteins passing through the first sheet. Their results suggest that PTA will be successful with most proteins. Other investigations have shown that probes can also be removed from nitrocellulose and the transfer reprobed several times (Legocki and Verma, 1981; Erickson *et al.*, 1982). Transfers may, in addition to their role in the identification and characterization of particular proteins and antibodies, be useful for the purification of antibody (Olmstead, 1981; Anderson *et al.*, 1982b).

II. Electrophoretic Transfer of Proteins

A. *Principle and Relevant Factors for Optimizing Transfer*

As in other types of electrophoresis of proteins the charges on the proteins or peptides allow a migration in an electric field. In a 2-D gel system usually SDS is used in a polyacrylamide gradient gel to obtain approximate molecular weights and for maximum resolution in the second dimension. In this case proteins are negatively charged and will migrate to the anode in an electric current. The molecules are not easily eluted from the gel so a relatively high current is needed for efficient transfer, particularly for larger molecules ($M_r > 100{,}000$). A solution of low ionic strength is used. Resolution is best on the transfers to diazo paper when they are performed as rapidly as possible, so that contact of protein to the paper will be made while the labile diazonium groups are still active. The first binding to diazo paper is probably noncovalent through ionic interaction between negative charges on the protein molecules and the positively charged diazonium groups, followed by covalent bond formation (Alwine *et al.*, 1977). Although the capacity for formation of covalent bonds is prolonged at low pH (Stellwag and Dahlberg, 1980) the binding of proteins transferred from SDS gels is better at a pH near neutrality or above (Reiser and Wardale, 1981; Symington *et al.*, 1981).

When nitrocellulose or nylon membrane is used the length of time of transfer can be extended as the binding to them is not dependent on a reaction with unstable groups. The electric field strength must not be too high or proteins pass through the sheet. Although attachment is noncovalent, with careful attention to transfer conditions and subsequent treatment, protein molecules will attach and remain bound throughout probing.

The ideal, of course, is to have quantitative elution of all proteins from the gel and to have them present on transfer sheets in an immunologically and functionally reactive state. Conditions most conducive to this may include cooling during prolonged transfers. Transfer solutions of various composition and pH can be tried for the transfer of different types of protein. For certain proteins a second dimension involving urea or electrofocusing may be the best solution.

When the proteins of interest are of high molecular weight one may consider improving elution by the use of a lower percentage of acrylamide or changing the amount of crosslinking. Partial digestion with protease has also been used to facilitate transfer of larger proteins. The addition of detergents such as SDS to the transfer buffer or the omission of methanol can be considered, although these changes may reduce binding of some proteins to nitrocellulose even though better elution is accomplished.

B. Preparation of Two-Dimensional Polyacrylamide Gels for Transfer

1. FIRST DIMENSION

The first dimension can be run by isoelectric focusing (O'Farrell, 1975), nonequilibrium pH gradient electrophoresis (O'Farrell *et al.*, 1977), or isotachophoresis (Nowinski *et al.*, 1979). The procedures for the preparation of these gels are presented in the citations given and discussed in the chapter by Bravo. For ribosomal proteins first-dimensional electrophoresis is performed using either acidic (Shen *et al.*, 1980) or basic (Nielsen *et al.*, 1982; Towbin *et al.*, 1982) gels.

Our technique for the preparation of the first-dimensional gels differs only in the use of a rapid method for making identical gels. It is particularly useful for very thin cylindrical gels. We make a bundle of isoelectric focusing tubes held together with a rubber band near the top. We insert the bundle into a heat sealable bag prepared to fit the tubes and clamp the bag to a stand making sure that the bottoms of the tubes are even. We then make the gel mixture and rapidly pipet it down the side of the bag, allowing the mixture to rise from the bottom to a premarked line on the tubes. The gel level can be adjusted by inserting solid glass rods into the bag. After polymerization has occurred, we cut the bag apart wearing gloves (acrylamide is toxic) and cut the excess gel from the ends of the tubes with a razor blade. Tubes are wiped with Kimwipes and rinsed in distilled water before use.

2. SECOND DIMENSION

Sodium dodecyl sulfate–5–20% polyacrylamide gradient gels are a good choice for general analysis of proteins in a complex preparation by PTA. These are cross-linked with *N,N'*-methylenebisacrylamide(bis) (0.13–0.54%), utilize the Laemmli (1970) buffer system, and are prepared and run according to O'Farrell (1975) (see also chapters by Bravo and by Garrels). Appropriate separative gel size is 13 cm × 15 cm × 1.5 mm. A 4.5% stacking gel with 0.12% bis is used and gels are run in an electrophoresis buffer consisting of 25 m*M* Tris base–192 m*M* glycine and 0.1% SDS.

An alternative system that has given good results (Reiser and Wardale, 1981) is the discontinuous sulfate borate buffer system of Neville (1971). Reiser and Wardale also altered cross-linking. In a 7–20% gel they used 0.09–0.15% bis. In addition, successful tranfers have been made from 2-D gels that included urea during the second-dimensional separation (Towbin *et al.*, 1979).

Lithium instead of SDS has also been used (Hawrot *et al.*, 1982). This has the advantage of not precipitating when cooling is required.

When the gel run is finished, cut one lower corner routinely for orientation. The gels should be preequilibrated if necessary and transferred immediately following the second-dimension electrophoresis.

3. PREEQUILIBRATION

It is important that gels do not change size during transfer. Smeared patterns will result. If size change can occur (a function of cross-linking and the compositions of solutions used for slab electrophoresis and transfer), the gel should be equilibrated in transfer buffer. For transfers to diazo paper it is also important that tris and glycine be removed to prevent their interaction with the diazonium groups. If slab gels contain urea it should be removed before electrophoretic transfer to diazo paper. Chilling the gel facilitates removal. If electrofocusing gels are used ampholytes must also be removed by washing before electrophoretic transfer. The stacking gel is removed prior to equilibration. Some investigators have suggested that 0.1% SDS be included in the wash buffers before transfer to avoid protein precipitation (Stellwag and Dahlberg, 1980). However, this may cause greater loss of proteins during the wash and reduced binding of the proteins to transfer sheets.

When nitrocellulose is used as transfer medium and 20% methanol is included in the transfer solution a brief equilibration period (15 min) is usually used. Whether or not size change will occur with particular systems should be determined empirically. If the particular protein being studied is very small the composition of the electrophoresis buffer and transfer solution should be such that the transfer can be made without equilibration to avoid loss.

C. Transfer Sheets and Apparatus

1. SELECTION OF TRANSFER SHEETS

The choice of transfer sheet depends on the type of probing desired. If repeated probing of a transfer is planned then diazo paper should probably be chosen. This type of sheet has been shown to be reprobable many times. Because proteins are not bound covalently to nitrocellulose, there is apt to be loss during erasure. The ability to reprobe is of particular importance if working with a limited sample.

In addition, reprobing of the identical 2-D transfer may allow proteins that are very similar in isoelectric points and molecular weight to be discriminated. Slight differences in location on different gel runs might otherwise be indiscernable. In order to compare DPT and DBM paper we analyzed a [^{35}S]methionine-labeled extract of human KB cells. We ran a 1-D SDS–5–18% polyacrylamide gradient gel with duplicate samples applied to two lanes. We transferred these proteins simultaneously to DPT and DBM paper strips for 2 h using conditions presented in Section II,D. Figure 1 shows that transfers to DPT and DBM paper have equal resolution. Small and large proteins have transferred to both types

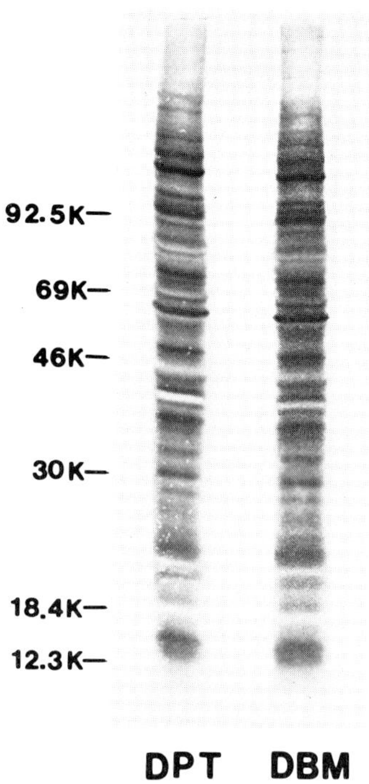

Figure 1. Electrophoretic transfer of proteins from an SDS–polyacrylamide gel to diazo paper. [^{35}S]Methionine-labeled polypeptides from KB cells (2.5 × 10^5 cells/lane) were transferred to DPT and DBM paper as described in the text. Transfers were autoradiographed. ^{14}C-labeled molecular weight markers were run on the same gel.

of sheet equally well. These papers have equal binding capacity, about 40 μg/cm^2 based on DNA (Seed, 1982). We recommend using DPT paper since it is more easily prepared and more stable.

About 50% of proteins M_r < 100,000 may be expected to be transferred and bound during a 2 h transfer (Symington *et al.*, 1981; Reiser and Wardale, 1981). Somewhat more complete transfer, particularly of larger proteins, can be accomplished with longer transfer times. Although detection of as little as 25 pg of antigen directly applied to diazo paper has been reported (Reiser and Wardale, 1981), one should in general expect to detect only about 1 to 10 ng of a particular protein (amount applied to first-dimensional gel) on either nitrocellulose or diazo paper 2-D transfers using the protocols that have been developed up to this time.

Nitrocellulose has several advantages. It is much cheaper than diazo paper and does not require chemical activation. In addition nitrocellulose transfers can be stained with amido black (Towbin *et al.*, 1979;

Gershoni and Palade, 1982) or toluidene blue (Towbin *et al.*, 1982), and immobilized proteins can be detected on nitrocellulose by means of an enzyme-linked immunoassay that requires no radiolabeled materials but uses a color-forming reaction for visualization. This detection system is comparable in sensitivity to the use of iodinated second antibody or ^{125}I-labeled PA (Hawkes *et al.*, 1982a). Extensive preequilibration is not required. Although transfers cannot be performed as rapidly with nitrocellulose, more complete transfer can be attained as no labile groups are involved. The capacity of nitrocellulose for protein binding is about 80 μg/cm^2 (Gershoni and Palade, 1982). More than 90% transfer of all proteins $M_r > 100{,}000$ is obtained in 22 h (Burnette, 1981), as determined by use of ^{14}C-labeled proteins. Care must be taken not to overload the membrane (Howe and Hershey, 1981), particularly when a transfer pattern is desired that accurately represents the original gel.

The mechanism of binding of proteins that have been run on SDS gels to nitrocellulose is unclear but hydrophobic forces as well as electrostatic must be involved. Legocki and Verma (1981) have reported erasure of probes from nitrocellulose at pH 2.2 and reuse of the transfers. However, they found that, unlike diazo paper transfers, the background became worse each time. In the erasure method of Erickson *et al.* (1982) 8 *M* urea is used. They were able to reprobe the blots with some kinds of antibody but not others. They suggest that some antigens may not remain reactive on nitrocellulose transfers during storage even without being exposed to erasure conditions. This recent progress toward satisfactory reuse of nitrocellulose transfers makes this medium increasingly attractive. Improvements in protocols for nitrocellulose transfers in this respect will very likely be forthcoming.

A positively charged nylon membrane, Zetabind, has been used successfully for electrophoretic transfer of protein (Gershoni and Palade, 1982). This material is notable for its high binding capacity (480 μg/cm^2 or about six times that of nitrocellulose). When it is used, methanol can be omitted. (Methanol presumably acts to strip some of the SDS from the protein molecules and prevent them from passing through the nitrocellulose sheet.) The use of Zetabind allows more rapid elution of proteins of all sizes and more complete elution of high molecular weight proteins. Most of the protein eluted in 2 h is bound (55% of the original 1.4 μg/lane load) and at least half of the bound protein is fixed by the first sheet. Under the same transfer conditions nitrocellulose binds less than 10% of the eluted protein to the first sheet. Transfer to this kind of sheet is more amenable to quantitation than transfer to other sheets. For use with a 2-D gel system the higher capacity would be most useful when studying those proteins present in high enough concentration to challenge the capacity of the transfer paper. However, achieving a low background is more difficult with this medium.

Although diethylaminoethyl (DEAE) membrane has not been probed, transfers to this type of sheet may have a role from a preparative standpoint.

2. TRANSFER CHAMBER AND POWER SOURCE

A schematic drawing of a typical loaded transfer chamber is shown in Fig. 2. The entire apparatus can be purchased as a unit specifically designed for this purpose (e.g. "Trans-Blot", Bio-Rad, Richmond, California; "Transphor", Hoefer Scientific Instruments, San Francisco, California) or adaptations can be made from available chambers. Details for the construction of a one-gel transfer unit similar to the one pictured are given in Bittner *et al.* (1980). Because high current is used to transfer proteins a safety lid should be used. The tank can be made of plexiglas and should be large enough to hold at least 3, and preferably 5, liters of

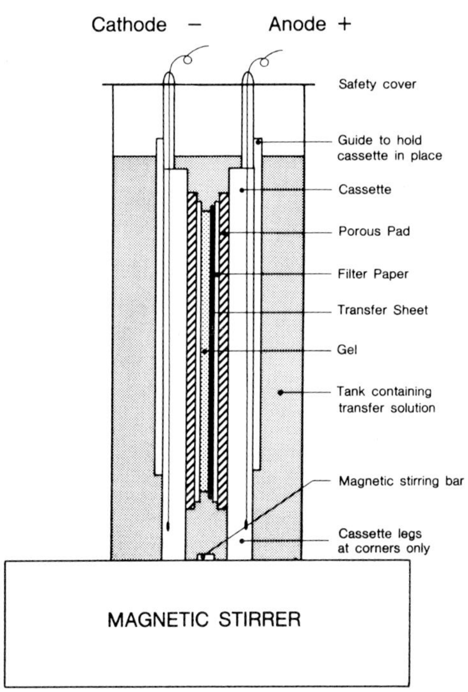

Figure 2. Schematic drawing of apparatus used for transfer of proteins from gels to sheets.

transfer solution in order to minimize the temperature rise during the run. A guide glued to the inside of the tank holds the cassette in place. The cassette can be of plexiglas also and the two halves held together by nylon hexbolts or a clamp after the "sandwich" is inserted. In the apparatus pictured platinum wire electrodes are strung back and forth lengthwise between the outside edges of each frame to achieve a uniform field. The electrodes are about 4 cm apart. Dialysis membrane or nylon fishing line can be affixed to the frame to hold the "sandwich" in place when it is inserted. The outside of the sandwich consists of two porous pads. Scotch-Brite made by 3M Company (Bio-Rad) is a good choice. These must be well washed before initial use. Nylon or Dacron sponge pads have also been used. The pads hold the gel firmly in contact with the transfer sheet without compressing it too much. Inside the pads are two sheets of heavy filter paper of high wet strength. The body of the cassette is held above the bottom of the tank on short legs to allow the stirring bar to rotate slowly to keep the pH even.

The gel pictured in the apparatus contains negatively charged proteins such as would be present in a SDS two-dimensional gel. Therefore the transfer sheet is located on the anodal side. The power source should be capable of providing at least 1.5 amps, preferably more. Field strengths used are usually 5–10 V/cm (relative to the distance between electrodes) (e.g., Hewlett–Packard DC 0–60 V/0–15 Amp; Bio-Rad 160/1.6).

3. ASSEMBLING THE "SANDWICH"

This is probably the aspect most responsible for difficulties in getting good transfers. If any bubbles are present between the gel and the transfer sheet the transfer will have poor resolution. It works well to perform the assembly under transfer buffer. This is particularly important with diazo paper because the diazonium groups are labile. Wearing gloves, place one Scotch-Brite pad in a plastic box full of transfer buffer. Pick up the gel from the equilibration buffer using one piece of equilibrated filter paper and lay them on top of the pad. Lay the transfer sheet on top of the gel starting at one end. Push out any bubbles and make sure contact is good. Add the other piece of filter paper and again push out any bubbles. Then add the other pad. The whole sandwich can be held together with large rubber bands. Place the sandwich in the cassette, secure the cassette, and slip it into the tank. If a cooling coil is to be used insert it now. Close the lid, turn on the magnetic stirrer, connect the electrode cables, and turn on the power.

When transferring to nitrocellulose some workers find it more convenient to assemble the sandwich by laying one saturated porous pad in a shallow layer of transfer buffer, moving the gel to this from the equilibration buffer with filter paper, pipetting a little buffer onto the gel, and rolling the wet nitrocellulose onto it. Bubbles can be easily seen through the nitrocellulose and removed by passing a gloved hand over the nitrocellulose or rolling a pipette across it. The wet final filter paper is added and rolled again. Then the saturated second pad is added and the package inserted into the cassette.

D. Transfer to Nitrocellulose Membrane

A general scheme for transfer and probing is shown in Table I. This is a protocol suitable for the analysis of total cellular proteins. For proteins of a particular size or nature adjustments in the protocol may be made to optimize it. Details of the protocol for nitrocellulose membrane follow. These transfer conditions are similar to those used by Burnette (1981).

1. MATERIALS AND REAGENTS

a. Nitrocellulose sheet 15 × 17 cm, 0.20 μm. (BA83; 0.45 μm, BA85 also acceptable for proteins of $M_r > 20{,}000$: Schleicher and Schuell, Keene, New Hampshire).
b. Two sheets of thick filter or chromatography paper of high wet strength 15 × 17 cm (Whatman 3MM).
c. Transfer solution for nitrocellulose 20 m*M* tris–base–154 m*M* glycine–20% methanol sufficient to cover electrode wires in tank (below the connections), fill assembly box, and use for preequilibration of gel before transfer.
d. Blocking and wash solution 50 m*M* Tris–HCl, pH 7.4–200 m*M* NaCl–0.05% Polyoxyethylene Sorbitan-Monolaurate (Tween-20; Sigma Chemical Co., St. Louis, Missouri)–3% gelatin (Laboratory Grade A, Fisher Chemical Co., St. Louis, Missouri) The Tween-20 should be added to the Tris–saline solution slowly with stirring. The gelatin is then added and the solution warmed to 37°C to dissolve the gelatin. Cool to room temperature before using. Store at 4°C. The Tween-20 is optional; the use of this or such detergents as NP-40 improves the background but may cause loss of some proteins from nitrocellulose (Lin and Kasamatsu, 1983).

TABLE I

General Scheme for Two-Dimensional Gel Protein Transfer Analysis

Nitrocellulose	Operation	Diazo paper
0–30 min, size equilibrate	Equilibrate gel in transfer solution	60 min, size equilibrate and remove components that react with diazonium group
10 min, wet in transfer solution	Pretreat transfer sheet	35 min, activate amino form to diazo form and equilibrate in transfer solution
2–22 h, 6–8V/cm	Electrophoretic transfer of proteins	2–4 h, 8–10V/cm; cooling for runs >2 h
1 h, 37°C	Incubate transfer sheet in blocking solution to saturate binding sites	2 h, 37°C
2 h, 22°C	Probe the transfer (e.g. by immunologic techniques.) Bind primary antibody (e.g. rabbit IgG)	2 h, 37°C
40 min: 22°C	Wash sheet	4 h, 37°C
1 h, 22°C e.g., Peroxidase-conjugated goat anti-rabbit IgG	Bind enzyme-conjugated second antibody or radiolabeled tracer	2 h, 37°C e.g., ^{125}I-labeled Protein A
40 min: 22°C	Wash sheet	2 h, 37°C
15–30 min: by reaction with chromogenic substrate	Detect specific protein	1–4 hr, autoradiography
2 or 3 times	Erase and reprobe	8 or more times

e. Platform rocker capable of gentle agitation (TEK-Pro V, Scientific Products, St. Louis, Missouri).

2. TRANSFER CONDITIONS

Equilibrate the gel with transfer solution to be used. Fifteen minutes is usually sufficient. Wear gloves and use forceps when handling nitrocellulose. Place the sheet in transfer buffer while allowing it and the filter papers to equilibrate 10 min. Assemble the sandwich and start the transfer.

With a field strength of about 6–8 V/cm most of the proteins separated on 2-D gels should be eluted within 22 h and good binding to the nitrocellulose should occur.

3. POSTTRANSFER TREATMENT OF NITROCELLULOSE MEMBRANE

After turning off the current and removing the cassette from the tank, open the cassette carefully and remove the filter paper from the gel side so that the position of the gel relative to the nitrocellulose sheet does not change. This is a precaution that is particularly important when strips or areas of the paper are to be cut and probed separately. Supporting the transfer and gel by the other filter paper, cut off the excess membrane and cut the lower right corner of the transfer sheet for orientation. Separate the sheet from the gel and label the side adjacent to the gel in soft lead pencil. Place the sheet in a 20 × 20 glass baking dish containing 200 ml of blocking solution. Rock the dish gently on a platform rocker for 1 h at 37°C to block residual protein binding sites.

4. OPTIONS

To elute high molecular weight proteins (>100,000) 0.01 to 0.1% SDS can be included during the transfer (Erickson *et al.*, 1982). A lower voltage gradient (3 V/cm) was used by these workers and good transfer was obtained in 21 h. If methanol is omitted from the transfer buffer (Anderson *et al.*, 1982b) much of the protein is eluted in 4 h. However, some protein may pass through the membrane (Gershoni and Palade, 1982). The use of 10% methanol might preserve efficient binding while hastening elution compared to transfers utilizing 20% methanol. Nitrocellulose membrane of 0.45 μm porosity has been widely used for transfers. This is satisfactory for large proteins but some loss occurs of proteins of molecular weight less than 20,000 (Burnette, 1981; Lin and Kasamatsu, 1983). A membrane of porosity 0.20 μm is preferable for the adsorption of smaller proteins.

The buffer concentration and the type of buffer may be varied to change the extent of transfer of particular proteins. Although smaller proteins elute more easily (Howe and Hershey, 1981) and transfer is generally better, differential transfer to nitrocellulose is not just a question of molecular weight differences (Hu *et al.*, 1981). Limited protease treatment of proteins in the gel may be done to aid transfer (Gibson, 1981). Transfer can also be done from urea-containing gels (Towbin *et al.*, 1979) using 0.7% acetic acid as transfer solution. In this case the nitrocellulose sheet should be positioned between the cathode and the gel during transfer. This transfer solution is suggested for basic proteins. Successful transfers of proteins to nitrocellulose have also been made using 25–50 m*M* sodium phosphate, pH 6.5. Some workers (Tsang *et al.*,

1983) recommend cooling for transfers to nitrocellulose but most find it unnecessary.

When gels have been removed from the transfer sheet, they may be stained to determine the extent of protein elution by comparison with an untransferred gel. The nitrocellulose transfer can be stained by immersing it about 1 min in 0.1% amido black 10B in 25% isopropanol–10% acetic acid in water. Destain about 30 min in 25% isopropanol–10% acetic acid. Wash in water before drying (Gershoni and Palade, 1982). Toluidene blue stain can also be used (Towbin *et al.*, 1982) or a very sensitive immunostain involving dinitrophenol derivitization of the proteins (Wojtkowiak *et al.*, 1983).

E. Transfer to Diazo Paper

A protocol suitable for transfer of total cellular proteins to diazo paper is outlined in Table I. It was used by Symington *et al.* (1981) for 2-D gels and is similar to that used by Bittner *et al.* (1980).

1. MATERIALS AND REAGENTS

a. Sheet 15 × 17 cm of aminophenylthioether cellulose (APT paper) or aminobenzyloxymethyl cellulose (ABM paper). Such sheets are available commercially (Bio-Rad, Schleicher and Schuell) or can be prepared in the laboratory. Aminophenylthioether paper can be prepared by the method of Seed (1982). Aminobenzyloxymethyl paper can be prepared by the method of Alwine *et al.* (1977). We have found that sheets can be stored fully activated in the acid form at −70°C for several weeks.
b. Activating Solution for Transfer paper (Alwine *et al.*, 1977):
   Ice-cold 1.2 *N* HCl
   2.5% $NaNO_2$ (freshly prepared)
   To 200 ml of 1.2 *N* HCl add 2 ml of sodium nitrite solution just before activation of paper.
c. Two sheets of Whatman 3 MM or equivalent paper.
d. Flat enamel pan (dissecting) slightly larger than the paper.
e. Transfer buffer for diazo paper: 25 m*M* sodium phosphate buffer, pH 6.5. Some of this should be stored cold for equilibration of the diazo paper after activation.
f. Blocking solution for diazo paper: 100 m*M* Tris, pH 9.0–0.25% gelatin–10% v/v ethanolamine.
g. Platform rocker.

2. TRANSFER CONDITIONS

Equilibrate gel for 3 × 20 min with 400 ml of transfer buffer per wash at room temperature with gentle rocking. Equilibrate filter papers also for the last 10 min. The activation of diazo paper is done in the cold, using all cold solutions in a precooled enamel pan according to the method of Alwine *et al.* (1977) except for the washes following activation. Time activation so that it will be finished as the gel equilibration is finished. Wear gloves when handling diazo paper. Label the paper in pencil in the lower right corner on the side to be in contact with the gel, put it into 200 ml of activating solution, and rock it gently 30 min. Then wash it 3 × 1 min in cold deionized distilled water and 2 × 1 min in cold transfer buffer (200 ml each wash). Assemble the sandwich and start the transfer.

The transfer should be stopped within 4 h as the reactivity of diazonium groups declines rapidly after that time. A voltage gradient of 8 to 10 V/cm is suggested. If more than 2 h are required to optimize transfer of a particular protein the tank should be cooled. For maximum recovery of proteins $M_r > 100{,}000$ 4 h is suggested.

3. POSTTRANSFER TREATMENT OF DIAZO PAPER

Turn off power, remove cassette from the tank, and disassemble the "sandwich". Mark the position of the gel, cut away excess paper, and cut the lower right corner for orientation. Be sure that the labeling is still legible so that the side to which proteins are attached is obvious. Immediately place the paper in 250 ml of blocking solution and rock for 2 h at 37°C to quench the unreacted diazonium groups.

After washing in blocking solution rinse the paper briefly in distilled water and proceed with the assay or store it. The paper may be stored for months at 4°C and still be immunologically reactive. We find it to be most reactive, however, if never allowed to dry out completely. Therefore storage in Saran Wrap and a sealed bag is suggested. An illustration of a two-dimensional transfer made under the above conditions is presented in Fig. 3. [^{35}S]Methionine-labeled proteins from human KB cells were run on identical 10–18% polyacrylamide gels and one was transferred to diazo paper. The dried gel (A) was autoradiographed simultaneously with the transfer sheet (B). We obtained transfers of all proteins that were detected on the 2-D gel with similar resolution. It is also of interest that radiographic analysis of the transfer is much more sensitive than that of dried gels.

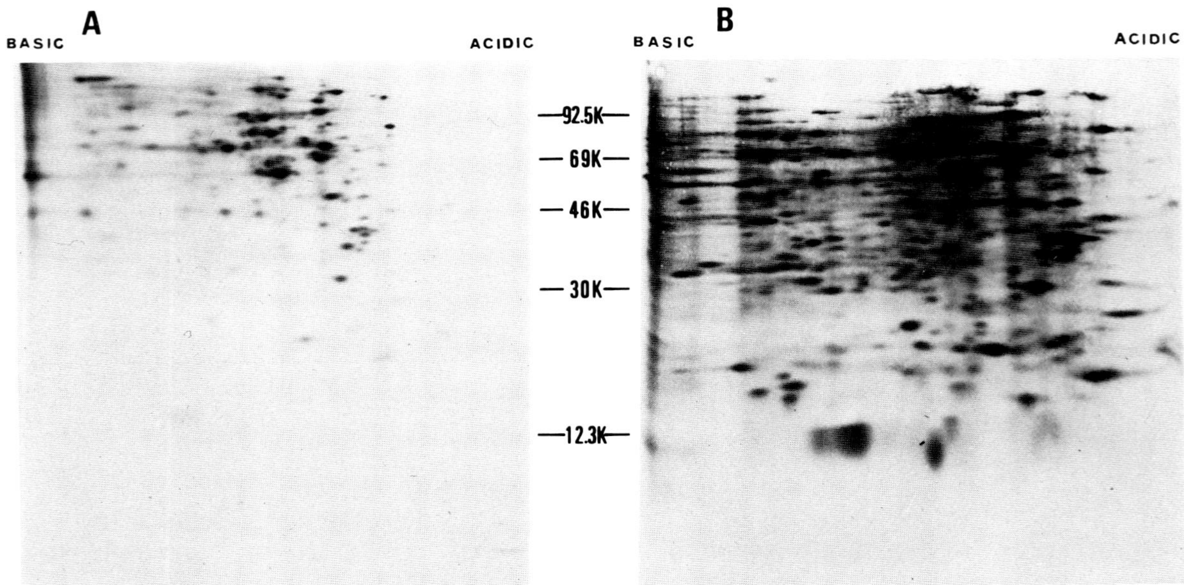

Figure 3. Electrophoretic transfer of proteins from 2-D SDS_4–polyacrylamide gel (A) to DBM paper (B). [^{35}S]Methionine-labeled polypeptides from KB cells were separated by two-dimensional gel electrophoresis. Molecular weights shown were determined by comparison with ^{14}C-labeled marker proteins in a parallel 1-D gel. Both the gel and the DBM paper autoradiograph were exposed for 7 days. From Symington *et al.* (1981), by permission of the authors.

4. OPTIONS

Another buffer system, 10 m*M* sodium borate buffer, pH 9.2, has also been found to give good transfers to diazo paper of proteins which remain immunologically reactive (Reiser and Wardale, 1981). In spite of the fact that diazonium groups are unstable at high pH values these authors have obtained good transfers using a voltage gradient of 32 V/cm. They started with ice-cold buffer but ran for 1 h without auxiliary cooling. This system may be preferable for some proteins. For particular proteins the relative amount of binding under different conditions will have to be determined.

A greater percentage of protein will be transferred from areas of a gel that have low protein loads than from those with higher loads (Symington *et al.*, 1981).

F. Transfer to Positively Charged Nylon Membrane

1. MATERIALS AND REAGENTS

a. Sheet of 15 × 17 cm positively charged nylon membrane ("Zetabind", AMF Specialty Materials CUNO Division, Meriden, Connecticut; "Zeta-Probe", Bio-Rad)
b. Two pieces of Whatman 3MM or equivalent paper
c. Transfer solution. That used for nitrocellulose is appropriate but methanol should be omitted.
d. Blocking solution for positively charged nylon: 0.01 *M* sodium phosphate buffer, pH 7.4–0.15 M NaCl–10% bovine serum albumin (BSA). This is a much higher concentration of protein than is needed to block nitrocellulose or diazo paper (Gershoni and Palade, 1982).

2. TRANSFER CONDITIONS

If any gel size change is expected during the run (determine empirically), the gel must be equilibrated in transfer solution. With transfer at 6 to 8 V/cm one can expect half of the protein to be eluted in a 2-h run. Longer runs are required for efficiently eluting higher molecular weight proteins.

3. POSTTRANSFER TREATMENT OF POSITIVELY CHARGED NYLON MEMBRANE

The run is stopped and the sandwich disassembled as in previous procedures. The paper is immersed in the blocking solution and incubated at 45 to 50°C for at least 12 h.

4. OPTIONS

An alternative transfer buffer is 41 m*M* Tris–40 m*M* boric acid, pH 8.3. Results are similar to those with the Tris–glycine buffer. Other blocking solutions may be used. See discussion in Section III,A.

III. Probing the Transfer

A. *Principle and Relevant Factors for Optimizing Detection*

The proteins that are immobilized on transfers are most frequently discriminated by use of antibody. Proteins are denatured after being separated on a SDS-polyacrylamide gel. Nevertheless, the conformation necessary for interaction with antibodies is, somewhat surprisingly, present in transferred protein in the majority of cases investigated (Anderson *et al.*, 1982b). Some sites that interact with antibodies may be reestablished during posttransfer treatment. As recent success in obtaining antibodies against very small peptides has verified, only a few amino acids are necessary for specific reactions with antibody. Particular folding may be unnecessary for the recognition of peptide sequences on a transfer. Some antisera have been produced using protein eluted from SDS gels as antigen (see Fig. 6 in chapter by Fey *et al.*). It is not surprising that the resulting antibodies can recognize protein transferred from SDS gels. It seems likely that the positive results achieved by this technique have been obtained without extensive recovery of the original configuration of the molecule, in view of the fact that the protein tightly bound to the sheets undoubtedly has difficulty in making major conformational rearrangements.

Monoclonal antibodies have been shown to react with proteins denatured in SDS-polyacrylamide gels and transferred to nitrocellulose by blotting. Shen and co-workers (1980) and Towbin and co-workers (1982) were able to detect specific proteins using monoclonal antibodies on 2-D gel transfers. Monoclonal antibodies have frequently been used successfully with electrophoretic transfers from 1-D gels to nitrocellulose (Dellagi *et al.*, 1982; Hawkes *et al.*, 1982a,b; Kasamatsu, 1982; Reinach *et al.*, 1982) and to diazo paper (Hogg *et al.*, 1981). Probing of 2-D gels with monoclonal antibodies may be an important specific technique of analysis in the future. Sites required for interaction with monoclonal antibodies may not always survive the 2-D gel separation process, however (Anderson *et al.*, 1982b).

The use of antibodies raised against small synthetic peptides will also allow very specific probing of antigenic relationships and steps in the processing and modification of proteins. They should be particularly useful as they recognize a short known sequence and will allow the detection of proteins previously mapped by genetic analyses to functional regions of the genome. They can be produced in rabbits and will therefore be good reagents for use with ^{125}I-labeled PA.

Blocking agents that have been used as an alternative to gelatin are 0.5–3% BSA, 5–10% calf serum or horse serum, 1% hemoglobin, and 1% ovalbumin. The choice must take into consideration the type of protein to be detected as well as the detecting agent. For example, glycoprotein contaminants in BSA may interfere with detection of specific glycoproteins by concanavalin A (Con A) and peroxidase activity in hemoglobin will affect the results if a peroxidase-conjugated enzyme system is used for detection. When radioactive tracers are used for detection instead of an enzyme conjugated system usually NP-40 has been used instead of Tween-20. Buffer containing several detergents has also been used (Feldman *et al.*, 1982). Several words of caution are in order. The strength of binding of the different proteins to nitrocellulose varies, so the practicality of including detergent in blocking and other solutions to minimize background without too much loss of protein will have to be determined experimentally. Palfree and Elliott (1982) present data showing a differential effect of several detergents on the binding of proteins to nitrocellulose. Batches of blocking agents may vary, so they should be tried out with a known system when changing. For example there may be immunoglobulin contaminants in some batches of BSA which react with ^{125}I-labeled PA so that specific complexes can not be distinguished.

After blocking of unreacted sites, Kakita *et al.* (1982b) used photolinking to keep insulin attached to nitrocellulose transfers during the antibody-mediated detection protocol. They again performed the crosslinking after antibody was attached to prevent it from coming off during the extensive washing required for a good background.

For good detection, of course, a crucial factor is the clarity of background for color visualization systems or a good signal to noise ratio for detection by radiographic methods. For each type of sample, transfer sheet, and antibody or other probe the best blocking and washing protocols must be determined experimentally. In Table I and the following sections we outline details of generally appropriate protocols for detection using a final color visualization (enzyme-conjugated antibody) (Section II,B) or a final autoradiographic visualization (Section II,C).

If radioactive marker proteins have been included in the gel the paper can be wraped in Saran Wrap and exposed before antibody treatment for exact orientation of proteins in later autoradiograms. If [^{35}S]methionine-labeled cellular proteins are used, exposure before antibody treatments may assist in the determination of which proteins are reacting later. After antibody probing with a ^{125}I-labeled detection agent, the ^{35}S decay can be blocked with four layers of aluminum foil before reexposure of the blot. This method is used by Cooper and Hunter (1983) with 2-D transfers to demonstrate the specificity of antisera raised against cellular targets for protein kinases.

The detection of antibody complexes to identify particular proteins is frequently done with an enzyme-linked second antibody system. For example, if peroxidase is the conjugate, visualization is accomplished by treating the sheet with a developing mixture consisting of H_2O_2 and a substrate that develops color when oxidized by the H_2O_2 in the presence of the enzyme. Color detection can be used for nitrocellulose transfers but not diazo paper. It is of the same order of sensitivity as radioimmune detection, that is 50–100 pg with some protein-antibody pairs. Amplification of sensitivity can be achieved by use of the peroxidase–antiperoxidase (PAP) method (Glass *et al.*, 1981a).

The ^{125}I-labeled PA detection system is very popular. Protein A binds to the Fc portion of many types of IgG (Kessler, 1975; Goding, 1978). If the specificity of the protein A is low for the type of immunoglobulin being used as the first antibody, rabbit antibody directed against the first antibody is used as second antibody followed by ^{125}I-labeled PA. Because the specificity of protein A is high for rabbit IgG this detection method is very sensitive. When monoclonal antibodies are used rabbit anti-mouse IgG second antibody can be used before the ^{125}I-labeled PA in order to get amplification (Nyari *et al.*, 1983). An alternative means of detection is the use of radiolabeled second antibodies.

Muilerman and colleagues (1982) have used both unspecific antiserum and a partially purified active enzyme to specifically detect inactive enzyme on nitrocellulose with electrophoretically transferred proteins. The transfer is treated with excess antibody raised against a crude enzyme preparation to get protein-IgG complexes that still have one combining site of the IgG molecule available. The sheet is then treated with the preparation of active enzyme and a chromogenic substrate.

See Section III,D for a discussion of nonimmunological probes.

B. Immunoprobing: Chromogenic System for Visualization

Reagents and materials are available in kit form (Bio-Rad) or separately from numerous laboratory supply houses. The protocol used is a modification of that used by Hawkes *et al.* (1982a) on dot blots.

1. REAGENTS AND MATERIALS

The use of Tween-20 in the following solutions is optional.

a. 50 m*M* Tris–HCl, pH 7.4–200 m*M* NaCl (TBS).
b. Blocking solution: TBS–.05% Tween-20–3% gelatin.
c. Binding solution: TBS–.05% Tween-20–1% gelatin.
d. First antibody solution: Antibody directed against the immobilized antigen (e.g., produced in rabbits). A 1:50 dilution in binding solution is suggested for initial trial. To obtain a better background, dilute the antiserum as much as possible.
e. Enzyme-conjugated second antibody solution, (e.g., horseradish peroxidase-conjugated antibody produced in goats directed against rabbit IgG, (Bio-Rad; Miles-Yeda, 30W475 North Aurora Road, Naperville, Illinois). The dilution depends on the titer but a dilution of 1:1000 or greater in binding solution is typical when using commercial preparations.
f. 30% H_2O_2 aqueous solution, stabilized (J. T. Baker Chemical Co., Phillipsburg, New Jersey).
g. Developing solution (prepared fresh): 30 mg 4-chloro-1-naphthol (Bio-Rad; Merck, P.O. Box 2000, Rahway, New Jersey, dissolved in 10 ml cold methanol. Keep cold and protected from light. To 50 ml TBS containing 0.05% Tween-20, add 30 μl of H_2O_2 just before mixing with the dye solution. Final concentrations are 0.05% 4-chloro-1-naphthol and 0.015% H_2O_2. Use immediately after mixing. Note that sodium azide, which is often added during immunological manipulations, will inhibit the peroxidase.
h. Platform rocker capable of gentle agitation.

2. TREATMENT WITH ANTIBODY

All antibody treatments and washes are performed with gentle rocking.

a. Nitrocellulose sheet has been rocking gently in blocking solution at 37°C for 1 h since transfer was finished (Section II,D,3).

b. Immerse sheet in 30 ml of first antibody solution in a 20 × 20 glass dish and gently rock 2 h at room temperature.
c. Wash at room temperature (200 ml of solution each change) as follows: 2 × 10 min TBS, 2 × 10 min blocking solution.
d. Immerse sheet in 30 ml of enzyme-linked second antibody in a 20 × 20 glass dish for 1 h at room temperature.
e. Wash (200 ml each) 1 × 10 min TBS, 2 × 10 min blocking solution, 1 × 10 min TBS.

3. DETECTION BY COLOR REACTION

f. Immerse sheet in 50 ml of freshly prepared developing solution in a 20 × 20 cm glass dish. Color should develop within 15 to 30 min.
g. Stop the reaction by washing the sheet in distilled water 3 × 5 minutes.
h. Photograph sheet while wet under reflected light using a yellow filter. Air dry and store in the dark.

4. OPTIONS

Other chromogenic substrates may be used but some are difficult to dissolve and some are mutagenic, possibly carcinogenic. Wear gloves when working with all such compounds. Other enzyme-conjugates can be used along with their appropriate substrates. Enzymes can be linked to a more specific second antibody, allowing greater sensitivity of detection. Karcher *et al.* (1981) use a 3-amino-9-ethylcarbazole as substrate for detection of proteins on 1-D strips. They dip the stained strips in paraffin oil and read optical densities in a photometer.

C. *Immunoprobing: Radiolabeled Components for Visualization*

The type of detection system, here described for ^{125}I-labeled PA, is appropriate for use with both diazo paper and other types of transfer sheets. The protocol used follows Symington *et al.* (1981) and is similar to that used by Renart *et al.* (1979).

1. REAGENTS AND MATERIALS

a. Blocking solution: 100 m*M* Tris–HCl, pH 9.0–10% v/v ethanolamine–0.25% gelatin. Store cold.

b. Binding solution: 50 m*M* Tris–HCl, pH 7.4–5m*M* EDTA–150 m*M* NaCl–0.25% gelatin–0.05% Nonidet P-40 (NP-40) (Bethesda Research Laboratories, Rockville, Maryland). Warm at 37°C to dissolve gelatin. Store cold.
c. Antibody solution: Microfuge (Beckman Inst. Co.) antiserum 4 min in cold before using it. A 1:30 dilution in binding solution is suggested for initial trial. The greater the dilution which can be used, the better the background.
d. Postiodination wash solution: 50 m*M* Tris–HCl, pH 7.4–5m*M* EDTA–1 *M* NaCl–0.25% gelatin–0.4% sodium *N*-lauroylsarcosinate (ICN Pharmaceuticals, Plainview, New York). Store cold.
e. ^{125}I-labeled PA stock solution, specific activity about 5 to 25 μCi/μg. (New England Nuclear, Boston, Massachusetts, or prepared by labeling *Staphylococcus* protein A (Pharmacia, Piscataway, New Jersey) by chloroglycouril or chloramine T procedures as described by Markwell and Fox (1978). As soon as a preparation of protein A has been ^{125}I-labeled it is tested on a small strip of diazo paper to which 100 ng of BSA has been transferred. The strip is reacted with rabbit anti-BSA and then ^{125}I-labeled PA. A dark band should be visible in 4 h. Preparations of ^{125}I-labeled PA can be used for about a month.
g. ^{125}I-labeled PA working solution: Stock solution is freshly diluted into binding solution to 2–5 $\times$ 10^5 cpm/ml. Thirty ml of the dilution is prepared for each run in a 50 ml disposable plastic tube (e.g., Falcon 2098).
h. Platform rocker; blotting paper; Saran Wrap

2. TREATMENT WITH ANTIBODY

a. Diazo transfer has been rocking in blocking solution at 37°C for 2 h.
b. Allow 0.02 ml of diluted antiserum per cm^2 of sheet (about 5 ml will be used). Cut a large enough piece of Saran Wrap to wrap tightly around the diazo paper. Put the antiserum dilution in the center. Use forceps when handling the transfer. Rinse the diazo paper rapidly in 3 $\times$ 500 ml distilled water, blot the back side with filter paper, and lay protein bound side down on top of antiserum. Make sure all of the paper is wet.
Wrap Saran Wrap around the sheet so that there are no trapped bubbles. Place the sheet on a glass plate in a plastic box lined with wet paper towels. Put on the lid and rock at room temperature 2 h or longer (overnight is permissible).
c. Rinse the paper with 50 ml binding solution. Wash it in 2 $\times$ 250 ml binding solution for 2 h each at 22°C, rocking.

d. Place 30 ml of ^{125}I-labeled PA in a 20 × 20 cm glass dish. Blot the back of the transfer sheet and put it into the dish right side up, making sure no bubbles of air are trapped beneath. Rock dish to wet transfer throughly across the top. Cover with Saran Wrap and rock on a platform rocker for 2 h.
e. Remove the Saran Wrap. With a 10 ml pipette, return the ^{125}I-labeled PA to its tube for proper disposal. Pass the transfer sheet rapidly through 3 × 500 ml distilled water and wash it 3 × 40 min in 250 ml postiodination wash solution.

3. DETECTION BY AUTORADIOGRAPHY

Take the transfer sheet out of the dish with forceps, pass it quickly through 3 × 500 ml distilled water, blot the back and wrap in Saran Wrap. Never let the paper dry completely. Have just one layer of Saran Wrap over the front and two overlapping on the back; turn ends toward back. This is important if you are using the transfer for simultaneous detection of ^{35}S or ^{14}C and ^{125}I disintegrations. Expose at −70°C with intensifying screens. Approximately 4 h is usually required for detection. The exposure may also be done at 4°C without intensifying screens exposing for a longer time.

4. OPTIONS

When this method of detection is used for transfers on nitrocellulose the solutions, incubations, and wash conditions should be the same as those presented in Section III,B,1 and 2. However, NP-40 is often used instead of Tween-20. Dilute the ^{125}I-labeled PA in 30 ml of binding solution to a concentration of 2 to 5 × 10^5 cpm/ml and substitute it for the enzyme-linked antibody. More extensive washes may be required to obtain a reasonable background. Feldman *et al.* (1982) use a solution containing several detergents. However, too much washing can cause loss of antigen and antibody; some workers prefer to omit detergent (Lin and Kasamatsu, 1983).

Antibody incubations can be done in a heat sealable bag (Dazey Seal-a-Meal). If so, 10 ml of solution suffices for a sheet. Be sure to exclude air bubbles and incubate with rocking. Such incubations can also be done in a 20 × 20 dish, using 30 ml of antibody dilution. Most antibody dilutions can be saved for reuse and kept at −20°C. When using the ab/^{125}I-labeled PA detection scheme 0.01% sodium azide will not interfere and it is a good precaution to use it. Radiolabeled antibodies, particularly ^{125}I-labeled, have also been used to probe transfers. The protocols are similar to that presented here.

D. *Alternative Probes*

Transfer sheets have been used with many types of affinity probes other than systems using antibodies. For example [^{32}P]DNA, ^{125}I-labeled RNA, and ^{125}I-labeled histone can be used to probe for DNA-binding proteins, RNA-binding proteins, or histone-binding proteins (Bowen *et al.*, 1980; Jack *et al.*, 1981; Aubertin *et al.*, 1983). ^{125}I-labeled lectins such as Con A, wheat germ agglutinin and lentil lectin have been used as a probe for glycoproteins, and ^{125}I-labeled avidin has been used to detect proteins labeled with *N*-hydroxysuccinimide biotin (Reiser and Wardale, 1981). Hawrot and co-workers (1982) have found that ^{125}I-labeled bungarotoxin can be used to probe for acetylcholine receptors transferred from gels to Zetabind. Gorelick and associates have used ^{125}I-labeled calmodulin as a probe to characterize calmodulin-enhanced protein kinase activity in the pancreas. They have used 2-D gels and have obtained good results with calmodulin overlay (J. Gershoni, personal communication).

Other options that do not involve the use of antibodies include the system of Bell and Engvall (1982) in which a fragment of fibronectin, which recognizes collagen, is linked to peroxidase. Yet another system which involves the use of enzyme for detection but does not rely on antibody depends on the binding of unlabeled Con A to immobilized glycoprotein and the location of bound Con A by treatment with unconjugated peroxidase, with which the free Con A sites can interact (Hawkes, 1982).

IV. Removal of Probes

A. *Removal of Probes from Nitrocellulose Membrane*

Legocki and Verma (1981) have been able to remove antibody and ^{125}I-labeled PA from ("erase") nitrocellulose transfers by treating the sheets in 100 m*M* glycine–HCl, pH 2.2–20 m*M* magnesium acetate–50 m*M* KCl for 1.5 h at room temperature. The membrane was then washed in 0.9%

NaCl–10 mM Tris-HCl, pH 7.4 and re-equilibrated in 0.9% NaCl–10 mM Tris, pH 7.4–3% BSA for 1 h at room temperature before reprobing with antibody and ^{125}I-labeled PA. Reprobing was successful although background increased with use. They find that transfers can be used at least three times. All antiserum and ^{125}I-labeled PA treatments were done at 30°C. Another method of removing probes from nitrocellulose has been used by Erickson *et al.* (1982). They use a treatment with 8 M urea–100 mM 2-mercaptoethanol–5 mg/ml BSA for 1 h at 60°C. Most transfers are reprobable at least once. However, they find that antigenicity of some proteins transferred to nitrocellulose may be lost by this treatment. For storage nitrocellulose sheets have been blotted dry and kept at −20°C, 4°C, or room temperature. Positive results have been obtained with probing of sheets stored many months at room temperature (Anderson *et al.*, 1982b). Legocki and Verma (1981) suggest that after erasure, transfers should be blotted dry and stored in Saran Wrap and a sealed bag at 4°C.

B. Removal of Probes from Diazo Paper

The advantage of diazo paper is the fact that it is easy to remove the probes from the transfers without removing the covalently bound proteins and the proteins remain antigenically probable. The background gets increasingly better with erasure and reprobing. We have used the same papers eight times. Protein can be removed by treatment in 50 mM–sodium phosphate buffer, pH 7.5–10 M urea/0.1 M 2-mercaptoethanol for 30 min at 60°C (Renart *et al.*, 1979). Erase in 10 ml of solution in a sealed plastic bag in a water bath. Agitate intermittently. Rinse with distilled water and equilibrate with binding solution for 2 h, rocking at 37°C before reacting with antibody. Probes have also been erased by treating with 2% SDS–50 mM sodium phosphate buffer, pH 7.5–100 mM 2-mercaptoethanol at 60°C for 30 min. However, the urea treatment seems to give a more reliably complete removal. Store the transfers moist in Saran Wrap in a sealed bag at 4°C.

C. Removal of Probes from Positively Charged Nylon Membrane

Probes can be removed from Zetabind by the urea protocol for nitrocellulose. Specific probes can be dissociated by particular treatments.

For example, Con A can be competed off by α-methylglucoside and α-methylmannoside, and calmodulin can be dissociated by a wash in 1 mM EDTA to remove calcium ions (Gershoni and Palade, 1983).

V. Examples and Applications of Protein Transfer Analysis

A. *Examples of the Detection of Specific Proteins on Transfers from Two-Dimensional Gels*

Detection by the immunoautoradiographic method is shown on a transfer to nitrocellulose membrane (Fig. 4) and to diazo paper (Fig. 5). The material used for the two-dimensional gels was similar, virus-infected whole cell lysates. Both are transfers from SDS–polyacrylamide gels. The conditions used were presented in Sections II,D and F. Both use specific antisera/^{125}I-labeled PA and autoradiography for detection. Figure 4 shows transfers of the proteins in lysates of two types of mouse cells infected with murine leukemia virus. The proteins were separated in the first dimension by cathode-directed isotachophoresis and in the second dimension by a SDS–10% polyacrylamide slab gel. A 1-D separation of authentic marker proteins is incorporated at the right. The transfers were probed with rabbit antiserum raised against p30, the major internal structural protein of murine leukemia virus and ^{125}I-labeled PA. P30 is a constituent of the M_r 65,000 polyprotein precursor ($Pr65^{gag}$) detected here. Major leukemia virus structural proteins and smaller processing intermediates are also detected, as well as an antigenically related larger protein. The resolution of proteins transferred to and detected on nitrocellulose is good, as is the background.

Figure 5 illustrates the use of immunoautoradiographic detection of proteins on diazo paper (DBM in this case). Proteins from a lysate of human KB cells infected with adenovirus 2 [Ad2] were separated in the first dimension by isoelectric focusing and in the second dimension by a SDS–10–18% polyacrylamide gel. Two identical gels were prepared. The gel pictured in Fig. 5A contains [^{35}S]methionine-labeled proteins and is an autoradiograph. Figure 5B is an immunoautoradiograph of proteins transferred to DBM paper and treated with goat anti-Ad2-virion serum followed by ^{125}I-labeled PA. Figure 5C shows [^{35}S]methionine-labeled purified Ad2 virion proteins for comparison. The Roman numerals refer

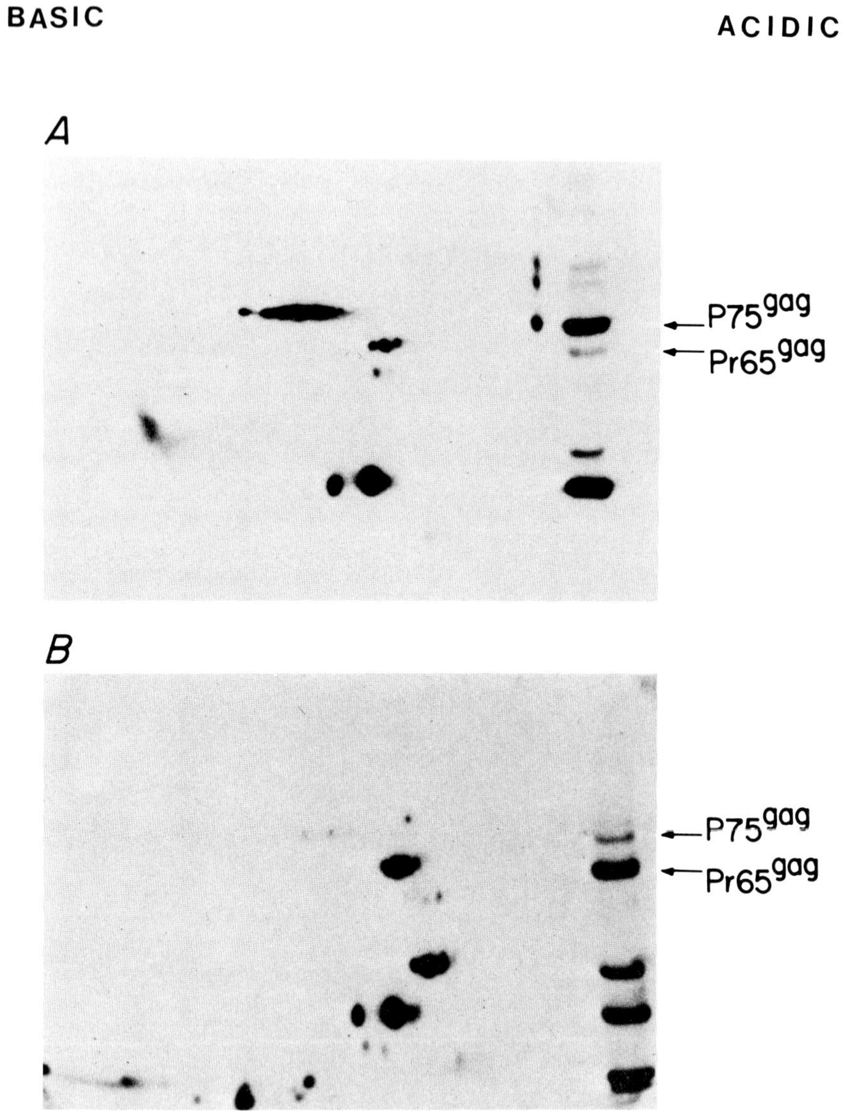

Figure 4. Immunoautoradiographic detection of murine leukemia virus-specific proteins transferred from a 2-D SDS–polyacrylamide gel to nitrocellulose. Polypeptides from (a) AKR murine leukemia virus-infected C57Bl/6 E♂G2 cells and from (B) Moloney murine leukemia virus-infected C3H cells were analyzed using anti-p30 serum and [^{125}I]PA as described in the text. Marker lanes to the right show the authentic samples run in a single dimension. Exposure time was 30 min. From Burnette (1981), by permission of the author and Academic Press.

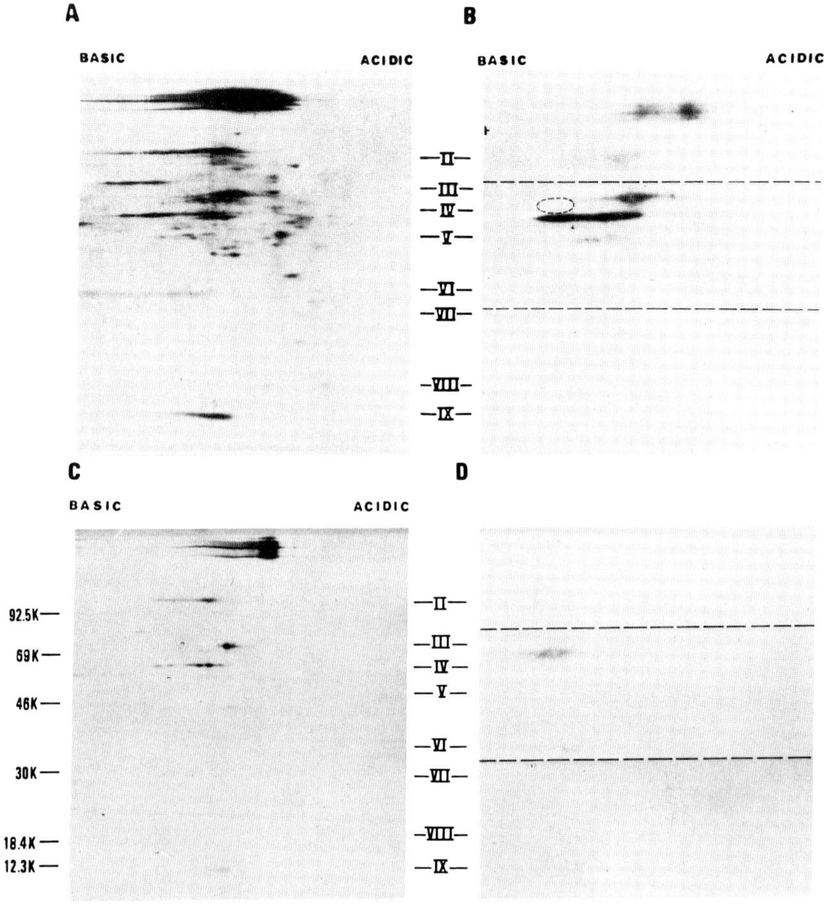

Figure 5. Immunoautoradiographic detection of adenovirus 2 proteins transferred from a 2-D gel to DBM paper. (A) 2-D gel autoradiograph of [^{35}S]methionine-labeled late Ad2-infected KB cell proteins; (B) 2-D immunoautoradiograph of late Ad2-infected KB cell proteins electrophoretically transferred to DBM paper and treated with anti-Ad2 virion serum and ^{125}I-labeled PA: (C) 2-D autoradiograph of purified [^{35}S]methionine-labeled Ad2 virion. The Roman numerals refer to Ad2-specific proteins. Molecular weights were determined by comparison with ^{14}C-labeled marker proteins in a parallel gel; (D) 2-D immunoautoradiograph of B after erasure and second treatment with antiserum to DNA-binding protein and ^{125}I-labeled PA. Only the portion of paper between the dotted bands was used for the second reaction; one spot was labeled. In the original reaction with antivirion serum this area was not labeled (indicated by dotted oval in B). From Symington *et al.* (1981). By permission of the authors.

to Ad2-specific proteins. The major virion proteins visualized in C are detected on the immunoautoradiograph in B. In addition there are several other spots seen on the immunoautoradiograph. Peptide maps and use of nonspecific sera are required to establish the identity of the other peptides visualized and to determine relationships. As with nitrocellulose, the resolution of proteins transferred to and detected on diazo paper is good, comparable to that found in a matching gel, and the background is also good.

The immunoautoradiograph shown in Fig. 5D illustrates the capacity of diazo paper for erasure and reuse. The same paper was used as for Fig. 5B (after erasure). The antiserum used was guinea pig anti-Ad2-coded early 73,000 dalton DNA-binding protein (DBP). Because the same transfer was used it is apparent that the DBP spot was not labeled when antivirion serum was used. Since 2-D gels may differ slightly it is important to be able to probe the same transfer with different sera in this way to distinguish or establish relationships between proteins that might have very similar isoelectric points and apparent molecular weight. It should be noted that although ^{125}I-labeled PA has been shown to have a lower affinity for goat IgG than for rabbit it can nevertheless be used with goat antiserum in this procedure with good results. It is effective also with guinea pig serum as demonstrated in this figure.

An important feature of the transfer analysis system is that the steady state concentrations of the intracellular proteins can be studied more easily than by metabolic labeling. Quantitation of particular proteins by use of PTA is receiving considerable attention (Vaessen *et al.*, 1981; Yen and Webster, 1982). Protocols that result in improvement in this aspect should be forthcoming. Modifications of proteins that would be reflected in a shift in mobility in an electric field or change in isoelectric point may be important in the function of these proteins. In this regard separation of proteins by 2-D gel electrophoresis combined with PTA can yield particularly important information.

B. Applications and Potential of the Method

Protein transfer analysis has become increasingly important as a way to probe infected and transformed cells for comparison with normal cells. PTA has become a frequently used tool in the analysis of blood and cell fractions. The specificity of antibodies can be precisely determined by the use of transfers and this has become an important use as well.

We present in Table II examples of systems that have been investigated using this method with 2-D gels. The diversity of proteins that have been studied is great, including viral structural and nonstructural, bacterial ribosomal and other bacterial proteins, avian transformation proteins, and mammalian proteins such as cytokeratin from rat cells and blood proteins. Polypeptides of the cytokeratin family were also studied by Franke *et al.* (1981) using 2-D gel capillary blots. Important procedural aspects of these studies are noted.

There have been several reports of the use of monoclonal antibodies with 2-D gel transfers. Shen *et al.* (1980) used diffusion blots. Other investigators report the use of monoclonals as probes on 2-D tranfers from urea gels (Nielsen *et al.*, 1982; Towbin *et al.*, 1982) Nielsen and coworkers investigated phosphorylation by this technique. Yurchenco *et al.* (1982) used monoclonals on electrophoretic transfers from second-dimensional gels containing SDS. They studied the subdomains of spectrin, utilizing tryptic peptides. Further important epitope mapping studies using PTA are to be expected. In addition, there have been numerous reports of successful use of monoclonals in probing 1-D transfers. For example, Hogg *et al.* (1981) have characterized a monoclonal antibody which reacts with a 45,000 dalton protein of human monocytes on transfers. Monoclonals have also been used successfully to probe transfers for viral peptides (Kasamatsu, 1982), intermediate filament proteins (Dellagi *et al.*, 1982), and a rat neuron-specific mitochondrial protein (Hawkes *et al.*, 1982b). Steitz *et al.* (1983) have used monoclonal and polyclonal antibodies and PTA to distinguish small ribonucleoproteins. Such studies may help to understand autoimmune disorders.

The possibilities for extension of the use of PTA to the greater discrimination possible with 2-D analysis are suggested by the range of successful 1-D studies that have been done using conventional polyclonal antisera and other probes. These include studies of the following: cytomegalovirus group common protein (Weiner and Gibson, 1981), cowpea mosaic virus precursor protein (Zabel *et al.*, 1982), antigens of subacute sclerosing panencephalitis (Karcher *et al.*, 1981), antigenically active material from *Schistosoma japonicum* eggs (Tsang *et al.*, 1982) and from *Mycoplasma pneumonia* (Hu *et al.*, 1981), ribosomal proteins (Towbin *et al.*, 1979; Stellwag and Dahlberg, 1980), DNA-binding proteins (Bowen *et al.*, 1980; Jack *et al.*, 1981); nonhistone nuclear proteins (Glass *et al.*, 1981a), receptor proteins such as the acetylcholine receptor (Hawrot *et al.*, 1982), and calmodulin-binding proteins (Gorelick *et al.*, 1982).

Enzymes are also being studied by PTA. These include adenosine deaminase (Wiginton and Hutton, 1982), 3-hydroxy-3-methylglutaryl-CoA reductase (Chin *et al.*, 1982), and phosphodiesterase (Muilerman *et al.*, 1982). A variety of glycoproteins have been detected on transfers (Glass *et al.*, 1981b; Hawkes, 1982; Hogan *et al.*, 1982; Langer *et al.*, 1982). Some of the diverse other proteins that have been studied by this

TABLE II

Applications of Protein Transfer Analysis Using Two-Dimensional Gels

Protein(s) studied	Material used	Amount loaded	Detection method	Author	Comments
Adenovirus 2 virion related and DNA-binding	Human KB cells	600 μg	Goat, Guinea pig ab/^{125}I-labeled PA	Symington *et al.* (1981)	Erasure and reprobing using DBM transfer
Simian virus 40 capsid	African green monkey kidney cells	5×10^4 cells	Rabbit ab–^{125}I-labeled PA	Reiser and Wardale (1981)	Erasure and reprobing using DPT transfer
Murine leukemia virus specific	Mouse cells	$1–3 \times 10^5$ cells	Rabbit ab–^{125}I-labeled PA	Burnette (1981)	Extended transfer time to 22 h to get >90% transfer of most proteins to nitrocellulose
Cross-reactive to strain 61A24 and yeast ATPase complex protein	*Rhizobium japonicium* strain 61A76	200 μg	Rabbit ab–^{125}I-labeled PA	Legocki and Verma (1981)	Multiple partial nitrocellulose transfers probed
Discrete ribosomal	*E. coli* ribosomes	100 μg	Monoclonal antibodies–^{125}I-labeled PA	Shen *et al.* (1980)	Use of monoclonal antibodies on diffusion transfers

Phosphorylated and protein kinase substrate (in normal and Rous Sarcoma-transformed cells)	Chicken embryo cells		Rabbit ab–^{125}I-labeled PA	Cooper and Hunter (1983)	[^{35}S]Methionine label, expose before and after probing with ^{125}I-labeled PA
Cytokeratin	Rat liver Novikoff ascites hepatoma cells	15 μg	Rabbit ab–Peroxidase-antiperoxidase	Schmidt *et al.* (1982)	Transfer identification confirmed by immunofluorescence
Cls piece of complement haptoglobin and 28 other components	Human plasma		Rabbit ab–peroxidase-ab	Anderson *et al.* (1982b)	Nitrocellulose transfer but methanol omitted to speed transfer
Heat-shock	Human mononuclear leukocytes, normal lymphoblastoid cells, testicular tissue		Rabbit ab–Peroxidase-ab	Anderson *et al.* (1982a)	Comparative probing transfers from different tissues
Spectrin α subunit	Human erythrocytes		Monoclonal antibodies–^{125}I-labeled PA	Yurchenco *et al.* (1982)	Use of monoclonal antibodies on electrophoretic transfers

method are surface layer proteins from *Caulobacter* (Smit and Agabian, 1982), human keratin (Moll *et al.*, 1982), myosin (Cavadore *et al.*, 1982), teratocarcinoma cell adhesion proteins (Yoshida and Takeichi, 1982), rat brain synaptosomal polypeptides (Erickson *et al.*, 1982), heat shock proteins from a wide variety of species (Kelley and Schlesinger, 1982), insulin (Kakita *et al.*, 1982a), and proteins from human placental coated vesicles (Pearse, 1982). Investigation of T cell suppressor factors is under way using PTA (C. Sorensen, personal communication).

Matus *et al.* (1980) investigated the proteins of brain postsynaptic densities. These authors point out that histochemical analysis should be used in addition to PTA to establish the location of a particular protein, especially when working with sticky proteins. This is far from a complete presentation of the proteins which are currently being studied by PTA but gives some idea of the scope.

One pertinent approach that we have used when the concentration of the protein of interest is low is immunoprecipitation followed by a transfer and detection on the transfer. We were able to detect the intracellular precursor of the Ad2 covalently bound terminal protein in this way (Green *et al.*, 1981).

VI. Concluding Remarks

The foregoing presentation gives some notion of the wide range of material and specific proteins that are amenable to protein transfer analysis. Two-dimensional transfer analysis is an exquisitely sensitive as well as highly discriminating approach to the characterization of proteins in cells, tissue, and fluids. It will be important in following the production of particular proteins by cells after insertion of genes using recombinant DNA techniques. Translation products from different cell systems can be compared in detail.

The proteins studied by this technique are both small and large. The technique must be varied by choosing transfer sheet and conditions suitable to the protein under study and the type of information sought. Transfer reuse is an important aspect of analysis of proteins from 2-D gels. By reprobing the same sheet proteins that are close together on a 2-D gel transfer may be distinguished or single proteins may be characterized by several kinds of probes such as antibody and lectin.

The technique allows characterization of monoclonal antibodies by specific reaction with particular proteins. The structure of antibodies and other proteins can be studied in detail by use of 2-D protein transfer analysis. Peptide fragments react in this system and extremely specific antibodies can be used. In this regard antibodies made to synthetic peptides (produced in rabbits, the IgG of which reacts well with ^{125}I-labeled PA) will help to determine details of structure of protein, precursors, and processing. Peptide antibodies will be useful in recognizing proteins coded by a particular part of a genome translated *in vitro* or present in a cell or tissue extract. This approach should be important in the detection of viral-related and transforming proteins.

At the present time an expectation of detection of 1 to 10 nanograms of protein (applied to the original first dimension) is reasonable, although sensitivity down to 25 pg has been reported based on detection of protein directly applied to a sheet. Greater sensitivity will depend on such factors as the use of antibodies with a high titer for proteins that have been transferred to sheets, improvement leading to better renaturation of bound proteins, improvements in washing protocols for better signal to noise ratio, and the use of probes of higher specific activity but undiminished specific binding capacity.

Protein transfer analysis has great potential for helping to solve many fundamental and practical biological problems when combined with 2-D gel electrophoresis. The specificity and sensitivity of protein detection is such that we can expect this technique to play an important role in the identification and characterization of particular proteins in the future.

Acknowledgments

I wish to thank all those who provided helpful information for their experience with this method, particularly George Bers, and those authors who made available preprints of unpublished articles. I am grateful to Stuart Symington, Jr. for his help in the preparation of the manuscript and to Ann Bergersen for typing it. I would also like to thank Maurice Green for insightful discussions and a stimulating atmosphere in which to work. My research was supported by grants CA-29561 and CA-21824 from the National Cancer Institute and a grant from The Council for Tobacco Research, U.S.A., Inc.

References

Adair, W. S., Jurivich, D., and Goodenough, U. W. (1978). Localization of cellular antigens in sodium dodecyl sulfate–polyacrylamide gels. *J. Cell Biol.* **79**, 281–285.

Alwine, J. C., Kemp, D. J., and Stark, G. R. (1977). Method for detection of specific RNAs in agarose gels by transfer to diazobenzyloxymethyl-paper and hybridization with DNA probes. *Proc. Natl. Acad. Sci. USA* **74**, 5350–5354.

Anderson, N. L., Giometti, C. S., Gemmell, M. A., Nance, S. L., and Anderson, N. G. (1982a). A two-dimensional electrophoretic analysis of the heat-shock-induced proteins of human cells. *Clin. Chem.* **28**, 1084–1092.

Anderson, N. L., Nance, S. L., Pearson, T. W., and Anderson, N. G. (1982b). Specific antiserum staining of two-dimensional electrophoretic patterns of human plasma proteins immobilized on nitrocellulose. *Electrophoresis* **3**, 135–142.

Aubertin, A. M., Tondre, L., Lopez, C., Obert, G., and Kirn, A. (1983). Sodium dodecyl sulfate–mediated transfer of electrophoretically separated DNA-binding proteins. *Anal. Biochem.* **131**, 127–134.

Bell, M. L., and Engvall, E. (1982). The specific detection of collagenous proteins after electrophoresis using enzyme-conjugated collagen-binding fibronectin fragments. *Anal. Biochem.* **123**, 329–335.

Bittner, M., Kupferer, P., and Morris, C. F. (1980). Electrophoretic transfer of proteins and nucleic acids from slab gels to diazobenzyloxymethyl cellulose or nitrocellulose sheets. *Anal. Biochem.* **102**, 459–471.

Bowen, B., Steinberg, J., Laemmli, U. K., and Weintraub, H. (1980). The detection of DNA-binding proteins by protein blotting. *Nucleic Acids Res.* **8**, 1–20.

Burnette, W. N. (1981). "Western Blotting:" Electrophoretic transfer of proteins from sodium dodecyl sulfate–polyacrylamide gels to unmodified nitrocellulose and radiographic detection with antibody and radioiodinated protein A. *Anal. Biochem.* **112**, 195–203.

Burridge, K. (1976). Changes in cellular glycoproteins after transformation: identification of specific glycoproteins and antigens in sodium dodecyl sulfate gels. *Proc. Natl. Acad. Sci. USA* **73**, 4457–4461.

Burridge, K. (1978). Direct identification of specific glycoproteins and antigens in SDS gels. *Methods Enzymol.* **50** (Part C), 54–64.

Cavadore, J.-C., Molla, A., Harricane M.-C., Gabrion, Y. B., and Demaille, J. G. (1982). Subcellular localization of myosin light chain kinase in skeletal cardiac and smooth muscles. *Proc. Natl. Acad. Sci. USA* **79**, 3475–3479.

Chin, D. J., Luskey, K. L., Anderson, R. G. W., Faust, J. R., Goldstein, J. L., and Brown, M. S. (1982). Appearance of crystalloid endoplamic reticulum in compactin-resistant chinese hamster cells with a 500-fold increase in 3-hydroxy-3-methylglutaryl-coenzyme A reductase. *Proc. Natl. Acad. Sci. USA* **79**, 1185–1189.

Cooper, J. A., and Hunter, T. (1983). Identification and characterization of cellular targets for tyrosine protein kinases. *J. Biol. Chem.* **258**, 1108–1115.

Dellagi, K., Brouet, J. C., Perreau, J., and Paulin, D. (1982). Human monoclonal IgM with autoantibody activity against intermediate filaments. *Proc. Natl. Acad. Sci. USA* **79**, 446–450.

Erickson, P. F., Minier, L. N., and Lasher, R. S. (1982). Quantitative electrophoretic transfer of polypeptides from SDS polyacrylamide gels to nitrocellulose sheets: a method for their re-use in immunoautoradiographic detection of antigens. *J. Immunol. Methods* **51**, 241–249.

Erlich, H. A., Cohen, S. N., and McDevitt, H. O. (1979a). Immunological detection and characterization of products translated from cloned DNA fragments. *Methods Enzymol.* **68**, 443–453.

Erlich, H. A., Levinson, J. R., Cohen, S. M., and McDevitt, H. O. (1979b). Filter affinity transfer: a new technique for the *in situ* identification of proteins in gels. *J. Biol. Chem.* **254**, 12240–12247.

Feldman, L. T., Imperiale, M. J., and Nevins, J. R. (1982). Activation of early adenovirus transcription by the herpesvirus intermediate early gene: Evidence for a common cellular control factor. *Proc. Natl. Acad. Sci. USA* **79**, 4952–4956.

Franke, W. W., Winter, S., Grund, C., Schmid, E., Schiller, D. L., and Jarasch, E.-D. (1981). Isolation and characterization of desmosome-associated tonofilaments from rat intestinal Brush border. *J. Cell Biol.* **90**, 116–127.

Gershoni, J. M., and Palade, G. E. (1982). Electrophoretic transfer of proteins from sodium dodecyl sulfate–polyacrylamide gels to a positively charged membrane filter. *Anal. Biochem.* **124**, 396–405.

Gershoni, J. M., and Palade, G. E. (1983). Protein blotting: principles and applications. *Anal. Biochem.* **131**, 11–15.

Gibson, W. (1981). Protease-facilitated transfer of high-molecular-weight proteins during electrotransfer to nitrocellulose. *Anal. Biochem.* **118**, 1–3.

Glass, W. F., III, Briggs, R. C., and Hnilica, L. S. (1981a). Identification of tissue-specific nuclear antigens transferred to nitrocellulose from polyacrylmide gels. *Science (Washington, D.C.)* **211**, 70–72.

Glass, W. F., III, Briggs, R. C., and Hnilica, L. S. (1981b). Use of lectins for detection of electrophoretically separated glycoproteins transferred onto nitrocellulose sheets. *Anal. Biochem.* **115**, 219–224.

Goding, J. W. (1978). Use of staphylococcal protein A as an immunological reagent. *J. Immunol. Methods* **20**, 241–253.

Gorelick, F. A., Freedman, S. D., Delahunt, N. G., Gershoni, J. M., and Jamieson, J. D. (1982). Calmodulin regulated protein kinase activity from exocrine pancreas. *J. Cell Biol.* **95** (Part 2), 401a (Abstr.).

Green, M., Symington, J., Brackmann, K. H., Cartas, M. A., Thornton, H., and Young, L. (1981). Immunological and chemical identification of intracellular forms of adenovirus type 2 terminal protein. *J. Virol.* **40**, 541–550.

Hawkes, R. (1982). Identification of concanavalin A-binding proteins after sodium dodecyl sulfate–gel electrophoresis and protein blotting. *Anal. Biochem.* **123**, 143–146.

Hawkes, R., Niday, E., and Gordon, J. (1982a). A dot-immunobinding assay for monoclonal and other antibodies. *Anal. Biochem.* **119**, 142–147.

Hawkes, R., Niday, E., and Matus, A. (1982b). MIT-23: a mitochondrial marker for terminal neuronal differentiation defined by a monoclonal antibody. *Cell* **28**, 253–258.

Hawrot, E., Gershoni, J. M., Burrage, T. G., Paladino, G. S., Lentz, T. L., and Chin, L. L. Y. (1982). Monoclonal antibodies to nicotinic acetylcholine receptor characterized by electro-transfer techniques. *Soc. for Neuroscience* **8**, Abstr. 91.7, 335.

Hoch, S. (1982). DNA-binding domains of fibronectin probed using Western blots. *Biochem. Biophys. Res. Comm.* **106**, 1353–1358.

Hogan, B. L. M., Taylor, A., Kurkinen, M., and Couchman, J. R. (1982). Synthesis and localization of two sulfated glycoproteins associated with basement membranes and the extracellular matrix. *J. Cell Biol.* **95**, 197–204.

Hogg, N., Slusarenko, M., Cohen, J., and Reiser, J. (1981). Monoclonal antibody with specificity for monocytes and neurons. *Cell* **24**, 875–884.

Howe, J. G., and Hershey, J. W. B. (1981). A sensitive immunoblotting method for measuring protein synthesis initiation factor levels in lysates of *Escherichia coli*. *J. Biol. Chem.* **256,** 12836–12839.

Hu, P. C., Huang, Y., Graham, J. A., and Gardner, D. E. (1981). Identification of immunogens for *Mycoplasma pneumoniae* by protein blotting. *Biochem. Biophys. Res. Commun.* **103,** 1363–1370.

Jack, R. S., Gehring, W. J., and Brack, C. (1981). Protein component from Drosophila larval nuclei showing sequence specificity for a short region near a major heat-shock protein gene. *Cell* **24**, 321–331.

Kakita, K., Giddings, S., and Permutt, M. A. (1982a). Biosynthesis of rat insulin I and II:

Evidence for differential expression of the two genes. *Proc. Natl. Acad. Sci. USA* **79**, 2803–2807.

Kakita, K., O'Connell, K., and Permutt, M. A. (1982b). Immunodetection of insulin after transfer from gels to nitrocellulose filters: A method of analysis in tissue extracts. *Diabetes* **31**, 648–652.

Karcher, D., Lowenthal, A., Thormar, H., and Noppe, M. (1981). Serological identification of viral antigens after electrophoretic transfer. *J. Immunol. Methods* **43**, 175–179.

Kasamatsu, H. (1982). Isolation of a monoclonal antibody against viral polypeptide VP2 of simian virus 40. *J. Virol.* **44**, 413–418.

Kelley, P. M., and Schlesinger, M. J. (1982). Antibodies to two major chicken heat shock proteins cross-react with similar proteins in widely divergent species. *Mol. Cell. Biol.* **2**, 267–274.

Kessler, S. W. (1975). Rapid isolation of antigens from cells with a Staphylococcal protein A-antibody adsorbent: Parameters of the interaction of antibody-antigen complexes with protein A. *J. Immunol.* **115**, 1617–1624.

Laemmli, U. (1970). Cleavage of structural proteins during the assembly of the head of bacteriophage T4. *Nature (London)* **227**, 680–685.

Langer, B. G., Leung, L. L. K., Gonnella, P. A., Nachmias, V. T., Nachman, R. L., and Pepe, F. A. (1982). α-Actinin and membrane glycoprotein IIIa are different proteins in human blood platelets. *Proc. Natl. Acad. Sci. USA* **79**, 432–435.

Langone, J. J. (1978). [^{125}I] Protein A: A tracer for general use in immunoassay. *J. Immunol. Methods* **24**, 269–285.

Legocki, R. P., and Verma, D. P. S. (1981). Multiple immunoreplica technique: Screening for specific proteins with a series of different antibodies using one polyacrylamide gel. *Anal. Biochem.* **111**, 385–392.

Lin, W., and Kasamatsu, H. (1983). On the electrotransfer of polypeptides from gels to nitrocellulose membranes. *Anal. Biochem.* **128**, 302–311.

Markwell, M. A. K., and Fox, C. F. (1978). Surface-specific iodination of membrane proteins of viruses and eukaryotic cells using 1,3,4,6-tetrachloro-3α,6α-diphenyl glycouril. *Biochemistry* **17**, 4807–4817.

Matus, A., Pehling, G., Ackermann, M., and Maeder, J. (1980). Brain postsynaptic densities: Their relationship to glial and neuronal filaments. *J. Cell Biol.* **87**, 346–359.

Moll, R., Franke, W. W., Volc-Platzer, B., and Krepler, R. (1982). Different keratin polypeptides in epidermis and other epithelia of human skin: A specific cytokeratin of molecular weight 46,000 in epithelia of the pilosebaceous tract and basal cell epitheliomas. *J. Cell Biol.* **95**, 285–295.

Muilerman, H. G., ter Hart, H. G. J., and Van Dijk, W. (1982). Specific detection of inactive enzyme proteins after polyacrylamide gel electrophoresis by a new enzyme-immunoassay method using unspecific antiserum and partially purified active enzyme: Application to rat liver phosphodiesterase 1. *Anal. Biochem.* **120**, 46–51.

Neville, D. M. Jr. (1971). Molecular weight determination of protein-dodecyl sulfate complexes by gel electrophoresis in a discontinuous buffer system. *J. Biol. Chem* **246**, 6328–6334.

Nielsen, P. J., Manchester, K. L., Towbin, H., Gordon, J., and Thomas, G. (1982). The phosphorylation of ribosomal protein S6 in rat tissues following cycloheximide injection in diabetes, and after denervation of diaphragm. *J. Biol. Chem.* **257**, 12316–12321.

Nowinski, R. C., Lostrom, M. R., Tam, M. R., Stone, M. R., and Burnette, N. (1979). The isolation of hybrid cell lines producing monoclonal antibodies against the p13(E) protein of ectotropic murine leukemia viruses. *Virology* **93**, 111–126.

Nyari, L. J., Tan, Y. H., and Erlich, H. A. (1983). Monoclonal antibodies directed against human fibroblast interferon: Characterization and functional studies. *Hybridoma*. **2**, 79–86.

O'Farrell, P. H. (1975). High resolution two-dimensional electrophoresis of proteins. *J. Biol. Chem.* **250**, 4007–4021.

O'Farrell, P. Z., Goodman, H. M., and O'Farrell, P. H. (1977). High resolution two-dimensional electrophoresis of basic as well as acidic proteins. *Cell* **12**, 1133–1142.

Olden, K., and Yamada, K. M. (1977). Direct detection of antigens in sodium dodecyl sulfate–polyacrylamide gels. *Anal. Biochem.* **78**, 483–490.

Olmstead, J. B. (1981). Affinity purification of antibodies from diazotized paper blots of heterogeneous protein samples. *J. Biol. Chem.* **246**, 11955–11957.

Palfree, R. G. E., and Elliott, B. E. (1982). An enzyme-linked immunosorbent assay (ELISA) for detergent solubilized Ia glycoproteins using nitrocellulose membrane discs. *J. Immunol. Methods* **52**, 395–408.

Pearse, B. M. F. (1982). Coated vesicles from human placenta carry ferritin, transferrin and immunoglobulin G. *Proc. Natl. Acad. Sci. USA* **79**, 451–455.

Reinach, F. C., Masaki, T., Shafig, S., Obinata, T., and Fischman, D. A. (1982). Isoforms of C-protein in adult chicken sheletal muscle: detection with monoclonal antibodies. *J. Cell Biol.* **95**, 78–84.

Reiser, J., and Wardale, J. (1981). Innumological detection of specific proteins in total cell extracts by fractionation in gels and transfer to diazophenylthioether-paper. *Eur. J. Biochem.* **114**, 569–575.

Renart, J., Reiser, J., and Stark, G. R. (1979). Transfer of proteins from gels to diazobenzyloxymethyl-paper and detection with antisera: A new method for studying antibody specificity and antigen structure. *Proc. Natl. Acad. Sci. USA* **76**, 3116–3120.

Schmidt, W. N., Pardue, R. L., Tutt, M. C., Briggs, R. C., Brinkley, B. R., and Hnilica, L. S. (1982). Identification of cytokeratin antigens in Novikoff ascites hepatoma. *Proc. Natl. Acad. Sci. USA* **79**, 3138–3142.

Seed, B. (1982). Diazotizable acrylamine cellulose papers for the coupling and hybridization of nucleic acids. *Nucleic Acids Res.* **10**, 1799–1810.

Shen, V. M., King, T. C., Kumar, V., and Daugherty, B. (1980). Monoclonal antibodies to *Escherichia coli* 50S ribosomes. *Nucleic Acids Res.* **8**, 4639–4649.

Smit, J., and Agabian, N. (1982). Cell surface patterning and morphogenesis: Biogenesis of a periodic surface array during *Caulobacter* development. *J. Cell Biol.* **95**, 41–49.

Steitz, J. A., Wolin, S. L., Rinke, J., Pettersson, I., Mount, S. M., Lerner, E. A., Hinterberger, M. and Gottlieb, E. (1983). *Cold Spring Harbor Symp. Quant. Biol.* **47**, 893–900.

Stellwag, E. J., and Dahlberg, A. E. (1980). Electrophoretic transfer of DNA, RNA, and protein onto diazobenzyloxymethyl (DBM)-paper. *Nucleic Acids Res.* **8**, 299–317.

Symington, J., Green, M., and Brackmann, K. (1981). Immunoautoradiographic detection of proteins after electrophoretic transfer from gels to diazo-paper: Analysis of adenovirus encoded proteins. *Proc. Natl. Acad. Sci. USA* **78**, 177–181.

Towbin, H., Staehelin, T., and Gordon, J. (1979). Electrophoretic transfer of proteins from polyacrylamide gels to nitrocellulose sheets: Procedure and some application. *Proc. Natl. Acad. Sci. USA* **76**, 4350–4354.

Towbin, H., Ramjoué, H.-P., Kuster, H., Liverani, D., and Gordon, J. (1982). Monoclonal antibodies against eucaryotic ribosomes: use to characterize a ribosomal protein not previously identified and antigenically related to the acidic phosphoproteins P1/P2. *J. Biol. Chem.* **257**, 12709–12715.

Tsang, V. C. W., Tao, Y., Qui, L., and Xue, H. (1982). Fractionation and quantitation of egg antigens from *Schistosoma Japonicum* by the single-tube kinetic-dependent–enzyme-linked immunosorbent assay (K-ELISA): Higher antigenic activity in urea-soluble than in aqueous-soluble fraction. *J. Parasitol.* **68**, 1034–1043.

Tsang, V. C. W., Peralta, J. M., and Simons, A. R. (1983). The enzyme-linked immuno-electro-transfer blot techniques (EITB) for studying the specificities of antigens and antibodies separated by gel electrophoresis. *Methods Enzymol.* **92** (Part E), 377–391.

Vaessen, R. T. M. J., Kreike, J., and Groot, G. S. P. (1981). Protein transfers to nitrocellulose filters: a simple method for quantitation of single proteins in complex mixtures. *FEBS Lett.* **124**, 193–196.

Weiner, D., and Gibson, W. (1981). Identification of a primate cytomegalovirus group-common protein antigen. *Virology* **115**, 182–191.

Wiginton, D. A., and Hutton, J. J. (1982). Immunoreactive protein in adenosine deaminase deficient human lymphoblast cell lines. *J. Biol. Chem.* **257**, 3211–3217.

Wojtkowiak, Z., Briggs, R. C., and Hnilica, L. S. (1983). A sensitive method for staining proteins transferred to nitrocellulose sheets. *Anal. Biochem.* **129**, 486–489.

Yen, T. S. B., and Webster, R. E. (1982). Translational Control of Bacteriophage f1 gene II and gene x proteins by gene V protein. *Cell* **29**, 337–345.

Yoshida, C., and Takeichi, M. (1982). Teratocarcinoma cell adhesion: Identification of a cell-surface protein involved in calcium-dependent cell aggregation. *Cell* **28**, 217–224.

Yurchenco, P. D. (1982). Monoclonal antibodies as probes of domain structure of the spectrin α subunit. *J. Biol. Chem.* **257**, 9102–9107.

Zabel, P., Moerman, M., van Straaten, F., Goldbach, R., and van Kammen, A. (1982). Antibodies against the genome-linked protein VPg of Cowpea Mosaic virus recognize a 60,000 dalton precursor polypeptide. *J. Virol.* **41**, 1083–1088.

CHAPTER 6

Correlation between Mouse and Human Two-Dimensional Gel Patterns: Peptide Mapping of Proteins Extracted from Two-Dimensional Gels

STEPHEN J. FEY
RODRIGO BRAVO*
PETER MOSE LARSEN
JULIO E. CELIS
Division of Biostructural Chemistry
Department of Chemistry
Aarhus University
Aarhus, Denmark

I. Introduction. 170
II. Methods. 170
A. Procedures for One-Dimensional Peptide Mapping. 171
B. Procedures for Two-Dimensional Peptide Mapping (Fingerprinting). 174
III. Correlation between the Mouse and Human Two-Dimensional Gel Patterns. 177
A. Updated Protein Catalog of Mouse Kidney Fibroblasts. . . . 177
B. Homologous Polypeptides in Mouse, Human, and Other Species. 177
IV. Concluding Remarks. 184
References. 187

* Present affiliation: European Molecular Biology Laboratory, 6900 Heidelberg, Federal Republic of Germany.

ISBN 0-12-164720-X

I. Introduction

Techniques such as two-dimensional gel electrophoresis have made possible the separation of thousands of proteins from a given cell type (O'Farrell, 1975; O'Farrell *et al.*, 1977). Protein catalogs have been prepared based on two-dimensional gel separations (Pedersen *et al.*, 1978; Bravo *et al.*, 1981a; Fey *et al.*, 1981; Bravo and Celis, 1982; J. I. Garrels, personal communication; N. L. Anderson and N. G. Anderson, personal communication; see also chapters by Neidhardt and Phillips and by Bravo and Celis), and there is now a pressing need to correlate the different catalogs through the identification of homologous polypeptides. In this laboratory we have been mainly concerned with studies of normal and transformed human and mouse cultured cells (for references see chapter by Celis *et al.*) and have made use of one- (Cleveland *et al.*, 1977; Fey *et al.*, 1981) and two-dimensional peptide mapping (Hutchinson *et al.*, 1978; Craig and Lancashire, 1980; Bravo *et al.*, 1981b) to identify homologous polypeptides between these species (Fey *et al.*, 1981).

It is the purpose of this chapter to describe peptide mapping techniques that have been adapted for the analysis of polypeptides recovered from two-dimensional gels (Fey *et al.*, 1981; Bravo *et al.*, 1981b). These techniques have been used to identify homologous polypeptides between the mouse and human protein catalogs. Homologous polypeptides have also been identified in other species such as duck, hamster, potoroo, and rat for which no numbering system is available at present.

II. Methods

Peptide mapping provides a further dimension for the study of proteins. Since its introduction by Ingram (1956, 1959), it has proved its utility in an ever-widening field of applications from the identification of related proteins (modified, degraded, or from different genes or species) to the determination of protein sequence.

In itself the technique consists of the specific proteolysis of a protein followed by fractionation and visualization of the resultant peptides. Specific proteolysis can be effected at a number of different sites by either chemical or enzymatic means (Table I) and can be partial (Cleveland *et al.*, 1977) or complete as required.

Many methods have been used to fractionate the peptides produced.

TABLE I

Sites for Protein Cleavage[a]

	Method	NH_2-Site-COOH
Chemical	CNBr	-Met↓-
	Formic acid[b]	-Asp↓-Pro-
Enzymatic	Trypsin[c]	-Lys↓-
		-Arg↓-
	Chymotrypsin	-Tyr↓-
		-Phe↓-
		-Trp↓-
	S. aureus V8[d]	-Asp↓-
		-Glu↓-
	Papain	Wide specificity

[a] See Smyth (1967) and Hirs and Timasheff (1977) for full details.
[b] Sonderegger *et al.* (1982).
[c] The specificity of trypsin can be restricted by blocking amino groups, Butler *et al.* (1969).
[d] The specificity can be restricted to glutamoyl bonds; Houmard and Drapeau (1972).

Initially these were paper chromatography and electrophoresis, but these have been largely superseded by more sensitive and convenient procedures such as thin-layer chromatography (TLC) (Watanabe and Yoshida, 1971), gel electrophoresis, (Laemmli, 1970) and high-performance liquid chromatography (HPLC) (Hearn, 1980).

There are important advantages to be gained by using two-dimensional gels to prepare polypeptides for peptide analysis. These are that many polypeptides are obtained reasonably pure and undegraded, and sufficient polypeptide is recovered to allow the peptide mapping procedures described below to be applied to many problems. Using radiolabeled samples and the more sensitive procedures, it is possible to peptide map (one-dimensionally) more than half of the discernible polypeptides from a total cell lysate (Fey *et al.*, 1981).

A. *Procedures for One-Dimensional Peptide Mapping*

This is essentially the method of Cleveland *et al.* (1977) modified to improve the resolution (Fey *et al.*, 1981).

1. The amount of polypeptide required depends upon the detection method used: about 10 μg if the peptide map is to be Coomassie Blue stained and about 50 ng if silver stained. For ^{3}H-labeled polypeptides 20,000 cpm are needed, whereas only 5,000–10,000 cpm are required in the case of ^{14}C- or ^{35}S-labeled polypeptides (corresponding to as little as 1–3 ng) if detected by fluorography.
2. Run two-dimensional gels as described in chapter by Bravo. Coomassie Blue stain the gel as usual. If the sample is radioactive, dry the gel without fixing and expose to X-ray film for 12–36 h. Gels treated for fluorography cannot be used.
3. Cut out the area containing the polypeptide and rehydrate if necessary.
4. Mash the gel between a folded piece of thin plastic using the straight edge of a scalpel handle and transfer to the bottom of a 1.5 ml centrifuge tube. This step is especially important for spots cut from two-dimensional gels as it markedly improves the resolution.
5. Add 20 μl of loading buffer (0.125 *M* Tris–HCl, pH 5.8, 0.5% SDS, 2.5% Dextran T-500, and 0.01% bromophenol blue) and boil for 5 min.
6. After cooling to room temperature, add 5 μl of protease dissolved in loading buffer and leave for 1 h to digest. The following amounts of protease per μg of polypeptide give satisfactory results: chymotrypsin 10–100 μg, elastase 20–200 μg, papain 2–20 μg, trypsin 10–100 μg, *Staphylococcus aureus* V8 protease 20–200 μg. The latter is the most widely used enzyme as it usually produces fewer peptides. The degree of digestion is relatively insensitive to the enzyme to substrate ratio facilitating the comparison of results from different experiments. Boil the sample again, cool, and then add 1 μl β-mercaptoethanol.
7. Prepare slab gels (15%) according to Laemmli (1970). For a standard slab gel with internal dimensions of 18 × 18 × 0.12 cm, mix the following solutions: 20 ml acrylamide : bisacrylamide (30 : 0.8%, w/v), 0.4 ml SDS (10% w/v Serva), 5 ml Tris–HCl (3 *M*, pH 8.8 at 20°C), 14.4 ml H_2O.
8. Degas and add 8 μl *N,N,N′,N′*-tetramethylethylenediamine (TEMED), 0.2 ml Ammonium persulfate, 10% w/v (APS).
9. Pour into assembled plates up to a mark 4.5 cm from the top.
10. Overlay with butanol saturated with slab gel buffer.
11. To prepare the stacking gel solution (5%), mix the following: 7.5 ml acrylamide–bisacrylamide stock (10 : 0.05% w/v), 0.15 ml SDS (10% w/v Serva), 3.6 ml Tris–HCl (0.5 *M*, pH 6.8), 3.6 ml H_2O.

12. When the slab gel has polymerized, wash the butanol away, degas the stacking gel solution, and add 7.5 μl TEMED, 0.225 ml APS (10% w/v).
13. Pour onto the slab gel and insert the well-forming comb.
14. Overlay with butanol saturated with the stacking gel buffer to minimize shrinkage during polymerization.
15. When the gel has polymerized, remove the comb and wash the wells with running buffer (6.06 g Tris–HCl, 28.8 g glycine, 10 ml of 10% SDS per liter).
16. Place gel in the electrophoresis chamber and add running buffer.
17. Load samples using a cut-off disposable pipette tip (diameter ~2 mm), and electrophorese at 11 mA for 14 h or until the tracking dye reaches 1 cm from the bottom of the gel.
18. After electrophoresis the gels are processed according to normal procedures for Coomassie Blue staining, silver staining (Mose Larsen, 1981; Mose Larsen *et al.*, 1983; see also chapters by Merril and Goldman and by Sammons *et al.*), or autoradiography–fluorography (Laskey and Mills, 1975; see also chapter by Bravo). If the gel is to be dried, wash first in 3% glycerol for 30 min to prevent it from cracking.

Note: Glass plates should be detergent washed and then heated to 140°C for 1 h to inactivate any traces of enzyme. Occasional chromic acid washing is also recommended, but the plates must be thoroughly rinsed before use.

Using this procedure it is possible to map polypeptides comprising less than 0.01% of the total cellular protein, depending on the amount of radioactivity applied to the two-dimensional gel. On a routine basis 40–50 polypeptides can be mapped per day.

If peptide mapping is to be used extensively, it is recommended to run a standard on each gel in order to facilitate comparison and to check the extent of digestion. We recommend the abundant protein actin for this purpose. Examples of one-dimensional peptide maps produced by partial digestion with *Staphylococcus aureus* V8 protease are given in Section III,B (Fig. 5).

In cases in which it is necessary to resolve small differences between homologous polypeptides it is advisable to use complete digestion and two-dimensional peptide mapping (fingerprinting). Using the protocol described in the next section about eight polypeptides can be fingerprinted per day.

B. *Procedures for Two-Dimensional Peptide Mapping (Fingerprinting)*

This procedure was modified from Craig and Lancashire (1980) and Hutchinson *et al.* (1978) for polypeptides extracted from two-dimensional gels (Bravo *et al.*, 1981b). Many other buffer systems are available (e.g., see Vandekerckhove and van Montagu, 1974; Elder *et al.*, 1977; Whittaker and Moss, 1981).

1. The amount of polypeptide required depends upon the detection method used: about 25 to 50 μg are needed if the TLC plate is to be developed with ninhydrin or fluorescamine; or about 25,000 cpm of ^{14}C- or ^{35}S-labeled proteins for fluorography (corresponding to 2 to 4 ng of the original protein).
2. Run the two-dimensional gels as described in the chapter by Bravo. For radioactive samples dry the unfixed gels and expose to X-ray films for 8 to 40 h.
3. Cut out the area containing the polypeptide and rehydrate in water if necessary.
4. Mash the gel between a folded piece of thin clear plastic using the straight edge of a scalpel handle and transfer to the bottom of a 1.5 ml centrifuge tube.
5. Add 0.5 ml of 100 m*M* ammonium bicarbonate and 3 μg of DPCC trypsin (Sigma type XI) and incubate for 10 h at room temperature.
6. Add another 3 μg of DPCC trypsin and continue incubation for 14 h to ensure complete digestion.
7. Pellet the gel fragment by low speed centrifugation and lyophilize the supernatant.
8. Rehydrate and lyophilize twice to remove traces of ammonium bicarbonate.

All subsequent steps should be carried out in a fume cupboard.

1. ELECTROPHORESIS: FIRST DIMENSION

9. Take up in 1 to 2 μl of running buffer of butanol : pyridine : acetic acid : H_2O (pH 4.7, v/v, 2 : 1 : 1 : 36), containing trace amounts of crystal violet, orange G, and xylene cyanole FF as tracking dyes.
10. Draw out a 10 μl glass microcapillary and break cleanly to use as an applicator.
11. Take sample up by capillary action and apply a little at a time onto

a corner (about 4 cm from two sides) of a cellulose thin-layer plate (20 × 20 × 0.01 cm 5716 Merck). The size of the spot should be less than 2 mm in diameter (smaller spots give better resolution).

12. Spray dampen the TLC plate with running buffer. The plate should become more translucent all over. Avoid over- or under-dampening.
13. Fill the flat-bed TLC chamber (Fig. 1) with running buffer. The supporting cooling plate should be horizontal and precooled by recirculating ice-cold water.
14. Wet the upper surface of the cooling plate with running buffer and lay the TLC plate onto it. Avoid trapping air bubbles, which will cause unequal cooling.
15. Connect the opposite sides of the TLC plate to the reservoirs using two pieces of folded presoaked Whatman 3 mm paper (20 × 10 cm).
16. Place two 10 μl glass microcapillaries along the remaining two sides and cover with a precooled glass "lid" (20 × 20 cm). This prevents evaporation and keeps a saturated atmosphere over the TLC plate.
17. Close TLC chamber and electrophorese towards the cathode for 40 min at 800 volts. The crystal violet (v) should migrate about 6 cm and the xylene cyanol FF (c) about 4 cm towards the cathode, and the orange G 1 cm towards the anode. All three should be in a straight line. Figure 1 shows the equipment after electrophoresis.
18. Dry the plate horizontally overnight.

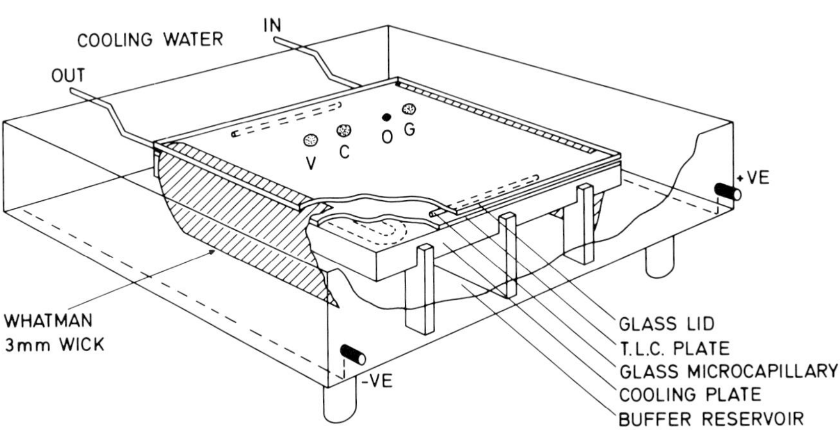

Figure 1. Apparatus for electrophoresis on thin-layer chromatography plates. +VE, positive; −VE, negative. G, Orange G; O, Origin; C, Xylene Cyanol; V, Crystal Violet.

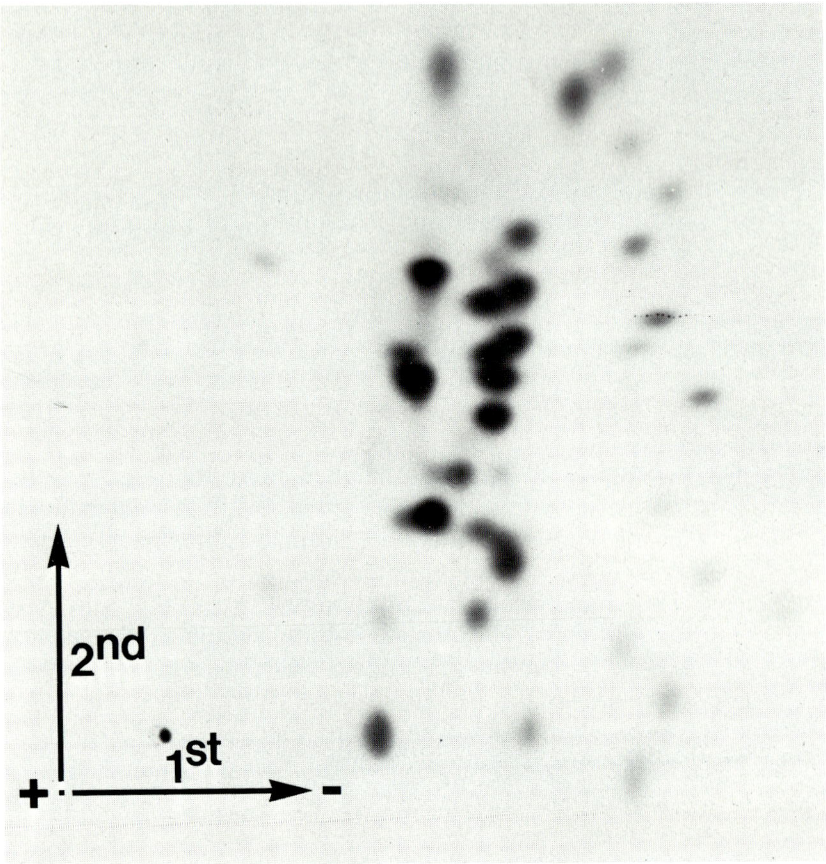

Figure 2. Two-dimensional tryptic peptide map of [^{14}C]β-actin from mouse Sarcoma 180. From Bravo *et al.* (1981b).

2. ASCENDING CHROMATOGRAPHY: SECOND DIMENSION

19. Stand the plate vertically in a preequilibrated ascending chromatography tank containing 200 ml of butanol : pyridine : acetic acid : H_2O (pH 5.3, v/v, 87 : 75 : 15 : 60) at room temperature and leave for 7 h to develop.
20. Dry the plates standing vertically.
21. If radioactive polypeptides are used, immerse the plates for 5 min in 0.4% diphenyloxazone in 2-methylnaphthalene and then leave to dry (Bonner and Stedman, 1978).
22. Mark the position of the origin and tracking dyes with radioactive ink (2000 cpm [^{14}C]/μl).

23. Wrap the plates in Saran Wrap.
24. Place the plate with film (Kodak XRP1) in a cassette and leave at −70°C.

A typical two-dimensional peptide map of [^{14}C]β-actin from mouse Sarcoma 180 is shown in Fig. 2.

Ninhydrin staining can be carried out according to Watanabe and Yoshida (1971) or Schiltz *et al.* (1977) and fluorescamine according to Vandekerckhove and van Montagu (1974; see also Stephens, 1978; Fishbein *et al.*, 1980). Both techniques can detect peptides at the nanomole level (25 μg protein) (Huisman *et al.*, 1978; Whittaker and Moss, 1981).

In situations in which it is not possible to obtain biosynthetically labelled proteins these can be radioiodionated. Iodination can be carried out within gel fragments at tyrosine residues with chloramine T (Elder *et al.*, 1977) or at the more common lysyl residues with the Bolton-Hunter reagent (Feitelson *et al.*, 1981; Sage *et al.*, 1981). With both procedures picomolar amounts of proteins can be analyzed.

III. Correlation between the Mouse and Human Two-Dimensional Gel Patterns

A. *Updated Protein Catalog of Mouse Kidney Fibroblasts*

We have presented a catalog of [^{35}S]methionine-labeled mouse kidney fibroblast polypeptides separated by high resolution two-dimensional gel electrophoresis (Fey *et al.*, 1981). A total of 1191 [^{35}S]methionine-labeled polypeptides (836 acidic, 365 basic) have been resolved and numbered as indicated in Fig. 3. The coordinates and identity of a few polypeptides are given in Tables II and III.

B. *Homologous Polypeptides in Mouse, Human, and Other Species*

Homologous polypeptides in mouse (kidney fibroblasts) (Fig. 3), human (HeLa) (see Fig. 1 of chapter by Bravo and Celis), duck (fibroblasts), hamster (BHK21), rat (L6), and Potoroo (PTK2) were tentatively identified on the basis of their position in two-dimensional gels. These were subsequently peptide mapped as described in Section II,A using *Staphylococcus aureus* V8 protease. The homologous polypeptides among the different species are indicated in Tables II and III. Both mouse and human polypeptides are identified with their corresponding protein catalog number (Bravo *et al.*, 1981a; Fey *et al.*, 1981; Bravo and Celis, 1982; see also Fig. 1 of chapter by Bravo and Celis). Because no numbering system is available for the duck, hamster, potoroo, and rat polypeptides, these are indicated with D, Ha, P, and R in Tables II and III, respectively.

A total of 119 homologous polypeptides (68 acidic, 51 basic) have been identified so far between mouse and HeLa, and 92 (52 acidic and 40 basic) among mouse, hamster, and human (HeLa). In a more limited study, 18 duck, 14 potoroo, and 8 rat polypeptides have been found to be homologous to mouse, HeLa, and hamster polypeptides. McConkey (1982) has reported that more than 185 out of 370 polypeptides analyzed coelectrophorese between hamster (CHO) and human (HeLa) proteins.

In most of our experiments, comigrating polypeptides gave similar one-dimensional peptide maps. For example, conserved proteins such as the actins (γ and β) and α- and β-tubulin comigrated in all species studied (Fig. 4) and showed similar one-dimensional peptide maps (Fey *et al.*, 1981). If necessary, subtle differences between this group of proteins could be further investigated using the two-dimensional peptide mapping procedure described in Section II,B (Bravo *et al.*, 1981b).

There are cases, however, in which peptide mapping revealed homologies between polypeptides that did not comigrate in all species. An example of such a situation is illustrated by the analysis of a quartet of cytoskeletal polypeptides (IEF 12, 24 (mitochondrial), 26 (vimentin) and 35; HeLa numbering system) found in a wide variety of cell lines, primary cultures, and tissues from several species (Fig. 4) (Bravo *et al.*, 1982; Mose Larsen *et al.*, 1982). Of these, the mitochondrial polypeptide

---

Figure 3. Two-dimensional gel electrophoresis of [^{35}S]methionine labeled polypeptides from asynchronous secondary (p4) mouse kidney fibroblasts labeled for 20 h. Both fluorograms (IEF and NEPHGE) have been overexposed in order to reveal minor polypeptides. In IEF the pH ranges from 7.5 (left) to 4.5 (right). In NEPHGE the pH ranges from 7.5 (right) to 10.5 (left). Following the convention used for HeLa cells (Bravo *et al.*, 1981a; Bravo and Celis, 982; chapter by Bravo and Celis), we have selected 101 original numbers (67 on the acidic and 34 on the basic side). Additional spots between two original numbers (large numbers) have been indicated with letters. In cases where there were more than 26 spots between two original spots we have indicated those spots with the letter z followed by a small number (Tables II and III). The spots have been numbered from higher to lower molecular weight and from the origin of the first dimension.

TABLE II

Updated Catalog of Mouse Fibroblast Acidic Polypeptides[a]: Homologous Polypeptides in Other Species[b]

Polypeptides	Coordinates molecular weight/ mobility relative to β-actin	Polypeptides with the same original number	Homologous[c] polypeptides in human HeLa cells	Homologous[c] polypeptides in other species
0	nd/1.29	35 (a–z8)		
1	132/0.20	55 (a–z28)		
			1	Ha
2	110/0.44	14 (a–m)		
			2	Ha
3	105/0.78	4 (a–c)		
4	100/0.83	11 (a–j)		
5	96/0.89	25 (a–x)	5	
5p	95/1.37		8n	
5t (α-Actinin)	94/0.97		8	Ha
5u	94/1.02		8i	Ha
6	95/0.10	1	6	Ha
7	95/0.12	5 (a–d)	7	Ha
7a	95/0.16		7a	Ha
8	94/0.73	44 (a–z17)	8a	Ha
8m	76/1.21 }		8z36	Ha
8n	76/1.24 }			
9	74/0.52	13 (a–l)	9	Ha
10	72/1.26	18 (a–q)	10	Ha, P
11	69/1.41	18 (a–p)		D
12	68/0.82	10 (a–i)	12	Ha, P, D, R
13	67/0.93	26 (a–y)	13	Ha, P
14	66/0.81	5 (a–d)	14	Ha, D
15	65/0.78	9 (a–h)	15	
15h	65/1.54		18f	Ha
16	65/0.76	5 (a–d)	16	Ha, P, D
17	64/0.26	2 (a)	17	Ha
18	63.5/0.20	20 (a–s)		
19 (α-Tubulin)	58/1.07	4 (a–c)	19	Ha, P, D
19c (α-Tubulin)	58/1.12		19b	Ha, P, D
20	59/0.78	10 (a–i)	21	Ha
			20	Ha
20h	57/0.63		23	Ha
21 (β-Tubulin)	56/1.27	16 (a–o)	25	Ha, P, D, R
21e	55/0.65		27	
22 (Vimentin)	54/1.18	9 (a–h)	26	Ha, P, D, R
22g (Phospho-vimentin)	54/1.22		26e	Ha
23	54/0.82	2 (a–b)		

(continued)

TABLE II (*continued*)

Polypeptides	Coordinates molecular weight/ mobility relative to β-actin	Polypeptides with the same original number	Homologous[c] polypeptides in human HeLa cells	Homologous[c] polypeptides in other species
24	53/0.88	9 (a–h)	24	Ha, P, D, R
25	53/0.37	7 (a–f)		
25f (vd1)	52.5/1.24		32	Ha
26	52/0.73	8 (a–g)		
26f (vd2)	50/1.31		33i	Ha
27	49.5/1.22	8 (a–g)		
27f (Skeletin)	49/0.99			
28	49.5/0.63	4 (a–c)	33	Ha
29	49/0.70	4 (a–c)		
30	49/1.12	9 (a–h)		
31	49/1.22	6 (a–e)	35	Ha, P, D, R
32	49/0.59	4 (a–c)		
33	49/0.16	6 (a–e)		
34	49/0.46	12 (a–k)	34	
34g (vd3)	47/1.32		41	Ha
35	45.5/0.83	4 (a–c)		
36	45.5/1.01	3 (a–b)	38	Ha
36b	46/1.20		35z3	Ha
37	46/0.22	2 (a)		
38	45.5/1.13	2 (a–b)		
39	45.5/0.32	1	39	Ha
40	45.5/0.19	14 (a–m)		
			40	Ha
41	45/0.69	7 (a–f)		
41d (vd4)	44/1.37		44i	Ha
41e	44/1.41		45	Ha, D
42	45/0.23	5 (a–d)		
43	44/0.47	7 (a–f)	42	
44	43.5/0.29	2 (a)		
45	43.5/0.44	3 (a–b)		
46	43.5/0.80	4 (a–c)		
46c (vd5)	43/1.44		47	Ha
47 (β-actin)	43/1.00	16 (a–o)	47	Ha, P, D, R
47g (γ-actin)	43/0.94		47a	Ha
48	42/0.09	63 (a–z36)		
			48r	Ha
48s	40/1.34		48z13	R
			48z30	Ha
48z8	39.5/1.26		48z67	
48z26	38/1.01 }		48z68	Ha, R
48z27	38/1.36 }			
			48z80	Ha
			48z90	Ha

TABLE II (*continued*)

Polypeptides	Coordinates molecular weight/ mobility relative to β-actin	Polypeptides with the same original number	Homologous[c] polypeptides in human HeLa cells	Homologous[c] polypeptides in other species
49	36/1.45	11 (a–j)		
49g	35.5/1.24		{48z99, 49z100}	R
50	35.5/0.64	9 (a–h)		
50h	35/0.11			Ha
51	35/0.24	8 (a–g)		
51f (cyclin)	35/1.43		49	Ha, P, D, R
52 (tropomyo-sin related)	35/1.45	21 (a–t)	52	Ha, P, D, R
52c	34.4/1.39		{49a, 49b}	
			52c	Ha, P
52t (tropomyo-sin related)	33/1.44		52x	D
53	32.5/1.36	10 (a–i)		Ha
53b	32.5/0.54		52r	
54	32/0.08	11 (a–j)		
54a	32/1.47		52m	Ha
55 (tropomyo-sin related	31.5/1.44	11 (a–j)	55	D
56 (tropomyo-sin related)	31/1.43	15 (a–n)	56	Ha, D
57	30/1.46	23 (a–v)	57	Ha
			57a	Ha
58	29/1.41	7 (a–f)	58	Ha
59	28/1.47	7 (a–f)		
59c	27/1.18		59a	Ha
60	26/1.45	3 (a–b)	60	Ha
61	25/0.37	15 (a–n)	60f	Ha
61d	24.5/0.62		59o	Ha
61k	24/0.41		60l	
62	23.5/1.43	31 (a–z4)	62	Ha, D
63	21/0.70	31 (a–z4)		
64	18/0.34	11 (a–j)	64	Ha
65	16/1.16	2 (a)	65	Ha
66	16/1.24	37 (a–z10)	66	Ha

[a] From Fey *et al.* (1981)

[b] Some polypeptides were found in human and hamster but not found in mouse. This does not mean that the corresponding polypeptide does not exist (and so it is simply left blank).

[c] Identified by one-dimensional peptide mapping; D, Duck; Ha, Hamster; P, Potoroo; R, Rat.

TABLE III

Updated Catalog of Mouse Fibroblast Basic Polypeptides[a]: Homologous Polypeptides in Other Species[b]

Polypeptide	Coordinates molecular weight/ mobility relative to NEPHGE 9	Polypeptides with the same original number	Homologous[c] polypeptides in human HeLa cells	Homologous[c] polypeptides in other species
0	nd/0.01	23 (a–v)		
+			0y	Ha
1	95/0.16	52 (a–z25)	1	Ha
1r	80/1.05		1r	Ha
1p	79/0.16		1w	Ha
1z22	69/0.25		1z31	Ha
2	67/0.35	4 (a–c)	2	Ha
3	66/0.77	3 (a–b)	3	Ha
4	65/0.28	1		Ha
5	64/0.74	2 (a)	5	Ha
5a	64/0.81		5a	Ha
6	62/0.91	13 (a–l)	6	Ha
7	61/0.53	11 (a–j)	7	Ha
7e	60/0.99		7j	Ha
7g	49/0.34		7a	
7h	59/0.10		7e	Ha
7i	59/0.20		7f	Ha
8	56/0.21	6 (a–e)	8	Ha
8d	55/1.09		7k	
8e	55/0.11		7m	
9	55/1.00	11 (a–j)	9	
9b	55/0.66		10a	Ha
10	54/0.53	9 (a–h)	10	Ha
10d	54/1.13		11a	
11	53/0.29	22 (a–u)	11	Ha
11b	52.5/0.15		11s	Ha
11c	52.5/0.50		11f	
11d	52.5/0.56		11e	
11g	52/0.44		11j	Ha
11h	52/0.33		11t	Ha
			11h	Ha
11o	51.5/0.60		12d	Ha
12	50/0.69	1	13	
13	50/1.51	5 (a–d)	12	Ha
14	50/0.15	13 (a–l)	14	Ha
14j	48/0.19		15b	Ha
15	47/1.01	13 (a–l)	15	Ha
15e	46.5/0.02		15j	Ha
			15f	Ha
15l	45/0.30		15m	Ha
16	45/0.66	38 (a–zll)	16	Ha

TABLE III (*continued*)

Polypeptide	Coordinates molecular weight/ mobility relative to NEPHGE 9	Polypeptides with the same original number	Homologous[c] polypeptides in human HeLa cells	Homologous[c] polypeptides in other species
16v	43/1.45		16k	Ha
			16w	Ha
17	41/1.16	3 (a–b)	17	Ha
18	40/0.31	1	19	Ha
19	39.5/0.88	11 (a–j)	18	Ha
19a	39.5/1.01		18a	
19c	39.5/1.18		19g	Ha
19i	37.5/0.93		19h	
20	37/0.69	12 (a–k)		Ha
20g	36/0.33			Ha
21	35/1.43	3 (a–b)	21	Ha
			21g	Ha
22	34.5/0.49	4 (a–c)		
23	33.5/0.33	12 (a–k)		
23j	33/0.68			Ha
24	33/1.43	4 (a–c)		Ha
			24e	Ha
25	32/0.08	7 (a–f)		
25e	31/0.04		26	Ha
26	31/0.20	32 (a–z5)		
			26j	Ha
26j	29.5/0.25		26g	Ha
27	27/0.75	9 (a–h)	27	Ha
28	25/0.62	4 (a–c)	28	Ha
29	23/0.37	6 (a–e)	29	Ha
30	20.5/0.64	2 (a)	30	Ha
31	20.5/0.93	4 (a–c)		
32	18.5/0.44	1	32	Ha
33	18/0.71	13 (a–l)	33	Ha
33h	16/0.76		33d	Ha

[a] From Fey *et al.* (1981).

[b] Some polypeptides were found in human and hamster but not in mouse. This does not mean that the corresponding polypeptide does not exist (and so it is simply left blank).

[c] Identified by one dimensional peptide mapping; D, Duck; Ha, Hamster; P, Potoroo; R, Rat.

IEF 24 (Mose Larsen *et al.*, 1982) showed the widest variation in mobility (Fig. 4) and one-dimensional peptide maps (Fig. 5). The homologous nature of these polypeptides has been subsequently confirmed by results showing that antibodies raised against the human variant (Mose Larsen *et al.*, 1982, 1983) cross-react with the other variants, staining a

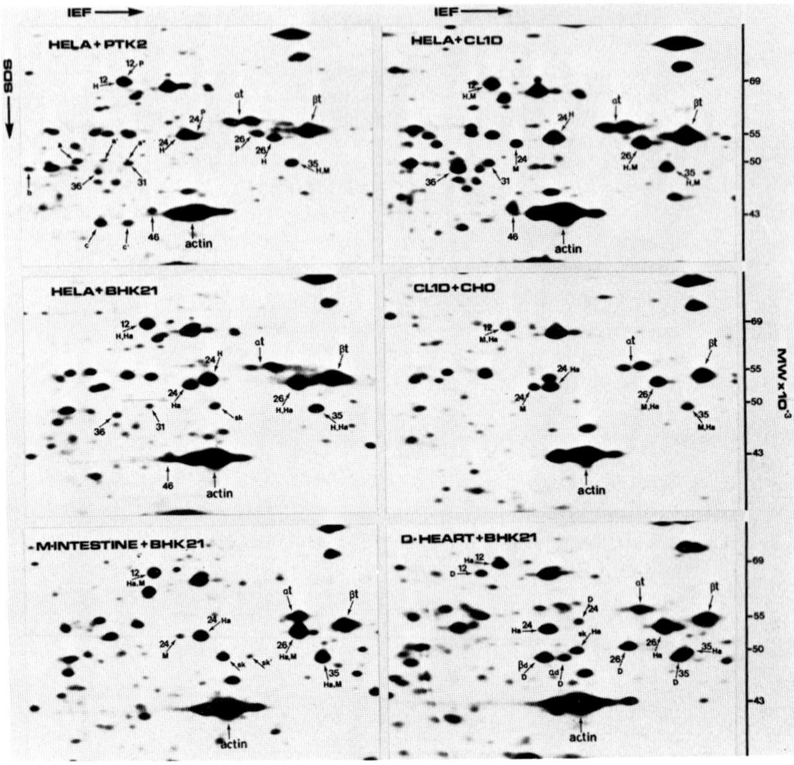

Figure 4. Two-dimensional gel coelectrophoresis of [^{35}S]methionine labeled polypeptides from pairs of different species. HeLa, human; CL1D, mouse; BHK21, hamster; CHO, Chinese hamster ovary; PTK2, potoroo; D-heart, duck; M-intestine, mouse. Arrows indicate cytoskeletal polypeptides. αt and βt, α- and β-tubulin. Polypeptides are numbered according to their homologous HeLa polypeptide. a, a′, a″, b, c, and c′, PTK2 keratins; sk and sk′, skeletin from hamster; αd and βd, α- and β-desmin from duck heart.

mitochondrial pattern in different species as seen by direct immunofluorescence (Fig. 6).

IV. Concluding Remarks

From the foregoing it is clear that peptide mapping, together with the use of specific antibodies, could be invaluable in the further identification of homologous polypeptides in various species. The data presented

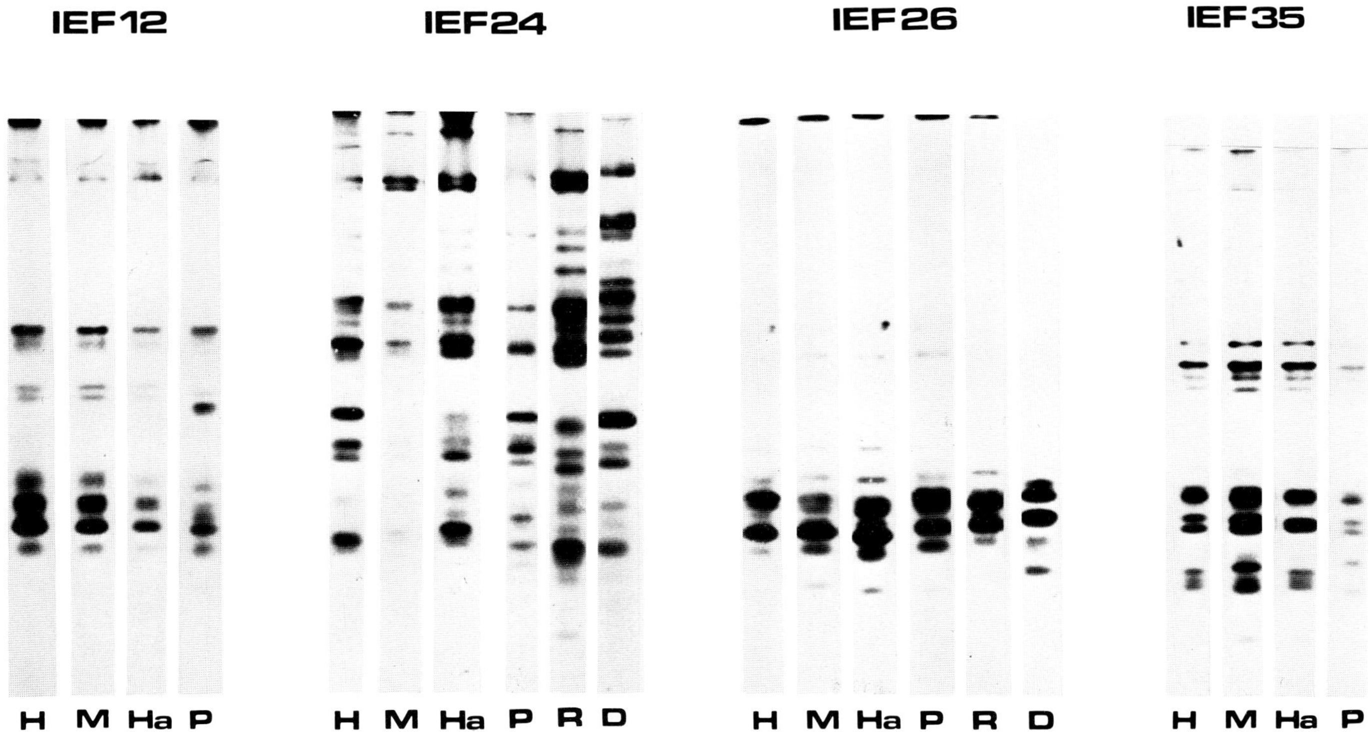

Figure 5. Peptide maps of some homologous cytoskeletal polypeptides from different species. Polypeptides are numbered according to the HeLa protein catalog (Bravo *et al.*, 1981a). H, human(HeLa); M, mouse(fibroblast); Ha, hamster(BHK21); P, Potoroo(PTK2); R, Rat(L6); D, Duck(fibroblast).

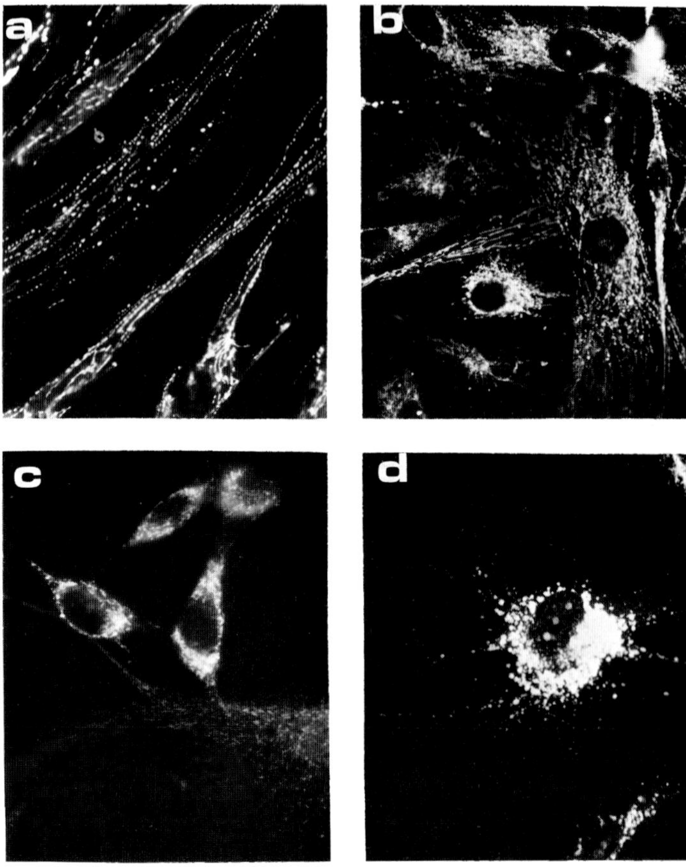

Figure 6. Mitochondrial localization of IEF 24 in different species. Immunofluorescence was carried out according to Mose Larsen *et al.* (1982, 1983) using anti-IEF 24 antibodies and methanol–acetone fixed cells. (a) Human skin fibroblasts; (b) Mouse 3T3; (c) Hamster BHK21; (d) Potoroo PTK2 cells. *Antibody production:* The cell content of approximately 30 confluent Nunc flasks (250 ml) seeded with transformed human amnion cells (AMA) were used to prepare IEF 24 for immunization (Mose Larsen *et al.*, 1982; Bravo *et al.*, 1983; Fey *et al.*, 1983). The cell monolayers were washed four times with sterile Hank's and scraped off with a rubber policeman in a final volume of 10 ml of Hank's. After centrifugation, the pelleted cells were treated with DNase and RNase, lyophilized and dissolved in IEF lysis buffer (O'Farrell, 1975). The procedure for two-dimensional gel electrophoresis (IEF) has been described in detail in the chapter by Bravo. Following electrophoresis (about 100 gels) of samples mixed with tracer radioactive proteins, IEF 24 was located using the autoradiograms and cut from the unfixed dried gels. The protein was eluted from the gels by shaking overnight at room temperature in 1 ml of 0.1% SDS (Celis *et al.*, 1973); lyophilized and redissolved in 1 ml of sterile Hank's. An equal volume of complete Freunds adjuvant was added to this solution. Two BALB/c mice were immunized intraperitoneally each with 0.3 ml of these suspensions and the injections repeated at 15 day intervals until a positive test was obtained by indirect immunofluorescence using methanol–acetone fixed cells. After a booster injection, the animals were bled from the heart. About 0.4 ml of sera was obtained in each case. The mice also produced about 2 ml of ascites fluid as a result of the repeated intraperitoneal injections. The specificity of these antibodies has been determined by immunoprecipitation (Mose Larsen *et al.*, 1982).

have allowed us to use the human and mouse systems interchangeably in our investigation of polypeptides whose rate of synthesis is sensitive to neoplastic transformation (see chapter by Celis *et al.*).

Acknowledgments

S. J. F. is a recipient of a fellowship from EMBO and the Danish Cancer Society. R. B. was a recipient of a fellowship from the Danish Medical and Natural Research Councils. P. M. L. is a recipient of a fellowship from the Aarhus University. This work was supported by grants from the Danish Medical and Natural Science Research Councils, Euratom, the Danish Cancer Society, Carlsberg, and Novo.

References

Bonner, W. M., and Stedman, J. D. (1978). Efficient fluorography of ^3H and ^{14}C on thin layers. *Anal. Biochem.* **89**, 247–256.

Bravo, R., and Celis, J. E. (1982). Up-dated catalogue of HeLa cell proteins: Percentage and characteristics of the major cell polypeptides labelled with a mixture of 16 [^{14}C]-labelled amino acids. *Clin. Chem.* **28**, 766–781.

Bravo, R., Bellatin, J., and Celis, J. E. (1981a). [^{35}S]-methionine labelled polypeptides from HeLa cells. Coordinates and percentage of some major polypeptides. *Cell Biol. Int. Rep.* **5**, 93–96.

Bravo, R., Fey, S. J., Small, J. V., Mose Larsen, P., and Celis, J. E. (1981b). Coexistence of three major isoactins in a single Sarcoma 180 cell. *Cell* **25**, 195–202.

Bravo, R., Small, J. V., Fey, S. J., Mose Larsen, P., and Celis, J. E. (1982). Architecture and polypeptide composition of HeLa cytoskeletons. Modification of cytoarchitectural polypeptides during mitosis. *J. Mol. Biol.* **154**, 121–143.

Bravo, R., Fey, S. J., Mose Larsen, P., Coppard, N., and Celis, J. E. (1983). Proteins IEF (isoelectric focussing) 31 and IEF 46 are keratin-type components of the intermediate-sized filaments. Keratins of various human cultured epithelial cells. *J. Cell Biol.* **96**, 416–423.

Butler, P. J. G., Harris, J. I., Hartley, B. S., and Leberman, R. (1969). The use of maleic anhydride for the reversible blocking of amino groups in polypeptide chains. *Biochem. J.* **112**, 679–689.

Celis, J. E., Smith, J. D., and Brenner, S. (1973). Correlation between genetic and translational maps of gene 23 in bacteriophage T4. *Nature New Biol.* **241**, 130–132.

Cleveland, D. W., Fischer, S. G., Kirschner, M. W., and Laemmli, U. K. (1977). Peptide mapping by limited proteolysis in sodium dodecyl sulfate and analysis by gel electrophoresis. *J. Biol. Chem.* **252**, 1102–1106.

Craig, S. W., and Lancashire, C. L. (1980). Comparison of intestinal brush-border 95-kDalton polypeptide and alpha-actinins. *J. Cell Biol.* **84**, 655–667.

Elder, J. H., Pickett, R. A., II, Hampton, J., and Lerner, R. A. (1977). Radioiodination of proteins in single polyacrylamide gel slices. Tryptic peptide analysis of all the major members of complex multicomponent systems using microgram quantities of total protein. *J. Biol. Chem.* **252**, 6510–6515.

Feitelson, M. A., Wettstein, F. O., and Stevens, J. G. (1981). Tryptic peptide mapping of picomolar quantities of protein labelled with the Bolton-Hunter reagent. *Anal. Biochem.* **116**, 475–479.

Fey, S. J., Bravo, R., Mose Larsen, P., Bellatin, J., and Celis, J. E. (1981). [^{35}S]Methionine labelled polypeptides from secondary mouse kidney fibroblast: Coordinates and one dimensional peptide maps of some major polypeptides. *Cell Biol. Int. Rep.* **5**, 491–500.

Fey, S. J., Mose Larsen, P., Bravo, R., Celis, A., and Celis, J. E. (1983). Differential immunological cross-reactivity of HeLa keratin antibodies with human epidermal keratins. *Proc. Natl. Acad. Sci. USA* **80**, 1905–1909.

Fishbein, J. C., Place, A. R., Ropson, I. J., Powers, D. A., and Sofer, W. (1980). Thin-layer peptide mapping. *Quant. Anal. Biochem.* **108**, 193–201.

Hearn, M. T. W. (1980). The use of reversed phase high performance liquid chromatography for the structural mapping of polypeptides and proteins. *J. Liq. Chromatogr.* **3**, 1255–1276.

Hirs, C. H. W., and Timasheff, S. W. (1977). Enzyme structure, part E. *Methods Enzymol.* **47**.

Houmard, J., and Drapeau, G. R. (1972). Staphylococcal protease: A proteolytic enzyme specific for glutamoyl bonds. *Proc. Natl. Acad. Sci. USA* **69**, 3506–3509.

Huisman, J. G., Stapel, S., and Geblink, M. G. T. (1978). Characterization of ferredoxins on a nanomole scale. *Anal. Biochem.* **90**, 501–509.

Hutchinson, M. A., Hunter, T., and Eckhart, W. (1978). Characterization of T antigens in polyoma-injected and transformed cells. *Cell* **15**, 65–77.

Ingram, V. M. (1956). A specific chemical difference between the globins of normal human and sickle-cell anæmia hæmoglobin. *Nature (London)* **178**, 792–794.

Ingram, V. M. (1959). Abnormal human hæmoglobin. III. The chemical difference between normal and sickle cell hæmoglobins. *Biochim. Biophys. Acta* **36**, 402–411.

Laemmli, V. (1970). Cleavage of the structural proteins during the assembly of the head of bacteriophage T_4. *Nature (London)* **227**, 680–685.

Laskey, R. A., and Mills, A. D. (1975). Quantitative film detection of ^{3}H and ^{14}C in polyacrylamide gels by fluorography. *Eur. J. Biochem.* **50**, 335–341.

McConkey, E. H. (1982). Molecular evolution, intracellular organization, and the quinary structure of proteins. *Proc. Natl. Acad. Sci. USA* **79**, 3236–3240.

Mose Larsen, P. (1981). An assessment of the potential offered by two dimensional gel electrophoresis and silver staining for developmental biology. Candidate Sci. Dissertion, Aarhus Univ., Aarhus, Denmark.

Mose Larsen, P., Bravo, R., Fey, S. J., Small, J. V., and Celis, J. E. (1982). Putative association of mitochondria with a sub-population of intermediate-sized filaments in cultured human skin fibroblasts. *Cell* **31**, 681–692.

Mose Larsen, P., Fey, S. J., Bravo, R., and Celis, J. E. (1983). Mouse mitochondrial protein IEF 24: Identification and immunohistochemical localisation of mitochondria in various tissues. *Electrophoresis* **4**, 247–256.

O'Farrell, P. H. (1975). High resolution two dimensional electrophoresis of proteins. *J. Biol. Chem.* **250**, 4007–4021.

O'Farrell, P. Z., Goodman, H. M., and O'Farrell, P. H. (1977). High resolution two dimensional electrophoresis of basic as well as acidic proteins. *Cell.* **12**, 1133–1142.

Pedersen, S., Block, P. L., Reeh, S., and Neidhardt, F. C. (1978). Patterns of protein synthesis in *E. coli:* A catalogue of the amount of 140 individual proteins at different growth rates. *Cell* **14**, 179–190.

Sage, H., Pritzl, P., and Bornstein, P. (1981). A new mapping technique for collagen chains. *Collagen Res.* **1**, 3–15.

Schiltz, E., Schnackerz, K. D., and Gracy, R. W. (1977). Comparison of Ninhydrin, Fluorescamine and O-Phthaldialdehyde for the detection of amino acids and peptides and their effects on the recovery and composition of peptides from thin-layer fingerprints. *Anal. Biochem.* **79**, 33–41.

Smyth, D. G. (1967). Techniques in Enzymic Hydrolysis. *Methods Enzymol.* **11**, 214–231.

Sonderegger, P., Jaussi, R., Gehring, H., Brunschweiler, K., and Christen, P. (1982). Peptide mapping of protein bands from polyacrylamide gel electrophoresis by chemical cleavage in gel pieces and re-electrophoresis. *Anal. Biochem.* **122**, 298–301.

Stephens, R. E. (1978). Fluorescent thin-layer peptide mapping for protein identification and comparison in the subnanomole range. *Anal. Biochem.* **84**, 116–126.

Vandekerckhove, J., and van Montagu, M. (1974). Sequence analysis of fluorescamine-stained peptides and proteins purified on a nanomole scale. Application to proteins of bacteriophage MS2. *Eur. J. Biochem.* **44**, 279–288.

Watanabe, S., and Yoshida, A. (1971). Micro-scale peptide mapping method for identification of variant proteins. *Biochem. Genet.* **5**, 541–547.

Whittaker, R. G., and Moss, B. A. (1981). Comparative peptide mapping at the nanomole level. *Anal. Biochem.* **110**, 56–60.

PART II

Applications of Two-Dimensional Gel Electrophoresis

CHAPTER 7

Clinical Applications of Two-Dimensional Gel Electrophoresis

RUSSELL P. TRACY*
DONALD S. YOUNG
Department of Laboratory Medicine
Mayo Clinic
Rochester, Minnesota

I. Introduction . . . 194
A. Electrophoresis in the Clinical Laboratory . . . 194
B. The Meaning of Clinical Application . . . 195
C. The "Team" Approach . . . 196
II. The Electrophoresis System . . . 196
A. The Need for High Throughput . . . 196
B. Specimen Storage and Management . . . 197
C. The Isoelectric Focusing Dimension . . . 198
D. The Dodecyl Sulfate Dimension . . . 201
E. Visualization . . . 202
F. Analysis . . . 206
G. Reproducibility and Quality Assurance . . . 207
H. Additional Techniques . . . 209
III. Serum and Plasma . . . 211
A. Specimen Procurement . . . 211
B. Sample Preparation . . . 211
C. Reference Serum Protein Map . . . 212
D. Alterations in the Reference Pattern . . . 214
IV. Urine . . . 220
A. Specimen Procurement . . . 220
B. Sample Preparation . . . 221
C. Reference Urine Map . . . 222
D. Alterations in the Reference Pattern . . . 225
V. Blood Cells . . . 226
A. Erythrocytes . . . 226
B. Lymphocytes . . . 227
C. Platelets . . . 229

* Present affiliation: Department of Pathology, Strong Memorial Hospital, University of Rochester Medical Center, Rochester, New York.

TWO-DIMENSIONAL GEL ELECTROPHORESIS OF PROTEINS

ISBN 0-12-164720-X

VI. Solid Tissue. 230
A. Sample Preparation. 230
B. The Need for Histological Analysis. 230
C. Colon and Colorectal Cancer. 231
D. Kidney. 231
E. Brain. 232
F. Muscle. 233
VII. Other Specimens. 233
A. Semen. 233
B. Pancreatic Fluid. 233
C. Spinal Fluid. 234
D. Amniotic Fluid. 234
E. Saliva. 234
VIII. Conclusions. 234
References. 235

I. Introduction

A. Electrophoresis in the Clinical Laboratory

Since its introduction (Tiselius, 1937), protein electrophoresis has undergone extensive evolution. Many of the modifications, such as isoelectric focusing and dodecyl sulfate electrophoresis, have paved the way for the development of two-dimensional gel electrophoresis (2-DGE) (see Table I in N. G. Anderson and N. L. Anderson, 1979). While most forms of electrophoresis have been used in clinical chemistry research at one time or another, the routine laboratory methods still produce the same five bands for serum that were seen originally: albumin, α_1-, α_2-, β- and γ-globulin (Tracy and Anderson, 1983). Although there is some impetus to replace cellulose acetate with agarose as the medium for routine serum electrophoresis, which would result in higher resolution (Sun *et al.*, 1978), there is no evidence that any other electrophoretic technique is needed in the routine laboratory at this time (Tracy *et al.*, 1983a). Nonetheless there is considerable potential for 2-DGE in the clinical laboratory, possibly as a routine test in the future but certainly even now as a research and development tool. This technique offers the researcher the opportunity to determine in a rational and methodical manner the appropriate proteins to be measured in a give clinical setting, which may help develop routine assays that are significantly better in specificity and sensitivity than those currently used.

B. The Meaning of Clinical Application

The term *clinical* may be interpreted in many ways. In the context of this chapter, we use it in the sense of *clinical chemistry,* and we will discuss applications of 2-DGE related to the analysis of (1) human body fluids; (2) blood cells, including erythrocytes, white cells, and platelets; and (3) solid tissue. We do not discuss cultured cells from either human or other sources, as this topic is covered in detail in other chapters. There are some obvious cases, however, in which such analyses may prove useful in a clinical laboratory, for example, the analysis of fibroblasts from patients with inborn errors of metabolism. Our aim is to review data that concern either proteins of clinical significance or studies directed toward the discovery of clinically significant proteins. By "clinically significant" we mean that measurements of the concentration of the protein in a given specimen are useful to the physician in the diagnosis of the disease or in the management of the patient.

We have defined several potential application areas for 2-DGE in the clinical laboratory now and in the future (Tracy *et al.*, 1983a; Tracy and Young, 1983). It is possible that 2-DGE may become a routine test for measuring serum, plasma, or urine proteins, but only if screens or panels of such proteins are shown to be clinically relevant (Tracy *et al.*, 1983a), because the power of 2-DGE as a quantitative procedure lies in its ability to analyze several proteins at once. Alternatively there may be a role for 2-DGE as a specialized test offered in a few centers to aid in the analysis of complex protein abnormalities, as in the case of multiple myeloma (Tracy *et al.*, 1982a). In the distant future 2-DGE may prove useful as the input device of a data-base protein index (Anderson and Anderson, 1982) in which gels may be automatically compared to standard normal gels to aid diagnosis.

It is now certain that 2-DGE is useful as a quality control tool, especially in evaluating the antigens and immunoglobulins used in routine immunologic assays (Swanson *et al.*, 1982). However, most importantly, at this time 2-DGE can be used in clinical chemistry research (Tracy and Young, 1983). The protein content of normal specimens and pathological specimens can be compared in detail and the appropriate proteins identified for further analysis. We anticipate that most, if not all, assays resulting from this type of research will require the isolation of this protein or proteins and the development of immunoreagents to be used for measurement. The actual assays may take several forms including radioimmunoassay, enzyme-linked immunosorbent assay (ELISA), histological staining of tissue sections, or radiographic imaging.

C. The "Team" Approach

We feel that for optimal use of 2-DGE in the clinical laboratory a team approach is necessary. Each member of the team has a specific function. The clinician is needed to obtain access to special patients as well as to provide insight into the relative importance of potential projects; the pathologist is essential for assurance as to the nature and quality of solid tissue and blood cell specimens. Specimens such as blood, serum, urine, and spinal fluid are most easily obtained by the clinical chemist, who along with the clinician also has access to other laboratory test results. The clinical immunologist is needed to produce immunoreagents (either antisera or hybridoma products) and to develop immunoassays. The necessary expertise may be available in fewer people than we have listed here, as in the case of the pathologist with specialty training in chemistry. The person responsible for running the gels might be any one of these members or someone else. We wish to emphasize that an interdisciplinary team approach is the most efficient way to use 2-DGE to achieve the goal of more specific and sensitive laboratory tests.

II. The Electrophoresis System

A. The Need for High Throughput

For a new test to have clinical utility considerable evaluation must be done concerning the sensitivity (i.e., true positive assay results/true positives plus false negatives), specificity (true negatives/true negatives plus false positives), and predictive value (true positives/all positives) of the test (Young and Tracy, 1983). Appropriate pathologic controls must be analyzed as well as many normal specimens. Even though we envisage most tests resulting from 2-DGE investigations to be immunologic in nature, much preliminary work must be done with 2-DGE. For this reason high throughput electrophoretic equipment must be used. Although several such devices have been described (Garrels, 1979; Jones *et al.*, 1980; see chapter by Garrels), the equipment designed by Anderson and Anderson (N. G. Anderson and N. L. Anderson, 1978; N. L. Anderson and N. G. Anderson, 1978) seems best suited for clinical chemistry research. Using a slightly modified version we can run on the order of 50 to 60 gels per week. An added advantage of using such equipment is that additional reproducibility is achieved by using large within-run batches (see Section II,C and D).

B. Specimen Storage and Management

Serum samples from normal donors may be adequately stored at −20°C for several months without major alterations of the pattern (Tracy *et al.*, 1982b), although several new minor spots are produced by this process. The authors believe that urine may also be stored in this manner but have not done a complete study. Spinal fluid also appears to be stable for several months at −20°C. Long-term storage for any of these specimens is best done at −70°C whenever possible. An alternative long-term storage method for urine is to prepare a lyophilized powder (Section IV,B) and store it at −20°C.

There are no major changes in the serum, plasma, or urine patterns that can be detected following incubation at room temperature for up to 3 h (Tracy *et al.*, 1982b). Even longer may be quite acceptable, especially in the case of normal specimens. Seminal plasma undergoes rapid proteolytic alteration and must be processed as quickly as possible (Edwards *et al.*, 1981). Standard dissociation buffers (e.g., Section III,B) seem adequate to arrest this process, however.

Erythrocytes, lymphocytes, and platelets may be stored at room temperature for several hours in heparin- or EDTA-anticoagulated blood or, in most cases, overnight at 4°C. Longer storage is not recommended. Red cell membranes are best prepared as rapidly as possible (Kasten *et al.*, 1982) and stored frozen at −70°C. Lymphocytes, other white cells, and platelets may be solubilized in a urea–NP-40 mix (Willard and Anderson, 1981) and stored frozen at −70°C for several months. Cells cannot be frozen, however, prior to solubilization (N. L. Anderson, personal communication).

Tissue specimens should be processed as quickly as possible; we attempt to have the surgical pathologist prepare the tissue (See Sections VI,A) and freeze it in liquid nitrogen within 30 min of excision (Tracy *et al.*, 1982c). Longer than this is not advised. Once frozen in liquid nitrogen and stored at −70°C, however, most tissues including tumors appear to be quite stable.

There are several reasons why high throughput 2-DGE requires a computerized sample index system for optimal efficiency. Our silver stain technique (Tracy *et al.*, 1982b) is most easily done over several days. This means that several experiments using different samples may

DISPLAY REC. 2D GEL P. 1 OF 3
RUN # ______________
SLOT # ______________
DATE ____ - ____ - ____
SPEC. # ______________
DUPLICATE? ______________
DISSOCIATION ______________
____
EQUILIBRATION ______________
____
COMMENTS 1 ______________
2 ______________
3 ______________
DALT RUN ____ - ____ - ____
DALT # ______________
VISUALIZATION ______________

DISPLAY REC. 2D GEL P. 2 OF 3
REVISION ____ - ____ - ____
SPEC. RECEIVED ____ - ____ - ____
SUPPLIER # ______________
TYPE # ______________
COMMENTS 1 ______________
2 ______________
3 ______________
AMT. LEFT ______________

Figure 1. Data base system for specimen organization. Our system generally uses 3 pages: the first two, which are standard, are shown here; the third is for data analysis and varies with the projects. Information is included on ISO run # and slot #, dissociation and equilibration, and DALT run and #, as well as the origin of the specimen.

be started between the time of a sample application and photography of a developed gel. Manual record keeping thus becomes difficult. Figure 1 illustrates the first two pages of our standard data base form, which allows rapid entry of critical data and easy retrieval of information. This type of system runs on a small microprocessor-based computer and may be purchased, set up, and run with minimal cost and effort.

C. The Isoelectric Focusing Dimension

The isoelectric focusing equipment we use has been made at our institution, essentially from the designs of N. G. Anderson and N. L. Anderson (1978).* The modifications that we made to ISO equipment include setting up the gels in units of 12, not 10, for quality assurance purposes (see Section II,G) and using removable tubes instead of those cemented in place (Tracy *et al.*, 1982b). Our standard equipment contains 24 tubes. This allows us to analyze ten samples in duplicate and still have four blank gels, from which we pick two to cut and equilibrate with small aliquots of degassed distilled water to establish the pH gradient, a qual-

* The term ISO-DALT (isoelectric focusing–SDS gel electrophoresis) refers to 2-DGE as described by N. L. Anderson and N. G. Anderson (1977). This technique utilizes equipment of standardized dimensions and allows multiple gels to be run at the same time.

TABLE I

Some Decision Criteria for Choosing Best ISO of Two Duplicates

Criteria	Comments
1. "Bulb" at acid end	Bulb is caused by dodecyl sulfate focusing
2. "Bulges" along the Iso	Bulges reveal the major serum proteins; does not apply to a tissue gel
3. Density line 2 mm below surface	Density line caused by lysine and arginine focusing (see Table II)
4. No breaks or fractures	May be caused by low conductivity, especially in center of gel
5. No "gritty" portions	—
6. No precipitate on the top	—

ity assurance parameter. From each set of duplicates the best gel is chosen to run in the DALT dimension. The decision criteria are outlined in Table I. This method of choosing the best gel out of two has greatly increased our reproducibility, even within a run.

The pH gradients we use have been described (Tracy *et al.*, 1982d, 1983a). The composition of our gels is shown in Table II, and the pH

TABLE II

Composition of ISO Gels[a]

Urea	11.0 g
Lysine	0.020 g
Arginine	0.033 g
Water	8.33 ml
Acrylamide solution[b]	2.37 ml
Ampholytes[c] 3.5–10	1.0 ml
5–7	0.25 ml
Degas	
NP-40[d]	0.4 ml
100 g/liter APS[d]	0.080 ml
TEMED[d]	0.013 ml
Catholyte : anolyte	1.0 *M* NaOH : 850 mg/liter H_3PO_4
Final volume:	21 ml

[a] Modified from Anderson and Anderson (1977).

[b] 30% acrylamide, 1.8% bisacrylamide.

[c] Value shown is for "narrow" range (MC-ISO 2) system; for broad range (MC-ISO 1), use 3.5–10, 1.2 ml–2.5–4, 0.10 ml–9–11, 0.24 ml (Tracy *et al.*, 1982b).

[d] NP-40, Non-ionic detergent; APS, Ammonium persulfate; TEMED, *N,N,N′,N′*-tetraethylmethylenediamine.

gradients that occur in these gels are shown in Fig. 2. Note that it is possible to extend the pH range to approximately 10 with the gels supplemented with lysine and arginine and by using 1 *M* NaOH as the catholyte.

Most of the specimens we analyze contain plasma proteins. This makes the assignment of apparent molecular mass for an unknown protein easy, using the known molecular masses of these internal standards (Tracy *et al.*, 1982d). These same plasma proteins, once characterized, may be used as orientation points in the charge dimension. Characterization is done relative to carbamylated hemoglobin charge standards (Tracy *et al.*, 1982a), which in turn have been characterized as to apparent isoelectric point (pI) by comparison with charge standard-containing gels that have been cut and used to establish the pH gradient.

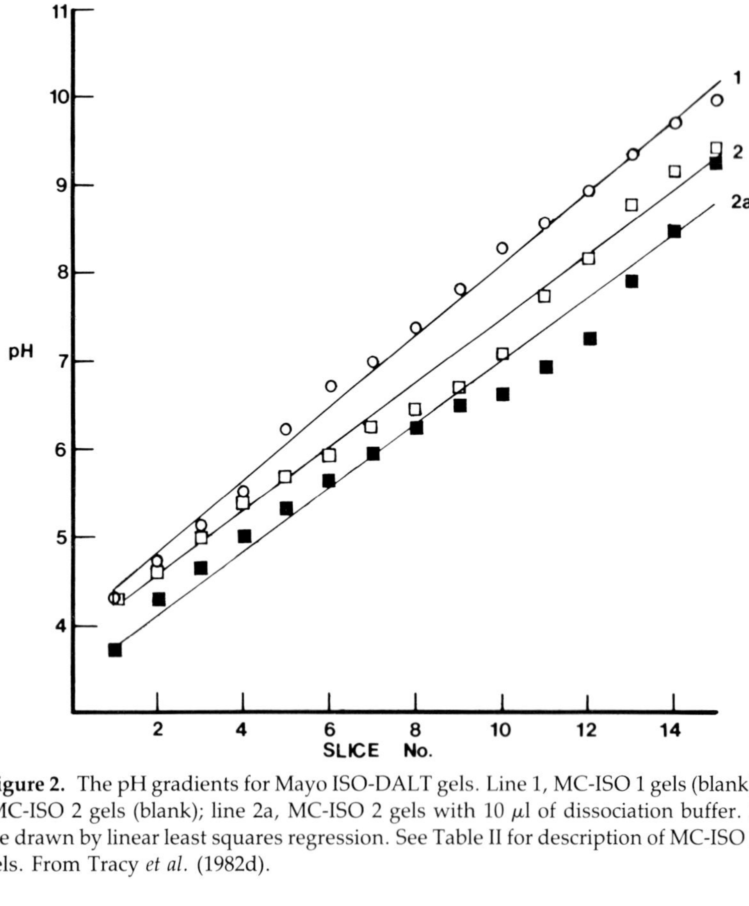

Figure 2. The pH gradients for Mayo ISO-DALT gels. Line 1, MC-ISO 1 gels (blank); line 2, MC-ISO 2 gels (blank); line 2a, MC-ISO 2 gels with 10 μl of dissociation buffer. Lines were drawn by linear least squares regression. See Table II for description of MC-ISO 1 and 2 gels. From Tracy *et al.* (1982d).

D. *The Dodecyl Sulfate Dimension*

The dodecyl sulfate (DALT) dimension system we use is essentially that of N. L. Anderson and N. G. Anderson (1978), with slight modification (Electro-Nucleonics, Oak Ridge, Tennessee). We use an acrylamide gradient that has been optimized for the separation of proteins in the 10–100 kdalton molecular mass range (Tracy *et al.*, 1982d). The recipe for this type of gel is listed in Table III, and the shape of the gradient is shown in Fig. 3. This type of DALT gel does not separate small proteins well. To do this we designed a gel that contains a gradient in acrylamide (15–25%), urea (6–8 *M*) and cross-linker (1 : 18.5–1 : 15). Proteins in the range of 1 to 10 kdaltons may be separated on this gel, with the limiting factor in visualization being the ability to fix small proteins in the gel and to stain them (Tracy *et al.*, 1983b).

ISO gels are prepared for the second dimension by equilibration in the standard dodecyl sulfate equilibration mix of Anderson and Anderson (N. G. Anderson and N. L. Anderson, 1978; N. L. Anderson and N. G. Anderson, 1978). We have modified this mixture to contain 5% dodecyl sulfate, and we have removed the thiol reagent from this mix because we have found that thiol-reducing agents will cause micro-streaking on the surface of silver-stained gels. The cause of this is apparently the

TABLE III

Composition of DALT Gels[a]

Component	10% Acrylamide	23% Acrylamide
Acrylamide solution[b]	28.5 ml	65.7 ml
Buffer[c]	25.3	25.5
Water	44.5	—
Glycerol	—	7.4
100 g/liter dodecyl sulfate	1.00	1.00
100 g/liter APS[d]	0.69	0.34
TEMED[d]	0.009	0.003
Total volume[e]	100 ml	100 ml

[a] Modified from Tollaksen *et al.* (1981).
[b] 350 g/liter acrylamide, 8 g/liter bisacrylamide
[c] 3 *M* Tris, pH 8.5 with HCl.
[d] APS, Ammonium persulfate; TEMED, *N,N,N',N'*-tetraethylmethylenediamine.
[e] The actual volumes used would depend upon such things as the type of gradient maker, the shape of the gradient, and the number of DALT gels to be made.

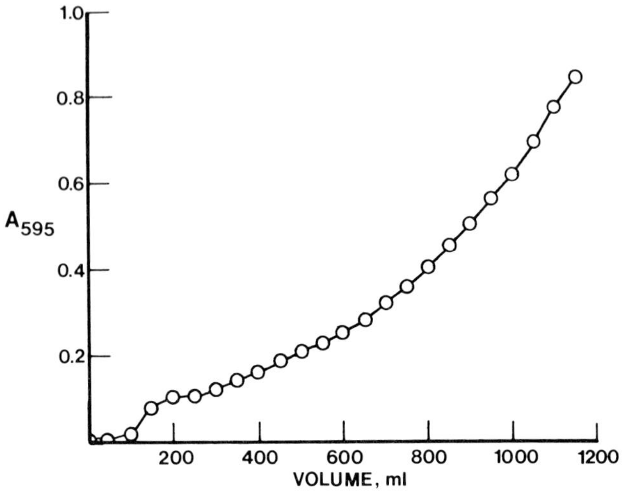

Figure 3. Acrylamide gradient for Mayo ISO-DALT gels. Gradient was measured by spiking "heavy" solution with bromophenol blue, collecting fractions, and measuring A_{595}. At 1200 ml of effluent the gradient concentration in the bottom of the DALT gels is approximately 200 g/liter. From Tracy *et al.* (1982d).

liberation of protein from macroparticles such as dust particles, which can be seen as micro-streaks with the sensitive silver stains. This is generally not a problem with Coomassie Blue staining procedures.

We also carefully time the equilibration step using rapid freezing in dry ice–ethanol as a convenient way to stop the process, because we have found this step to be the cause of significant protein loss and staining variability (see Section II,G).

E. Visualization

1. COOMASSIE BLUE

Coomassie Blue staining techniques have been extensively discussed in the literature. We use the technique devised for the ISO–DALT system, which includes five washes of ethanol of successively lower concentration varying from 50 to 5% (Tollaksen *et al.*, 1981) to effect destaining. Our motorized stainer (see Section II,E,2) allows the semiautomated staining of ten gels simultaneously, although Coomassie Blue staining and destaining may be readily done in plastic boxes (1 liter of solution/10

TABLE IV

Silver Stain Procedure[a]

Step	Quantity	Time
1. Fix in glutaraldehyde	100 g/liter	30 min
2. Wash (with water)	Several liters	Extensive
3. Stain in:		15 min
Water	21 ml	
10 *M* NaOH	0.19 ml	
NH_4OH	1.4 ml	
200 g/L $AgNO_3$	4.0 ml	
total volume (with water)[b]	100 ml	
4. Wash (with water)		5 min, plus a rapid rinse
5. Develop in:		Generally, developing done as 5–10 min
Water[b]	100 ml	
1.0 g/liter citric acid	11 ml	
HCO_3	0.1 ml	

[a] Modification of Oakley *et al.* (1980); see Tracy *et al.* (1982b) for details of the procedure.

[b] The actual volumes used will depend upon the gel size, container size, and number of gels; e.g., for 10 standard DALT gels, we use 750 ml of stain when done in boxes and 3000 ml when done in automatic stainer.

gels) on a reciprocal shaker set at approximately 60–70 cycles/min. The sensitivity of this method of staining with respect to visual spot detection is of the order of 50 to 100 ng/spot (Anderson and Anderson, 1977).

2. SILVER STAIN

Since the appearance of the initial publications concerning the use of silver nitrate to visualize proteins in gels (Switzer *et al.*, 1979), several other methods and modifications have been reported (Oakley *et al.*, 1980; Sammons *et al.*, 1981). Silver staining is discussed in more detail elsewhere (see chapters by Merril and Goldman and by Sammons *et al.*), but we will comment on its use to stain multiple gels simultaneously. The stain we have modified for staining 10 gels at a time (Tracy *et al.*, 1982b) is that of Oakley *et al.* (1980), employing alkaline ammoniacal silver nitrate as the stain and citric acid–formaldehyde as the developer. The recipe is listed in Table IV. We originally used plastic boxes on a reciprocal shaker for staining (750 ml/10 gels). Development of each gel was then done individually. A motorized unit has been designed to conserve technician time, decrease breakage and markings on the gels due to handling, and increase within-run reproducibility (R. Tracy, unpublished). Gels are held vertically between removable plexiglas separators; solution flows past the gels through the porous top and bottom as the container is cycled up and down inside the larger solution-filled container. The volumes of the solutions are increased over the manual method, but the silver (the major expense) may be precipitated ultimately as silver chloride and recycled. We find the use of such a device greatly facilitates high throughput silver staining.

We should note that other staining procedures, such as that of Sammons *et al.* (1981) develop colors in the spots (see chapter by Sammons *et al.*). This technique is useful in certain cases although we do not currently recognize the need to use it on a routine basis in the clinical laboratory.

3. RADIOLABELING, AUTORADIOGRAPHY, AND FLUOROGRAPHY

Radiolabeling of proteins is most often done by the incorporation of [^{35}S]methionine into actively growing cells in culture. A discussion of this is out of the scope of this article (see chapter by Bravo); however, there are other methods of radiolabeling that are germane. The most common is to label tyrosine residues with ^{125}I by the method of Hunter and Greenwood (1962).

These techniques are commonly used in the clinical laboratory to prepare antigens for radioimmunoassay; these antigens may then be analyzed by 2-DGE for quality control. The second stage of electroblotting techniques (see Section II,H,4; see also chapter by Symington) often use ^{125}I-labeled protein A as the signal molecule. ^{125}I may be readily visualized by direct autoradiography using intensifying screens (Laskey and Mills, 1977) with exposure at −70°C. An example of using 2-DGE as a quality control tool for radioimmunoassay is shown in Fig. 4.

Other types of radiolabeling include reduction and alkylation with [^{14}C]iodoacetamide, phosphorylation with ^{32}P, and reductive methylation with $CH_2{}^3H$. Gels containing proteins labeled by these methods are usually visualized by fluorography at −70°C following precipitation of a fluor compound in the gel (Bonner and Laskey, 1974; see also chapter by Bravo and Celis). These techniques are not commonly used in the clinical laboratory.

4. PHOTOGRAPHY

Regardless of the method of visualization we routinely photograph each gel, keeping the negative as a permanent record. A large format film such as 4 × 5 in (10 × 12.5 cm) is recommended for clarity and

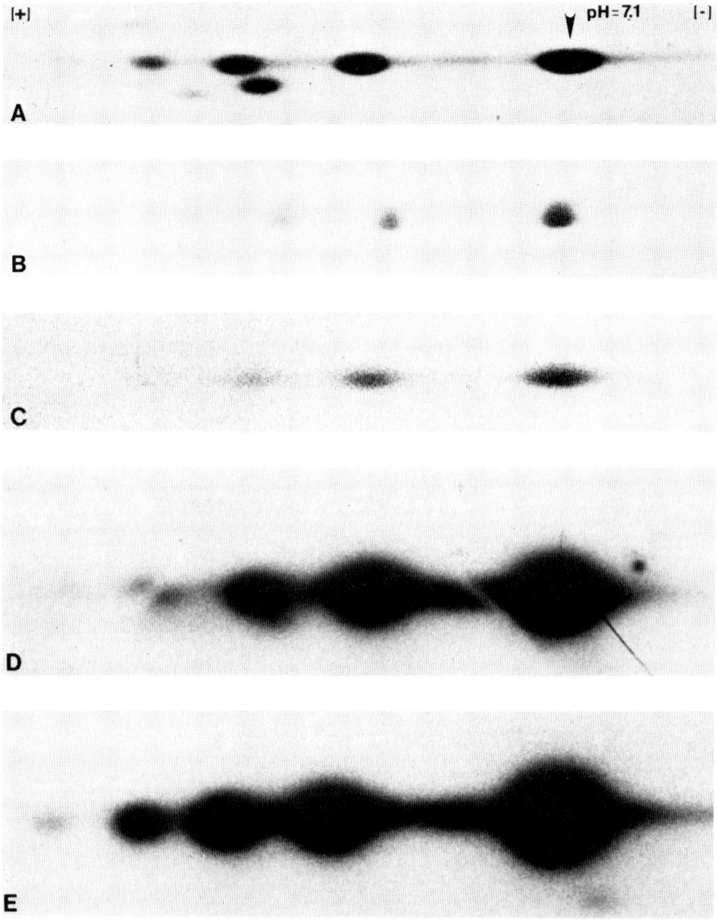

Figure 4. Low molecular mass regions of two-dimensional gel electrophoretograms of β_2-microglobulin. No proteins were seen on the other parts of the gels. A–C are stained with Coomassie Blue; D and E are autoradiographs. (A) Conventionally purified β_2-microglobulin; (B) conventionally purified β_2-microglobulin, additionally purified by affinity chromatography; (C) β_2-microglobulin from whole serum of an anephric patient after single pass affinity chromatography; (D) conventionally purified β_2-microglobulin labeled with ^{125}I; (E) ^{125}I-labeled β_2-microglobulin immunoprecipitated with assay monoclonal antibody. From Swanson *et al.* (1982).

resolution. A film with a wide dynamic range such as Kodak Super-XX maintains the relationship between density and amount of protein (N. G. Anderson, personal communication). Photographs as well as transparencies may be made from the negatives. Transparencies are useful for visually comparing several similar gels (see Section II,F,1).

F. *Analysis*

Two-dimensional gels may be analyzed in many ways depending upon the analytical equipment available and the intent of the researcher. The authors believe that a minimal system in the clinical laboratory should include the ability to make transparencies for overlay comparison, and some sort of densitometric device for (at least) crude quantitation. Manually overlaying transparencies is an efficient method to compare 2-DGE patterns qualitatively but not quantitatively. Individual spots identified as interesting by overlay comparison may be quantitated somewhat laboriously by using a scanning densitometer (Tracy *et al.*, 1982b). However, the researcher must be aware of several caveats: (1) the linearity of the visualization technique, (2) the linearity of the densitometer, (3) the need for careful background corrections, and, (4) if photographic transparencies are used, the need to correct for the photographic process (such as including a photographic step wedge in the original negative along with the gel).

Beyond the minimal system computer-aided devices have been and are being developed at various levels of sophistication. Most systems are intended to analyze entire gels or sections of gels at one time, although we have developed a TV camera, microcomputer-based system that allows rapid quantitation of individual spots picked out by the operator (R. Tracy, unpublished). Systems aimed at analyzing patterns range from relatively simple micro- or small minicomputer-based devices that use TV cameras (Aycock *et al.*, 1981; Mariash *et al.*, 1982) to highly sophisticated equipment that is capable of removing streaks, directly comparing gels on a spot by spot basis, and quantitating every spot in a pattern (Garrels, 1979; Anderson *et al.*, 1981; see also chapter by Garrels *et al.*). These elaborate systems have uses in long range projects such as human protein indexing (Anderson and Anderson, 1982) and estimation of human mutation rate (Skolnick *et al.*, 1982) but are not currently needed in any project of immediate clinical significance. Certainly their use will become more important, especially in the clinical laboratory, as data bases are enlarged.

Because plasma proteins are present in virtually every sample we analyze we have developed a coordinate system for all of our gel patterns based on the location of several abundant plasma proteins. This system facilitates the comparison of various specimens, for example, serum and urine and is used as the basis of our numbering system. Each gel is divided into six parts; a vertical line bisects the large albumin spot and two horizontal lines, one through albumin and one just above apo

A-I lipoprotein, complete the sectioning. The sections are lettered A to F, left to right, top to bottom, and spots are numbered randomly within the sections. Plasma is used as the primary reference so that a protein in any specimen that has been verified as a plasma protein receives the prefix letter "P" (or "S" if present only in serum). All other proteins receive a different prefix letter identifying the type of specimen (e.g., UA-20; U, Urine, A-20, Location; see Section III,A for example). It should be noted that this system, albeit useful, is interim in nature and will be replaced later by a computer-based system such as TYCHO (Anderson *et al.*, 1981).

G. Reproducibility and Quality Assurance

As a prelude to the use of 2-DGE in the clinical laboratory, the authors have done extensive work to determine reproducibility and protein loss (Tracy *et al.*, 1982d). The intrinsic reproducibility of the system was estimated in a two-step manner. First, ISO gels were used to establish pH gradients over several runs. The slopes and intercepts of the gradients were analyzed statistically: within-run CV (coefficient of variation) = 5.2% (slope), 3.0% (intercept); between-run CV = 4.6% (slope), 4.3% (intercept) (data averaged from Table V in Tracy *et al.*, 1982d).

Second, serum protein DALT gels were used to establish log molecular weight versus R_m curves over several runs. Again, slopes and intercepts were analyzed statistically: within-run CV = 3.2% (slope), 3.2% (intercept); between-run CV = 3.4% (slope), 2.7% (intercept) (data from Table VII in Tracy *et al.*, 1982d).

These estimations of error do not require a sample; for example, the ISO gels used were blank gels. The actual position of a given spot is a function of many variables, only one of which is the intrinsic variability of the system. Protein load, the presence of salt, and pH and density of the sample can all affect the absolute position of a given protein spot (Tracy *et al.*, 1982d). Nonetheless these data indicate that 2-DGE as currently done has sufficient intrinsic reproducibility for exploration in a clinical setting.

The question of protein loss is different from that of reproducibility of spot density for a given amount of protein. The first question should be independent of visualization technique whereas the second question is heavily dependent on it. We have explored the first question, protein loss, in depth using ^{125}I-radiolabeled purified plasma proteins α_2-macroglobulin, albumin, and β_2-microglobulin (Tracy *et al.*, 1982d). Our

TABLE V

Quantitation and Reproducibility in the Steps of the ISO-DALT Procedure[a]

		Percentage in ISO gel after focusing[b]		Percentage in ISO gel after equilibration[b]		Percentage lost during DALT[c]	
		MC-Iso 1	MC-Iso 2	MC-Iso 1	MC-Iso 2	MC-Iso 1	MC-Iso 2
α_2M[d]	$\bar{X}$	75.1	67.6	48.4	51.7	2.5	6.6
	SD	6.1	5.5	3.9	4.3	0.9	2.0
	CV, %	8.1	8.2	8.1	8.3	n.a.	n.a.
HSA[d]	$\bar{X}$	88.3	72.5	66.2	61.1	4.8	22.2
	SD	5.7	3.5	6.7	3.1	2.2	2.9
	CV, %	6.5	4.9	10.1	5.0	n.a.	n.a.
β_2M[d]	$\bar{X}$	91.4	79.4	56.0	46.9	0.5	0.5
	SD	6.6	4.3	6.7	4.6	0.1	0.1
	CV, %	7.2	5.3	11.9	9.8	n.a.	n.a.

[a] Reproduced with permission from Tracy *et al.* (1982d).

[b] Results expressed as (TCA-precipitable cpm in gel/TCA-precipitable cpm in starting sample) × 100%; $n = 6$.

[c] Results expressed as (TCA-precipitable cpm in gel after DALT)/(TCA-precipitable cpm in gel before DALT) × 100%; $n = 3$.

[d] M_r (reduced) and approximate serum concentration of α_2-macroglobulin (α_2M), albumin (HSA), and β_2-microglobulin (β_2M) are 185,000 (200 mg/dliter), 67,000 (3.5 g/dliter), and 12,000 (2.0 mg/liter). n.a., Not applicable due to the low values for the mean.

studies, shown in Table V, indicate that the major sources of protein loss during the procedure are in the ISO step and in the equilibration step (see also chapter by Bravo). A careful examination of the factors affecting protein loss in the ISO step indicates that the pH of the sample and the concentration of catholyte are most important: the higher the pH of the sample and the lower the concentration of sodium hydroxide used as catholyte, the greater the percentage of the sample protein that enters the gel (Tracy *et al.*, 1982d). Of course these two parameters may have to be adjusted independent of this consideration for any given sample.

The protein loss during equilibration occurs as a function of time with as much as 15% of a small protein loss in under 1 min of equilibration (Tracy *et al.*, 1982d). This is presumably because the protein has focused around the outside of the tube gel in a disproportionate manner, possibly because of lower acrylamide concentration at the interface with the glass tube. From 40–50% may be lost in 45 min. To reduce this variability we routinely time the equilibration step at 10 min (see Section II,D). In our routine system 50–60% of the starting protein is present on the DALT gel and the overall CV for the system is 8–12% (Tracy *et al.*,

1982d). These data agree well with those of Peterson and McConkey (1978), who used a different system (for discussion see Tracy *et al.*, 1982d).

Reproducibility of spot density for a given amount of protein is more difficult to determine as many more variables are present. Two estimates are available based upon autoradiographs produced by ^{35}S-labeled tissue culture cells and sophisticated computer-based scanning systems. Garrels (1979) has estimated that the error in scanning two autoradiographs from two gels of a single sample is 20%, 15% of which is due to the densitometry and 5% due to electrophoresis. Anderson *et al.* (1981) estimate that the average error in their system for a similar test was 8–10%, half of which they feel is electrophoretic error. These error estimates, however, are general in nature and contain no information about the reproducibility of spot density for a given protein over several runs.

H. Additional Techniques

There are several ancillary techniques that have proved useful in our laboratory and are reported here briefly.

1. PROTEIN CONCENTRATION

We frequently need to concentrate protein specimens. We have found that dialysis against a volatile buffer such as ammonium formate or ammonium carbonate, followed by lyophilization, concentrates proteins well in a manner suitable for 2-DGE. The buffer is removed during lyophilization and no significant residue that might perturb the ISO step remains. When desalting of a specimen is needed, as in the case of urine, high-pressure gel permeation chromatography using FRACTOGEL HW-40F (EM Reagents, Gibbstown, New Jersey) and 0.1 *M* ammonium formate, pH 6.5, is rapid and allows immediate concentration of protein by lyophilization (see Section IV,B).

2. PROTEIN ASSAY

We often determine the protein concentration of our samples after solubilization, just prior to 2-DGE. When using the dodecyl sulfate dissociation buffer of Anderson and Anderson (1977) this is easily done using the Coomassie Blue–Bradford assay (Bio-Rad, Richmond, California). We incorporate 5 μl of the dissociation buffer into the blank, the standards, and any controls; this allows us to measure up to 5 μl of sample. Amounts greater than this adversely affect the assay.

3. RADIOLABELING

As mentioned previously, the chloramine T method of iodination (Hunter and Greenwood, 1962) is routinely used in our laboratory and does not seem to affect the apparent molecular mass or pI of the proteins we have examined or induce any charge heterogeneity when gentle conditions are used. We have started using an improved form of this process in which the oxidizing agent is coupled to a solid phase, thereby limiting the detrimental effects and allowing for rapid and easy termination of the iodination (Iodo-Gen, Pierce Chemical Company, Rockford, Illinois).

4. ELECTROBLOTTING

Electroblotting or so-called Western blotting is the powerful "third-dimension" technique whereby the proteins separated in the two-dimensional gel are electrophoresed out of the gel and onto a solid phase, usually nitrocellulose, where they can be exposed to immunoglobulins and other detection molecules for specific identifications. This technique is addressed in detail in the chapter by Symington and therefore will be discussed only briefly here. We use the method of Anderson *et al.* (1982) with some modification. Proteins are electrophoresed onto nitrocellulose. The nitrocellulose sheet is incubated with a bovine serum albumin-containing buffer to block extra adsorption sites and then with a solution containing detection molecules, usually antibody. After excess detection solution is washed off the blot is incubated with a signal molecule, either peroxidase coupled second stage antibody or, generally, ^{125}I-labeled protein A, which binds to most imunoglobulins. We use three steps during each wash (i.e., after incubation with detection molecule and after incubation with signal molecule): phosphate buffered saline (PBS), PBS with 0.05% nonionic detergent (NP-40), and PBS again (T. Skully, personal communication). This significantly improves the quality of the final image. We have found immunoglobulin preparations to be superior to whole antisera.

III. Serum and Plasma

A. Specimen Procurement

Serum and plasma samples may be obtained by standard venipuncture techniques or even by "finger-stick" methods (Tracy *et al.*, 1982b). Serum is obtained by allowing blood to clot for 30 min. Various types of standard anticoagulants have been evaluated in the preparation of plasma (Tracy *et al.*, 1982b). These appear to yield equivalent results. The researcher should be aware of the many variables that can affect the concentration of macromolecules in serum or plasma such as tourniquet use, posture, and meals (Young, 1979). Although the effect of most of these variables has not been determined in detail, any abnormal results should be evaluated with them in mind, before assuming that abnormalities are disease related.

B. Sample Preparation

The initial publication of Anderson and Anderson (1977) illustrating the use of 2-DGE for the examination of plasma described the use of a dodecyl sulfate dissociation buffer (20 g/liter dodecyl sulfate, 50 ml/liter 2-mercaptoethanol, 100 ml/liter glycerol, 50 m*M* cyclohexylamine sulfonic acid). The authors have found this buffer to be useful for all specimens containing proteins soluble in aqueous medium and use it routinely for serum and plasma. The ratio of the amount of serum or plasma to dissociation buffer depends upon the visualization method ultimately to be used. For Coomassie Blue, 10 μl of serum or plasma are mixed with 30 μl of buffer; for silver staining 5 μl of specimen are mixed with 100 μl of buffer. In both cases the capped samples are heated to 95°C for 5 min, cooled, and briefly centrifuged to eliminate loss by evaporation and condensation on the upper walls of the tube. Following this, 10 μl (20 μl or more for visualizing proteins present in low relative abundance) are applied to the gels.

There are potentially many ways to pretreat plasma prior to electrophoresis to highlight certain aspects of the sample. The clotting process is the most widely used of these. We have explored the use of two of the pretreatments in an effort to remove major serum proteins, thereby clarifying the 2-DGE pattern: affinity chromatography of serum on Affi-Gel Blue (Bio-Rad, Richmond, California) and Sepharose-bound protein

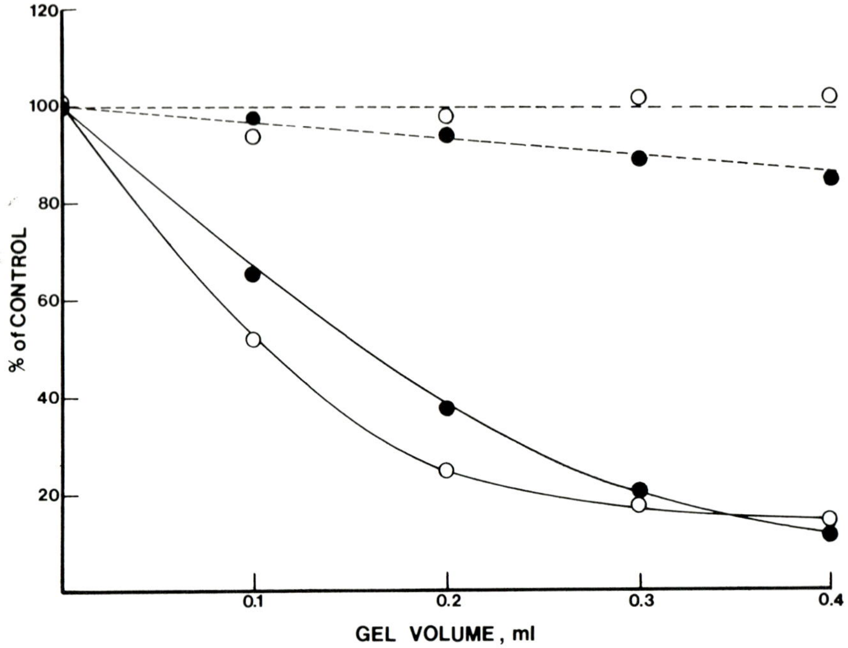

Figure 5. Removal of albumin and IgG by solid-phase adsorbants, Cibachron Blue-Sepharose (Affi-Gel Blue) and protein A-Sepharose. Cibachron Blue-Sepharose experiment: solid dots, solid line, albumin; solid dots, dotted line, IgG; protein A-Sepharose experiment: open dots, solid line, IgG; open dots, dotted line, albumin. A swollen gel, 0.25 ml, was used with 100 μl of serum. From Tracy *et al.* (1982b); reprinted by permission.

A. The first method mainly removes albumin when done under appropriate conditions and the second method mainly removes immunoglobulins (Tracy *et al.*, 1982b). The pattern of removal of these proteins is shown in Fig. 5. It should be noted that these methods are not entirely specific. The method of choice would make use of specific antibodies in high titer, most likely monoclonal in nature.

C. Reference Serum Protein Map

Figure 6 illustrates our current reference map of serum. Figure 7 is the schematic drawing of this gel with the numbering system described in Section II,F. Table VI is the key to the numbers and contains our current knowledge of the 2-DGE pattern. The schematic drawing also contains

Figure 6. Two-dimensional gel electrophoresis map of normal serum. Approximately 0.5 μl of serum have been applied to the ISO gel. The DALT gel was stained with silver nitrate. Compare to Fig. 7 for identification.

the pH gradient from the blank quality control gel, the location of hemoglobin β chain charge standards, and a molecular mass scale based upon the location of serum proteins.

Many serum proteins appear as *charge trains* due to glycosylation. The additional sialic acid residues cause the spots to extend upward to the left (N. L. Anderson and N. G. Anderson, 1979). In addition, heterogeneity is seen due to polymorphic forms as in the case of haptoglobin. Not all serum proteins stain with silver to the same degree but for α_1-antitrypsin at least (and probably for the others) densitometric analysis reveals reasonable linearity between total density and amount of protein in silver-stained gels (Tracy *et al.*, 1982b).

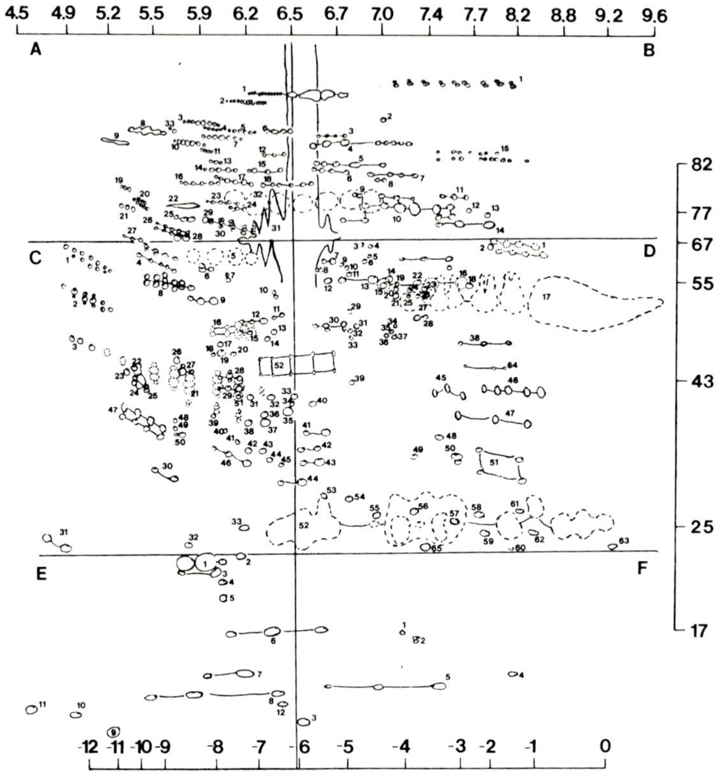

Figure 7. Reference schematic of human serum. This map should be compared to Fig. 6 and Table VI for identification of the spots. The number and letter system is explained in the text. Scales: top, pH gradient in a blank gel of the same run; side, molecular mass of several prominent proteins; bottom, location of hemoglobin charge standards; points 0 and −1 refer to alpha chain; −2 to −12 refer to beta chain.

D. Alterations in the Reference Pattern

There are many clinical settings in which the concentration of various serum proteins is known to be altered. Inflammation causes rapid increases in the acute phase reactant proteins, which include the complement components, plasminogen, α_1-antitrypsin, ceruloplasmin, and C-reactive protein among others (Whicher, 1981).

Tissue damage causes leakage of tissue proteins into the blood. One example is a rise in the serum concentration of creatine phosphokinase (CK) due to muscle trauma or myocardial infarction; Fig. 8 illustrates circulating CK in the serum of a patient with extensive muscle damage.

TABLE VI

List of Spots Observed in Serum and Plasma (Revision: 11/82)

Number	Name	Comments
A 1	α_2-Macroglobulin	*a,b*
3	Ceruloplasmin	*a,b*
10	Ceruloplasmin	*a,b*
14	Unknown plasma protein	Protein in freeze/thaw plasma
19	C_1S	*a*
22	Prothrombin	*b,k*
25	α_1B-glycoprotein	*a*
30	Hemopexin	*a*
31	Albumin	Two genetic variants; *a,b*
32	IgM heavy chain	*b,c*
33	α_1-antitrypsin dimer	*a*
B 1	IgG, polymers	*c*
5	C_3 activator	*a*
6, 7	Unknown plasma protein	Protein in serum, not plasma
10	Transferrin	Two genetic variants; *a*
15	Plasminogen	Glu form is larger, lys form smaller; *a*
C 1	α_1-antichymotrypsin	*a*
2	α_2HS-glycoprotein	Two genetic variants; *a*
5	IgA, heavy chains	*b,c*
6	Antithrombin III	*a,b*
8	α_1-antitrypsin	Three genetic variants; *a*
9	GC-globulin	Two genetic variants; *a*
16	Fibrinogen, γ chain	*a,d*
21	Haptoglobin, β chain	Three genetic forms of holohaptoglobin; *a,e*
23, 24, 26	Unknown plasma protein	Protein in freeze/thaw serum or plasma
27	Unknown plasma protein	Unknown genetic variant; two forms seen
30	Unknown plasma protein	Becomes a major urine protein; *f*
31	J-chain	*b,k*
46	Apolipoprotein E	Four forms seen; *a*
47	G4-glycoprotein	*a*
52	IgM, heavy chain frags.	*b*
D 1	Fibrinogen, α chain	*a,d*
16	Fibrinogen, β chain	*a,d*
17	IgG, heavy chains	*a,b,c*
45–47	IgG, heavy chain frags.	*g*
50	$C_4\ \gamma$	*a*
52	κ, λ light chains	*a,b,c*
53–58, 61, 63	Nonlight chains	*h*
59, 62	Nonlight chains	pH dependent; *h;i*
64	Creatine phosphokinase	Present in patients with muscle damage; *j*
65	Unknown plasma protein	Unknown nonlight chain present in patient with idiopathic urticaria

(*continued*)

TABLE VI (*continued*)

Number	Name	Comments
E 1	Apolipoprotein A-I	*a*
3	Retinol binding protein	*c*
6	Haptoglobin, α_2 chain	*a,e*
7	Prealbumin	*a,c*
8	Haptoglobin, α_1 chains	*a,e*
9	Apolipoprotein A-II	*a*
10, 11	LDL-related lipoproteins	*a*
4	Hemoglobin, β chain	*a*
5	β_2-microglobulin	Only seen in cases of renal transplant rejection; *b*

[a] Comparison to map in Anderson and Anderson (1977).

[b] Coelectrophoresis with a pure protein in our laboratory.

[c] Immunologic identification (precipitate/electrophoresis or electroblot) in our laboratory.

[d] Comparison of 2-DGE analysis of serum and plasma.

[e] 2-DGE analysis of ahaptoglobinemic serum.

[f] Comparison of 2-DGE analysis of serum and urine.

[g] These spots were adsorbed onto protein A, but did not react with antilight chain serum.

[h] These spots in the light chain region were not adsorbed onto protein A.

[i] These spots were present when samples were prepared at pH 6.5, but absent at 9.5.

[j] Comparison to map in Giometti *et al.* (1979).

[k] N. L. Anderson, personal communication.

Cancer also causes changes in serum 2-DGE patterns, but these changes appear to be nonspecific and restricted mainly to the immunoglobulin region (see Fig. 9).

The most extensively studied plasma protein disorder is multiple myeloma (and the other monoclonal gammopathies; Latner *et al.*, 1980; Thorsrud *et al.*, 1980; Kyle, 1980; Tracy *et al.*, 1982b). Two-dimensional gel electrophoretic examination of sera from patients with monoclonal gammopathies and electroblotting experiments using heavy chain-specific antisera have revealed that the heavy chains of immunoglobulins are highly restricted in terms of map location (Tracy *et al.*, 1982b). In fact, IgG subclasses exhibit characteristic electrophoretic behavior on 2-DGE, although the location of these proteins do overlap somewhat (Tracy *et al.*, 1982b). Figure 10 illustrates examples of IgG, IgM, IgA, and IgD myeloma sera. The monoclonal heavy chains always exhibit charge heterogeneity presumably due to glycosylation (Latner *et al.*, 1980). The immunoglobulin light chains are also restricted in terms of location but may not be classified by spot location with greater than 70% accuracy

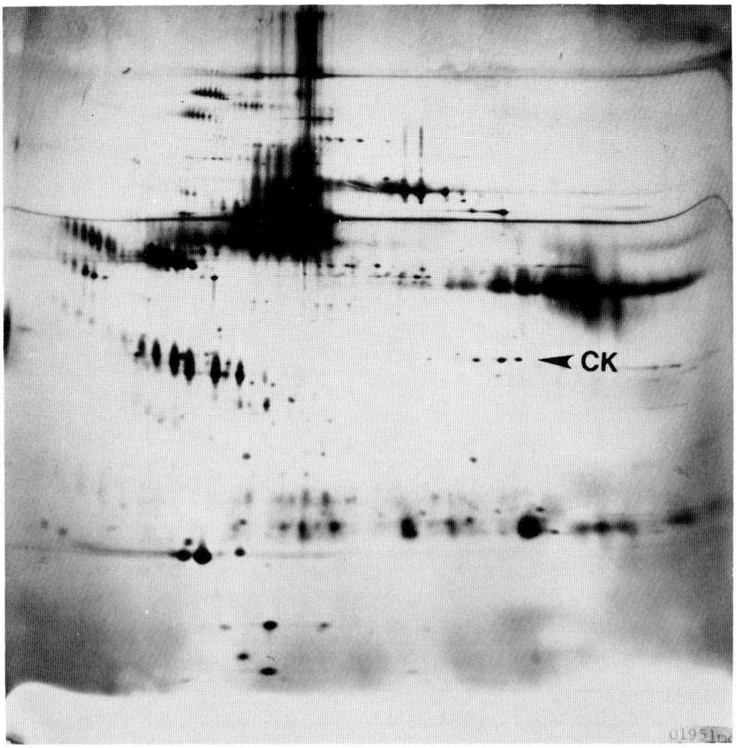

Figure 8. Two-dimensional gel electrophoresis map of serum from a patient with severe muscle damage. CK, creatine phosphokinase, identified by comparison to the map in Giometti *et al.* (1979).

because of location overlap of kappa and lambda light chains (Tracy *et al.*, 1982b).

There are many genetic variants detectable in serum with 2-DGE. These include, but are not limited to transferrin, α_2-HS-glycoprotein, and haptoglobin (Anderson and Anderson, 1977; N. L. Anderson and N. G. Anderson, 1979); α_1-antitrypsin, GC-globulin, and an unknown serum protein (Tracy *et al.*, 1982b); apolipoprotein E (Zannis *et al.*, 1981; Borresen and Berg, 1981); and albumin and β-globin (Wanner *et al.*, 1982). The most clinically significant of these is α_1-antitrypsin; the PiZZ-phenotype is associated with hepatic cirrhosis in children and pulmonary insufficiency in adults (Morse, 1978). Figure 11 depicts the phenotypic information concerning this protein obtainable from 2-DGE. It is clear that the normal phenotype PiM is clearly distinguishable from both the PiS form and the undesirable PiZ form. Currently, however, 2-DGE

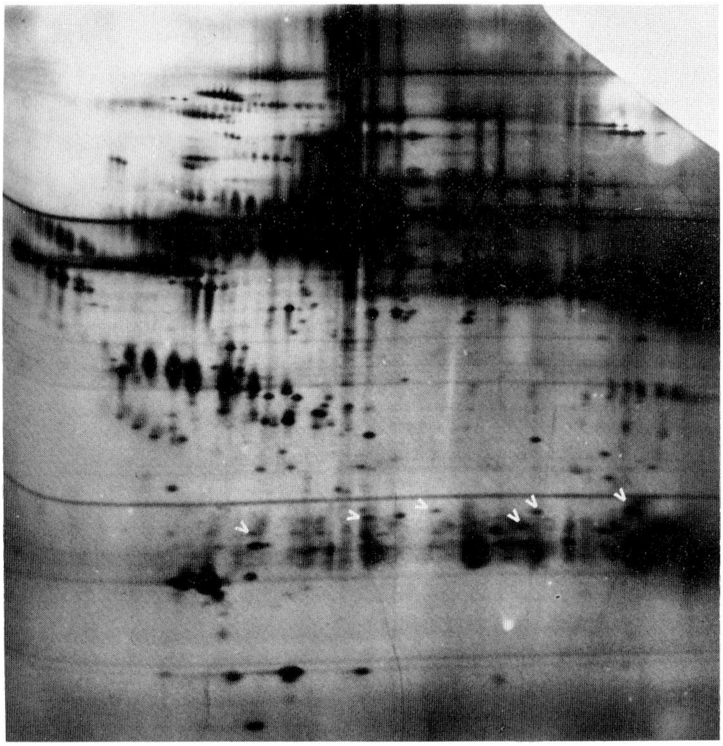

Figure 9. Two-dimensional gel electrophoresis map of serum from a patient with colorectal cancer. White arrows point to proteins, most likely "oligoclonal" light chains indicating increased immunologic activity. This pattern has been seen in many of these patients.

does not appear to distinguish the many various subforms that may be seen with high voltage, thin film analytical isoelectric focusing.

There are many clinical settings, some of which we have mentioned, where 2-DGE analysis of serum or plasma proteins may be a useful adjunct to more traditional methods. As mentioned before 2-DGE may be accepted as a method for measuring several serum proteins at once, should serum protein panels become accepted. For the near future, however, we feel that 2-DGE analysis of serum and plasma in the clinical laboratory will not become a routine procedure at least until the data base is expanded to a much greater degree and automated analysis systems become more readily available. There will, of course, be an ongoing role for 2-DGE analysis in clinical chemistry research.

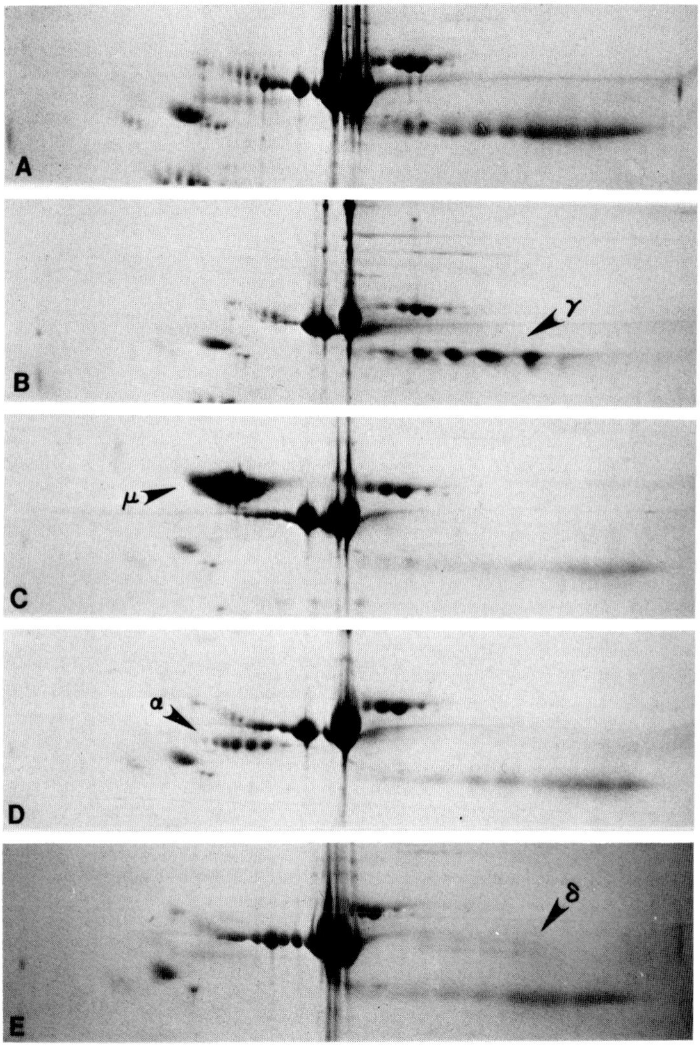

Figure 10. Heavy chain locations on two-dimensional gels. Sections through the heavy chain region of five gels are shown. (A) Normal serum; (B–E) myeloma sera containing γ, μ, α, and δ heavy chains, respectively. The large spots on the center of each gel are albumin. All gels are Coomassie Blue stained. From Tracy *et al.* (1982a).

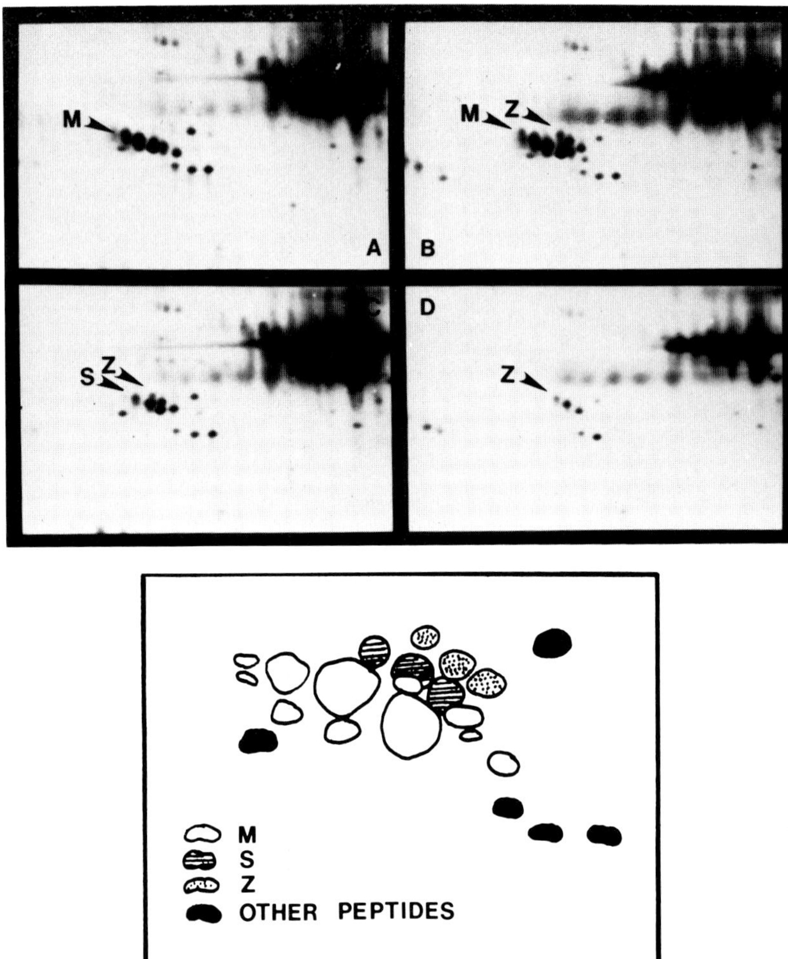

Figure 11. Genetic variation in α_1-antitrypsin. Phenotyping was done by nondenaturing isoelectric focusing. (A) Homozygous PiMM; (B) heterozygous PiMZ; (C) heterozygous PiSZ; (D) homozygous PiZZ. The drawing is a composite of forms identified so far. From Tracy *et al.* (1982b).

IV. Urine

A. Specimen Procurement

There are three major types of urine collection specimens: the first morning specimen, the random "spot" specimen, and the 24-h collection (Free and Free, 1975). The first two are generally obtained without preservatives and processed in the urinalysis laboratory within 2 h. The third type, the 24-h collection, may have one of several preservatives present such as glacial acetic acid. Alternatively a collection vessel may be kept in the refrigerator or on ice. The authors prefer the latter procedure for 24-h collections because we have not yet determined in detail the effect of the various preservatives on the proteins of urine.

There is also no information yet on the effect of delay between collection and analysis, storage temperature, etc. For the time being we advocate collection of a spot or first morning specimen using an appropriate midstream technique. The specimen is then processed or frozen at −70°C (as rapidly as possible, certainly within 3 h).

B. Sample Preparation

Urine may contain much particulate matter including cells, casts, and crystals. If turbidity persists when the urine has equilibrated to room temperature it must be eliminated before the urine is processed further. We have found that centrifugation for 10 min at 2000 × *g* is adequate for most samples.

Urine contains thousands of compounds, the vast majority of which are nonprotein in nature. The ISO dimension is particularly susceptible to the effect of salts so that simple concentration that would not dilute and might even concentrate the nonprotein constituents is unadvisable. Therefore the protein of urine, which amounts to only about 0.1 g/liter in healthy adults, must be separated from the majority of the nonprotein components either before or after the concentration but prior to 2-DGE. Two major methods discussed in detail by Andersons' group (Anderson *et al.*, 1979a,b; Edwards *et al.*, 1982b) utilize either dialysis followed by lyophilization or gel permeation chromatography followed by lyophilization. The most practical of these seems to be the latter technique. We have slightly modified this technique (Tracy *et al.*, 1983b) and offer the following as a routine procedure for sample preparation:

1. Apply 10 ml of urine to a 1.5 × 60 cm column containing Fracto-Gel HW40-F (EM Reagents, Gibbstown, New Jersey) equilibrated in 0.1 *M* ammonium formate, pH 6.5, pumped at approximately 120 psi (0.82 MPa).
2. Collect the protein peak, letting the remainder flow to waste (see Fig. 12).
3. Lyophilize the sample, then weigh it.

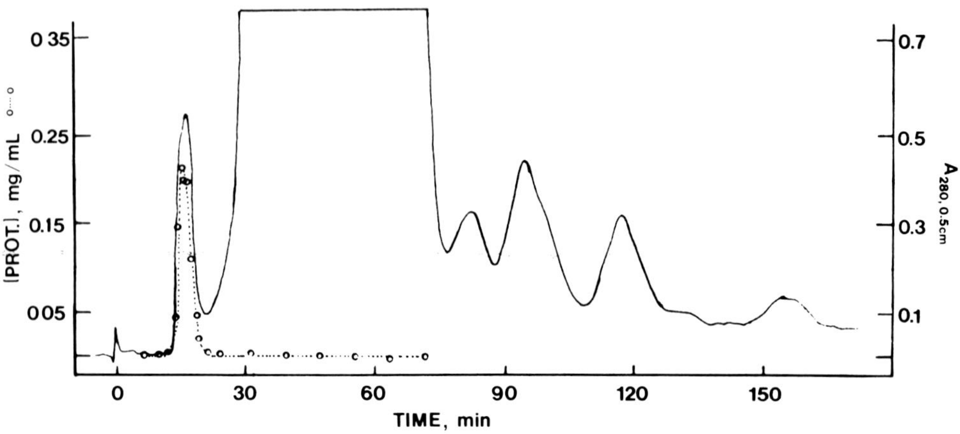

Figure 12. Fractionation of urine by HPLC gel permeation. Procedural details are given in the text. Solid line, absorbance at 280 nm; dotted line, protein concentration determined by Coomassie Blue dye-binding assay. The urine used in this experiment was graded as 1+ in our routine Urinalysis Laboratory.

4. This sample contains variable amounts of nonprotein components as a function of the grade of proteinuria and efficiency of chromatography. We routinely weigh 1 mg and dissolve it in 50 μl dodecyl sulfate dissociation buffer (see Section III,B), then analyze the sample for protein (see Section II, H,1). We attempt to apply 20–30 μg/gel for silver staining using a bovine serum albumin-based standard curve in the protein assay.

The specimens that cause most difficulties are those from normal, healthy subjects. In a single chromatography step under these conditions too much nonprotein material often remains in the protein peak even though 99% has been removed. Depending upon the chromatogram we will often redissolve the initial lyophilizate in 1 ml buffer and rechromatograph it on a 0.9 × 50 cm column. The protein peak of the smaller column elutes in approximately 10 min. The sample must then be relyophilized.

C. Reference Urine Map

Figure 13 illustrates our reference urine map, in this case from a 35-year-old female. We have noticed several differences between urine

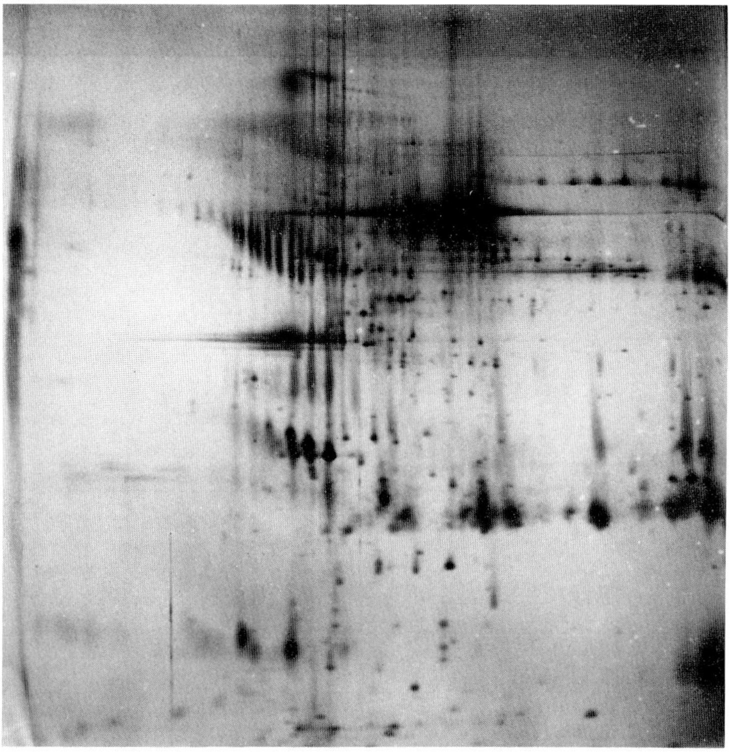

Figure 13. Two-dimensional gel electrophoresis map of normal human urine. This gel is oriented with acid to the left and represents approximately 20 μg of total urine protein. Compare to Fig. 14 for identifications. The pH gradient has been enhanced in the acid region, but much protein is still too acid to remain on the ISO gel.

from men and women, but we have not analyzed enough samples to know if these differences are sex-linked or, more probably, related to diet, exercise patterns, etc. Figure 14 depicts our reference urine schematic, established after examining approximately 20 normal urines, both male and female. The most common male and female proteins have been included. We have modified the ISO system in the acid region to illustrate the detail present there.

Certain spots on this schematic are keyed to the analogous spots in plasma; there may be other identical spots but we are taking a conservative approach to claiming identity. Anderson *et al.* (1979a) have discussed the possibilities of origin for a given urinary protein; in a slightly modified form these possibilities appear in Table VII. The spots in urine

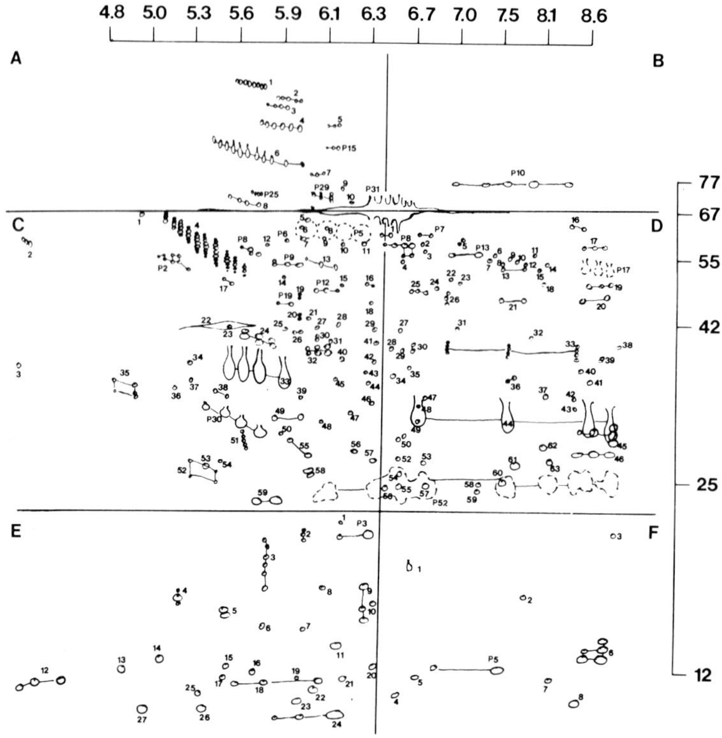

Figure 14. Reference schematic of human urine. This map should be compared to Figs. 6, 7, and 13 and Table VI for identification of spots. The number and letter system are explained in the text. Numbers preceded by "P" are known plasma proteins. Scales: top, pH gradients; side, molecular mass, based on known proteins present in the sample.

TABLE VII

Origin of Urinary Proteins[a]

Plasma Proteins

1. Major constituents: e.g., albumin
2. Minor constituents: e.g., β_2-microglobulin
3. Hormones: e.g., hCG in pregnancy urine
4. Tissue proteins: e.g., myoglobin in muscle damage
5. Tumor-associated antigens: e.g., CEA
6. Products of infection

Proteins of the genito-urinary tract

1. Kidney: e.g., *N*-acetyl-glucosaminidase in renal tubular disease
2. Other parts of urinary tract
3. Male accessory glands, vaginal secretions
4. Genito-urinary tract tumor-associated antigens: e.g., prostatic acid phosphatase
5. Products of genito-urinary tract infection

---

[a] Modified from N. G. Anderson *et al.* (1979b).

not keyed to the plasma map must originate from one of the alternative categories. The authors are currently engaged in a project to evaluate the exact nature of many of the proteins in urine by immunologic methods. Other maps of urine based on Coomassie Blue staining have been published (Anderson *et al.*, 1979a; Edwards *et al.*, 1982b).

The normal urinary pattern appears quite reproducible, at least qualitatively, although the number of specimens we have examined is small. Nonetheless we feel that there is sufficient evidence to warrant the large scale analysis of urine, especially because it is the easiest human specimen to obtain and therefore the prime candidate for any screening procedure.

D. Alterations in the Reference Pattern

The most obvious clinical settings to investigate concerning 2-DGE of urine are those that involve renal disease. In this regard we have investigated renal transplant rejection (R. Tracy, unpublished). We have used 2-DGE to assess the relative abundance of large and small molecular mass serum proteins and have noted striking differences during rejection (Tracy *et al.*, 1982f) (Fig. 15). We are currently developing immune reagents to the appropriate proteins to analyze the quantitative aspects of this phenomenon in more detail.

Multiple myeloma and the other monoclonal gammopathies cause large changes in urine protein excretion in addition to their effects on serum proteins. Most striking are the Bence Jones proteins, which often accompany these diseases. These are detected in the immunological light chain region but are easily studied by other means (Kyle, 1980). The 2-DGE technique may pick up more subtle effects but they are yet to be delineated.

Both glomerular and tubular diseases cause striking alterations in the urine pattern but again the exact meaning of these changes needs to be evaluated in the context of a large study.

A potential marker for prostatic carcinoma has been detected by 2-DGE (Edwards *et al.*, 1982a) as has a potential marker for muscular dystrophy (Frearson *et al.*, 1981). Both of these potential markers, although highlighting the ability of 2-DGE to compare complex protein mixtures, must be rigorously evaluated before acceptance.

Finally, urine from subjects with rheumatoid arthritis has been studied and potential markers have been seen, but again the study was small and needs to be expanded (Clark *et al.*, 1980).

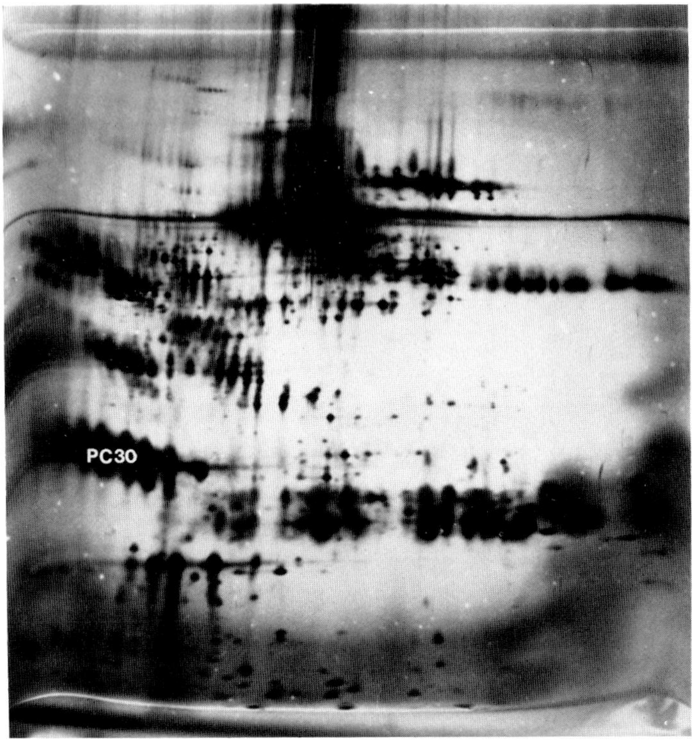

Figure 15. Two-dimensional gel electrophoresis of urine from a patient undergoing renal transplant rejection. Approximately 40 μg of protein were applied to the gel. Orientation is the same as in Fig. 6. Many normal serum proteins may be seen (compare to Figs. 6, 7); protein PC-30 is present in especially high relative abundance.

V. Blood Cells

A. Erythrocytes

Erythrocytes have been extensively studied by two-dimensional electrophoretic techniques (for review see Tracy and Anderson, 1983) but not very often by ISO–DALT or similar methods. Most procedures are based on two-dimensional dodecyl sulfate electrophoresis, especially those used to analyze "ghosts".

The initial 2-DGE study of erythrocyte membranes was done by Bhakdi *et al.* (1975). The resolution they obtained was not impressive, however. Bienzle and Pjura (1977), using this same system, examined Triton-X-100 extracts of fresh and aging erythrocytes. They noted that a 75 kdalton protein disappeared within a few hours and several other major proteins exhibited quantitative changes.

Rubin and Milikowski (1978) published the first high resolution gel maps of erythrocyte membranes and used their method to compare normal and sickle membranes. They failed to demonstrate any unique proteins in either preparation, but they did demonstrate differences in the cross-linking patterns (Rubin *et al.*, 1980).

Rosenblum *et al.* (1982) have published high resolution maps of erythrocyte membranes using a different solubilization solution from that of Rubin and using silver stain techniques. Their procedure eliminates spectrin from the gels (a major cause of streaking) but causes many other changes as well. Kasten *et al.* (1982) in our laboratory have analyzed membranes prepared by both methods (see Fig. 16) and documented the differences.

Edwards *et al.* (1979) have studied red cell lysates in some detail. Many of the major enzymes have been identified. A standard Coomassie Blue-stained map has proved useful in evaluating the effects of hemolysis on serum and urine (Tracy *et al.*, 1982b).

B. Lymphocytes

The Andersons' group has extensively studied human peripheral lymphocytes (Anderson *et al.*, 1981; Gemmell and Anderson, 1982; Willard, 1982). Many subsets of lymphocyte proteins have been identified that are susceptible to various perturbants such as interferon, phorbol esters, and urine from patients with renal transplant rejection (Anderson, 1981) or are connected with subcellular fractions such as the mitochondrion and the cytoskeleton (Anderson, 1981). Markers for subset populations have been identified (Gemmell and Anderson, 1982) and have been used recently in a pilot study to confirm the identification of leukemias made by monoclonal antibody histochemical techniques (N. L. Anderson, *et al.*, 1983). During the course of their work an excellent map of human peripheral lymphocytes has been published (Fig. 17). A preliminary report has appeared from Willard (1982) indicating that a marker protein has been detected for the early detection and diagnosis of infectious mononucleosis.

Lester *et al.* (1981, 1982), Goldman *et al.* (1982), and Hanash *et al.* (1982) have also studied the protein expression of lymphocytes.

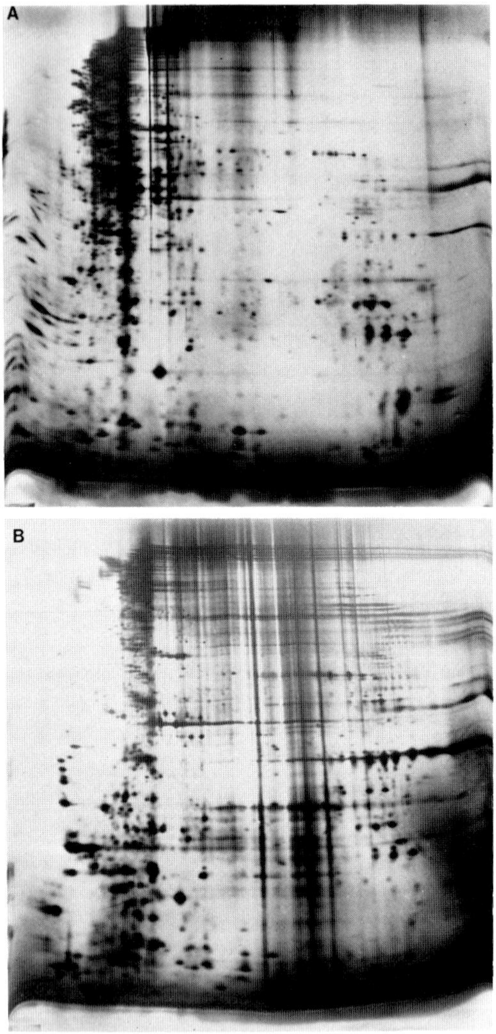

Figure 16. Two-dimensional gel electrophoresis maps of human red cell membranes. Red cell "ghosts" were prepared and solubilized as described in Kasten *et al.* (1982). (A) Prepared to exclude spectrin; (B) prepared to include spectrin. The crosshatched pattern in the upper right of (B) is produced by spectrin. Gels were stained with silver nitrate.

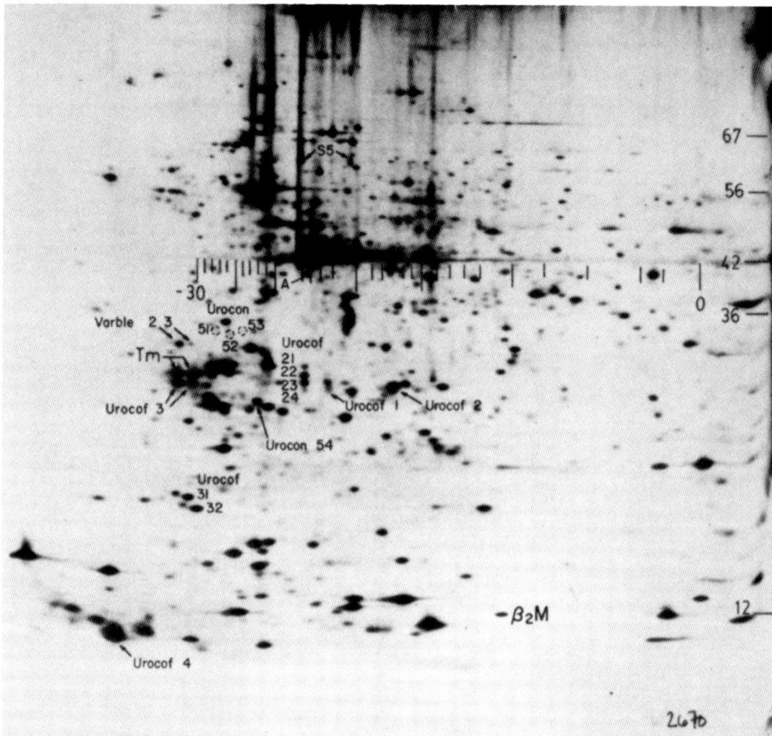

Figure 17. Two-dimensional gel electrophoresis map of human peripheral lymphocytes labeled with [^{35}S]methionine in overnight culture. An autoradiograph was made from dried gel, which is oriented with acid side to the left. Scales: middle, CK-charge standards; side, molecular mass. A, actin, β_2M, β_2-microglobulin. The other names refer to the TYCHO data base system used at Argonne National Laboratory. From Willard and Anderson (1981).

C. Platelets

Platelets have been extensively studied by several groups. Clemetson, McGregor and co-workers (Clemetson *et al.*, 1979; McGregor *et al.*, 1980) have proposed a revised nomenclature for platelet membrane proteins based upon reproducible, complex patterns achieved with 2-DGE. A new nomenclature has also been proposed by Sixma and Schiphorst (1980).

Clemetson *et al.* (1980) and Holahan and White (1981) have compared platelets from normals with those from patients with Glanzmann's thrombasthenia, and have noted several consistent changes. The data gathered so far seem to indicate that the disease is heterogeneous, the various forms possibly resulting from genetic variability of the patient.

VI. Solid Tissue

A. Sample Preparation

Various methods have been advocated for the analysis of solid tissue. We have found the method outlined below to be effective for most samples we have analyzed.

1. Frozen tissue is broken into very small pieces by mechanical smashing (without thawing).
2. Tissue, 150 mg, is homogenized in a Polytron homogenizer (4 bursts, 16 sec/burst) with 600 μl of 8 *M* urea, 50 ml/liter 2-mercapto-ethanol, 50 ml/liter 3.5–10 ampholytes, 2m*M* phenylmethyl-sulfonyl fluoride (PMSF), 2m*M* benzamidine, 20 μM pepstatin; adjusted to pH 9.5 with 1 *M* sodium hydroxide.
3. Most of the sample is frozen at −70°C; 100 μl is mixed with 10 μl 100 mg/liter dodecyl sulfate, 33 μl of a solution containing 8 *M* urea, 8% NP-40; then adjusted to pH 9.5.
4. Sample is centrifuged for 3 min at 8000 $\times$ *g* (in a microfuge).
5. Apply from 10 to 20 μl to the gel.

This method (Tracy *et al.*, 1982c) is a modification of that of Wilson *et al.* (1977). We have found that many tissues contain proteases that can be inactivated by the protease inhibitors and/or the alkaline pH (L. Anderson, personal communication).

B. The Need for Histological Analysis

Specimens of most solid tissues must be prepared very carefully by a surgical pathologist prior to analysis (muscle is an exception because of the homogeneity of large pieces). It is critical that fatty tissue, ligaments, and necrotic tissue be trimmed away and that the homogeneity of the piece of tissue is confirmed by frozen section analysis (Tracy *et al.*, 1982c).

C. Colon and Colorectal Cancer

Tracy *et al.* (1982c) and Thorsrud *et al.* (1982) have compared normal colonic mucosa and colon adenocarcinoma by 2-DGE; the first group, using silver stain and the dissociation scheme shown above, detected very little difference between normal tissue and tumor. However, 5 proteins out of approximately 1000 seem to be present in the tumor but not in the normal. There were many quantitative changes. Figure 18 illustrates the differences seen.

Thorsrud *et al.* (1982), using Coomassie Blue staining and a simpler urea–NP-40 dissociation, detected many more differences. Because the authors have seen incomplete dissociation of some tissues in a preparation of this nature, the meaning of the differences is not clear.

D. Kidney

We have reported (Tracy *et al.*, 1982e) on a comparison of normal renal cortex (approximately 70% tubular epithelial cells) and renal cell carci-

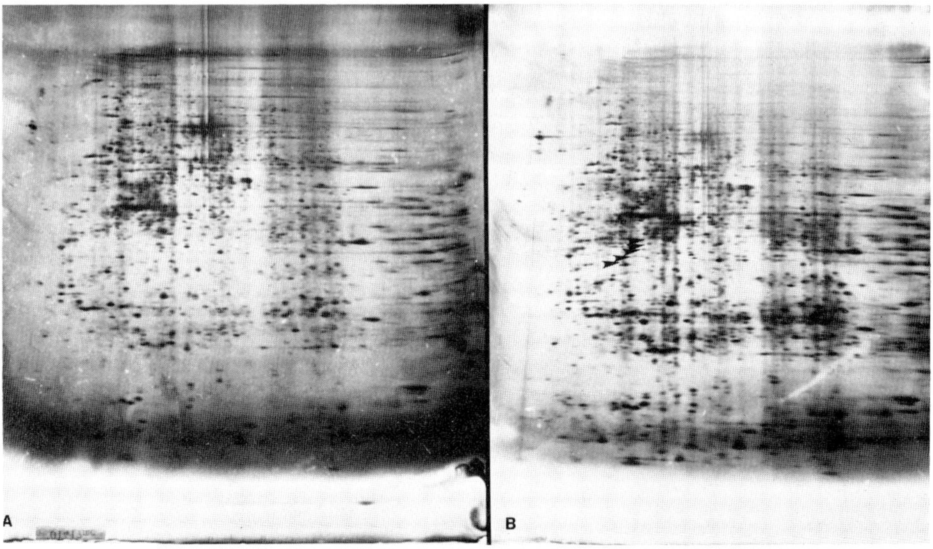

Figure 18. Two-dimensional gel electrophoresis maps of human colon mucosa and colon adenocarcinoma. (A) Normal mucosa; (B) adenocarcinoma from the same patient. Arrows in B point out apparantly tumor-specific proteins.

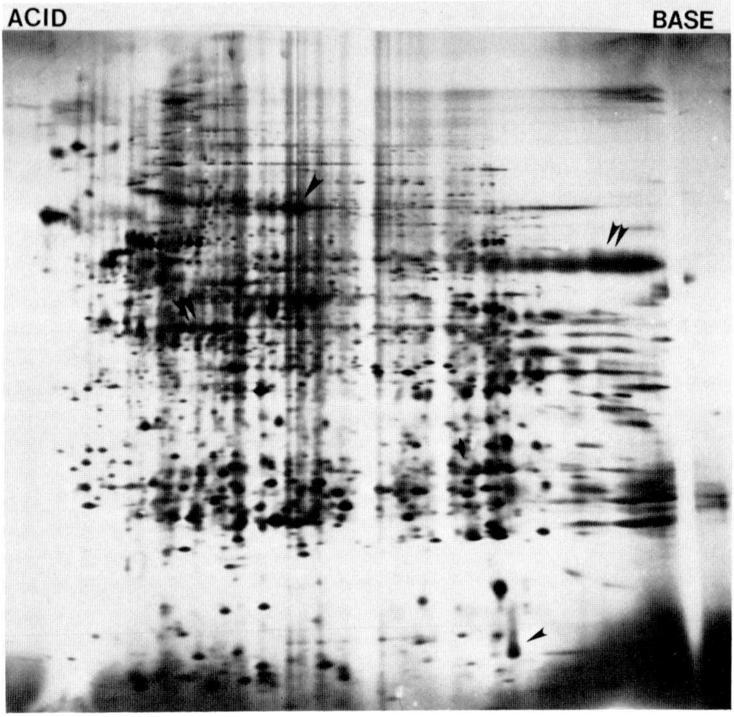

Figure 19. Two-dimensional gel electrophoresis map of normal human renal cortex. Arrows: large, in center, albumin; double large, actin; medium, lower right, hemoglobin β chain; double medium, IgG heavy chains. Sample was prepared as described for tissue; the gel was stained with silver nitrate.

noma (approximately 90% transformed tumor cells). As was the situation with colon, most of the proteins are identical although there are again many changes in relative abundance, several of which are striking. A major normal protein is absent from most tumor tissues; this protein has been isolated by cutting out spots from preparative gels and we are currently attempting to develop monoclonal immune reagents. Figure 19 illustrates a typical pattern of normal renal cortex.

The kidney has also been studied by Smith *et al.* (1980) although the differences in their technique and staining make comparisons difficult.

E. Brain

The human brain has been extensively studied by Comings and coworkers (Comings, 1982a,b,c; Comings *et al.*, 1982; Comings and Pekkula-Flagan, 1982). They have examined nearly 200 cadaver brains and detected 11 putative mutations from the 300 spots/gel that they analyzed. These mutations include a polymorphism of glial fibrillary protein associated with many neurological disorders, a mutation of myelin basic protein in one patient with schizophrenia, and a possible mutation of protein 17:1 in Joseph's disease.

Jackson and co-workers examined thirteen human tissues including brain and determined eight proteins that are apparently brain-specific (Jackson and Thompson, 1981; Jackson *et al.*, 1982). Four have been identified as 14-3-2 protein, aldolase C_4, creatine kinase BB, and 14-3-3 protein. The others may be potential markers for neurological disease.

F. *Muscle*

Giometti *et al.* (1979, 1980) have studied human skeletal muscle in detail. These workers have compared specimens from patients with adult-onset dematomyositis to normal specimens and detected differences in the myosin light chain and tropomyosin patterns (Danon *et al.*, 1981). They have shown that single histological sections can be used for analyses of this kind (Giometti and Anderson, 1981).

Whalen and Ecob (1982) have examined the contractile proteins of muscle from normal individuals and patients with idiopathic scoliosis. In several patients they identified alterations in the myosin light chain patterns on the different sides of the scoliotic region of the spine.

VII. Other Specimens

A. *Semen*

Edwards *et al.* (1981) have analyzed human seminal plasma by 2-DGE. Their map illustrates 200 proteins, some of which have been characterized by concanavalin A binding or absence from semen of vasectomized subjects. A number of basic proteins are rapidly proteolysed during the liquefaction process.

B. *Pancreatic Fluid*

Scheele, who has published extensively on the 2-DGE separation of pancreatic fluid, was one of the first investigators to use isoelectric focusing and dodecyl sulfate in combination (Scheele, 1975, 1980, 1981). He has carefully mapped the major exocrine enzymes (Scheele, 1980) and has determined that pancreatic fluid from a patient with pancreatic cancer contains several spots not seen in normals (Scheele, 1981).

C. Spinal Fluid

Goldman *et al.* (1980) have studied human spinal fluid using sensitive silver stain and established the identity of several proteins apparently missing in plasma. Tracy *et al.* (1982e) have also examined spinal fluid from normal individuals and from patients with multiple sclerosis. The authors have demonstrated several new spots in the light chain region from specimens that had so-called oligoclonal banding upon Coomassie Blue staining of agarose gels. We feel that 2-DGE may be a fruitful method for the examination of spinal fluid from patients with demyelinating disease.

D. Amniotic Fluid

Jones *et al.* (1981) have analyzed 48 specimens of amniotic fluid by Coomassie Blue-stained 2-DGE. The majority of the proteins present are serum proteins although there appear to be several proteins specific to the amniotic fluid. The origin of these proteins is as yet unknown.

E. Saliva

One report on the 2-DGE analysis of human saliva has appeared in the literature (Giometti and Anderson, 1980). Approximately 100 protein spots were seen, most of the protein present being amylase. Several of these proteins have been identified as originating in the parotid gland.

VIII. Conclusions

High resolution, two-dimensional gel electrophoresis has been developed to the point at which it holds great promise for the clinical laboratory. Current techniques are characterized by a high degree of reproducibility, throughput, and sensitivity, as well as, in some cases, quantitative information. The techniques may be used effectively in the clinical laboratory with appropriate collaboration and have already been instrumental in elucidating differences between some normal and diseased specimens that may prove useful either in diagnosis or in our understanding of underlying disease processes. It is not yet certain that 2-DGE will ever be utilized on a routine basis in the clinical laboratory, but it clearly has already achieved some value as a research and development tool and will continue to be used in this manner for some time.

References

Anderson, N. G., and Anderson, N. L. (1978). Analytical techniques for cell fractions. XXI. Two-dimensional analysis of serum and tissue proteins: multiple isoelectric focusing. *Anal. Biochem.* **85**, 331–340.

Anderson, N. G., and Anderson, N. L. (1979). Molecular anatomy. *Behring Inst. Mitt.* **63**, 169–210.

Anderson, N. G., and Anderson, N. L. (1982). The human protein index project and the molecular pathology data base. *Med. Lab.* **11**, 75–94.

Anderson, N. G., Anderson, N. L., Tollaksen, S. L., Hahn, H., Giere, F., and Edwards, J. (1979a). Analytical techniques for cell fractions. XXV. Concentration and two-dimensional electrophoresis of human urinary proteins. *Anal. Biochem.* **95**, 48–61.

Anderson, N. G., Anderson, N. L., and Tollaksen, S. L. (1979b). Proteins of human urine. I. Concentration and analysis by two-dimensional electrophoresis. *Clin. Chem.* **25**, 1199–1210.

Anderson, N. L. (1981). Studies of gene expression in human lymphocytes using high-resolution two-dimensional electrophoresis. *In* "Electrophoresis '81" (R. Allen and P. Arnaud, eds.), pp. 309–316. de Gruyter, New York.

Anderson, N. L., and Anderson, N. G. (1977). High resolution two-dimensional electrophoresis of human plasma proteins. *Proc. Natl. Acad. Sci. USA* **74**, 5421–5425.

Anderson, N. L., and Anderson, N. G. (1978). Analytical techniques for cell fractions. XXII. Two-dimensional analysis of serum and tissue proteins: multiple gradient-slab electrophoresis. *Anal. Biochem.* **85**, 341–354.

Anderson, N. L., and Anderson, N. G. (1979). Microheterogeneity of serum transferrin, haptoglobin and α_2HS-glycoprotein examined by high resolution two-dimensional electrophoresis. *Biochem. Biophys. Res. Commun.* **88**, 258–265.

Anderson, N. L., Taylor, J., Scandora, A. E., Coulter, B. P., and Anderson, N. G. (1981). The TYCHO system for computerized analyses of two-dimensional gel protein mapping data. *Clin. Chem.* **27**, 1807–1820.

Anderson, N. L., Nance, S., Pearson, T., and Anderson, N. G. (1982). Specific antiserum staining of two-dimensional electrophoretic patterns of human plasma proteins immobilized on nitrocellulose. *Electrophoresis* **3**, 135–142.

Anderson, N. L., Wiltsie, J. J., Li, C. Y., Willard-Gallo, K. E., Tracy, R. P., Young, D. S.,

Powers, M. T., and Anderson, N. G. (1983). Analysis of human Leukemic cells using two-dimensional gel electrophoresis. I. Results of a pilot study. *Clin. Chem.* **29**, 762–767.
Aycock, B. F., Weil, D. E., Sinicropi, D. V., and McIlwain, D. L. (1981). Television-based densitometric analysis of proteins separated by two-dimensional gel electrophoresis. *Comput. Biomed. Res.* **14**, 314–326.
Bhakdi, S., Knüfermann, H., and Wallach, D. F. H. (1975). Two-dimensional separation of erythrocyte membrane proteins. *Biochem. Biophys. Acta* **394**, 550–557.
Bienzle, U., and Pjura, W. J. (1977). Alteration of membrane proteins during erythrocyte aging. *Clin. Chim. Acta* **76**, 183–186.
Bonner, W. M., and Laskey, R. A. (1974). A film detection method for tritium-labelled proteins and nucleic acids in polyacrylamide gels. *Eur. J. Biochem.* **46**, 83–88.
Borresen, A.-L., and Berg, K. (1981). The apo E polymorphism studied by two-dimensional, high-resolution gel electrophoresis of serum. *Clin. Genet.* **20**, 438–448.
Clark, P. M. S., Kricka, L. J., and Whitehead, T. P. (1980). Pattern of urinary proteins and peptides in patients with rheumatoid arthritis investigated with the ISO–DALT technique. *Clin. Chem.* **26**, 201–204.
Clemetson, K. J., Capitanio, A., and Lüscher, E. F. (1979). High resolution two-dimensional gel electrophoresis of the proteins and glycoproteins of human blood platelets and platelet membranes. *Biochem. Biophys. Acta* **553**, 11–24.
Clemetson, K. J., Capitanio, A., Pareti, F. I., McGregor, J. L., and Lüscher, E. F. (1980). Additional platelet membrane glycoprotein abnormalities in Glanzmann's Thrombasthenia: A comparison with normals by high resolution two-dimensional polyacrylamide gel electrophoresis. *Thromb. Res.* **18**, 797–806.
Comings, D. E. (1982a). Two-dimensional gel electrophoresis of human brain proteins. I. Techniques and nomenclature of proteins. *Clin. Chem.* **28**, 782–789.
Comings, D. E. (1982b). Two-dimensional gel electrophoresis of human brain proteins. III. Genetic and non-genetic variations in 145 brains. *Clin. Chem.* **28**, 798–804.
Comings, D. E. (1982c). Two-dimensional gel electrophoresis of human brain proteins. IV. Disorders of glial proliferation and a polymorphism of glial fibrillary acid protein–GFAP–Duarte. *Clin. Chem.* **28**, 805–812.
Comings, D. E., and Pekkula-Flagan, A. (1982). Two-dimensional gel electrophoresis of human brain protein. V. Non-equilibrium gel electrophoresis, with detection of a myelin basic protein mutation–MBL–Duarte. *Clin. Chem.* **28**, 813–818.
Comings, D. E., Carraway, N. G., and Pekkula-Flagan, A. (1982). Two-dimensional gel electrophoresis of human brain proteins. II. Specific proteins and brain subfractions. *Clin. Chem.* **28**, 790–797.
Danon, M. J., Giometti, C. S., Manaligod, J. R., Perurena, O. H., and Skosey, J. L. (1981). Adult-onset nemaline rods in a patient treated for suspected dermatomyositis. *Arch. Neurol. (Chicago)* **38**, 761–766.
Edwards, J. J., Anderson, N. G., Nance, S. L., and Anderson, N. L. (1979). Red cell proteins. I. Two-dimensional mapping of human erythrocyte lysate proteins. *Blood* **53**, 1121–1132.
Edwards, J. J., Tollaksen, S. L., and Anderson, N. G. (1981). Proteins of human semen. I. Two-dimensional mapping of human seminal fluid. *Clin. Chem.* **27**, 1335–1340.
Edwards, J. J., Anderson, N. G., Tollaksen, S. L., von Eschenbach, A. C., and Guevara, J. (1982a). Proteins of human urine. II. Identification by two-dimensional electrophoresis of a new candidate marker for prostatic cancer. *Clin. Chem.* **28**, 160–163.
Edwards, J. J., Tollaksen, S. L., and Anderson, N. G. (1982b). Proteins of human urine. III. Identification and two-dimensional electrophoretic map positions of some major urinary proteins. *Clin. Chem.* **28**, 941–948.

Frearson, N., Taylor, R. D., and Perry, S. V. (1981). A protein in urine associated with muscle disease and muscle damage. *Br. Med. J.* **282**, 2002–2003.

Free, A. H., and Free, H. M. (1975). "Urinalysis in Clinical Laboratory Practice." CRC Press, West Palm Beach, Florida.

Garrels, J. I. (1979). Two-dimensional electrophoresis and computer analysis of proteins synthesized by clonal cell lines. *J. Biol. Chem.* **254**, 7961–7977.

Gemmell, M. A., and Anderson, N. L. (1982). Lymphocyte, monocyte and granulocyte proteins compared by use of two-dimensional electrophoresis. *Clin. Chem.* **28**, 1062–1066.

Giometti, C. S., and Anderson, N. G. (1980). Two-dimensional electrophoresis of human saliva. *In* "Electrophoresis '79" (B. Radola, ed.), pp. 395–404. de Gruyter, New York.

Giometti, C. S., and Anderson, N. G. (1981). Muscle protein analysis. III. Analysis of solubilized frozen-tissue sections by two-dimensional electrophoresis. *Clin. Chem.* **27**, 1918–1921.

Giometti, C. S., Anderson, N. G., and Anderson, N. L. (1979). Muscle protein analysis. I. High-resolution two-dimensional electrophoresis of skeletal muscle proteins for analysis of small biopsy samples. *Clin. Chem.* **25**, 1877–1884.

Giometti, C. S., Bárány, M., Danon, M. J., and Anderson, N. G. (1980). Muscle protein analysis. II. Two-dimensional electrophoresis of normal and diseased human skeletal muscle. *Clin. Chem.* **26**, 1152–1155.

Goldman, D., Merril, C. R., and Ebert, M. H. (1980). Two-dimensional gel electrophoresis of cerebrospinal fluid proteins. *Clin. Chem.* **26**, 1317–1322.

Goldman, D., Merril, C. R., Polinsky, R. J., and Ebert, M. H. (1982). Lymphocyte proteins in Huntington's disease: Quantitative analysis by use of two-dimensional electrophoresis and computerized densitometry. *Clin. Chem.* **28**, 1021–1025.

Hanash, S. M., Tubergen, D. G., Heyn, R. M., Neel, J. V., Sandy, L., Stevens, G. S., Rosenblum, B. B., and Krzesicki, R. F. (1982). Two-dimensional gel electrophoresis of cell proteins in childhood leukemia, with silver staining: A preliminary report. *Clin. Chem.* **28**, 1026–1030.

Holahan, J. R., and White, G. C. (1981). Heterogeneity of membrane surface proteins in Glanzmann's Thrombasthenia. *Blood* **57**, 174–187.

Hunter, W. M., and Greenwood, F. C. (1962). Preparation of ^{131}I-labeled human growth hormone of high specific activity. *Nature (London)* **194**, 495–496.

Jackson, P., and Thompson, R. J. (1981). The demonstration of new human brain-specific proteins by high-resolution two-dimensional polyacrylamide gel electrophoresis. *J. Neurol. Sci.* **49**, 429–438.

Jackson, P., Thompson, V. M., and Thompson, R. J. (1982). Demonstration of basic human-brain-specific proteins by the BASO-DALT system. *Clin. Chem.* **28**, 920–924.

Jones, M. I., Massingham, W. E., and Spragg, S. P. (1980). Electrophoresis tank for running multiple slab gels. *Anal. Biochem.* **106**, 44–449.

Jones, M. I., Spragg, S. P., and Webb, T. (1981). Detection of proteins in human amniotic fluid using two-dimensional gel electrophoresis. *Biol. Neonate* **39**, 171–177.

Kasten, W. R., Tracy, R. P., Currie, R. M., and Young, D. S. (1982). Two-dimensional gel electrophoresis of red cell membranes. *Clin. Chem.* **28**, 1628 (Abstr.).

Kyle, R. (1980). Classification and diagnosis of monoclonal gammopathies. *In* "Manual of Clinical Immunology" (N. R. Rose and H. Friedman, eds.), pp. 135–150. Amer. Soc. Microbiol., Washington, D.C.

Laskey, R. A., and Mills, A. D. (1977). Enhanced autoradiographic detection of ^{32}P and ^{125}I using intensifying screens and hypersensitized film. *FEBS Lett.* **82**, 314–316.

Latner, A. L., Marshall, T., and Gambi, M. (1980). Microheterogeneity of serum im-

munoglobulins revealed by a technique of high resolution two-dimensional electrophoresis. *Electrophoresis* **1**, 82–89.
Lester, E. P., Lemkin, P., Lipkin, L., and Cooper, H. L. (1981). A two-dimensional electrophoretic analysis of protein synthesis in resting and growing lymphocytes in culture. *J. Immunol.* **126**, 1428–1434.
Lester, E. P., Lemkin, P., and Lipkin, L. (1982). A two-dimensional gel analysis of autologous T and B lymphoblastoid cell lines. *Clin. Chem.* **28**, 828–839.
Mariash, C. N., Seelig, S., and Oppenheimer, J. H. (1982). A rapid, inexpensive, quantitative technique for the analysis of two-dimensional electrophoretograms. *Anal. Biochem.* **171**, 388–394.
McGregor, J. L., Clemetson, K. J., James, E., Lüscher, E. F., and Dechavanne, M. (1980). Characterization of human blood platelet membrane proteins and glycoproteins by their isoelectric point (pI) and apparent molecular weight using two-dimensional electrophoresis and surface-labelling techniques. *Biochim. Biophys. Acta* **599**, 473–483.
Morse, J. O. (1978). Alpha$_1$-antitrypsin deficiency. *New Engl. J. Med.* **299**, 1045–1048.
Oakley, B. R., Kirsch, D. R., and Morris, N. R. (1980). A simplified ultrasensitive silver stain for detecting proteins in polyacrylamide gels. *Anal. Biochem.* **105**, 361–363.
Peterson, J., and McConkey, E. H. (1978). Non-histone chromosomal proteins from HeLa cells. *J. Biol. Chem.* **251**, 548–554.
Rosenblum, B. B., Hanash, S. M., Yew, N., and Neel, J. V. (1982). Two-dimensional electrophoretic analysis of erythrocyte membranes. *Clin. Chem.* **28**, 925–931.
Rubin, R. W., and Milikowski, C. (1978). Over two hundred polypeptides resolved from the human erythrocyte membrane. *Biochim. Biophys. Acta* **509**, 100–110.
Rubin, R. W., Milikowski, C., and Wise, G. E. (1980). Organizational differences in the membrane proteins of normal and irreversibly sickled erythrocytes. *Biochim. Biophys. Acta* **595**, 1–8.
Sammons, D., Adams, L. D., and Nishizawa, E. E. (1981). Ultrasensitive silver-based color staining of polypeptides of polyacrylamide gels. *Electrophoresis* **2**, 135–141.
Scheele, G. A. (1975). Two-dimensional gel analysis of soluble proteins. *J. Biol. Chem.* **250**, 5375–5385.
Scheele, G. A. (1980). Biosynthesis, segregation and secretion of exportable proteins by the exocrine pancreas. *Am. J. Physiol.* **238**, G467–G477.
Scheele, G. A. (1981). Human pancreatic cancer: Analysis of proteins contained in pancreatic juice by two-dimensional isoelectric focusing/sodium dodecyl sulfate gel electrophoresis. *Cancer (Philadelphia)* **47**, 1513–1515.
Sixma, J. J., and Schiphorst, M. E. (1980). Identification of ectoproteins of human platelets. Combination of radioactive labelling and two-dimensional electrophoresis. *Biochim. Biophys. Acta* **603**, 70–83.
Skolnick, M. M., Sternberg, S. R., and Neel, J. V. (1982). Computer program for adapting two-dimensional gels to the study of mutation. *Clin. Chem.* **28**, 969–978.
Smith, S. C., Racine, R. R., and Langley, C. H. (1980). Lack of genetic variation in the abundant proteins of human kidney. *Genetics* **96**, 967–974.
Sun, T., Lien, Y. Y., and Gross, S. (1978). Clinical application of a high-resolution electrophoresis system. *Ann. Clin. Lab. Sci.* **8**, 219–227.
Swanson, R. A., Tracy, R. P., Katzmann, J. A., Wilson, D. M., and Young, D. S. (1982). β_2-microglobulin determined by radioimmunoassay with monoclonal antibody. *Clin. Chem.* **28**, 2033–2039.
Switzer, R. C., Merril, C. R., and Shifrin, S. A. (1979). A highly sensitive silver stain for detecting proteins and peptides in polyacrylamide gels. *Anal. Biochem.* **98**, 231–237.

Thorsrud, A. K., Haugen, H. F., and Jellum, E. (1980). High resolution two-dimensional electrophoresis (ISO–DALT) in clinical chemistry: Comparison with routine electrophoretic and immunologic methods. *In* "Electrophoresis '79" (B. Radola, ed.), pp. 425–435. de Gruyter, New York.

Thorsrud, A. K., Vatn, M. H., and Jellum, E. (1982). Two-dimensional electrophoresis patterns of proteins of normal mucosa, polyps and carcinomas of the large intestine. *Clin. Chem.* **28**, 884–889.

Tiselius, A. (1937). A new apparatus for electrophoretic analysis of colloidal mixtures. *Trans. Faraday Soc.* **33**, 524–531.

Tollaksen, S. L., Anderson, N. L., and Anderson, N. G. (1981). "Operation of the Iso-Dalt system," 6th ed., Argonne Natl. Lab. Publ. No. ANL–BIM–81–1. Argonne Natl. Lab., Argonne, Illinois.

Tracy, R. P., and Anderson, N. L. (1983). Applications of two-dimensional gel electrophoresis in the clinical laboratory. *In* "Clinical Laboratory Annual" (J. Batzakis and H. Homburger, eds.), pp. 101–130. Appleton-Century-Crofts, New York.

Tracy, R. P., and Young, D. S. (1983). Two-dimensional gel electrophoresis: methods and potential application in the clinical laboratory. *J. Clin. Lab. Auto.,* **3**, 235–244.

Tracy, R. P., Currie, R. M., Kyle, R. A., and Young, D. S. (1982a). Two-dimensional gel electrophoresis of serum specimens from patients with monoclonal gammopathies. *Clin. Chem.* **28**, 900–907.

Tracy, R. P., Currie, R. M., and Young, D. S. (1982b). Two-dimensional gel electrophoresis of serum specimens from a normal population. *Clin. Chem.* **28**, 890–899.

Tracy, R. P., Wold, L. E., Currie, R. M., and Young, D. S. (1982c). Patterns of normal colon mucosa and colon adrenocarcinoma compared by two-dimensional gel electrophoresis. *Clin. Chem.* **28**, 915–919.

Tracy, R. P., Currie, R. M., and Young, D. S. (1982d). Reproducibility and quality assurance of two-dimensional gel electrophoresis of serum specimens. *Clin. Chem.* **28**, 908–914.

Tracy, R. P., Wold, L. E., Currie, R. M., and Young, D. S. (1982e). Analysis of renal cortex and renal cell carcinoma by two-dimensional gel electrophoresis. *Proc. Acad. Clin. Lab. Physicians Sci.* Abstr. No. 15.

Tracy, R. P., Hill, H. D., Currie, R. M., and Young, D. S. (1982f). Two-dimensional gel electrophoresis of urinary proteins from renal transplant recipients. *Clin. Chem.* **28**, 1580 (Abstr.).

Tracy, R. P., Currie, R. M., Hill, H. D., and Young, D. S. (1983a). Two-dimensional gel electrophoresis in clinical chemistry. *Colloq. Int. Biol. Prospect. 5th,* (in press).

Tracy, R. P., Currie, R. M., Hill, H. D., and Young, D. S. (1983b). Two-dimensional gel electrophoresis for the clinical laboratory. *In* "Protides of the Biological Fluids" (H. Peeters, ed.), Pergamon Press, Oxford, pp. 581–586.

Wanner, L. A., Neel, J. V., and Meisler, M. H. (1982). Separation of allelic variants by two-dimensional electrophoresis. *Am. J. Hum. Genet.* **34**, 209–215.

Whalen, R. G., and Ecob, M. S. (1982). Two-dimensional electrophoretic analysis of muscle contractile proteins in patients with idiopathic scoliosis. *Clin. Chem.* **28**, 1036–1040.

Whicher, J. T. (1981). Inflammation: The role of the acute phase proteins. *Diagn. Med.* No. 3, 52–69.

Willard, K. E. (1982). Two-dimensional analysis of human lymphocyte proteins. III. Preliminary report on a marker for the early detection and diagnosis of infectious mononucleosis. *Clin. Chem.* **28**, 1031–1035.

Willard, K. E., and Anderson, N. G. (1981). Two-dimensional analysis of human lymphocyte proteins. I. An assay for lymphocyte effectors. *Clin. Chem.* **27**, 1327–1334.

Wilson, D. L., Hall, M. E., Stone, G. C., and Rubin, R. W. (1977). Some improvements in two-dimensional gel electrophoresis of proteins. *Anal. Biochem.* **83**, 33–44.

Young, D. S. (1979). Biological variability. *In* "Chemical Diagnosis of Disease" (S. Brown, F. L. Mitchell, and D. S. Young, eds.), pp. 1–114. Elsevier/North-Holland, Amsterdam.

Young, D. S., and Tracy, R. P. (1983). Two-dimensional electrophoresis in the development of laboratory tests. *Electrophoresis* **4**, 117–121.

Zannis, V. I., Just, P. W., and Breslow, J. L. (1981). Human apolipoprotein E isoprotein subclasses are genetically determined. *Am. J. Hum. Genet.* **33**, 11–24.

CHAPTER 8

Two-Dimensional Gel Electrophoresis for Studies of Inborn Errors of Metabolism

DAVID GOLDMAN
Laboratory of Clinical Science
National Institute of Mental Health
Bethesda, Maryland

CARL R. MERRIL
Laboratory of General and Comparative Biochemistry
National Institute of Mental Health
Bethesda, Maryland

I. Introduction . . . 241
II. Detection of Protein Polymorphism . . . 242
III. Protein Mutation by Two-Dimensional Gel Electrophoresis . . . 248
IV. Characteristic Patterns of Polypeptide Modulation . . . 251
V. Conclusion . . . 254
References . . . 255

I. Introduction

Application of two-dimensional gel electrophoresis (2-DGE) (O'Farrell, 1975) in studies of heritable diseases will extend our perception of the individual at the molecular level. This will allow increasingly accurate recognition, classification, understanding and treatment of pathology. There are four major areas in which a contribution to human medical genetics will be made: (1) the identification of primary mutations .in genetic diseases, (2) the study of molecular alterations in inborn errors of metabolism that may provide disease-specific and disease-associated

TWO-DIMENSIONAL GEL
ELECTROPHORESIS
OF PROTEINS

ISBN 0-12-164720-X

markers, (3) observations of metabolic perturbations that will lead to a better understanding of normal metabolism and to the elucidation of metabolic patterns and pathways of diseases, and (4) the identification of protein polymorphisms and use of these polymorphisms in gene mapping. In this chapter, each of the above approaches will be described (see also chapter by Neel *et al.*).

Because genetic variants with metabolic consequences usually alter the quantity or quality of proteins, identification of alterations in protein quantity, composition, functional activity, localization, immunological determinants, or electrophoretic mobility have all contributed to our understanding of particular genetic diseases. In addition these methods have helped reveal extensive, apparently selection-neutral, protein genetic variation (Harris, 1980). An unknown fraction of protein genetic variants confer gross structural and phenotypic variation and are the biochemical basis of phenotypic individuality. These are the genetic factors that help define us as individuals both in health and in disease.

Two-dimensional gel electrophoresis extends the capacity for detecting and studying the extent, mechanisms, and consequences of genetic variation, chiefly because it enables more than 1000 proteins to be visualized from a single cell type (O'Farrell, 1975) and can be applied with little modification to different cells and tissue fluids and their subfractions. Sensitive visualization methods allow largely indiscriminate detection so that the high resolving power of 2-DGE can be fully exploited. These methods include silver stains and autoradiography or fluorography after appropriate labeling. Discriminate detection, as with radiolabeling of phosphorylated proteins or detection with specific antibody, is also useful. The large number of proteins resolved and the ready-made two-dimensional matrix make possible the systematized mapping and cross-correlation of new data, especially when 2-DGE is combined with computer-assisted measurement and data reduction (Lutin *et al.*, 1978; Garrels, 1979; Miller *et al.*, 1979; Bossinger *et al.*, 1979; Merril *et al.*, 1981a; Goldman *et al.*, 1982; see also chapter by Garrels *et al.*).

II. Detection of Protein Polymorphism

Individual variation has both genetic and environmental origins. Genetic and environmental variation are manifested grossly or only on the molecular level, but all genetic phenotypic variation is determined by molecular variation and is most precisely studied at the molecular level.

The study of the subtle molecular differences will lead to the dissection and understanding of human individuality. A category of normal genetic variation is polymorphism. Polymorphisms are normal, genetic phenotypic variants found with an allelic frequency of >0.01. Although usually thought of as selectively neutral, an unknown fraction produce subtle metabolic and other phenotypic effects. As described in Section III protein profiling by 2-DGE may help uncover those which do so.

Of more immediate interest is the role molecular polymorphisms play in linkage analysis and in construction of a map of the human genome. For *Drosophila*, an effective genetic map was pieced together by Morgan and his students by 1926. This map was based solely on external characteristics and omitted only the Y chromosome (Morgan, 1926). The process of mapping the human genome has lagged because of its greater size, the long species generation time, and the fact that crosses cannot be arranged for experimental purposes and mutants cannot be deliberately induced in our society. However, the availability of molecular polymorphisms, in combination with linkage analysis and the use of somatic cell genetics, has brought this goal within reach. Molecular polymorphisms that are generally useful for assigning new loci by linkage analysis are those whose frequency is great enough that there is a reasonable likelihood that they will be present in a family under study.

Currently fewer than 50 commonly polymorphic loci are available, including allozyme polymorphisms detectable by one-dimensional electrophoresis, serological antigen loci, and DNA restriction fragment length polymorphisms (RFLPs). The probability analysis presented in Fig. 1 shows that the discovery of new polymorphisms results in the probability that some additional fractions of the genome will become available for genetic linkage studies. Approximately 26% of the human genome can be analyzed by linkage studies at a recombination frequency of 10%. However, utilization of the polymorphisms discoverable by two-dimensional electrophoresis will greatly increase the fraction that can be mapped if it can be shown that the polymorphisms being discovered are new ones.

Table I shows the polymorphisms that have been demonstrated through 2-DGE. Comings described a polymorphism of a perchloric acid–soluble brain protein, Pc 1-Duarte, in 1977 (Comings, 1977). Comings has since described two other 2-DGE polymorphisms of human brain proteins (Comings, 1982a,b). Giometti and Anderson (1981) showed that a nonmuscle tropomyosin of human fibroblasts was polymorphic by 2-DGE. In a computer-assisted analysis of 28 individuals, Goldman and Merril found 19 polymorphisms among 186 proteins of the human lymphocyte (Goldman *et al.*, 1982; Goldman and Merril,

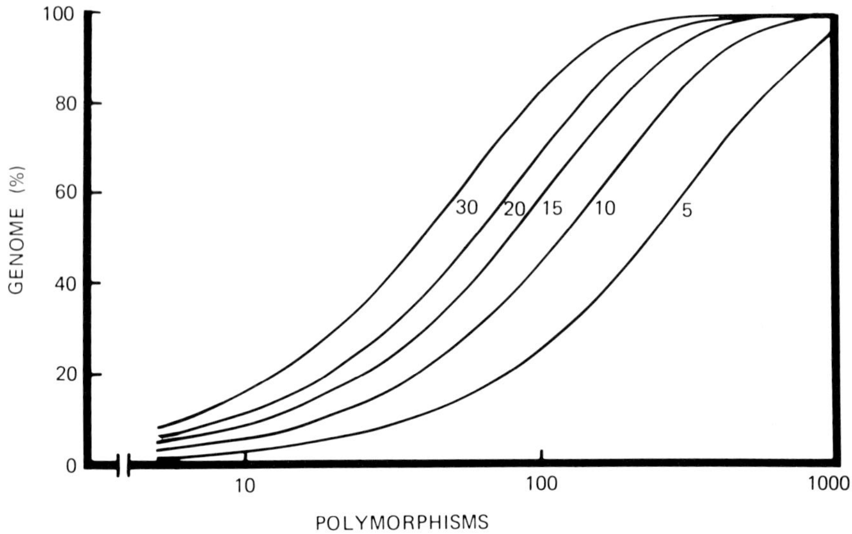

Figure 1. Theoretical projection of the fraction of the genome within which genetic linkage can be recognized at different recombination distances (see also Elston and Lange, 1975; Goldman and Merril, 1983). The fraction of the genome that can be analyzed is plotted against the logarithm of the number of polymorphisms discovered. The projection was carried out with the assumptions that the length of the human genome is approximately 3300 centimorgans (Renwick, 1971) and that the distribution of the polymorphisms is random. The genome was divided into 23 chromosomal segments to take into account boundary effects. Each new polymorphism was assigned randomly and the percentage of genome within which linkage could be assessed within different recombination distances was then increased according to how much new area was covered. The means of 100 trials are plotted.

1983). The locations of these 19 polymorphisms are shown on the two-dimensional gel in Fig. 2. Hamaguchi *et al.* (1981) have independently described three of these lymphocyte protein polymorphisms in a Japanese population.

In each case the polymorphism is detected by a charge shift. In a general fashion the extent of this shift in charge varies inversely with molecular weight, as predicted (O'Farrell, 1975). Within the pH range examined the isoelectric focusing stage of 2-DGE resolves all charge variants (O'Farrell, 1975; Steinberg *et al.*, 1977). Two-dimensional gel electrophoresis is thus capable of detecting almost all of the allozyme variants detected by one-dimensional gel electrophoresis because these variants generally involve charge shifts. Wanner *et al.* (1982) found that 90% of the known variants they examined were detectable by 2-DGE.

TABLE I

Protein Polymorphism by Two-Dimensional Electrophoresis

Polymorphism	Number[a]	References
In man lymphocyte	<u>19</u>	
	19	Goldman and Merril (1983)
	7	Goldman *et al.* (1982)
	3	Hamaguchi *et al.* (1981)
Fibroblast	<u>1</u>	
Nonmuscle tropomyosin		Giometti and Anderson (1981)
Brain	<u>3</u>	
Pc 1-Duarte		Comings (1977)
MBL-Duarte		Comings (1982a)
GFAP-Duarte		Comings (1982b)
Pancreas	<u>1</u>	
Amylase		Wanner *et al.* (1982)
Serum	<u>7</u>	Anderson and Anderson (1977)
Haptoglobin		
Gc-globulin		
α_1-antitrypsin		
Transferrin		
Arginine-rich lipoprotein		
α_2-HS glycoprotein		
G-4		
Total	<u>31</u>	
In mouse lymphocyte	<u>8</u>	Elliot (1979)

[a] Underlined numbers are totals for unique polymorphisms in a tissue or overall.

In serum, seven polymorphisms were shown to be detectable by 2-DGE (Anderson and Anderson, 1977). One of these, G-4, represents a newly described polymorphism. Rosenblum *et al.* (1982) reported that 5 out of 20 serum protein spots representing unknown polypeptides were polymorphic. Some or all of these are new serum polymorphisms. In all at least 31 polymorphisms including 24 that may be new ones have now been characterized by 2-DGE.

In addition to the shift in pI there is often a slight shift in apparent molecular mass (see Fig. 3). This shift in SDS–polyacrylamide gel mobility may be caused by altered conformation of the denatured protein in the second-dimension gel and has been frequently observed for charge-substituted variants (Vandekerckhove *et al.*, 1980). Some serum polymorphisms showed a shift in apparent molecular mass but no shift in

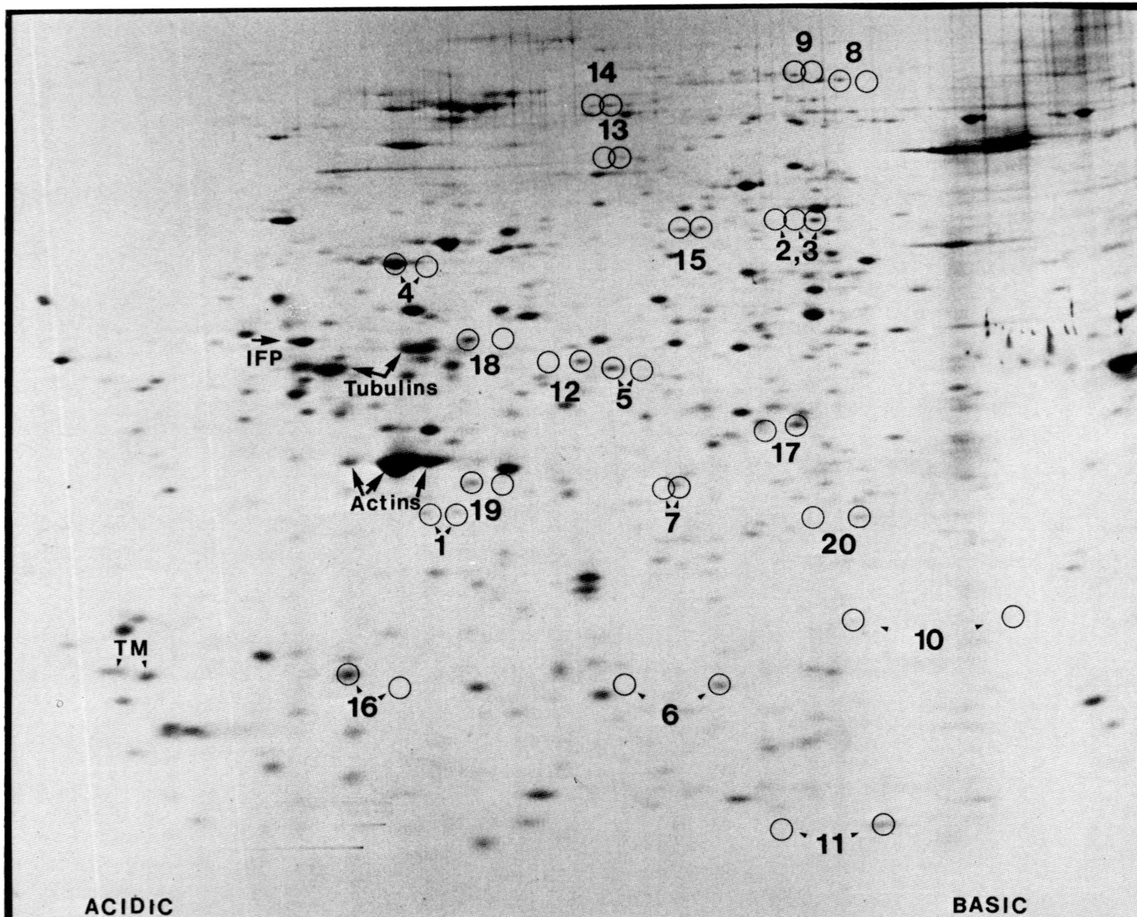
9
8
14
13
15
2,3
4
IFP
Tubulins
18
12
5
17
Actins
19
1
7
20
10
TM
16
6
11
ACIDIC
BASIC

charge (Rosenblum *et al.*, 1982). Such variants could involve charge-neutral amino acid substitutions that result in a conformational change or alter posttranslational processing.

Gene dosage for 2-DGE protein polymorphisms has generally been observed when it has been looked for quantitatively (Goldman *et al.*, 1982; Goldman and Merril, 1983). Gene dosage is evident for the polymorphism displayed in Fig. 3. The heterozygote (Fig. 3B) displays two molecular forms, each present in half the density seen for the corresponding molecular form in the homozygotes (Fig. 3A and 3C).

In the area of human population genetics 2-DGE has enabled investigators to expand greatly the number of loci that can be assessed for variability. The overall level of genetic variability in a population is expressed as average heterozygosity, which is computed by averaging the heterozygosity observed for all loci scored. Average heterozygosity detectable by allozyme electrophoresis was 6.3% for man (Harris, 1980), but has generally been estimated to be lower by 2-DGE. McConkey *et al.* (1979) and Walton *et al.* (1979) found levels of less than 1%. Goldman and Merril (1983) have found an average heterozygosity of 2.4% for 186 proteins of human lymphocyte. However, Rosenblum *et al.* (1982) found an average heterozygosity of 6.5% for 20 randomly selected serum proteins.

Although considerable progress has been made in the art of detecting 2-DGE polymorphisms, extensive family studies have not been reported and none have been genetically mapped. In the mouse lymphocyte Elliot (1979) has demonstrated eight polymorphisms by 2-DGE and by using recombinant inbred strains was able to assign five to a chromosome and show all eight were independent of previously known polymorphisms. Human 2-DGE polymorphisms will generally have to be assigned by linkage analysis and somatic cell genetics.

Figure 2. Autoradiogram of human lymphocyte proteins to identify the positions of 19 polymorphic proteins. Lymphocytes were isolated from peripheral blood, stimulated with phytohemagglutinin, radiolabeled for 3 h with [^{14}C]leucine and electrophoresed using the procedure of O'Farrell with minor modifications. A 4:1 mixture of pH 5–7 and pH 3–10 ampholytes was used for isoelectric focusing and a 10% polyacrylamide gel was used for second stage electrophoresis in the presence of SDS.

Polymorphisms were identified among 186 proteins in a population of 28 individuals. Criteria for assigning proteins as polymorphic included the detection of a discrete charge-shifted variant and the observation of heterozygotes who showed gene dosage: the two electromorphs were approximately equal in density when present in the same individual. Limited family data support these as genetic variants. The population distribution of alleles was in each case consistent with Hardy-Weinberg expectations. Population data, molecular weights and isoelectric points, nomenclature, and methods are described more fully elsewhere (Goldman *et al.*, 1982; Goldman and Merril, 1983).

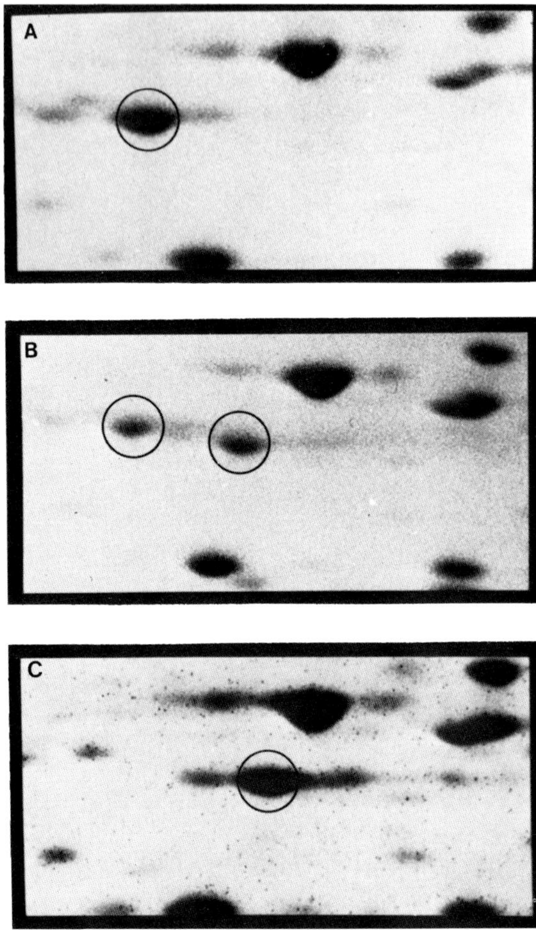

Figure 3. Illustration of the three observed phenotypes for the polymorphism designated NIMH-4 (Goldman *et al.*, 1982; Goldman and Merril, 1983), a polymorphism of a 64 kdalton, pI 5.2 polypeptide of human lymphocytes. (A) Acidic homozygote; (B) heterozygote; (C) basic homozygote.

III. Protein Mutation by Two-Dimensional Gel Electrophoresis

Discovery of new protein mutations and primary protein abnormalities in genetic disease by 2-DGE is an exciting prospect because no other technique allows so many gene products to be surveyed.

TABLE II

Hypoxanthine-Guanine Phosphoribosyl Transferase in Lesch–Nyhan Patients and Controls[a]

	Enzyme activity[b]	Immuno-precipitable protein	2-DGE Density[b]	
			Silver	Autoradiography
Patients	86.7 ± 76.8	–	25.1 ± 1.2	18.2 ± 3.7
Controls	1748.7 ± 99.2	+	38.6 ± 3.2	19.0 ± 3.3

[a] From Merril *et al.* (1981a).
[b] Mean ± S.E.M. (n = 3).

Steinberg *et al.* (1977) examined cAMP-resistant S49 mouse lymphoma cells. Of 26 clones that showed a higher activation constant for the cAMP-dependent protein kinase, 16 showed charge mutations in the regulatory subunit of the kinase whereas in 10 this protein electrophoresed in the normal position. Regulation of the synthesis and phosphorylation of a number of other proteins by cAMP (Steinberg and Coffino, 1979) and trans-dominant mutations (Steinberg *et al.*, 1979) were also studied in this experimental system.

Some of the first clinical studies of inborn errors of metabolism were done on the Lesch-Nyhan syndrome. Lesch-Nyhan syndrome is an X-linked recessive trait caused by a deficiency in hypoxanthine–guanine phosphoribosyl transferase (HGPRT) activity; it results in spasticity, hyperuricemia, and self-mutilation. Because HGPRT activity was often undetectable and because some antibodies prepared against normal HGPRT did not recognize mutant HGPRT proteins, it was hypothesized that most of these patients totally lacked the protein (Ghangas and Milman, 1977). The mutations responsible for total absence of a protein could be mutations in a control protein or mutations at the structural locus that resulted in either a failure in transcription or early termination of translation (amber mutation). In *E. coli,* 2-DGE was successfully used to demonstrate amber mutations in three enzymes of the galactose metabolic pathway: uridine diphosphate galactose 4-epimerase, galactokinase, and galactose 1-phosphate uridyl transferase (Merril *et al.*, 1981b). However, the Lesch–Nyhan syndrome Merril *et al.* (1981b) resolved, visualized, and measured HGPRT protein in each of three Lesch–Nyhan patients with virtual absence of enzyme activity and virtual absence of immunoprecipitable HGPRT. In each case, there was only a slight reduction in HGPRT protein on 2-DGE gels (see Table II).

Strauss *et al.* (1981), using antibody prepared in a different manner, found cross-reacting material in three out of three patients they examined. Hypoxanthine–guanine phosphoribosyl transferase was detectable in the correct position by SDS–PAGE but isoelectric focusing results were not presented. Charge-altered HGPRT variants are detectable by 2-DGE (Ghangas and Milman, 1977; Zannis *et al.*, 1980). Thus the 2-DGE studies indicate that Lesch–Nyhan disease usually results from a point mutation at the structural locus.

However, the detection of primary mutation remains the most difficult task in the investigation of genetic diseases. The major reasons for this are as follows:

1. Two-dimensional gel electrophoresis probably surveys only a fraction of the total cellular proteins, although this has been questioned (Duncan and McConkey, 1982). On the basis of the apparent number of mRNA species the number of active cellular structural genes in somatic tissues is between 10,000 and 20,000 (Williams and Penman, 1975; Lewin, 1980), and the number is perhaps somewhat higher in brain, for which the data suggest between 12,000 and 70,000 active structural genes (Young *et al.*, 1976; Bantle and Hahn, 1976; Hastie and Bishop, 1976). If the mutation involves a minor polypeptide or one that is not visualized, it is likely to be missed. To overcome this difficulty one may study tissues and subcellular fractions that are more likely to contain the protein whose mutation is responsible for the disease.
2. Even if the mutant protein is detectable, it may not show a charge alteration and may be present in nearly normal concentration, although its functional level is greatly reduced (Merril *et al.*, 1981a). One way of overcoming this difficulty is to study a sufficient number of subjects with a disease so that genetic heterogeneity will make likely the study of some individuals who manifest a charge mutation.
3. Charge variants discovered may be coincidental polymorphisms and must be proven to be otherwise.
4. Protein alterations not detected as charge variants are likely to be secondary alterations as described below. Such proteins can be useful as trait- or state-specific molecular markers and may give clues to pathogenesis (Merril *et al.*, 1981a).

Although detection of the specific mutation in a disease of unknown molecular origin is still unlikely, 2-DGE is far superior to any other method for this purpose. Two-dimensional gel electrophoresis also holds considerable promise for detecting the effects of mutagens both *in vitro* and in the whole organism. Two-dimensional gel electrophoresis and computer-assisted analysis are being used to search for new mutations in parent-child trios (Skolnick *et al.*, 1982; see also chapter by Neel *et al.*).

IV. Characteristic Patterns of Polypeptide Modulation

Genetic traits and biological states may be associated with characteristic patterns of protein modulations. Some diseases that are phenocopies on a gross level may reveal trait-specific molecular markers due to secondary effects of the primary mutation.

The effects we are concerned with detecting are due to feedback control mechanisms, posttranslational modification, altered protein stabilities, or disrupted compartmentation. Patterns of protein modulation can be state-specific or associated or trait-specific or associated. The development of a correlative catalog of protein modulations in different diseases and protein identifications and associations with cell structures and functions will greatly assist in defining the specificity of protein modulations or patterns of modulations and will help dissect pathogenic and normal processes.

Patterns of state-associated protein modulations have been quantitatively characterized by exposing cells or tissues to a variety of treatments. These include heat shock (Miller *et al.*, 1979; Anderson *et al.*, 1982), nerve growth factor (Garrels, 1979), dibutyryl cAMP (Steinberg and Coffino, 1979), calcium (Willard and Anderson, 1981), and thyroid hormone (Nikodem *et al.*, 1981). Proteins have also been shown to vary with the cell cycle (Bravo and Celis, 1980) and after neoplastic transformation (Leavitt and Kakunaga, 1980; Leavitt *et al.*, 1982; Bravo and Celis, 1982; Celis *et al.*, this volume). In most cases, none of the observed modulations can be said to be state-specific because few treatments have been examined and because the identity and biology of modulation of most of the altered polypeptides is not understood.

Trait-associated protein alterations have now been identified in a number of human diseases. Merril *et al.* (1981a) found eleven quantitative alterations among 400 proteins of Lesch–Nyhan syndrome lymphocytes (Fig. 4). These differences were present in the patterns of all Lesch–Nyhan patients examined and were not found in any normal individuals. Secondary protein modulations have also been found in alterations of chromosome number, as in trisomy 21 (Weil and Epstein, 1979; Van Keuren *et al.*, 1981). Proteins from aneuploid individuals may also display increased or decreased quantities that are secondary to gene

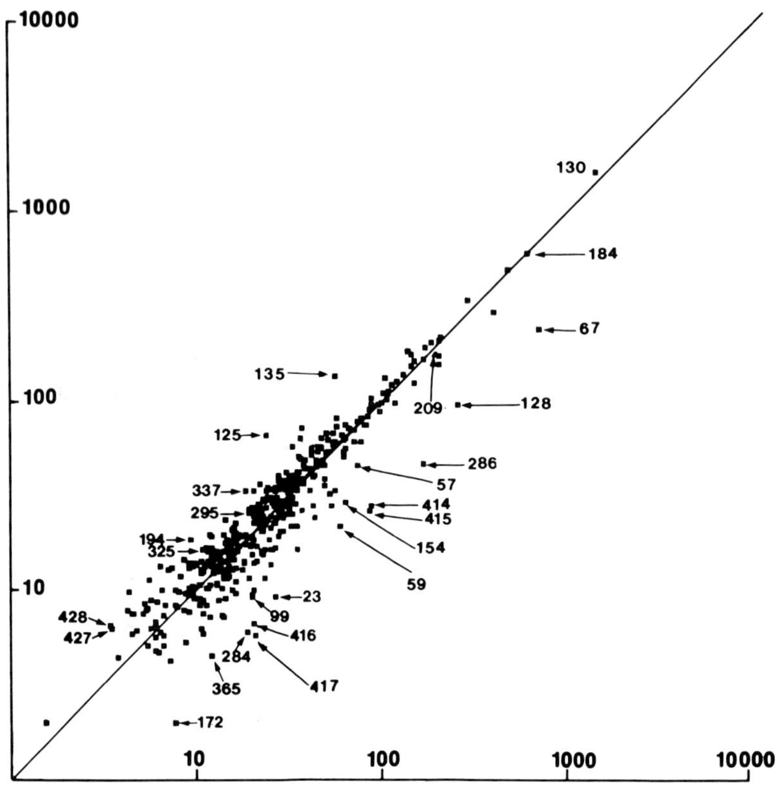

Figure 4. Analysis of 400 lymphocyte proteins in Lesch-Nyhan syndrome. Mean densities of proteins measured from autoradiograms of phytohemagglutinin-stimulated, [^{14}C]leucine-labeled proteins from three Lesch-Nyhan patients are plotted against the mean densities in three controls. Outliers were found to be consistently different in the Lesch-Nyhan lymphocytes. From Merril *et al.* (1981a).

dosage. In this context such alterations might also be classified as trait-associated modulations. Cytoplasmic superoxide dismutase, which is known to map to chromosome 21 and which shows enzyme activity proportional to the number of copies of chromosome 21 (Schitiu *et al.*, 1974), shows a density proportional to gene dosage when examined on 2-DGE gels of cells mono-, di-, and trisomic for chromosome 21 (Van Keuren *et al.*, 1981; Brown *et al.*, 1981) (see Table III). It should be noted that altered copy number for one chromosome may alter expression of genes on other chromosomes. For instance, activities of several enzymes that map to chromosomes other than 21 have been shown to be elevated in Down's syndrome (Francke, 1981). Nevertheless, using this approach

TABLE III

Cytoplasmic Superoxide Dismutase (SOD-1) and Protein 377 in Fibroblasts Mono-, Di-, and Trisomic for Chromosome 21[a]

	Silver Stain	Autoradiography
SOD-1 Density		
Monosomy 21	68.2	33.0
Disomy 21	100.0	100.0
Trisomy 21	161.7	164.7
Protein 377 Density[b]		
Monosomy 21	24.2	67.7
Disomy 21	100.0	100.0
Trisomy 21	137.2	182.3

[a] From Van Keuren *et al.* (1981).

[b] Density of SOD-1 and protein 377 were measured from silver-stained gels and autoradiograms of cells monosomic (GM230), disomic (GM10), and trisomic (GM2504) for chromosome 21. Densities were normalized using densities of other proteins that did not vary with chromosome number. Values are expressed as the percentage of the SOD or protein 377 density in disomic cells.

investigators have tentatively assigned unknown 2-DGE proteins to chromosome 21 (see Table III) and the X chromosome (Van Keuren *et al.*, 1981; Devine *et al.*, 1982).

Studies of the tissue that is primarily involved in a disease process have usually yielded extensive differences when compared to the tissue in the normal state. These differences are at times due to cell death and altered cellular composition of the tissue. Brains of patients with Huntington's disease, Joseph's disease, and multiple sclerosis show extensive protein alterations, but most of these are secondary to gliosis (Comings, 1977; Rosenberg *et al.*, 1982). When peripheral fibroblasts and lymphocytes have been examined in patients with Huntington's disease, no polypeptide alterations were detected (Comings, 1977; Goldman *et al.*, 1982; Rosenberg *et al.*, 1982). In one study, although more than 300 proteins were quantitatively analyzed in 13 patients and 15 controls and individuals at risk for Huntington's disease, no abnormalities were observed in the patterns examined (Goldman *et al.*, 1982).

Another example of how an alteration in cellular composition of a disease-affected structure can lead to an alteration in 2-DGE protein profile is provided by a study of muscle proteins in idiopathic scoliosis (Whalen and Ecob, 1982). By comparing paraspinous muscle samples from the convex and concave sides of the curved spine, they identified variations in muscle contractile proteins that were probably due to differences in the ratio of fast to slow muscle fibers.

A potential trait-specific muscle protein disease marker is the lack or deficiency of two myosin light chains in nemaline rod myopathy (Giometti *et al.*, 1980; Danon *et al.*, 1981). The abnormality occurs in both congenital and adult-onset forms of the disease but was not observed in patients with Duchenne's and myotonic dystrophy and dermatomyositis.

The grey platelet syndrome involves an inherited deficiency of the alpha granule, one of the three types of platelet granules. Nurden *et al.* (1982) have shown by 2-DGE that a number of proteins that are associated with the alpha granule are deficient. The specific molecular defect has not yet been pinpointed.

Bernard–Soulier disease is an inherited platelet disorder in which the defect apparently is in the membrane and in which there are giant morphologically abnormal platelets. Electrophoretic analysis reveals a deficiency of an approximately 140 kdalton membrane glycoprotein (Nurden *et al.*, 1981; Peterson *et al.*, 1982; Clemetson *et al.*, 1982) in patients but not in heterozygotes (Peterson *et al.*, 1982). Other membrane proteins have also been found to be deficient in this disease (Clementson *et al.*, 1982), indicating that these defects may be secondary.

V. Conclusion

Clinical studies utilizing 2-DGE may contribute knowledge on polymorphism, linkage of polymorphic loci to disease, mutations (both new and disease-associated), polypeptide modulation, and regulation associated with normal processes and disease. In the area of protein polymorphism 31 human polymorphisms, including 24 which may be new ones, have been described and better estimates of human protein heterozygosity have been provided. For primary mutations 2-DGE has been successfully used to identify charge-shift and amber mutations. For secondary protein modulations patterns of these have been demonstrated in a number of genetic diseases. The specificity of these alterations as markers will be established as additional diseases and metabolic alterations are studied. The prospect of correlating the information that is being accumulated is good owing to the suitability of the high resolution 2-DGE matrix to cataloging and the application of the method to a growing variety of questions.

Acknowledgment

This work was supported in part by the National Institute on Alcohol Abuse and Alcoholism.

References

Anderson, N. L. and Anderson, N. G. (1977). High resolution two-dimensional electrophoresis of human plasma proteins. *Proc. Natl. Acad. Sci. USA* **74**, 5421–5425.

Anderson, N. L., Giometti, C. S., Gemmell, M. A., Nance, S. L., and Anderson, N. G. (1982). A two-dimensional electrophoretic analysis of the heat-shock-induced proteins of human cells. *Clin. Chem.* **28**, 1084–1092.

Bantle, J. A., and Hahn, W. E. (1976). Complexity and characterization of polyadenylated RNA in the mouse brain. *Cell* **8**, 139–150.

Bossinger, J., Miller, M. J., Vo, K. P., Geiduschek, E. P., and Xuong, N. H. (1979). Quantitative analysis of two-dimensional electrophoretograms. *J. Biol. Chem.* **254**, 7986–7998.

Bravo, R., and Celis, J. E. (1980). A search for differential polypeptide synthesis throughout the cell cycle of HeLa cells. *J. Cell Biol.* **48**, 795–802.

Bravo, R., and Celis, J. E. (1982). Human proteins sensitive to neoplastic transformation in cultured epithelial and fibroblast cells. *Clin. Chem.* **28**, 949–954.

Brown, W. T., Dutkowski, R., and Darlington, G. T. (1981). Localization and quantitation of human superoxide dismutase using computerized 2-D gel analysis. *Biochem. Biophys. Res. Commun.* **102**, 675.

Clemetson, K. J., McGregor, J. L., James, E., Dechavanne, M., and Luscher, E. F. (1982). Characterization of the platelet membrane glycoprotein abnormalities in Bernard–Soulier syndrome and comparison with normal by surface-labelling techniques and high resolution two-dimensional electrophoresis. *J. Clin. Invest.* **70**, 304–311.

Comings, D. E. (1977). Pc 1 Duarte, a common polymorphism of a human brain protein, and its relationship to manic–depressive disease and multiple sclerosis. *Nature (London)* **277**, 28–32.

Comings, D. E. (1982a). Two-dimensional gel electrophoresis of human brain proteins. V. Non-equilibrium gel electrophoresis, with detection of a myelin basic protein mutation—MBL-Duarte. *Clin. Chem.* **28**, 813–818.

Comings, D. E. (1982b). Two-dimensional gel electrophoresis of human brain proteins. IV. Disorders of glial proliferation and a polymorphism of glial fibrillary acidic protein—GFAP Duarte. *Clin. Chem.* **28**, 805–812.

Danon, M. J., Giometti, C. S., Manaligod, J. R., Perurena, O. H., and Skosey, J. L. (1981). Adult-onset nemaline rods in a patient treated for suspected dermatomyositis. *Arch. Neurol. (Chicago)* **38**, 761–766.

Devine, E., Brown, T., Krueger, L., Jenkins, E., Dubkowski, R., and Sammons, D. (1982). Assignment of proteins to human chromosomes 21 and X using two dimensional gel electrophoresis. *Am. J. Hum. Genet.* **34**, A170.

Duncan, R., and McConkey, E. H. (1982). How many proteins are there in a typical mammalian cell? *Clin. Chem.* **28**, 749–755.

Elliot, R. W. (1979). Use of two-dimensional electrophoresis to identify and map new mouse genes. *Genetics* **91**, 295–308.

Elston, R. C., and Lange, K. (1975). The prior probability of autosomal linkage. *Ann. Hum. Genet.* **38**, 341–350.

Francke, U. (1981). *In* "Trisomy 21 (Down Syndrome) Research Perspectives" (F. F. de la Cruz and P. S. Gerald, eds.), pp. 237–251. Univ. Park Press, Baltimore, Maryland.

Garrels, J. (1979). Modulation of protein synthesis by nerve growth factor, *J. Biol. Chem.* **254**, 7961–7977.

Ghangas, G. S., and Milman, G. (1977). Hypoxanthine phosphoribosyltransferase two-dimensional gels from normal and Lesch Nyhan hemolysates. *Science (Washington, D.C.)* **196**, 1119–1120.

Giometti, C. S., and Anderson, N. L. (1981). A variant of human nonmuscle tropomyosin found in fibroblasts by using two-dimensional electrophoresis. *J. Biol. Chem.* **256**, 11840–11846.

Giometti, C. S., Barany, M., Danon, M. J., and Anderson, N. G. (1980). Muscle protein analysis. II. Two-dimensional electrophoresis of normal and diseased human skeletal muscle. *Clin. Chem.* **26**, 1152–1155.

Goldman, D., and Merril, C. R. (1983). Detection of human lymphocyte polymorphisms by two-dimensional electrophoresis. *Am. J. Hum. Genet.* **35**, 827–837.

Goldman, D., Merril, C. R., Polinsky, R. J., and Ebert, M. H. (1982). Lymphocyte proteins in Huntington's disease: Quantitative analysis by use of two-dimensional electrophoresis and computerized densitometry. *Clin. Chem.* **28**, 1021–1025.

Hamaguchi, H., Ohta, A., Mukai, R., Yabe, T., and Yamada, M. (1981). Genetic analysis of human lymphocyte proteins by two-dimensional gel electrophoresis. *Hum. Genet.* **59**, 215–220.

Harris, H. (1980). "The Principles of Human Biochemical Genetics," 3rd ed. Elsevier-North Holland, New York.

Hastie, N. D., and Bishop, J. D. (1976). The expression of three abundance classes of mRNA in Mouse Tissues. *Cell* **9**, 761–774.

Leavitt, J., and Kakunaga, T. (1980). Expression of a variant form of actin and additional polypeptide changes following chemical induced *in vitro* neoplastic transformation of human fibroblasts. *J. Biol. Chem.* **255**, 1650–1661.

Leavitt, J., Goldman, D., Merril, C. R., and Kakunaga, T. (1982). Changes in gene expression accompanying chemically-induced malignant transformation of human fibroblasts. *Carcinogenesis (N.Y.)* **3**, 61–70.

Lewin, B. (1980). "Gene Expression," 2nd ed. Eucaryotic Chromosomes, p. 178, Wiley, New York.

Lutin, W. A., Kyle, C. F., and Freeman, J. A. (1978). Developments in biochemistry, quantitation of brain proteins by computer-analyzed two dimensional electrophoresis. *Electrophoresis* **2**, 93–106.

McConkey, E. H., Taylor, B. J., and Phan, D. (1979). Human heterozygosity: a new estimate. *Proc. Natl. Acad. Sci. USA* **76**, 6500–6504.

Merril, C. R., Goldman, D., and Ebert, H. (1981a). Protein variations associated with the Lesch–Nyhan syndrome. *Proc. Natl. Acad. Sci. USA* **78**, 6471–6475.

Merril, C. R., Gottesman, M. E., and Adhya, S. L. (1981b). *Escherichia coli gal* operon proteins made after prophage lambda induction. *J. Bacteriol.* **147**, 875–887.

Miller, M. J., Xuong, N. H., and Geiduschek, E. P. (1979). A response of protein synthesis to temperature shift in the yeast *Saccharomyces cerevisiae*. *Proc. Natl. Acad. Sci. USA* **76**, 5222–5225.

Morgan, T. H. (1926). "The Theory of the Gene." Yale Univ. Press, New Haven, Connecticut.
Nikodem, V. M., Trus, B. L., and Rall, J. E. (1981). Two-dimensional gel analysis of rat liver nuclear proteins after thyroidectomy and thyroid hormone treatment. *Proc. Natl. Acad. Sci. USA* **78**, 4411–4415.
Nurden, A. T., Dupuis, D., Kunicki, T. J., and Caen, J. P. (1981). Analysis of the glycoprotein and protein composition of Bernard-Soulier platelets by single and two-dimensional SDS-polyacrylamide gel electrophoresis. *J. Clin. Invest.* **67**, 1431–1440.
Nurden, A. T., Kunicki, T. J., Dupuis, D., Soria, C., and Caen, J. P. (1982). Specific protein and glycoprotein deficiencies in platelets isolated from two proteins with the gray platelet syndrome. *Blood* **59**, 709–718.
O'Farrell, P. H. (1975). High resolution two-dimensional electrophoresis of proteins. *J. Biol. Chem.* **250,** 4007–4021.
Peterson, D. M., Hirst, A., and Wehring, B. (1982). Comparison of normal and Bernard-Soulier platelet membrane glycoproteins. *J. Lab. Clin. Med.* **100**, 26–36.
Renwick, J. H. (1971). The mapping of human autosomes. *Annu. Rev. Genet.* **5,** 81–120.
Rosenberg, R. R., Ivy, N., Kirkpatrick, J., Bay, C., Nyhan, W. L., and Baskin, F. (1982). Joseph disease and Huntington disease: protein patterns in fibroblasts and brain. *Neurology* **31**, 1003–1014.
Rosenblum, B. B., Hanash, S. M., and Neel, J. V. (1982). Heterozygosity in plasma proteins as analyzed by two-dimensional gels. *Am. J. Hum. Genet.* **34**, A192.
Schitiu, S., Sinet, P. M., Lejeune, J., and Frezal, J. (1974). Dosage effects for superoxide dismutase-1 in nucleated cells aneuploid for chromosome 21. *Humangenetik* **23,** 65
Skolnick, M. M., Steinberg, S. R., and Neel, J. V. (1982). Computer programs for adapting two-dimensional gels to the study of mutation. *Clin. Chem.* **28**, 969–978.
Steinberg, R. A., and Coffino, P. (1979). Two-dimensional gel analysis of cyclic AMP effects in cultured S49 lymphoma cells: protein modifications, inductions and repressions. *Cell* **18**, 719–723.
Steinberg, R. A., O'Farrell, P. H., Friedrich, U., and Coffino, P. (1977). Mutations causing charge alterations in regulatory subunits of the cAMP-dependent protein kinase of cultured S49 lymphoma cells. *Cell* **10**, 381–391.
Steinberg, R. A., Vandaale, T., and Coffino, P. (1979). Kinase-negative mutants of S49 mouse lymphoma cells carry a trans-dominant mutation affecting expression of cAMP-dependent protein-kinase. *Cell* **15**, 1351–1361.
Strauss, M., Lubbe, L., and Geissler, E. (1981). HGPRT structural gene mutation in Lesch-Nyhan-Syndrome as indicated by antigenic activity and reversion of the enzyme deficiency. *Hum. Genet.* **57**, 185–188.
Vandekerckhove, J., Leavitt, J., Kakunaga, T., and Weber, K. (1980). Coexpression of a mutant beta-actin and the two normal beta- and gamma- cytoplasmic actins in a stably transformed human cell line. *Cell* **22**, 893–899.
Van Keuren, M. L., Goldman, D., and Merril, C. R. (1981). A quantitative two-dimensional electrophoretic survey of proteins affected by chromosome 21. "Electrophoresis '81" *In* (R. C. Allen and P. Arnaud, eds.), pp. 355–369. de Gruyter, Berlin.
Walton, K. E., Styer, D., and Gruenstein, E. I. (1979). Genetic polymorphism in normal human fibroblasts as analyzed by two-dimensional gel electrophoresis. *J. Biol. Chem.* **254**, 7951–7960.
Wanner, L. A., Neel, J. V., and Meisler, M. H. (1982). Separation of allelic variants by two-dimensional electrophoresis. *Am. J. Hum. Genet.* **34**, 209–215.

Weil, J., and Epstein, C. (1979). The effect of trisomy 21 on the patterns of polypeptide synthesis in human fibroblasts. *Am. J. Hum. Genet.* **7**, 478–488.
Whalen, R. G., and Ecob, M. S. (1982). Two-dimensional electrophoretic analysis of muscle contractile proteins in patients with idiopathic scoliosis. *Clin. Chem.* **28**, 1036–1040.
Willard, K. E., and Anderson, K. E. (1981). Two-dimensional analysis of human lymphocyte proteins. I. An assay for lymphocyte effectors. *Clin. Chem.* **27**, 1327–1334.
Williams, J. G., and Penman, S. (1975). The messenger RNA sequences in growing and resting mouse fibroblasts. *Cell* **6**, 179–206.
Young, B. D., and Birnie, G. D., and Paul, J. (1976). Complexity and specificity of polysomal poly(A)+ RNA in mouse tissues. *Biochemistry* **15**, 2823–2828.
Zannis, V. I., Gudas, L. J., and Martin, D. W., Jr. (1980). Characterization of the subunit composition of HGPRTase from human erythrocytes and cultured fibroblasts. *Biochem. Genet.* **18**, 1–19.

CHAPTER 9

Adapting Two-Dimensional Gel Electrophoresis to the Study of Human Germ-Line Mutation Rates*

J. V. NEEL
B. B. ROSENBLUM
C. F. SING
M. M. SKOLNICK
S. M. HANASH
S. STERNBERG
University of Michigan Medical School, Ann Arbor, Michigan

I. Introduction. 260
II. Some Basic Facts Concerning Mutation. 260
III. Why Attempt to Use 2-D Gels to Study Mutation? 262
IV. Some General Considerations in the Design of a Program to Detect Mutations. 263
V. Selection of the Most Appropriate Material for Study. 265
VI. Technical Considerations. 269
VII. A Statistical Analysis of Factors Influencing Spot Location. . . . 272
VIII. Ability of the System to Detect Genetic Variants. 274
IX. Nomenclature. 282
A. Basic Considerations in Nomenclature. 282
B. The Subdivision of the Gel into Constellations. 283
C. Designation of Variants. 283
X. Gel Analysis. 288
XI. Steps Necessary to Identify a Putative Mutant. 292
XII. Data Management. 296
A. A Friendly Interface. 297
B. Interactive Data Acquisition. 297
C. Networking. 298
XIII. The Selection of Populations for Study. 299
XIV. Concluding Comments. 302
References. 303

* Research supported in part by U.S. National Cancer Institute Grant O1CA2 6803-02 and Dept. of Energy Contract E(11-1)(2828).

TWO-DIMENSIONAL GEL
ELECTROPHORESIS
OF PROTEINS

ISBN 0-12-164720-X

I. Introduction

The question of the sensitivity of the human genetic material to the mutagenic effects of ionizing radiation and a variety of natural and man-made chemical agents has for some time been very much on the collective minds of geneticists, regulatory agencies, and the informed public. The past difficulties in reaching a precise evaluation of the genetic effects of a human mutagenic exposure (e.g., to ionizing radiation) have been widely discussed and are well known. Although a variety of convenient tests of somatic cell genetic damage in exposed persons have recently been developed (e.g., increases in chromosomal aberrations, sister-chromatid exchanges, and/or DNA adducts), the relevance of these findings to genetic damage transmitted to the next generation remains to be clarified. There is an urgent need for better methods for the study of germ line mutation. In recent years we have devoted a major effort to employing one-dimensional gel electrophoresis (1-DGE) in the study of spontaneous and induced mutation (see Neel, 1971, 1980, 1983; Neel *et al.*, 1979, 1980a,b, 1983a). In this chapter we will summarize the current status of our efforts to develop second-generation technology, that is, to adapt the two-dimensional polyacrylamide gel (2-D PAGE) technologies to this objective. It must be emphasized that this effort is still in its early stages and much of the material presented is in the nature of a progress report.

II. Some Basic Facts Concerning Mutation

The classic definition of a *point mutation* (as opposed to mutation involving all or a portion of a chromosome) demands the appearance in an offspring of a trait not present in either parent, which trait is then transmitted to subsequent generations according to Mendelian principles. Given the resolving power of current biochemical techniques, this definition must now be expanded to encompass any transmitted alteration in the sequence of nucleotides comprising an organism's DNA even though some of these alterations, as in "pseudogenes" or third positions in the triplet code, may not be immediately attended by what is usually perceived as phenotypic effects. In this presentation we will consider that a sequence of nucleotides discharging a specific function is a genetic locus; the alternative nucleotide sequences that occur at this locus are alleles. As a generalization, these alleles express themselves through the sequence of amino acids that comprise a polypeptide, the principal exception to that statement being mutations involving RNA functions. Even in this case the ultimate effect is on a polypeptide. Thus studies of the rate at which mutation alters protein structure or synthesis can often be related to the level of the basic phenotype, that is, the DNA level.

Two-dimensional electrophoresis of the O'Farrell type (1975) can in theory detect two kinds of mutations: (1) those in which the result of one or more nucleotide substitutions, losses, or duplications is a change in the amino acid composition of a polypeptide that alters the charge or apparent molecular weight of the polypeptide and (2) those in which the nucleotide change results in the failure to elaborate any of the customary gene product (i.e., true nulls). The ability to detect this second type of mutational event is not as yet well developed. In addition, as two-dimensional electrophoretic techniques not requiring denaturation are developed (as is true for 1-DGE), one could also under favorable circumstances (a combination of protein and activity stains) detect mutations involving enzymes that result in the elaboration of a protein of normal charge and weight that lacks enzyme activity.

Mutation is a rare event. Studies utilizing 1-DGE have suggested that spontaneous mutations in humans resulting in an electrophoretic variant that in the case of enzymes retains some activity have a frequency of approximately 2×10^{-6}/locus/generation (see review in Neel, 1983). On the basis of the composition of the average polypeptide, for each amino acid substitution that alters molecular charge there should also be two or three uncharged substitutions not detected by most screening electrophoresis. Furthermore, although as yet no spontaneous loss-of-activity mutations have been detected in relatively limited studies of humans and mice (Soares, 1979; Mohrenweiser, 1981; Johnson and Lewis, 1983; Satoh *et al.*, 1983b), for chemically induced mutations in mice the ratio of enzyme loss-of-activity mutations to electrophoretic mutations is 2 : 1 (Soares, 1979; Johnson *et al.*, 1981; Johnson and Lewis, 1981, 1983). Among spontaneous mutations in *Drosophila* the ratio of loss-of-activity to electrophoretic mutations was 6 : 1 (Mukai and Cockerham, 1977). Studies on the relative frequency of these two types of mutations in mammalian somatic cells in culture suggest an even higher relative frequency of loss-of-activity mutations than in *Drosophila* (Milstein *et al.*, 1974; Arthur *et al.*, 1975; Adetugbo *et al.*, 1977; Nelson and Harris, 1978), but one must be cautious in extrapolating from somatic cell to germ line events.

Let us assume that in humans the relative frequency of enzyme loss-of-activity mutants is similar to that observed in induced mutations in the mouse (a conservative estimate). It is not known what fraction of these mutations involve failure to form any polypeptide gene product, which in theory should result in a 50% reduction in polypeptide staining intensity on a 2-D gel, and what proportion result in a normal amount of inactive gene product with normal mobility not detectable with polypeptide stains in a 2-D gel. Let us assume that these two types of loss-of-activity mutants occur with equal frequency. Summing all the previously mentioned types of mutational events, we set the total frequency of mutation per locus per generation in humans at about 1×10^{-5}, but even with 2-D gels detectable and potentially detectable events will be about 0.4×10^{-5}/locus/generation. Thus any program using 2-D gels for the study of mutation must be able to detect very rare events with high accuracy. The attempt to use 2-D gels in this context is surely one of the most demanding applications of this technique to date.

III. Why Attempt to Use 2-D Gels to Study Mutation?

Early studies of both spontaneous and induced mutation rates in humans relied of necessity on morphological traits—medical syndromes attributable to dominant mutation (see review in Vogel and Rathenberg, 1975)—or to population indices to which fresh mutation contributed to some extent: congenital defect, early death, or impaired somatic development (see Neel and Schull, 1956). These suffered from the disadvantages that for the former, the precise genetic events underlying the phenomenon could not be specified, and that for the latter, in addition to this disadvantage, factors other than mutation contributed to the phenomenon under study.

With the development of convenient techniques for karyotyping and for one-dimensional electrophoresis, much more exact approaches to the question of the frequency and kinds of both point and chromosomal mutations were at hand. This presentation will not deal with the latter. With respect to the former, given that the ultimate basis of gene expression is mainly through protein structure, the ability to detect protein variants provides a more basic approach than previously available, often permitting extrapolation to events at the DNA level. Furthermore, although a minority of the electrophoretic variants are harmful (but recall hemoglobins S, C, and E or some of the glucose-6-phosphate dehydrogenase variants), electrophoretic mutants can serve as indices to more serious mutational changes, that is, loss of proteins. There is reason to believe some success will attend efforts to detect this latter type of mutation by 2-DGE in which case the frequency can be estimated directly rather than indirectly.

Because mutation is a rare event the numbers of electrophoretic observations necessary to detect an altered mutation rate are quite large, so large as to be (until now) an almost prohibitive consideration to the use of electrophoresis in this context. The principal such attempt to date for humans is in the study of the potential genetic effects of the atomic bombs, where because of the extreme interest in the question and the presumed high genetic risk, observations employing 1-DGE have been initiated (Neel *et al.*, 1980a). Given the extent to which concern over the possibility of genetic damage enters into the discussion of industrial, medical, and military developments, there is an urgent need for the development of more efficient approaches to generating the necessary number of observations.

At present, 2-D PAGE is one of the two best prospects for improving the efficiency of the study of mutation. The other involves studies at the DNA level itself, but aside from the fact that a suitably efficient technology is not yet at hand, these have the drawback that unless probes for defined DNA sequences involving translated information are employed, there is no way to extrapolate from an observed mutation to the protein (or grosser) levels. Despite the promise of 2-D PAGE, many developments and cross-verifications are necessary before the approach can be certified as useful; these will be discussed in this chapter.

IV. Some General Considerations in the Design of a Program to Detect Mutations

Given the requirement for very large numbers of highly accurate observations in the study of mutation, our challenge is the development of a system that will yield the maximum amount of reliable information per unit effort. This system begins with the choice of sample and the manner in which it is collected and processed and terminates with a format for data management that ideally not only permits ready retrieval of the findings with respect to any specific protein but—because a body of data of general genetic interest will be generated—permits relating the findings on any particular protein to the findings on any other protein in the study.

The strategies for studying mutation differ in many ways from those appropriate to attacking many of the other problems in whose solution 2-D gels will be helpful. For many of these problems the findings on a relatively few critical gels will be decisive, and the investigator runs and reruns his gels until the desired result is achieved, often slightly varying technique from run to run in order to achieve better resolution. By contrast, in an ongoing study of two groups of children, one born to parents exposed to a possibly mutagenic experience and one born to parents not so exposed (the controls), once the study is initiated the technique must remain rigidly standardized to avoid time-dependent biases as, for example, the proportion of children born to the two groups changes.

If one is interested in visualizing "all" the protein constituents of a cell following some experimental manipulation, it will usually be necessary to run several gels, each under somewhat different constraints (isoelectric focusing gradients, electrophoretic conditions). If, however, a single set of gel conditions yields 80% of the information that could be obtained from multiple gels of the same source material, then in a search for mutations it may be more efficient to run a single gel on many people than several gels on half as many people. As a further example of the strategic considerations that enter into the study of mutation, consider that in some studies one scans a gel for any detectable departure from a standard. By contrast, in the study of mutation one must work only with that subset of the many protein moieties displayed on a gel in which genetic variation can be detected with high accuracy.

In the more familiar morphological approach to mutation, such an event is detected through the occurrence of a child with an abnormal phenotype, which phenotype is then transmitted to the next generation. Figure 1 illustrates several examples from previous studies by our group. In such studies one selects phenotypes whose diagnosis is minimally ambiguous. Once the dominant inheritance of the trait has been established one can score as mutations isolated instances of the phenotype even without the test of genetic transmission. In the present context, a mutation will be manifest as a polypeptide variant in a child not present in either parent. The demonstration of this fact requires a family study; if both parents of a child with a polypeptide variant of interest cannot be studied, that variant does not contribute information regarding mutation. If the 2-D PAGE approach is to be successful, one must have the same level of familiarity with protein phenotypes that is necessary in the study of morphological phenotypes. Thus one must be able to identify and exclude from consideration fetal proteins that might be present in a premature child but not in a full-term infant or older child. One must also have a knowledge of epigenetic phenomena, such as

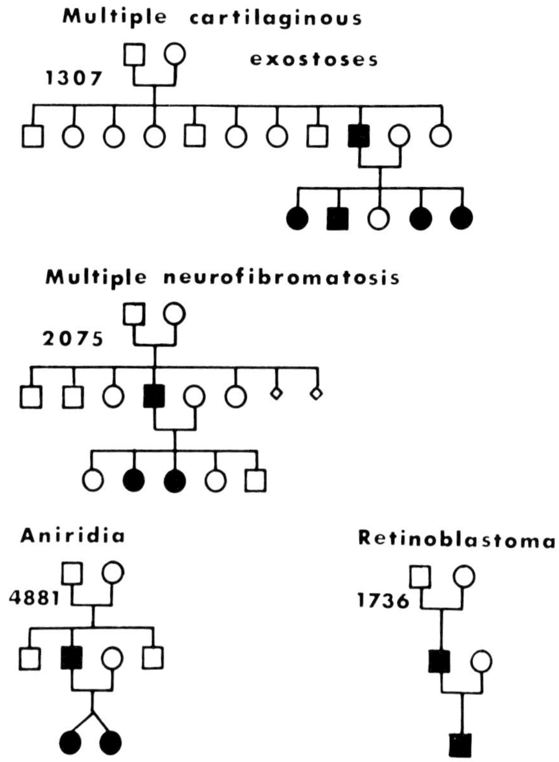

Figure 1. Four examples of mutation resulting in a dominantly transmitted trait in humans. ■, affected males; ●, affected females; □, unaffected males; ○, unaffected females. The numbers attached to each pedigree indicate the kindred number in the files of the University of Michigan Heredity Clinic.

posttranslational modifications, so that these are excluded from consideration. For any putative mutation the possibility of a discrepancy between legal and biological parentage must be considered. Finally, in the ultimate a protein moiety thought to result from a mutational event should be characterized to the extent that its affinities (i.e., the protein of which it is a variant) can be identified (see Section XI).

V. Selection of the Most Appropriate Material for Study

The most readily accessible sources of data concerning human protein phenotypes are urine, saliva, sweat, hair follicles, and blood. Of these, blood is the most informative, and it is this tissue that forms the mainstay of our studies. The principal components of interest are the plasma and four cell types (or fractions thereof): erythrocytes, polymorphonuclear leucocytes (including eosinophils and basophils), nonpolymorphonuclear leucocytes (lymphocytes of various types, monocytes), and platelets. In the interest of cross-preparation comparisons, in our laboratory all routine 2-D preparations are silver-stained; isotopic labeling of the proteins is reserved for special studies. For all of the four cell types, one can elect to study cellular membrane proteins separately from the cytosol.

Studies in this laboratory suggested a high overlap among the proteins of polymorphonuclear leucocytes, nonpolymorphonuclear leucocytes, and platelets, based on the criterion of amount and position in the gel (S. Hanash, personal communication). The term *position* refers not simply to x and y coordinates, which for a given protein vary from gel to gel, but to position within a constellation of protein moieties. In time this "guesstimate" of the amount of overlap should be subject to precise verification by a variety of polypeptide characterizations. This observation suggests that it would not be efficient to run preparations of these three types of cells routinely. Platelets seem to yield better patterns than either of the other two cell types as judged by dispersal over the gel and resolution of the individual spots. Therefore for the present we have elected to study platelet preparations routinely but harvest and hold the leucocytes with no separation of membrane and cytosol. In the event of an interesting finding, not only can the platelet preparation be rerun but, because of the apparent overlap in proteins among the different cell types, there is the possibility of confirming the finding in preparations based on the other two cell types. Figure 2 illustrates what appears to be the same variant in gels of solubilized platelets and of polymorphonuclear leucocytes from the same individual. The relative ease and quantity with which erythrocyte membranes can be harvested has led us to consider these as separate sources of material for the study of genetic variation (Rosenblum *et al.*, 1982). Thus it is our current practice to run erythrocyte membranes and erythrocyte cytosols separately.

Because of the frequency with which they are subject to posttranslational modification, plasma proteins constitute the least desirable set among the four sets of proteins with which we have elected to work. They have, however, been more extensively studied than the other sets of proteins both from the biochemical and the genetic viewpoints. In our work the availability from past studies of a considerable number of variants of such proteins as albumin, group-specific component, transferrin, and ceruloplasmin has been an important asset in testing the resolving power of our 2-D PAGE procedures.

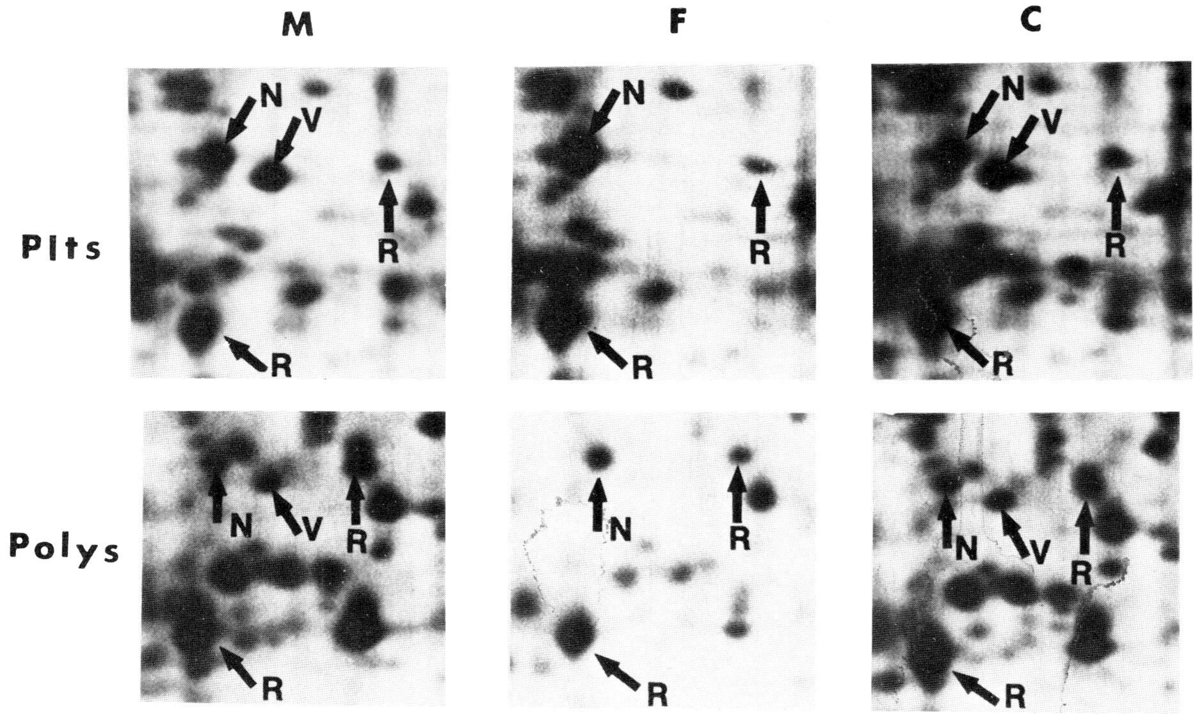

Figure 2. Enlarged sections of corresponding regions in gels of human platelets (Plts) and polymorphonuclear leucocytes (Polys), illustrating the inheritance of a variant protein (V). N, normal protein; M, mother; F, father; C, child. The variant in the child is inherited from the mother. Identity of the variable protein in the two preparations is judged not only by position in gel but by position with reference to certain invariant, apparently homologous spots in the two types of preparations, labeled *R*.

Not all the protein moieties that can be visualized on a gel are suitable for the study of mutation. For each type of preparation we are in the process of selecting a subset of spots meeting two criteria: (1) they are of sufficient intensity (silver staining) that if genetic variation were to result in two spots instead of one spot (i.e., a heterozygote), each of the two would have a staining intensity well above background and (2) there is a sufficient free area surrounding the spot in both axes that a variant would be unlikely to be superimposed on another protein moiety and so be missed. It must be admitted that this criterion is not always absolutely maintained. If the spot of interest is deeply staining but there are a few faintly staining spots in the immediate vicinity, the deeply staining spot will be included with those to be studied.

With these restrictions we are currently estimating that the minimal number of moieties suitable for the study of mutation in each type of preparation is at this stage the following: plasma, 30; erythrocyte membranes, 45; erythrocyte cytosols, 45; and platelets, 60, for a total of 180 polypeptides, the product of 360 alleles at 180 loci. (Only experience will reveal exactly how many loci are involved.) With further familiarity with the gel patterns, improvement in gel technology, and increasing sophistication in the computer alogrithms, this number may increase by at least 50%.

In the context of a study of mutation genetic variants are of two types: variants corresponding to common polymorphisms (allele frequency $>$ 0.01) and rare variants with lesser allele frequencies. Although concealed among the former there could be an occasional mutant, the likelihood of encountering a mutant is substantially higher among the rare variants. This is because the common polymorphisms may be maintained by selection, whereas the rare variants are more apt to reflect mutation pressure (detailed argument in Neel *et al.*, 1983a). Therefore we limit our search for mutation to the rare variants. The experience with 1-D gels suggests that among Caucasoid populations nonpolymorphic electrophoretic variants of the type requiring family studies occur with a frequency approximating 1–2 per 1000 protein determinations, the number varying widely from system to system (see review in Harris *et al.*, 1974; Neel and Rothman, 1981). It is not clear whether this figure can be extrapolated to 2-D gels in view of the apparently lower levels of heterozygosity being encountered in these gels (discussion in Section VIII). If we assume the extrapolation to be proper, then with 180 spots being scored per individual, approximately one in three individuals will exhibit a variant requiring family studies as a possible mutant. However, at this stage there is still a great deal of normative work necessary to distinguish common polymorphisms from rare variants, and this also requires family studies. With the various preparations we make, every child exhibits some finding of interest. Thus we find it efficient at this stage of our program to make preparations of child, mother, and father initially. Indeed, the probable necessity for so many family studies suggests that in a study in which the focus is strictly mutation, that is, surveys of children born to parents who are thought to have sustained a mutagenic exposure, it will be desirable to collect blood samples from both the parents at the time the sample is obtained from the child, as more economical of effort than attempting to contact and obtain samples from the parents only in the event of an interesting finding in a child. In a survey to be described briefly later, we only make preparations from the child when blood from both parents is at hand.

VI. Technical Considerations

As indicated earlier, for the present we are attempting to build this study of mutation around the use of blood samples. From each sample of blood six fractions are obtained for analysis by two-dimensional electrophoresis. The whole blood sample is first subjected to low speed centrifugation and the platelet-rich plasma removed (see Holmsen *et al.*, 1972). The platelets are then pelleted from the plasma by high speed centrifugation and the plasma removed and saved. The platelet pellet is washed with phosphate-buffered saline (PBS) and is now ready for solubilization. Following removal of the platelet-rich plasma, the remaining red and white cells are diluted with PBS to the original volume and the nonpolymorphonuclear cells separated by centrifugation on Ficoll-Hypaque and recovered (Boyum, 1968). The remaining mixture of polymorphonuclear cells and erythrocytes is then separated into the two cellular components by sedimentation on 3% dextran sulfate (see Tauber and Goetzel, 1979). These steps result in the isolation of five fractions: plasma, platelets, nonpolymorphonuclear cells, polymorphonuclear cells, and red cells. The latter fraction is lysed and cytosol and membranes are processed separately.

For each fraction specific techniques have been developed for the solubilization of the sample and for the first and second dimension of the electrophoresis, techniques that yield maximum reproducibility of the gel patterns in our hands. Following isolation and washing, platelets, polymorphonuclear cells and nonpolymorphonuclear cells are treated identically. If samples cannot be solubilized the day of preparation, they are held at −80°C until further processing is convenient. To limit proteolysis, phenylmethylsulphonyl fluoride is added at the time of solubilization. The amount of solubilization solution (9 *M* urea, 2% Nonionic detergent P-40 (NP-40), 2% β-mercaptoethanol, and 2% LKB pH 3.5–10 Ampholines) added to the cell pellet is determined by the size of the pellet. Following the solubilization, the samples are centrifuged at 13,000 × *g* for 2 min and the supernatant applied to the first-dimension gel (LKB pH 3.5–10 Ampholines). Electrophoresis is for 16 h at 700 V and then for 2 h at 1200 V. The gels are then extruded from the tubes and equilibrated for approximately 5 min in the solution described by O'Farrell (1975). Substantial losses of protein from the first-dimension gels may occur with longer equilibration times (Tracy *et al.*, 1982a; B. B. Rosenblum and P. E. Daddona, unpublished). First-dimension gels may be frozen and held indefinitely. Second-dimension SDS gels are prepared with the DALT apparatus (Anderson and Anderson, 1978); we use a 11.25–13.75% acrylamide gradient (Rosenblum *et al.*, 1982). Following electrophoresis, gels are fixed in 50% ethanol–5% acetic acid and then washed with two changes of ethanol–acetic acid in which the ethanol concentrations are 25 and 10%. Finally the gels are soaked in distilled water. Gels are routinely stained with silver by either the technique of Merril *et al.* (1981) or Sammons *et al.* (1981) (see also chapters by Merril and Goldman and Sammons *et al.*). Silver staining by the technique of Sammons *et al.* (1981) is more sensitive and results in differences between spots in coloration that appear to be at least in part polypeptide-specific.

Plasma samples are treated with two volumes of a 2% SDS, 5% β-mercaptoethanol, and 10% glycerol solution as described by Anderson and Anderson (1977) and Rosenblum *et al.* (1983a). First-dimension gels (LKB pH 3.5–10 Ampholines) are run for 15 h at 1200 V and 2 h at 1500 V. Gels are extruded and equilibrated as described above. Second-dimension SDS gels are prepared as described for white cells. Following electrophoresis, plasma patterns are initially stained with 0.2% Coomassie Brilliant Blue in 50% ethanol–5% acetic acid. Gels are destained with decreasing concentrations of ethanol in 5% acetic acid and finally distilled water. Several proteins including transferrin, Gc-globulin, α-fibrinogen and α_1-antitrypsin are more easily scored for genetic variation on Coomassie Blue–stained gels than following silver staining. All gels found to exhibit satisfactory Coomassie Blue patterns are subsequently silver-stained by the technique of Sammons *et al.* (1981).

The amount of albumin and immunoglobulin have significant effects on the pattern of proteins from plasma samples. Proteins are amphoteric and hence function as ampholytes in the first-dimension (focusing) gels.

Specific proteins appearing in high concentrations, such as albumin, therefore significantly affect the pH gradient in the vicinity of the specific protein. The immunoglobulins, in contrast to albumin, which focuses in a relatively narrow pH range, stretch out over a major portion of the gel (see Fig. 6). They may obliterate and interact with part of the nonimmunoglobulin plasma pattern because of the large portion of the gel they cover. Previous studies have suggested that Cibachrome Blue coupled to Sepharose may be used to remove albumin and Protein A may be used to remove the IgG fraction of immunoglobulin (Tracy *et al.*, 1982b). We have found that both Cibachrome Blue and Protein A are nonspecific and remove many additional proteins from the plasma pattern (B. B. Rosenblum and M. Gagnon, unpublished). Some of these additional proteins were removed as efficiently or more efficiently than either albumin or IgG. One of the reasons for our finding is the use of a more sensitive (silver) stain than employed by other investigators.

The red cell lysate is prepared by diluting the washed packed red cells with two volumes of 5 m*M* phosphate buffer, pH 8.0. For first-dimension electrophoresis the sample is solubilized in an equal volume of 9 *M* urea–2% NP-40–2% β-mercaptoethanol–2% pH 3.5–10 Ampholines (LKB) (Edwards *et al.*, 1979). First-dimension gels (LKB pH 3.5–10 Ampholines) are focused at 1200 V for 15 h and then at 1500 V for 2 h. Gels are extruded and equilibrated as previously described. Following the second-dimension electrophoresis, the gels are stained with Coomassie Blue as described for plasma as a check on the quality of the gel. Carbonic anhydrase yields a diffuse pattern after silver staining and therefore must be scored from Coomassie Blue-stained gels. The hemoglobin chains are also better evaluated from the Coomassie Blue-stained gels. Satisfactory gels are subsequently silver-stained by the technique of Sammons *et al.* (1981).

Red cell membranes are washed until they are colorless with 5 m*M* phosphate buffer, pH 8.0 (see Rosenblum *et al.*, 1982), then lyophilized. Membranes are solubilized in a solution of 4 *M* urea, 2% NP-40, 2% β-mercaptoethanol, and 2% pH 3.5–10 Ampholines (LKB) at a concentration of 40 mg dried membrane per ml of solubilization solution. First-dimension gels are prepared as described by Rosenblum *et al.* (1982); the Ampholines (LKB) are added in a ratio of 62.5% pH 5–7 and 37.5% pH 3.5–10. Electrophoresis is performed at 1200 V for 15 h and then for 2 h at 1500 V. Gels are extruded and equilibrated. Second-dimension gels are prepared and electrophoresed as described above. Gels are screened for quality by staining with Coomassie Blue as described above. Streaking in the high molecular weight range and a large actin spot usually indicate poor gels, which will not be useful for genetic analysis. Acceptable gels are further stained with silver by the technique of Sammons *et al.* (1981).

For the present, as already noted we prefer to work with father–mother–child trios for any specific cell type or fraction. The probability of obtaining a research-quality gel on the first run now varies (according to preparation) from 50 to 90%. Working with trios rather than singletons thus requires not only running three instead of one gels originally but an additional 10–100% more gels to ensure satisfactory preparations. The total number of gels would be reduced if one ran gels on the parents only after the demonstration of an interesting variant in a child. While we are still defining the frequency and kinds of variation in these preparations, we find that the convenience of immediate decisions concerning the genetic nature of a finding outweighs the added amount of work involved.

We do not routinely include molecular or charge standards in the gels because sufficient internal control landmarks have been established in the gels. Such standards are employed in detailed studies of specific proteins.

VII. A Statistical Analysis of Factors Influencing Spot Location

An important consideration in the evaluation of 2-D PAGE data is an evaluation of the experimental and biological sources of variability among gels. We have chosen to begin with a consideration of the factors other than allelic variation influencing the dispersion of spot locations. The objective is to estimate the range of variation in the x and y dimension in order to set statistical limits on the contributions of technical and biological factors to shifts in spot location and by identifying the experimental sources of such variation to direct our attention to those aspects of laboratory and image evaluation processing procedures that might be improved. We briefly summarize here our first attempts to understand variability in spot location.

Our initial study (Sing *et al.*, 1983) involved a sample of 21 gels that were evaluated for spots that were unambiguously identifiable and present on all gels. Multiple gels were run on each of the blood samples drawn from six individuals. Seven other individuals were represented by only one gel. Each of the gels was evaluated by three observers, resulting in 63 observations on each spot. We estimated the variation among observers of the same gel, among gels within individuals, and among individuals for the x and the y coordinates of spots defined by two different geometric transformations. The first transformation involved establishing a common coordinate system based on the location of β-actin and the dye front. This treatment results in the rescaling and rotating of all spots to a standard coordinate system with actin at the origin and the y axis perpendicular to a line drawn between the ends of the dye front. The effects of local distortion were investigated by considering a second transformation that further modified the location of a spot according to the definition of a standard experimental unit or field corresponding in most respects to what elsewhere we have termed a constellation. This was carried out by designating three spots that bounded an area that defined the region of the gel being studied. The linear transformation that mapped these three spots on each gel into their respective average location based on all 63 observations was used to transform the x and y coordinates of the remaining study spots on each gel. The effect of this second transformation was to remove variation due to distortion of the gel pattern among the sample of gels being considered.

Table I summarizes the analysis of variation in four spots, R1–R4, transformed to the actin-dye front coordinate system and in two spots, β-actin and R4, transformed to remove the variation in the shape of the

TABLE I

An Analysis of the Sources of Variability in 2-D Gel Spot Coordinates

Spot Number		Mean (mm)	Total variance	% Variance due to Individual	Replication	Observer	C.V.[a]
R1	X	18.2	4.1	68.2	7.7	24.1	6.3
	Y	18.8	0.9	34.9	14.7	50.4	4.1
R2	X	0.6	0.7	32.3	37.4	30.3	112.7
	Y	62.6	5.9	62.8	3.4	33.8	2.4
R3	X	−14.4	1.1	43.0	28.7	28.3	5.5
	Y	39.6	3.9	62.1	5.9	32.0	3.1
R4	X	33.0	5.3	45.0	16.8	38.2	5.2
	Y	57.3	5.7	53.4	4.0	42.6	2.8
Analysis after transformation based on experimental unit							
β-Actin	X	−18.2	1.5	55.1	3.7	41.2	4.5
	Y	−18.8	1.0	42.0	0.0	58.0	4.1
R4	X	14.8	1.3	22.6	15.4	62.0	6.7
	Y	38.5	2.0	14.8	15.4	69.8	3.3

[a] C.V. is coefficient of variation based on the sum of replication and observer error variances, i.e., the square root of the percentage of variance due to replication and observer times the total variance divided by the mean.

region defined by the triangular experimental unit bounded by the vectors between R1, R2, and R3. For each actin-corrected spot, the appropriate adjusted variance reveals that approximately 95% of these data fall in a radius of 4 mm of its average location. When distortion is removed by the method considered, the 95% confidence region about spot R4 is reduced from 9.5 mm to approximately 5.6 mm. Failure to detect a statistically significant correlation between the coordinates for any spot suggests a symmetrical confidence interval.

Statistically significant variability attributable to differences between individuals was detected at the 0.05 level of probability in both the x and y dimensions for every spot studied. There was greater replicate variability in the x dimension and greater observer variability in the y dimension in every case. The coefficient of variation based on these two sources of variance was less than 7% in all but one case where the spot was near the origin of the x axis. The increase in the fraction of the observer variance upon correction for distortion suggests that there may be individual and gel effects on the relationships among spots in this series of gels.

This preliminary study documents as much as a 5 mm radial shift in the location of a given spot when replicated gels are available on a dozen individuals. Correction for distortion in the immediate vicinity of the spot of interest may reduce this by one-half. A significant fraction of the variation in the location of a spot may be attributable to effects, either intrinsic or acquired, that are small in size and specific to the individual.

VIII. Ability of the System to Detect Genetic Variants

As a prerequisite to evaluating genetic data derived from the use of 2-D PAGE, it is necessary to know the resolving power of this technique compared to standard (1-D) techniques. To this end we first tested the ability of a single set of gel conditions to resolve previously detected variants of four human proteins (β-globin, carbonic anhydrase I and II, and albumin) and one mouse protein (pancreatic amylase) (Wanner *et al.*, 1982). Because β-globin focuses closely to the basic edge on a standard O'Farrell type gel, a gel gradient with a range of pH 5.5–8.5 was used in the study of variants of this protein. Out of a total of twelve varients, nine were resolved unambiguously. Two of the albumin variants behaved inconsistently, being well resolved on some gels but not on others run under presumably identical conditions. One of the amylase variants was not resolvable with the gel conditions employed. Figure 3 illustrates the degree of resolution obtained for several of the variants. The actual physical distance in the electrophoretic axis occasioned by a gain or loss of one charge of the β-globin or carbonic anhydrase I polypeptides (charge change determined by sequence data) will depend on molecular weight, pH gradient of the gel, and other factors. Our experience with this limited number of known charge changes suggests that with a standard gel (pH gradient 4–7) a gain or loss of one charge alters the mobility of polypeptides such as these by about 7 mm and two charges, as expected, double that. It should be recalled that in most of the 1-D electrophoretic systems that have been developed an effort has been made to optimize the electrophoretic conditions in respect to detecting variants of a specific protein. Sometimes several 1-D systems have been employed simultaneously in the detection of variants of a particular protein because, as is well known, it is possible to find a pH at which two proteins differ in net charge even though their isoelectric points are identical.

Employing the gel conditions described earlier, we have studied the ability of the technique to resolve variants of a series of plasma proteins whose identity was unknown at the time they were selected for study (Rosenblum *et al.*, 1983a,b). For this purpose samples have been available for newborn children (placental blood) and their mothers and fathers. One investigator not involved in running the gels selected from typical silver-stained gels of a small series of children a battery of polypeptides whose identity was unknown to him, all polypeptides approximately meeting the criteria laid out in Section IV. ("Unknown" proteins were used to avoid the bias that might occur if one knew the nature of the proteins and/or had previously worked with the gels extensively.) Several of the apparently suitable spots selected (primarily those focusing in the albumin region but of lower molecular weight) exhibited excessive gel-to-gel variation and had to be eliminated from consideration. There were left 20 spots, which have been analyzed in 56 gels (with parental material). The results are shown in Table II. All variants scored in a child were observed in one parent or the other. Variants of five of these proteins are figured in Rosenblum *et al.*, (1983b). Not all the proteins could be scored unambiguously in all of the gels. Because ambiguity may indicate a variant, in a study of mutation a repeat run is indicated for a sample yielding a gel in which a protein could not be scored. However, two of the proteins (D-005 and D-013) could not be scored in 15 of the 112 opportunities. Unless technical improvements permit better scoring of these proteins on the first gel run, they will not be consid-

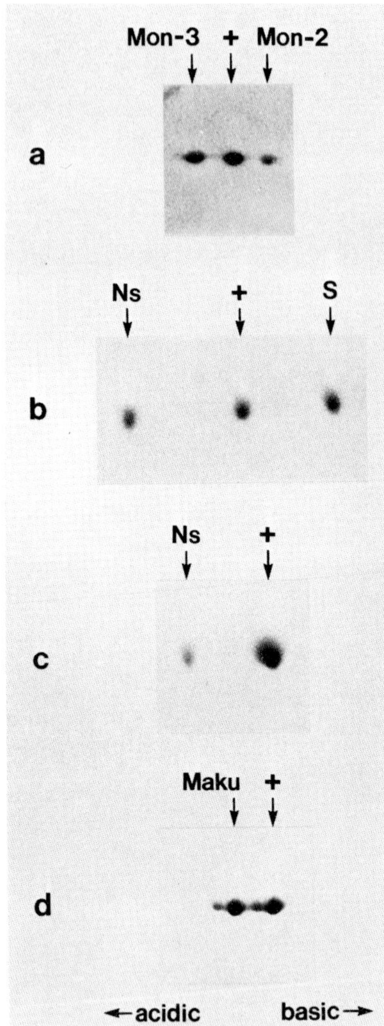

Figure 3. Demonstration of the effectiveness of detection of known protein variants with 2-D PAGE. (a) Comparison of migration of normal carbonic anhydrase I with Montreal-3 (−1 charge) and Montreal-2 (+1 charge); gel gradient pH 4–7. (b) Comparison of normal β-globin with sickle cell β-globin (+1 charge) and globin $N_{Seattle}$ (−2 charges); gel gradient pH 5.5–8.5. (c) Separation of these same three β-globins on a gel with the "standard" pH gradient (4–7). Now globin $N_{Seattle}$ separates less well and sickle cell β-globin either does not separate from normal, because of the marginal position of normal β-globin, or else runs off the gel. (d) Separation of albumin Maku from normal albumin (+). The charge difference is unknown. Further details in Wanner *et al.* (1982).

ered appropriate for the study of mutation. Five of the 20 polypeptides proved to be variable; of a total of 1091 polypeptides scored, 70 exhibited inherited variants. This corresponds to a frequency of heterozygotes (index of heterozygosity) of 6.4 ± 0.7%.

We have also scored these same gels from Coomassie blue-stained preparations for variants with respect to 11 previously identified additional proteins, most of which we do not regard as suitable for studies of mutation because their position in the gel suggests that a considerable fraction of the variants could not be readily visualized. These include many well-known proteins. The results are given in Table III. Five of these 11 proteins exhibited genetic variation (confirmed by studies of the father and mother); of a total of 591 polypeptides scored, 30 exhibited heterozygosity, for a heterozygosity index of 5.1%. Table II also presents data on two other unknown proteins (B-003 and C-011) scored from Coomassie Blue preparations, not included in the battery of 20 identified as suitable for mutation studies, which were incidentally observed to be polymorphic in the course of this study. The C-011 polymorphism was previously observed by Tracy *et al.* (1982a); we have identified an additional variant in this system. Figure 4 illustrates the appearance in 2-D PAGE of variants of five known proteins, four from the group scored in Table II (transferrin, group-specific component protein, α-fibrinogen, and prealbumin) and one protein selected (when its identity was still unknown to us) as suitable for mutation studies (apolipoprotein E). This figure also presents the results of limited family studies (father, mother) of the same variants. Figure 5 illustrates variants of the two additional currently unknown proteins (B-003 and C-011). As the figure demonstrates, the genetic nature of these variants has also been confirmed by limited family studies.

The variants detected in this study of plasma proteins differed from normal in migration in the isoelectric focusing (IEF) axis by 1 to 8 mm and in the molecular axis by approximately 2 mm. It is of course possible that we have failed to detect some variants with even greater mobility shifts. Nevertheless, the data suggest that the ideal spot for scoring for genetic variants should be surrounded by an "unpopulated" radius of some 10 mm, an ideal which will not even be approximated by most of the spots on a gel under what currently are standard conditions for most laboratories. For critical genetic investigations it may be necessary to resort to procedures that fractionate the solubilized mixture by some criterion other than IEF or molecular weight before the sample is analyzed.

The frequency with which heterozygotes for polypeptide variants were encountered in this study is in good agreement with the index of

TABLE

Occurrence of Genetic Variation in 20 Proteins Scored in

	A-001	C-001	C-007	C-008	C-009	C-012	C-013	C-015	C-016	D-004	D-005
Normal homozygotes	55	53	55	24	56	56	56	56	56	56	15
Heterozygotes	0	3	0	18	0	0	0	0	0	0	26
Variant homozygotes	0	0	0	13	0	0	0	0	0	0	7
Total	55	56	55	55	56	56	56	56	56	56	48
p[b]	1.0	0.946	1.0	0.6	1.0	1.0	1.0	1.0	1.0	1.0	0.583
q[b]	0	0.054	0	0.4	0	0	0	0	0	0	0.417
χ^{2}[c]				5.57[d]							0.63

[a] The identity of the proteins was unknown to the individual who selected them for study. Criteria for selection were as described in text; they are condensed to be suitable for the study of mutation (with two possible exceptions). The genetic nature of the variation was in all instances confirmed by the presence of the same variant in the father or mother.

[b] p, frequency of normal allele; q, frequency of variant$_1$ allele.

[c] Test for agreement with Hardy-Weinberg equilibrium proportions.

[d] D.F. = 1, .05 > p > .01.

TABLE

Occurrence of Genetic Variation in 13 Well-Known Proteins

	A-002 Hemopexin	A-003 Albumin	B-001 Plasminogen	B-002 Transferrin	B-003 (?)	C-003 α_1-Antitrypsin
Normal homozygotes	51	56	47	47	36	48
Heterozygotes	0	0	0	3	18	5
Variant homozygotes	0	0	0	0	0	0
Total	51	56	47	50	54	53
p[b]	1.0	1.0	1.0	0.970	0.833	0.953
q[b]	0	0	0	0.030	0.167	0.047
χ^{2}[c]						

[a] Identity of the proteins is given in the legend for Fig. 4. These proteins did not meet the criteria as objects for the study of mutation. The genetic nature of all variants was confirmed by the presence of the same variant in the father or the mother.

[b] p, frequency of normal allele; q, frequency of variant allele.

[c] Test for agreement with Hardy-Weinberg equilibrium proportions.

II

56 Persons from Silver-Stained Gels (Sammon's Technique)[a]

D-006	D-007	D-008	D-009	D-010	D-011	D-012	D-013	C-014	
52	55	52	55	55	54	55	49	Normal homozygote	35
								N/V_1, Normal/variant$_1$ heterozygote	6
0	0	3	0	0	0	0	0	N/V_2, Normal/variant$_2$ heterozygote	13
0	0	0	0	0	0	0	0	V_1/V_2, Variant$_1$/variant$_2$ heterozygote	1
—	—	—	—	—	—	—	—	V_1/V_1, Variant$_1$ homozygote	1
52	55	52	55	55	54	55	49	V_2/V_2, Variant$_2$ homozygote	0
									—
1.0	1.0	0.973	1.0	1.0	1.0	1.0	1.0	Total	56
0	0	0.027	0	0	0	0	0		
								p^b	0.795
								q^b	0.080
								r^e	0.125
								χ^{2c}	0.55

III

Scored in 56 Persons from Coomassie Blue–Stained Gels[a]

C-004 Gc-Globulin	C-005 α-Fibrinogen	C-006 α_1-Acid glycoprotein	C-011 (?)	D-001 γ-Fibrinogen	D-002 β-Fibrinogen	E-002 Prealbumin
29	54	56	48	56	55	55
20	1	0	8	0	0	1
6	0	0	0	0	0	0
—	—	—	—	—	—	—
55	55	56	56	56	55	56
0.709	0.991	1.0	0.929	1.0	1.0	0.991
0.291	0.009	0	0.071	0	0	0.009
0.77						

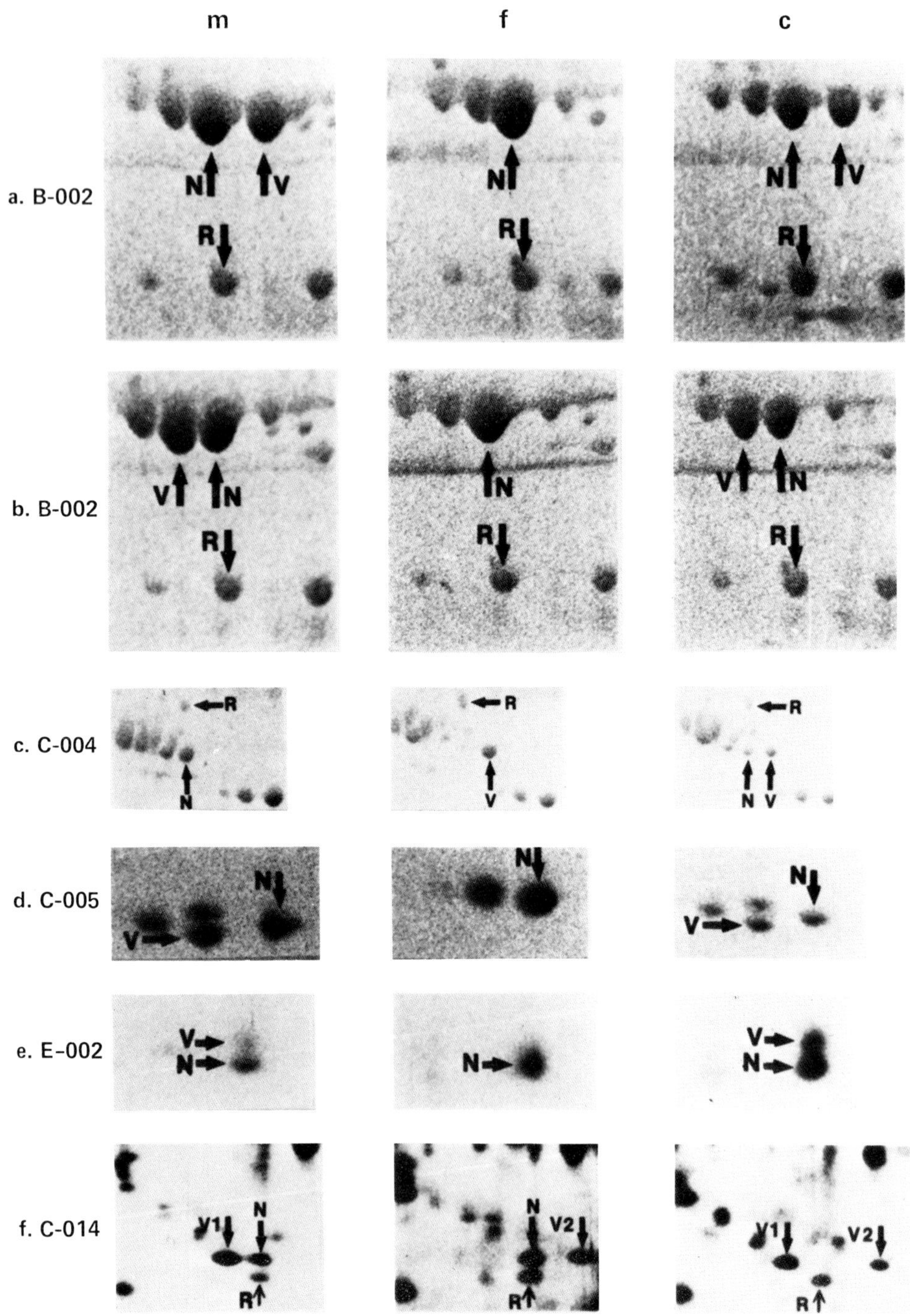
m
f
c
a. B-002
b. B-002
c. C-004
d. C-005
e. E-002
f. C-014
N
V
R
V1
V2

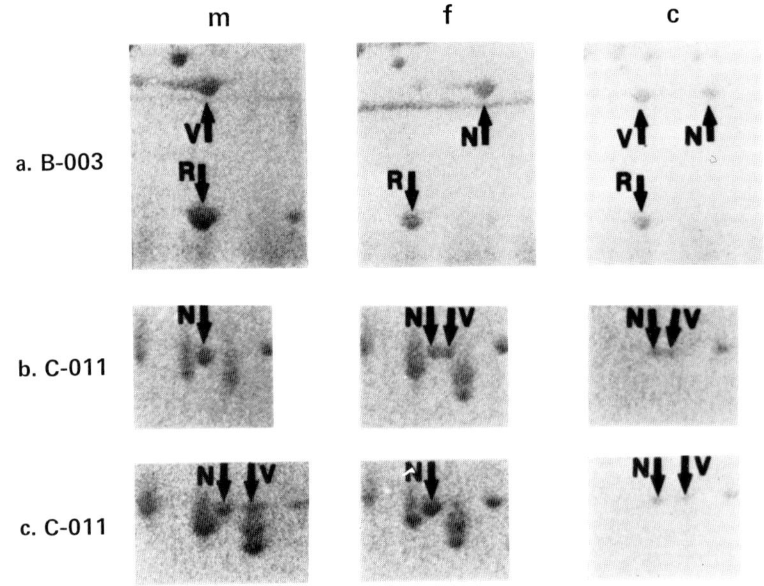

Figure 5. Illustration of the appearance of variants of two presently unidentified proteins (B-003 and C-001) not included in the batteries listed in either Table I or Table II. (a) A highly polymorphic protein best scored in the Coomassie Blue–stained gels. One parent is homozygous for the allele encoding the variant, the other for the normal allele, and the child is a heterozygote. (b) and (c) Two different variants of protein C-011. The variant illustrated in (b) apparently corresponds to one identified by Tracy *et al.* (1982a). Note the persistent reference spot to the far right of each field but the absence in both children of two polypeptides well represented in their parents; these are the multiple β-haptoglobin polypeptides.

heterozygosity (6–7%) based on 1-D studies of a variety of proteins (see review in Harris, 1980). This observation is, however, in sharp contrast to the findings of others regarding apparent heterozygotes in 2-D PAGE preparations of human lymphocytes, kidney and brain samples, and cultured fibroblasts, studies that also involved unknown polypeptides

---

Figure 4. Illustration of appearance of variants of five proteins whose identities are known. Given our criteria, only C-014 is thought to be suitable for the study of mutation. R, a convenient nearby reference spot (where one is available); N, the normal position of the spot being scored; and V, a variant of that spot. (a) and (b) "Fast" basic (D type) and "slow" acidic (B type) variants of transferrin. (c) The common polymorphism of group-specific component protein (Gc 1 and 2). (d) A variant of the α-chain of fibrinogen. The normal pattern consists of two polypeptides, which vary concurrently in the mother and child. (e) A molecular weight variant of prealbumin. (f) Two variants of what we have tentatively identified as apolipoprotein E, the child being a double heterozygote.

in which heterozygote frequencies of less than 2% have been reported (McConkey *et al.*, 1979; Walton *et al.*, 1979; Hamaguchi *et al.*, 1981; Smith *et al.*, 1981; Comings, 1982b). Elucidating the cause of these differing estimates of heterozygosity must have high priority. It has been suggested that the proteins visualized in these 2-D PAGE preparations are for the most part structural (rather than enzymatic) and that these are less free to vary than the proteins on which estimates of heterozygosity have thus far been based (see previous citations). This reservation would not apply to the plasma proteins. This explanation implies remarkably strong stabilizing selection against variants of these cellular proteins. (For the present, we assume essentially similar mutation rates at all the loci coding for proteins.) An alternative (and not exclusive) explanation is that other investigators have not placed the same constraints on the proteins they scored as we have. We hesitate to compare our results with those of any study that did not employ the same scoring criteria as we have; parental material should be available for the verification of the genetic nature of subtle alterations. We propose in the near future to analyze for the frequency of genetic variation the other types of preparations that are routine in this laboratory by the criteria employed in the study of plasma proteins. The issue of the amount of heterozygosity in populations is central to population genetics.

IX. Nomenclature

Any system of nomenclature for the proteins visualized on a 2-D gel advanced at this time must be considered temporary and stopgap; as familiarity with 2-D gels increases it will undoubtedly be superseded by a more precise system of nomenclature just as the hemoglobin variants have progressed from an alphabetical–geographical type nomenclature to one which identifies the precise amino acid substitution in the α, β, δ, or γ hemoglobin polypeptides.

A. Basic Considerations in Nomenclature

Two considerations have emerged as basic in developing this system of nomenclature. First, because variants of a given protein may differ from normal by only a few millimeters in their mobility but in both the isoelectric and electrophoretic axes the absolute distance between two widely spaced spots may vary from gel to gel by as much as 0.5 cm, a system of absolute *x*, *y* coordinates has to us always appeared unworkable. Second, whereas in 1-D electrophoresis variants are distributed along a single axis, rendering the terms *fast* or *slow* convenient and meaningful, in 2-D electrophoresis not only may variants occur in the charge or molecular size axes but at all possible positions with reference to these two axes. This statement is illustrated in Fig. 4, wherein we picture an inherited variant in the isoelectric axis (B-002), one in the molecular axis (E-002), and one which departs from the normal pattern in both axes (C-005).

B. The Subdivision of the Gel into Constellations

To meet the fact that gels are not completely reproducible, we find it convenient to subdivide a gel into a series of constellations defined by readily recognized "landmark spots" (or streaks), the boundaries of the constellation then being defined by lines connecting the most external of these landmark spots. These compare to the *natural groupings* of Comings (1982a). However, because a genetic variant may be separated in either axis by as much as a centimeter from its normal counterpart, the boundaries of a constellation must be somewhat imprecise and indeed, the operational boundaries of two constellations might overlap. Within each constellation the spots of interest are numbered roughly from left to right and top to bottom. As noted previously, with our interest in genetic variation no effort is made to designate all the protein moieties found within a given constellation. Rather, after inspection of a series of gels assumed to be representative we select spots meeting the criteria mentioned earlier as well as certain reference spots (e.g., actin, albumin, transferrin, the tubulins) and simply arbitrarily number as just described. Figure 6 illustrates the result of applying this system of nomenclature to the plasma proteins and explains the nomenclature of the preceding section.

C. Designation of Variants

The variants of interest can for present purposes be classified as qualitative or quantitative.

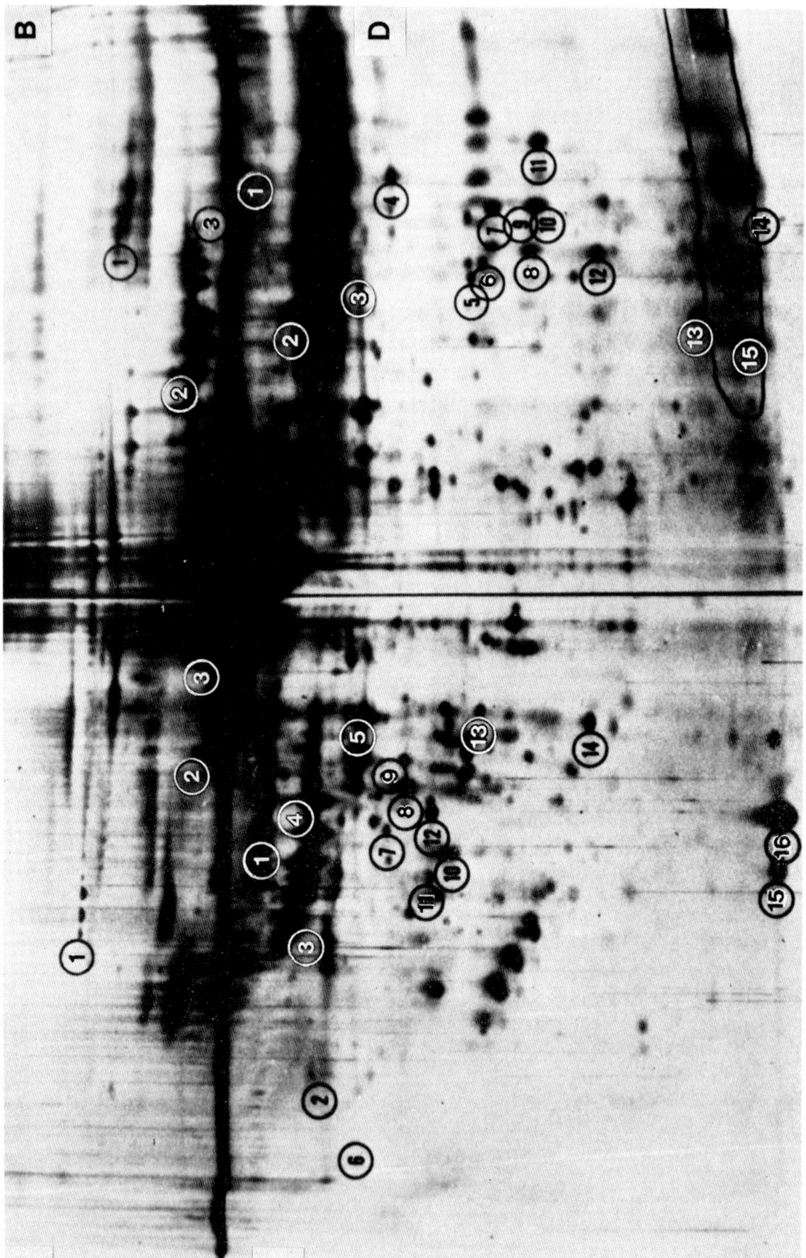
A
B
C
D

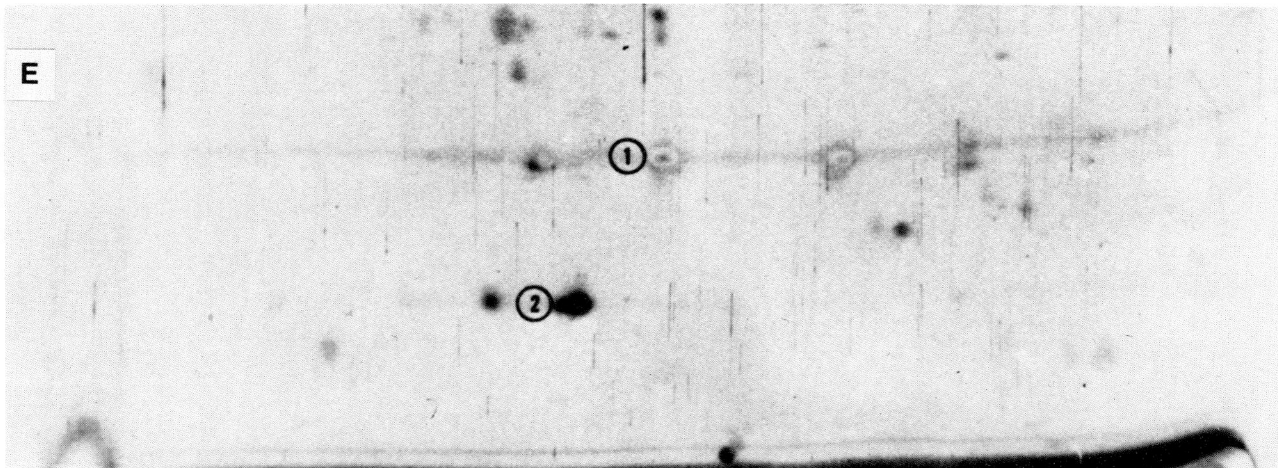

Figure 6. Subdivision of 2-D PAGE gel into convenient regions and designation of proteins within these regions that are scored for genetic variants. Where possible the label is to the left of the spot. Proteins whose identities are known include the following: A-002, hemopexin; A-003, albumin; B-001, plasminogen; B-002, transferrin; B-003, polymorphic protein scored in Coomassie Blue-stained gels; C-002, α_2-HS glycoprotein; C-003, α_1-antitrypsin; C-004, Gc globulin; C-005, fibrinogen α chain; C-006, α_1-acid glycoprotein; C-010, haptoglobin β chain; C-011, protein identified as variable by Tracy *et al.* (1982a); C-014, apolipoprotein E; C-016, apolipoprotein A-1; D-001, fibrinogen γ chain; D-002, fibrinogen β chain; D-003, IgG α chain; D-014, pregnancy-specific protein; D-015, IgG κ and λ light chains; E-001, haptoglobin α chain; E-002, prealbumin.

1. QUALITATIVE VARIANTS

These are as illustrated in Figs. 4 and 5. There is, however, a quantitative aspect even to variants such as these because in the presence of a variant the amount of protein (read *staining intensity*) in the normal position is diminished and indeed, in cases of doubt concerning the protein moiety homologous to a variant, a decrease in staining intensity should often help identify that homologue. For qualitative variants step 1 in our system of nomenclature is based upon measured distance from the presumed homologue and angular vector by degrees from 0° to 359°. Thus with reference to the presumed normal spot, the three variants referenced above would be designated as follows:

a. horizontal variant: 9 mm, 0°
b. vertical variant: 2 mm, 270°
c. off-axis variant: 3 mm, 165°

It should be noted that with the use of suitable computer subroutines, obtaining these measurements from a gel (actual, photograph, or displayed on screen) can be accomplished with much less effort than if protractor and ruler were employed. It is recognized that there is a degree of approximation in assigning the angular vector.

The next step in developing a nomenclature can only occur when one has accumulated a sufficient number of variants to make a comparison between them profitable. Let us assume that in the course of 1000 gels one has encountered some 60 variants of a given protein. Figure 7 is a hypothetical plot of findings. The origin of this figure corresponds to the position of the normal spot. This too is subject to variation in exact location as described earlier but for these purposes can be regarded as fixed. The variants for the most part fall into two clusters, which can be assumed to define two polymorphisms. In addition, however, there are three outliers. These all need to be rerun with suitable comparison gels. On the rerun in our hypothetical example, outlier 1 now joins cluster 1 and is assumed to be that variant. However, variants 2 and 3 still maintain their positions and, as rare variants, would require family studies as possible mutations.

At this stage one may elect either to continue to carry the variants under the strictly descriptive nomenclature or for ease of tabulation and statistical analysis one can give them arbitrary designations, such as V_1, V_2, ..., V_n. An alternative, which might facilitate comparison between tabulations, would be to derive the average position of the variant. Thus V_1 would become V_{X001} 315°; 6 mm, the subscript indicating the spot designation. Comparison across laboratories would of course assume

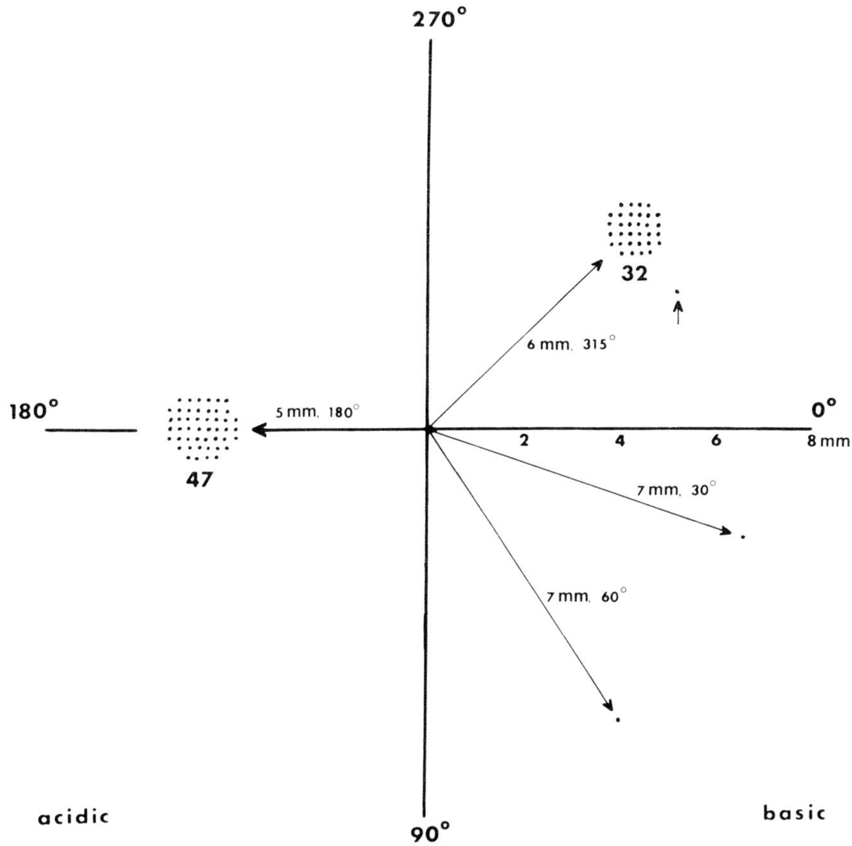

Figure 7. A diagrammatic representation of the distribution in the isoelectric focusing (IEF) and molecular weight axes of a group of variants of a hypothetical protein. It is suggested that, for the present, variants are most conveniently described in terms of an angular vector with reference to the normal spot and a distance from the spot.

some standardization in the designation of constellations and numbering of spots. This may be difficult to maintain, because as sophistication in spot analysis improves additional spots within the constellation may be designated for analysis and it will be difficult to maintain adequate communication between laboratories.

2. QUANTITATIVE VARIATION

A second type of genetic variant detectable in a 2-D system is characterized by failure of formation of a protein product. Individuals homozygous for such variants should lack specific protein moieties. These should be relatively easy to detect but quite rare. Heterozygotes should be characterized by half-normal staining intensities of these spots. Preliminary observations lead us to believe that under favorable circumstances such heterozygotes can also be detected in silver-stained gels. To this end we must obtain the optical density (absorbance units) of each of the spots in a carefully selected set of spots and then compute the contribution of each spot to the optical density of the entire set. On the basis of observations involving known heterozygotes (whose normal spot will reflect the presence of only one allele), it appears that it will be possible to detect with high accuracy a spot whose contribution to the density of the set is only half normal.

The ability to detect heterozygotes for true nulls with high accuracy would be a development of great usefulness in the study of mutation because on the basis of the data on kinds of mutations reviewed earlier, it should double or treble the amount of information concerning mutation to be extracted from a gel. It is also useful in the presence of a variant whose "mother" spot is not immediately obvious, because one criterion of the mother spot should be half-normal intensity. For those spots for which quantification is feasible we propose to make this observation a regular entry into our data files. The basic entry will be a percentage figure based on the contribution of that spot's density to the density of the battery of spots being evaluated. Verification of the real usefulness of this approach will depend on the accumulation of data on multiple proteins, similar to the accumulation of data with respect to quantitative enzyme variation (Mohrenweiser, 1981; Satoh *et al.*, 1983b).

Ultimately, of course, this rather cumbersome terminology should be unnecessary. It seems unlikely, however, that a more rational approach based on a high level of gel standardization and knowledge of the exact nature of the various spots will be available within the next several years. By that time image analysis algorithms should have made substantial progress. The Gabriel graph approach that our group is attempting to develop for the identification of newly arisen mutations (see Section X) must, of course, have the ability to deal with genetic variation. This system of nomenclature is consistent with our projections for the use of this graph structure in the analysis of gels.

X. Gel Analysis

The analysis of gels for genetic variants with particular reference to the possibility of mutation can be conducted by eye, as illustrated by many of the previously cited studies and some of our own work. Visual scoring is not only monotonous and laborious; the eye-brain axis tires and inaccuracies creep in. Accordingly we have devoted a major effort to developing computer programs that efficiently and reliably first locate the individual protein spots that appear on each gel and then compare the spot patterns encountered in family members (currently, mother–father–child trios) in such a way as to gather information about events that may have genetic significance, for example, a spot appearing in a child's gel that has no counterpart in preparations from either parent. These two tasks of spot identification and cross-gel spot comparison are very different in nature and thus require different approaches.

As soon as a gel is digitized by computer one is faced with the problem of what to do with the large amount of information that goes into making an image. It is not unusual to have one million pixels per image, each with 256 grey levels. This presents a fundamental problem for most serial computer architectures because they have the restriction that only one pixel of information can be examined at a time. Thus it is highly desirable to use some form of parallel data processing in which more than one pixel is examined per computer clock cycle.

A parallel computer called the Cytocomputer, which is especially designed for image processing, has been developed by Sternberg (1979). Basically, the Cytocomputer through its parallel "pipeline" architecture is able to examine simultaneously 3 × 3 pixel "windows" and transform the center pixel of the window (based upon the values of its neighboring pixels) according to a large set of functions that can be specified in an appropriate programming language. A typical Cytocomputer program is composed of a sequence of different neighborhood transformations and, in addition to evaluating each 9-pixel neighborhood transformation in parallel, the Cytocomputer is evaluating each of the different transformations in parallel. The operations we describe in this section are conveniently implemented by a Cytocomputer (or similar device) coupled to a VAX 11/780 but can be executed by the VAX alone.

Window transformations, in addition to being amenable to parallel implementation, provide a very concise and elegant conceptual framework for image processing. The mathematical foundations of such transformations have been formalized with the development of what has been termed *mathematical morphology* (Serra, 1982). A basic idea of mathematical morphology is that appropriate sequences of window transformations can be used to perform various geometric operations on an image. Thus one can conceptualize a sphere of a radius larger than the largest spot and effectively "roll" the sphere "underneath" the image surface in order to define the background levels of the image (i.e., nongenetic noise). This allows us to remove the variable background levels encountered in most images in order to get estimates of spot volumes that are not affected by large variations in background intensity due to conditions of illumination or gel preparation. By choosing appropriate sequences of window operations, we can perform other geometric operations that aid in filtering noise (both that caused by imperfections in the digitization process and the streaking that is peculiar to gels) and that eventually lead to the detection of spot locations. A more detailed description of the transformations of mathematical morphology as they have been applied to the problem of spot detection is given by Skolnick (1982, 1983).

Once the spot locations have been determined, presumably homologous spots must then be compared in order to arrive at the desired genetic decisions. The process of gel generation cannot be standardized to the extent that it would be feasible simply to superimpose the images of the detected peaks for the different family members, as is done in the flicker technique of astronomy. Because of presently unpredictable distortions in a gel preparation, there is no guarantee that the coordinates of homologous spots on different gels will be the same. However, within a local region of gel the distortions are less of a problem and thus it is possible to superimpose spot patterns locally. This is possible only if spots common to the trio are identified on each family member's gel based on their membership in shared constellations so that each common spot can serve as an origin point from which to do comparisons on the surrounding local region. Although it is possible for an operator to identify each common spot, such an approach would be prohibitive given the volume of data to be processed for genetic monitoring. Another possible procedure would be to determine prior to processing the characteristic constellations of spots that would then be used to automate the search for spots that specify a superimposition origin point for each local region of each family member. This a priori approach to gel comparison is useful and, in general, any system for genetic monitoring should have the capability of searching for known spots. A problem is that when we are starting our investigations it is difficult to have precise expectations as to what spot patterns will appear, and furthermore, there is the problem of how to detect and not be confused by variations from the expected patterns.

With these issues in mind we have approached the gel comparison problem from a broader perspective that includes the possibility of embedding in the algorithm an a priori approach to a given constellation but that does not require a priori information to perform comparisons. This has been achieved by imposing a graph structure of the Gabriel type (Gabriel and Sokal, 1969) on the spot patterns of each gel and

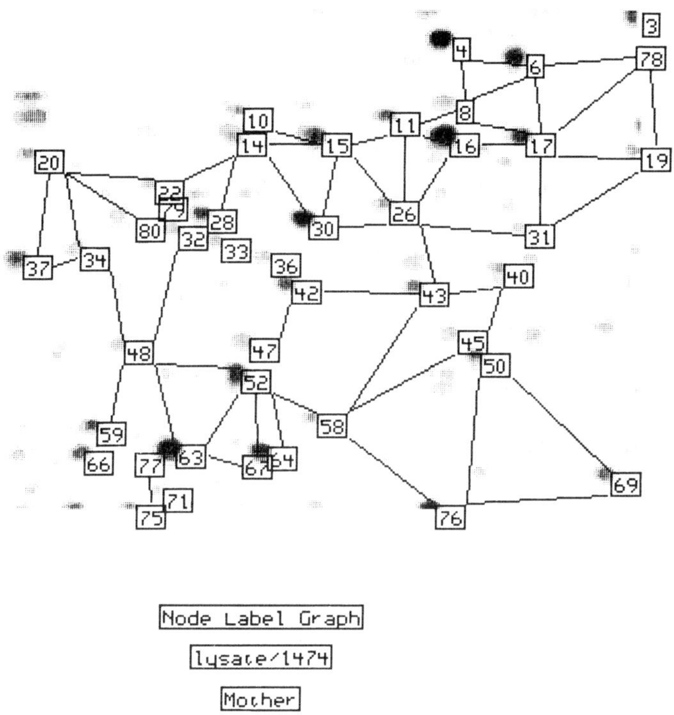

Figure 8. An illustrative Gabriel-type graph superimposed on the processed image of an area in a 2-D PAGE gel prepared from an erythrocyte cytosol. A computer program will compare this graph with corresponding graphs from the father and mother, and classify the protein moieties visualized in all three images according to the genetic pattern that they exhibit. In the event that a node present in one gel of the trio is lacking in a gel from another member of the trio, the algorithm creates a "virtual node" in the latter. This is illustrated by node 78. Although the algorithm assigns numbers to all discernible densities (polypeptides), for clarity only the 44 most densely staining polypeptides are shown.

performing comparisons of graphs as opposed to comparisons of spot locations. Figure 8 illustrates a Gabriel graph derived from a 2-D gel of a child's erythrocyte lysate. Similar graphs are derived for the child's father and mother. These graphs bring out enough of the communalities across spot patterns that a completely automatic comparison across gels without any a priori expectations is possible. As the comparison program processes unknown gels it can develop a model of the spot patterns it has encountered. The model can be thought of as consisting of prototypic intersecting spot constellations that are embedded in the graph structure. As the comparison program is exposed to new gels it can report on the variance between the new information and the learned prior structure. The investigator can also embed his expectations and nomenclature within the comparison program in terms of the constellations within the graph structure. The output of the program consists of a plot of the final graph superimposed on the processed gel images and a table that indicates the comparison result for the final graph. This table indicates which spots were found in the preparations of one or both parents but not the child, which in one or both parents and the child, and which in the child only. The latter, after other explanations are excluded, are considered as possibly resulting from mutation. A more detailed description of the basic mechanisms of the comparison program can be found in Skolnick (1982).

Currently an operator compares the hard copy output of the Gabriel graph algorithm with a digitized, processed image of the corresponding portion of the father's, mother's, and child's gels, makes a decision as to the status of each spot being scored, and enters this into the computer. The computer output serves as a "prompter" directing the operator to regions of interest, but the output, although filed, is not routinely entered into the computer.

XI. Steps Necessary to Identify a Putative Mutant

When a protein moiety in a child is detected that is not present in either parent, that is, a putative mutant, the first step is a repeat run on stored samples from the three individuals concerned. If the finding is confirmed, the next step is to perform on these samples the various typings brought to bear in medico-legal parentage questions. We customarily store an aliquot of glycerolized cells. In our laboratory, given the availability of these cells plus hemolysate and plasma, typings relevant to resolving questions of parentage can be performed for the following systems exhibiting genetic polymorphisms useful in paternity issues: A_1A_2BO, MNSs, Rh, K, Fy, Jk, AcP, ADA, AK, Cp, EsD, GALT, GPT, PGM_1, and G6PD. We have estimated that in the situation in which the legal father is not the biological father, with this battery the fact will be detected at the 90% level (H. Gershowitz, personal communication). However, we are reluctant to accept a single system exclusion if the critical determination depends on an indirect (Coombs) reaction. In addition the polymorphisms present in the various 2-D PAGE preparations should be useful with respect to questions of parentage. If there is no exclusion, we would then attempt to repeat the observation on fresh samples. Given the need to bring as much power as possible to bear on the question of parentage discrepancies, typings for the major histocompatibility complex are indicated.

The question of a discrepancy between legal and biological parentage can never be resolved with absolute certainty. Thus in any collection of putative mutants some fraction might indeed derive from such discrepancies. However, a probable value can be assigned to this fraction (Neel, 1971; Rothman *et al.*, 1981). This requires that in the population under study one determines for a randomly selected subset of the children the frequency of discrepancies between legal and biological parentage. Let I = the frequency of such discrepancies, W = average frequency at the genetic loci under study of the alleles responsible for the kinds of rare variants being studied as to their mutation origin, and D = the probability of detecting nonpaternity with the available laboratory tests. The frequency with which undetected parentage discrepancies result in an apparent mutation will be equal to $IW(1 - D)$. The corrected mutation rate μ_{EST} is derived from the apparent mutation rate μ_{OBS} simply by subtracting this term, that is

$$\mu_{\text{EST}} = \mu_{\text{OBS}} - IW(1 - D).$$

An expression for the standard error of μ_{EST} can be obtained from the formula for the variance of the sum of two random variables. We suggest the following:

$$\text{SD } \mu_{\text{EST}} = \sqrt{\text{var } \mu_{\text{OBS}} + [W(1 - D)]^2 V(I) - 2(1 - D) \operatorname{cov}(\mu_{\text{OBS}}, I)},$$

where V is the variance of I.

A worked example is provided by Rothman *et al.* (1981). If one were contrasting two groups as to mutation rates, hopefully identical except for the exposure of the one set of parents to a mutagenic influence, one should still derive an independent estimate of parentage exclusions (I) for both groups.

Unfortunately, with this approach one cannot specify among an apparent collection of mutations which are the result of parentage discrepancies and which are true mutations. However, we have also derived an expression for the probability *for* paternity in any instance of apparent mutation in which the child and its parents have been subjected to extensive genetic typings and the frequency in the population of the alleles whose products are being typed is known (Rothman *et al.*, 1981). With this one can rank a series of apparent mutations with reference to the probability for paternity in the children exhibiting the mutations and so form an opinion as to which of the putative mutants are most likely true mutants. We have also shown that this odds–ratio formula may be used to obtain an independent direct estimate of the mutation rate.

An equally important question in using 2-D gels to study mutations, which we have not yet been forced to confront, is to determine with some certainty the polypeptide of which the variant is a mutant. The position of the variant will serve to suggest a probable origin, but quite conceivably a chain termination mutation affecting a large molecule could result in a smaller molecule that fortuitously may migrate in a position suggesting affinities with a completely unrelated protein. In this situation, we envision these possible steps.

1. *Densitometric studies.* The polypeptide of origin should exhibit half its normal total density (absorbance).
2. *Use of specific protein stains.* The mutant polypeptide can usually be expected to exhibit whatever specific staining properties characterize the protein of origin, (i.e., those of a lipoprotein, glycoprotein, etc).
3. *Use of activity stains.* If it is possible to recover the protein in question and permit it to renature, it may exhibit the same activity as the protein of origin.
4. *Parallel studies with other biochemical techniques.* For putative mutants that appear to involve well-known proteins, one can turn for their characterization to studies utilizing other established techniques such as 1-DGE or high-performance liquid chromatography (HPLC). The latter technique is proving highly effective in the separation of closely related proteins. For instance, Fig. 9 illustrates the resolution of transferrin variants by HPLC in our laboratory (Strahler *et al.*, 1983). High-performance liquid chromatography should be useful in the isolation of sufficient quantities of both known and unknown proteins for full or partial characterization. The most obvious such characterization is the amino acid sequence of the polypeptide; establishing the sequence of only the first of 10 to 20 amino acids of a polypeptide should provide a "signature" for comparison with other possible homologous polypeptides.
5. *Immunological studies.* Both the mutant protein and the suspected protein(s) of origin can be recovered from the gel or from other types of preparations as just mentioned and used as antigens to elicit antibodies, polyclonal or monoclonal. Hopefully, these antibodies will cross-react with the parent or variant proteins. The feasibility of this technique has been demonstrated by Mose Larsen *et al.* (1982), who have also shown that by adding a fluorescent tag to the antibody one can localize the position of the antigen within the cell (Bravo *et al.*, 1983).

Figure 9. Demonstration of the ability of anion exchange HPLC to discriminate plasma transferrin variants (D_1, D_3, B_1, B_2) from normal (C). The various types are defined by the time at which they come off the column, as follows: Tf C, C at 25.3 min (top panels); Tf CD_1, C at 25.3 min; D_1 at 23.2 min (middle left); Tf CD_3, C at 25.3 min; D_3 at 18.9 min (bottom left); Tf CB_1, C at 25.5 min; B_1 at 29.9 min (middle right); Tf CB_2, C at 25.3 min; B_2 at 26.8 min (bottom right). Only that portion of the chromatogram between 15 and 40 min is shown. Further details regarding the procedure are given in Strahler *et al.* (1983).

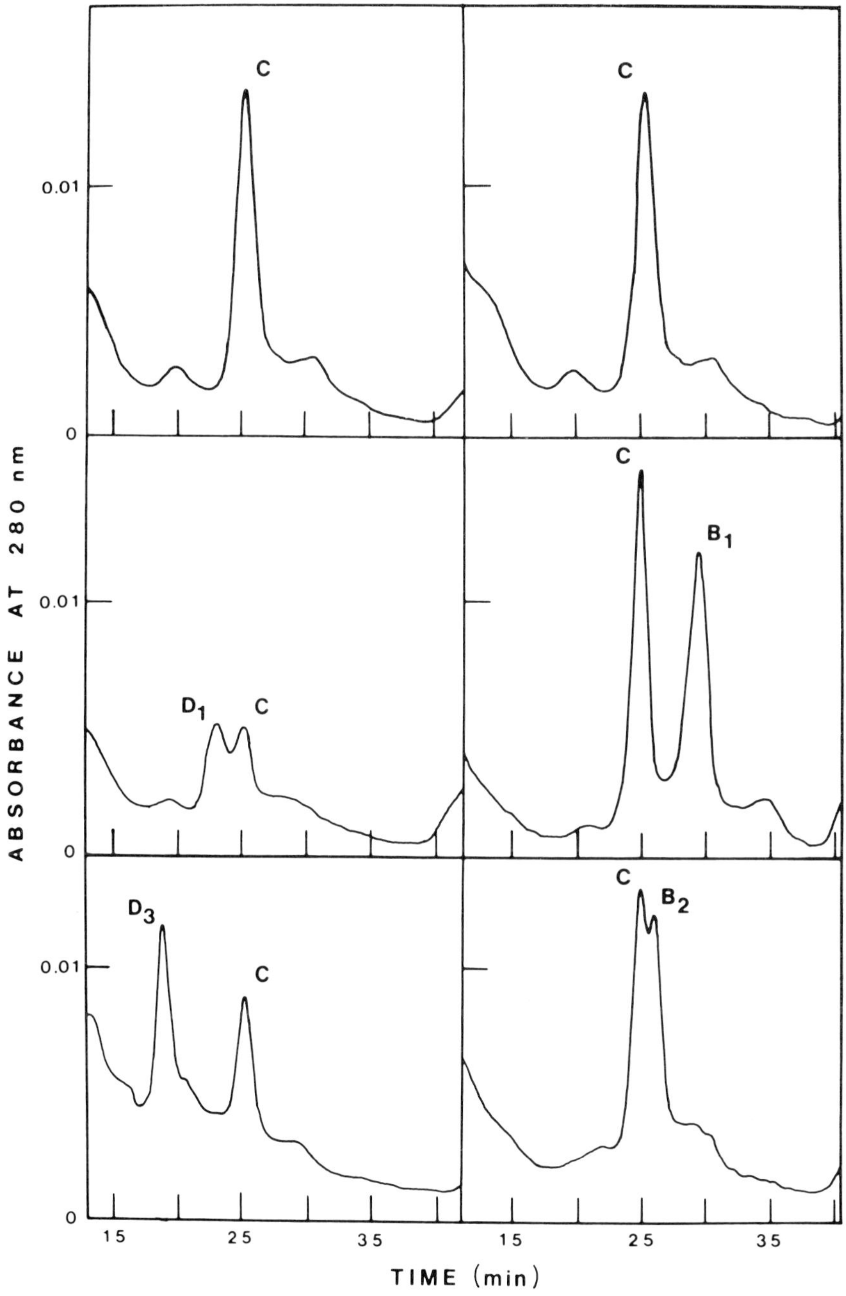
ABSORBANCE AT 280 nm
0.01
0
C
D1
D3
B1
B2
15
25
35
TIME (min)

The ultimate test of mutation is the transmission of the phenotype to the next generation according to Mendelian principles, and this test has been employed in many of the studies of the dominantly inherited medical syndromes. This test is more easily implemented in mutation rate studies based on gross phenotypes than in those based on biochemical findings because the population surveys necessary to the former will often have located sporadic examples of the syndrome who are in their reproductive years, many of whom will already have produced children among whom the transmission of the defect can be evaluated. By contrast, studies based on 2-D PAGE will usually be directed towards younger children. There will thus be a delay between detection and the opportunity to study transmission. It is important that the need for such studies not be lost sight of, because transmission is the ultimate genetic test. In this context we recognize that a somatic mutation in an early embryonic cell might result in sufficient variant protein to be detectable with 2-D PAGE; such an occurrence would not result in a transmitted trait. The occurrence of this phenomenon should be recognizable by a departure from the expected 1:1 ratio in the offspring of a series of apparently mutant individuals (and usually by the small amount in which the variant polypeptide was represented).

XII. Data Management

The application of computer technology to the management of 2-D gel data is an integral part of our effort to establish a mutation monitoring system. We discussed the power of the computer as a tool for the acquisition and processing of digitized images in Section X. Here we address the application of the computer to the management and analysis of data generated by research using 2-D gel methods, including data generated by computer-assisted methods.

Each research question defines a subset of variables from the large number of possibilities that may be generated from the processing of a 2-D gel image. The identification of those appropriate for the efficient detection and characterization of germinal mutations has been a primary concern in the early stages of adapting the 2-D gel technology to mutation research. Although the details are not yet in place, knowledge about the general nature of the variables to be expected and how they will be generated in a mutation-monitoring setting has enabled us to proceed to develop a general strategy to assemble information about the 2-D gel from a number of sources into a centralized data base. Because ours is a research environment it is imperative that the strategy be flexible enough to accommodate changes in the definition of variables and their sources and that the plan that we develop be adaptable to those situations in which the mutation-monitoring technologies developed may be implemented.

Computer hardware and programs that will permit various members of the research team to interface with a central data management system are being developed for the purpose of conveniently (1) transmitting data directly from the source to a central data storage facility, (2) manipulating data (either single variables, a record, or an entire file of records) that has been collected, and (3) assembling subsets of the total data base to be used to generate reports or to carry out statistical analyses necessary to answer research questions. To accomplish these goals we have set three criteria that must be satisfied. The system must be "friendly" to the user, be on-line and interactive, and take advantage of the technological benefits of distributed networking of shared computing resources. Access to both the computer hardware and the programming to control the computers are involved in each.

A. A Friendly Interface

With regard to the friendliness of the interface with the user, we have designed all computing functions to operate in an interactive mode from a terminal device in a time-sharing environment. Operator convenience has been given the highest priority in the design of the computer programs that support this interface. Two categories of programs are being developed. First, there are the programs that enable the user to interact with the system to control the execution of the programs that have been written to carry out data manipulations and statistical computations and to generate tabular summaries. Second, there is the set of programs being developed to communicate between the devices that are being developed to collect data on mutation and the central (host) computer.

B. Interactive Data Acquisition

We have developed a computerized method to manage the specification of a format that enables the user to function in an interactive mode without the assistance of a computer programmer. By calling on a program the user can interact with the computer to build the definition of the data files that are appropriate for his aspect of the project. This results in the cataloging of all the information required to set up the entering, editing, and listing of data by persons with little previous experience with a computer language. A special feature of the strategy is the ability to edit the format of these functions to include the addition, deletion, or redefinition of variables. Two kinds of computer files store the information necessary to control the interaction of the user with the data base he has defined. The first type of file stores information about the questions that define a data collection instrument (question or variable name, permissible responses, the description of any codes, format for the output, and the position in the data file where the responses are to be stored). The second type of file contains the data records themselves. If the format of the data collection should change during the course of a study, an interactive program allows one to modify the computer files that define the form and then update the data files to reflect changes in the format of the form.

Once the data record is defined for each source of data that relates to a common project, information may be entered from any terminal device by a technician or from another computer, for those data records set up to accept the output of an automated data acquisition device. Our program has evolved a need to input data according to both approaches.

C. Networking

The distribution of computing resources among the nodes of a collaborative research activity that typifies a project such as the one outlined in this chapter is made possible by the changing nature of computers, computer programming, and communications technology. Individual computer installations connected by communication links form a network of computers. A network enables one to select a computer appropriate for the specific tasks that are defined for each of the activities in a multidisciplinary effort to estimate mutation rates. We have developed a network to facilitate the collection of mutation data that involves data acquisition computers at remote sites and a PDP 11/70 central data-handling device. In one example involving the 2-D gel laboratory, we have selected a combination of hardware and communication software to facilitate an on-line interactive mode for acquiring data that involves the simultaneous control of the operations of a VAX 11/780 and a PDP 11/70. The investigator interacts with these two computers from a laboratory located at a distance from either of these separate installations. Once data is collected he may proceed to route subsets of data to a third computer, an Amdahl 470 V/8, for the execution of statistical analyses.

XIII. The Selection of Populations for Study

There are in principle two general approaches to the selection of study populations. One can select a geographic area of interest, such as one in which the population is entering into a new industrial exposure, and over an extended period of time follow the children being born with reference to the possibility of a changing mutation rate. Because the increase to be expected (if any) will almost certainly be small, the numbers of observations necessary for a significant study can be enormous. Let us assume, for a simple statistical example, that one wishes to compare the data collected on the population in two consecutive 5-year periods and would like to be able to detect an increase in the mutation rate of 50% or greater with type I (α) and type II (β) errors of 0.05 and 0.20, respectively. A conservative formula for estimating the two samples sizes (Neel, 1980) is

$$n = \frac{2(2.487)^2 P_{2i}(1 - P_{2i})}{(P_1 - P_{2i})^2},$$

where n = sample size, P_1 = assumed baseline mutation rate, and P_{2i} = assumed present rate.

The constant 2.487 is based on a one-tailed test at level $\alpha = 0.05$ with probability of a type II error (β) equal to 0.20; the usual variance of the difference of two proportions $(p_1q_1 + p_2q_2)/n$ has been replaced by the more conservative value of $2p_2q_2/n$. With this formula and assuming the detectable baseline rate suggested earlier (4×10^{-6}/locus/generation), the necessary sample sizes are 18,000,000 observations each. This is the number of discrete data points. Given that each protein determination is a test of two alleles and our suggestion that a minimum of 180 proteins can be tested per child, these numbers would necessitate two samples each involving 50,000 children. Because of the exponential nature of this formula the necessary number is less than half if one wished within the same constraints to detect an increase of 100% or more; approximately 18,000 children would be needed in each of the samples. Even with this reduction in numbers only the 2-D PAGE offers the possibility of the biochemical approach becoming feasible in this setting.

The alternative approach is to select for study populations thought to be at high risk of exposure to a mutagen, a "worst-cases" approach. Since fortunately such populations are usually small, the results of several such studies will undoubtedly have to be combined in order to amass the numbers necessary to a significant test of hypothesis. Since (also fortunately) such high risk populations are rare in any single country, a significant effort to evaluate the possibility of an increased mutation rate will generally call for international collaboration. Assume again that we wished to be able to detect any increase greater than 100% with the errors discussed above. Control data would be collected simultaneously and in an identical fashion. If six study populations each with an appropriate control of equal size could be defined in six different countries, then meeting the sample sizes of 18,000 mentioned above would entail for any single country the study of 3000 children of exposed parents and 3000 controls or, over a 5-year period 1200 subjects to be studied per year.

A worst-cases approach conducted on the necessary scale either will or will not yield a significant finding within the limits mentioned. If there is a significant increase, not only is there a base for dealing with the concerns of the individuals involved but also the first basis for cautious extrapolation to lower doses. If there is no significant increase, there is the basis for some reassurance to individuals exposed to lower doses, the reassurance inversely proportional to the actual exposure.

With the very small numbers of positive events involved, regression techniques would bring greater statistical power to bear on the data analysis. However, although such statistical techniques might be possible in the case of radiation exposures, in which cases gonadal dose can be calculated with some accuracy, they are not yet applicable with respect to the chemical mutagens, for which post-hoc evaluations of gonadal dose remain extremely difficult.

Should an international collaborative effort be pursued, strenuous efforts at standardizing the techniques of the various laboratories involved are indicated. This standardization is necessary not only for the study of mutation but because these investigations should result in large amounts of data concerning genetic variation, which one would hope to be able to compare between laboratories.

Because of the costs inherent in such studies, it will be highly desirable either to incorporate these studies into the fabric of medical practice or, if a "high risk" group is to be studied, to incorporate the genetic studies into a study that is already projected, which in most instances will be directed at the somatic effects of the exposure.

We are currently involved in three pilot study attempts to explore the feasibility of applying 2-D PAGE to the study of mutation.

1. In Ann Arbor, Michigan, placental cord blood samples from newborn infants and venous samples from their parents are being collected at two local hospitals with the informed consent of the parents involved.

This study could be regarded as a prototype for a continuous monitoring program, although on so small a scale that a significant body of data will not result for many years. The study was originally initiated to gather with 1-DGE data on the rate at which mutation alters protein structure (Neel *et al.*, 1979), and a strength of the 2-D effort is that 1-DGE studies are made of a subsample of the enzymes and serum proteins visualized in the 2-D gels. A drawback of such a study as this is that if the repeat blood sample is necessary from the child within a reasonable interval after the first sample, the child will still be a small baby, of whom the parents are properly very protective.

2. In Hiroshima a small pilot study is being initiated on children born to atomic bomb survivors who received ≥ 1 rad at the time of the detonations, as well as a suitable group of children whose parents are thought to have received essentially no radiation at that time. The pilot study is within the framework of the program of the Radiation Effects Research Foundation, jointly funded by the U.S. Department of Energy through the National Academy of Sciences and the Japan Ministry of Health and Welfare. These children are already being surveyed for electrophoretic variants by 1-DGE techniques and for activity variants by centrifugal fast analyzer techniques (Neel *et al.*, 1980a; Satoh *et al.*, 1983b). Any data collected with the 2-D PAGE approach can be treated as additive to the other electrophoretic data.

3. In Hiroshima a second small pilot study is also being initiated in collaboration with Y. Nishimoto and colleagues on a sample of children born to wartime workers engaged in the manufacture of sulfur mustard gas on Ohkuno Island, some 50 miles from Hiroshima. The parents are already under long term surveillance in a study also supervised by Nishimoto. This is anticipated to be a small sample of children but their parents, to judge by the frequency of a variety of neoplasms (Wada *et al.*, 1963, 1968; Shimura *et al.*, 1978), must be considered at relatively high risk from a chemical mutagen. Further details concerning this study will be found in Neel *et al.* (1983b).

It should be noted that all three of these studies are to a considerable extent integrated into ongoing aspects of the practice of medicine or special studies. It seems clear that studies of mutation rates employing 2-D PAGE will be sufficiently expensive that a major effort must be made to combine such studies with other medical activities.

In the application of 2-D PAGE to the study of possible increases in mutation rates, sociopolitical considerations arise that are absent in almost all other studies involving this technology. For instance, in the

United States there is intense concern on the part of former military personnel who participated in atomic bomb tests and residents in the area of the tests concerning the possible somatic and genetic effects of the resultant exposures, and a variety of legal actions have been initiated by the individuals concerned. The magnitude of the exposures remains highly debatable (see Smith, 1982) but they were undoubtedly well below the average sustained by the survivors of Hiroshima and Nagasaki. At present 2-D PAGE offers the most hopeful and objective approach to evaluating the questions of whether these persons have sustained transmitted genetic damage. Given the results of the studies in Hiroshima and Nagasaki (Schull *et al.*, 1981), however, it is highly unlikely from what has been published concerning the radiation exposures that any study of these groups will yield statistically significant findings, but the numbers necessary to exclude a 100% increase in the mutation rate could probably be amassed. Should such studies be initiated as a component of a more general effort to put in perspective the entire issue of potentially mutagenic experiences? Is this same line of reasoning relevant to similar issues being raised in other countries? How far should society go to meet the concerns of its citizens, concerns that actions of the society have created? What level of "exclusion-of-effect" constitutes a useful service?

XIV. Concluding Comments

It is clear that the application of the 2-D PAGE technique to the study of mutation should result in substantial improvements in the efficiency with which this phenomenon can be studied. Progress has been made in a number of respects but much remains to be done before the technique will have fully realized its potential in this context. On the one hand, even if the application of the technique is limited to the visual reading of a few of the more prominent spots on assorted gels, a new source of information concerning mutational events will have been gained. If, on the other hand, the reading of as many as 500 polypeptides could be both accurate and almost fully automated, this would represent a spectacular advance in our ability to study human mutational rates and monitor for genetic damage. Given the rapid advances in 2-D PAGE, there are grounds for reasonable optimism that the larger objective will be accomplished.

References

Adetugbo, K., Milstein, C., and Secher, D. S. (1977). Molecular analysis of spontaneous somatic mutants. *Nature (London)* **265**, 299–304.

Anderson, N. L., and Anderson, N. G. (1977). High resolution two-dimensional electrophoresis of human plasma proteins. *Proc. Natl. Acad. Sci. USA* **74**, 5421–5425.

Anderson, N. L., and Anderson, N. G. (1978). Analytical techniques for cell fractions. XXII. Two-dimensional analysis of serum and tissue proteins. Multiple gradient-slab gel electrophoresis. *Anal. Biochem.* **85**, 341–354.

Arthur, E., Steel, C. M., Evans, H. J., Povey, S., Watson, B., and Harris, H. (1975). Genetic studies on human lymphoblastoid cell lines: Isozyme and cytogenetic heterogeneity in a cell line, with evidence for localization of the Pep A locus in man. *Ann. Hum. Genet.* **39**, 33–42.

Boyum, A. (1968). Isolation of mononuclear cells and granulocytes from human blood. *Scand. J. Clin. Lab. Invest.* **21** (Suppl. 97), 77–89.

Bravo, R., Fey, S. J., Mose-Larson, P., Coppard, N., and Celis, J. E. (1983). Proteins IEF 31 and IEF 46 are keratin-type components of the intermediate-sized filaments: Keratins of various human cultured epithelial cells. *J. Cell Biol.* **96**, 416–423.

Comings, D. E. (1982a). Two-dimensional gel electrophoresis of human brain proteins. I. Technique and nomenclature of proteins. *Clin. Chem.* **28**, 782–789.

Comings, D. E. (1982b). Two-dimensional gel electrophoresis of human brain proteins. III. Genetic and non-genetic variations in 145 brains. *Clin. Chem.* **28**, 798–804.

Edwards, J. J., Anderson, N. G., Nance, S. L., and Anderson, N. L. (1979). Red cell proteins. I. Two-dimensional mapping of human erythrocyte lysate proteins. *Blood* **53**, 1121–1132.

Gabriel, K. R., and Sokal, R. R. (1969). A new statistical approach to geographic variation analysis. *Syst. Zool.* **18**, 259–270.

Hamaguchi, H., Ohta, A., Mukai, R., Yabe, T., and Yamada, M. (1981). Genetic analysis of human lymphocyte proteins by two-dimensional gel electrophoresis. I. Detection of genetic variant polypeptides in PHA-stimulated peripheral blood lymphocytes. *Hum. Genet.* **59**, 215–220.

Harris, H. (1980). "The Principles of Human Biochemical Genetics," 3rd revised ed. Elsevier-North Holland, Amsterdam.

Harris, H., Hopkinson, D. A., and Robson, E. B. (1974). The incidence of rare alleles determining electrophoretic variants: Data on 43 enzyme loci in man. *Ann. Hum. Genet.* **37**, 237–253.

Holmsen, H., Day, H. J., and Setkowsky, C. A. (1972). Secretory mechanisms. Behavior of adenine nucleotides during the platelet release reaction induced by adenosine diphosphate and adrenaline. *Biochem. J.* **129**, 67–82.

Johnson, F. M., and Lewis, S. E. (1981). Electrophoretically detected germinal mutations induced in the mouse by ethylnitrosourea. *Proc. Natl. Acad. Sci. USA* **78**, 3138–3141.

Johnson, F. M., and Lewis, S. E. (1983). The detection of ENU-induced mutations in mice by electrophoresis. *In* "Utilization of Mammalian Specific Locus Studies in Hazard Evaluation and Estimation of Genetic Risk" (W. Sheridan and F. deSerres, eds.), Plenum, New York. In press.

Johnson, F. M., Roberts, G.T., Sharma, R. K., Chasalow, F., Zweidinger, R., Morgan, A., Hendren, R. W., and Lewis, S. E. (1981). The detection of mutants in mice by electrophoresis: Results of a model induction experiment with procarbazine. *Genetics* **97**, 113–124.

McConkey, E. H., Taylor, B. J., and Phan, D. (1979). Human heterozygosity: a new estimate. *Proc. Natl. Acad. Sci. USA* **76**, 6500–6504.

Merril, C. R., Goldman, D., Sedman, S. A., and Ebert, M. H. (1981). Ultrasensitive stain for proteins in polyacrylamide gels shows regional variation in cerebrospinal fluid proteins. *Science (Washington, D.C.)* **211**, 1437–1438.

Milstein, C., Adetugbo, K., Cowan, N. J., and Secher, D. S. (1974). Clonal variants of myeloma cells. *Prog. Immunol.* **1**, 157–168.

Mohrenweiser, H. W. (1981). Frequency of enzyme deficiency variants in erythrocytes of new born infants. *Proc. Natl. Acad. Sci. USA* **78**, 5046–5050.

Mose Larsen, P., Bravo, R., Fey, S. J., Small, J. V., and Celis, J. E. (1982). Putative association of mitochondria with a sub-population of intermediate-sized filaments in cultured human skin fibroblasts. *Cell* **31**, 681–692.

Mukai, T., and Cockerham, C. C. (1977). Spontaneous mutation rates at enzyme loci in *Drosophila melanogaster*. *Proc. Natl. Acad. Sci. USA* **74**, 2514–2517.

Neel, J. V. (1971). The detection of increased mutation rates in human populations. *Perspect. Biol. Med.* **14**, 522–537.

Neel, J. V. (1980). Some considerations pertinent to monitoring human populations for changing mutation rates. *Proc. Int. Congr. Genet. 6th* **1**, 225–238.

Neel, J. V. (1983). The frequency of spontaneous and induced "point" mutations in higher eukaryotes. *J. Hered.* **74**, 2–15.

Neel, J. V., and Rothman, E. (1981). Is there a difference between human populations in the rate with which mutation produces electrophoretic variants? *Proc. Natl. Acad. Sci. USA* **78**, 3108–3112.

Neel, J. V., and Schull, W. J. (1956). "The Effects of Exposure to the Atomic Bombs on Pregnancy Termination in Hiroshima and Nagasaki." Natl. Acad. Sci. Natl. Res. Council, Washington, D.C. (Publ. No. 461).

Neel, J. V., Mohrenweiser, H., Satoh, C., and Hamilton, H. B. (1979). A consideration of two biochemical approaches to monitoring human populations for a change in germ cell mutation rates. *In* "Genetic Damage in Man Caused by Environmental Agents" (K. Berg, ed.), pp. 29–47. Academic Press, New York.

Neel, J. V., Satoh, C., Hamilton, H. B., Otake, M., Goriki, K., Kageoka, T., Fujita, M., Neriishi, S., and Asakawa, J. (1980a). Search for mutation affecting protein structure in children of atomic bomb survivors: preliminary report. *Proc. Natl. Acad. Sci. USA* **77**, 4221–4225.

Neel, J. V., Mohrenweiser, H. W., and Meisler, M. H. (1980b). Rate of spontaneous mutation at human loci encoding for protein structure. *Proc. Natl. Acad. Sci. USA* **77**, 6037–6041.

Neel, J. V., Mohrenweiser, H., Hanash, S., Rosenblum, B., Sternberg, S., Wurzinger, K., Rothman, E., Satoh, C., Goriki, K., Krasteff, T., Long, M., Skolnick, M., and Krzesicki, R. (1983a). Biochemical approaches to monitoring human populations for germinal mutation rates. I. Electrophoresis. *In* "Utilization of Mammalian Specific Locus Studies in Hazard Evaluation and Estimation of Genetic Risk" (W. Sheridan and F. deSerres, eds.), pp. 71–93. Plenum, New York.

Neel, J. V., Nishimoto, Y., Goriki, K., Satoh, C., Fujita, M., and Yoshimoto, Y. (1983b). Epidemiological studies of germ cell mutation in human populations, with particular reference to groups with unusual chemical exposures. *In* "Methods for Estimating Exposure, Quantifying Risks to Human Health, and Measuring Injury to Ecosystems" (N. Nelson and V. Vonk, eds.), Scientific Group on Methodologies for the Safety Evaluation of Chemicals, Rome, in press.

Nelson, R. L., and Harris, H. (1978). The detection of mutation in human diploid fibroblasts after mutagen treatment using non-selective cloning and enzyme electrophoresis. *Mutat. Res.* **50**, 277–283.

O'Farrell, P. H. (1975). High resolution two-dimensional electrophoresis of proteins. *J. Biol. Chem.* **250**, 4007–4021.

Rosenblum, B. B., Hanash, S. M., Yew, N., and Neel, J. V. (1982). Two-dimensional electrophoretic analysis of erythrocyte membranes. *Clin. Chem.* **28**, 925–931.

Rosenblum, B. B., Hanash, S. M., and Neel, J. V. (1983a). High-resolution separation of plasma proteins, application of genetic analysis. *In* "Biologie Prospective, Tome 1" (M. M. Galteau, G. Siest, and J. Henny, eds.) pp. 85–90. Masson, Paris.

Rosenblum, B. B., Neel, J. V., and Hanash, S. M. (1983b). Genetic variation in polypeptides visualized with two-dimensional gel electrophoresis. *Proc. Natl. Acad. Sci. USA,* in press.

Rothman, E. D., Neel, J. V., and Hoppe, F. M. (1981). Estimating a probability for paternity in apparent cases of mutation. *Am. J. Hum. Genet.* **33**, 617–628.

Sammons, D. W., Adams, L. D., and Nishizawa, E. E. (1981). A silver-based color development system for staining of polypeptides in polyacrylamide gels. *Electrophoresis* **2**, 135–141.

Satoh, C., Awa, A. A., Neel, J. V., Schull, W. J., Kato, H., Hamilton, H. B., Otake, M., and Goriki, K. (1983a). Genetic effects of atomic bombs. *Proc. Int. Congr. Hum. Genet. 6th,* Part A, pp. 267–276.

Satoh, C., Neel, J. V., Yamashita, A., Goriki, K., Fujita, M., and Hamilton, H. B. (1983b). The frequency among Japanese of heterozygotes for deficiency variants of 11 enzymes. *Am. J. Hum. Genet.,* in press.

Schull, W. J., Otake, M., and Neel, J. V. (1981). A reappraisal of the genetic effects of the atomic bombs: Summary of a thirty-four year study. *Science (Washington, D.C.)* **213**, 1220–1227.

Serra, J. P. (1982). "Mathematical Morphology and Image Analysis." Academic Press, New York and London.

Shimura, K., Nomura, H., Niimoto, M., Hattori, T., Kamitsuna, A., Shigenobu, T., Nishimoto, Y., Hayashi, Y., and Tokuoka, S. (1978). Pathological review of four cases of stomach cancer which developed in former workers of poison gas factory in Ohkunojima. *Hiroshima Igaku (J. Hiroshima Med. Assoc.)* **31**, 277–281.

Sing, C. F., Gallarda, H. S., and Hanash, S. M. (1983). Statistical sources of 2D gel variability. *Comput. Biol. Med.,* in press.

Skolnick, M. M. (1982). An approach to completely automatic comparison of two-dimensional electrophoresis gels. *Clin. Chem.* **28**, 979–986.

Skolnick, M. M. (1983). Application of morphological transformations to the analysis of two-dimensional electrophoretograms of biological materials. *Comput. Vision, Graphics and Image Process.,* in press.

Skolnick, M. M., Sternberg, S. R., and Neel, J. V. (1982). Computer program for adapting two-dimensional gels to the study of mutation. *Clin. Chem.* **28**, 969–978.

Smith, C. R., Racine, R. R., and Langley, C. H. (1981). Lack of genic variation in the abundant proteins of human kidney. *Genetics* **96**, 967–974.

Smith, R. J. (1982). Atomic bomb tests leave infamous legacy. *Science (Washington, D.C.)* **218**, 266–269.

Soares, E. R. (1979). TEM-induced gene mutations at enzyme loci in the mouse. *Environ. Mutat.* **1**, 19–25.

Sternberg, S. R. (1979). Parallel architectures for image processing. *Proc. IEEE Comput. Soc. Int. Software Applic. Conf.* 3rd (Catalog #7990968).

Strahler, J., Hanash, S., Gupta, G., Rosenblum, B., and Butkunas, R. (1983). Separation of transferrin variants in plasma by ion exchange HPLC. *J. Chromatogr.,* in press.

Tauber, A. I., and Goetzl, E. J. (1979). Structural and catalytic properties of the solubilized superoxide-generating activity of human polymorphonuclear leukocytes. Solubilization, stabilization in solution and partial characterization. *Biochemistry* **18**, 5576–5584.

Tracy, R. P., Currie, R. M., and Young, D. S. (1982a). Two-dimensional gel electrophoresis of serum specimens from a normal population. *Clin. Chem.* **28**, 890–899.

Tracy, R. P., Currie, R. M., and Young, D. S. (1982b). Reproducibility and quality assurance of two-dimensional gel electrophoresis of serum specimens. *Clin. Chem.* **28**, 908–914.

Vogel, F., and Rathenberg, R. (1975). Spontaneous mutation in man. *Adv. Hum. Genet.* **5**, 223–318.

Wada, S., Yamada, A., Nishimoto, Y., Tokuoka, S., Miyanishi, M., Katsuta, S., and Umisa, H. (1963). Neoplasms of the respiratory tract among poison gas workers. *Hiroshima Igaku (J. Hiroshima Med. Assoc.)* **16**, 728–745.

Wada, S., Miyanishi, M., Nishimoto, Y., Kambe, S., and Miller, R. W. (1968). Mustard gas as a cause of respiratory neoplasia in man. *Lancet* **i**, 1161–1163.

Walton, K. E., Styer, D., and Guenstein, E. I. (1979). Genetic polymorphism in normal human fibroblasts as analyzed by two-dimensional polyacrylamide gel electrophoresis. *J. Biol. Chem.* **254**, 7951–7960.

Wanner, L. A., Neel, J. V., and Meisler, M. H. (1982). Separation of allelic variants by two-dimensional electrophoresis. *Am. J. Hum. Genet.* **34**, 209–215.

CHAPTER 10

Expression of Cellular Proteins in Normal and Transformed Human Cultured Cells and Tumors: Two-Dimensional Gel Electrophoresis as a Tool to Study Neoplastic Transformation and Cancer

JULIO E. CELIS
RODRIGO BRAVO*
PETER MOSE LARSEN
STEPHEN J. FEY
JAIME BELLATIN
ARIANA CELIS
Division of Biostructural Chemistry
Department of Chemistry
Aarhus University
Aarhus, Denmark

I. Introduction. 308
II. Transformation-Sensitive Polypeptides in Cultured Epithelial and Fibroblast Human Cells: Work in This Laboratory 311
A. Transformation-Sensitive Polypeptides: Working with a Small Number of Cells. 311
B. The Synthesis of Most Transformation-Sensitive Polypeptides Is Sensitive to Changes in the Growth Properties of the Cells. 320
III. Transformation-Sensitive Polypeptides of Known Identity. . . . 323
A. Vimentin. 323
B. Cyclin. 326
C. Tropomyosin. 332
IV. Variation in the Expression of Mutant Actins in Chemically Transformed Human Diploid Fibroblasts: Actin in General. . . . 334

* Present affiliation: European Molecular Biology Laboratory, 6900 Heidelberg, Federal Republic of Germany.

TWO-DIMENSIONAL GEL
ELECTROPHORESIS
OF PROTEINS

ISBN 0-12-164720-X

V. Phosphoproteins. 337
A. Transformation-Sensitive Phosphoproteins in Cultured Epithelial and Fibroblast Cells 337
B. Tyrosine Phosphorylation of Proteins in A431 Human Tumor Cells. 341
VI. Polypeptide Synthesis in Human Tumors and Normal Tissue. . 343
A. Polypeptide Synthesis in Human Sarcomas and Normal Tissue. 343
B. Emerging Technology. 346
VII. Conclusions. 347
References. 353

I. Introduction

The molecular mechanisms underlying malignant transformation and cancer are for the most part unknown (for references see Foulds, 1969; Cairns, 1975, 1978, 1981; Peto, 1977; Klein, 1981; Bishop, 1982; Radman *et al.*, 1982; Franks, 1983). In general the study of cancer has been the domain of tumor virologists and various types of oncogenic viruses have provided useful experimental models for the study of neoplastic transformation in cultured cells.

Perhaps the most comprehensive view of viral oncogenesis has come from studies of retroviruses, a group of viruses that are the etiological agents of sarcomas, carcinomas, and leukemias in many avian and mammalian species (for references see Weiss et al, 1982). Genetic as well as biochemical studies of these viruses and especially of the Rous sarcoma virus (RSV) have led to the discovery that some of these viruses carry host genetic information ("oncogenes") which has been implicated in the process of neoplastic transformation (for reviews see Bishop, 1981, 1982; Klein, 1981; Bishop and Varmus, 1982; Weinberg, 1982a,b; Cooper, 1982). Sequences homologous to the viral oncogenes (v-*onc* genes) have been detected in the genome of vertebrate cells (c-*onc* genes or proto-oncogenes), and in many cases the cellular oncogenes have been shown to be conserved throughout evolution and sometimes expressed in a tissue-specific manner (reviewed by Coffin *et al.*, 1981). To date, 17 retrovirus oncogenes have been identified and 16 are known to have close relatives in the genome of normal vertebrate cells (Bishop, 1982). It has been proposed that the activity of the proto-oncogene(s) is important in the control of cell division and/or cell proliferation and that transformation results in the abnormal expression of these genes.

The best known of these genes is the RSV *src* gene, which codes for a phosphoprotein ($pp60^{src}$) of molecular weight of 60,000 that has protein kinase activity and that specifically phosphorylates tyrosine residues (Hanafusa, 1977; Vogt, 1977; Brugge and Erikson, 1977; Purchio *et al.*, 1978; Collet and Erikson, 1978; Erikson *et al.*, 1979; Radke and Martin, 1979; Collet *et al.*, 1980; Hunter and Sefton, 1980; Levinson *et al.*, 1980; Radke *et al.*, 1980). Tyrosine phosphorylation of cellular proteins has also been implicated in the transformation mechanism of retroviruses other than RSV (Beemon, 1981; Sefton *et al.*, 1981a; for reviews see Hunter, 1982; Ghysdael, 1983). The kinase activity of these proteins is thought to affect the activity of one or more cellular proteins, but even though various substrates have been identified there is no direct experimental proof correlating the modification of these substrates and the altered growth properties of transformed cells. We will return to this point later because the identification of these substrates has been aided by the use of high resolution two-dimensional gel electrophoresis.

Tyrosine-specific protein kinases are also involved in the metabolism of normal cells (Ushiro and Cohen, 1980; Hunter and Cooper, 1981; Cooper and Hunter, 1981b; Ek *et al.*, 1982; Hunter, 1982; Cooper *et al.*, 1982; Ghysdael, 1983). For example, it has been shown that binding of polypeptide growth factors such as epidermal growth factor (EGF) and platelet-derived growth factor (PDGF) to various cells stimulates the phosphorylation of tyrosine residues of cytoplasmic and membrane proteins. There is now evidence, however, that tyrosine phosphorylation is not a universal mechanism of transformation, because cells transformed with Kirsten and Moloney sarcoma virus as well as SV40 do not show an increased level of phosphotyrosine (Sefton *et al.*, 1980).

Studies of DNA tumor viruses, especially of the small papovaviruses SV40 and polyoma, have revealed other proteins (large T antigen alone or with small t in SV40. Middle T antigen alone or together with small t in polyoma) as the principal candidates for transforming proteins (Schaffhausen and Benjamin, 1979; Howley, 1980; Ito, 1980; Martin, 1981; Segawa and Ito, 1982). The study of these proteins has been difficult, however, because they are also involved in the late replication cycle of the virus. Although the SV40 large T antigen is found complexed to a cellular nuclear 53-kdalton protein, the polyoma middle T antigen is not (Lane and Crawford, 1979). Interestingly, the 53-kdalton protein has been detected in small amounts in some normal mouse tissues (Linzer and Levine, 1979) but is not present in human tumor cell lines HeLa, EJ, and T24 (Benchimol *et al.*, 1982). A similar 53-kdalton protein has been found to be complexed with the Epstein–Barr virus nuclear antigen (EBNA) of Epstein–Barr virus (EBV) (Klein *et al.*, 1979).

Even though most naturally occurring tumors are not of viral etiology, there is now a unifying view as to how viruses, chemicals, and radiation induce carcinogenesis (see Klein, 1981; Bishop, 1982; Radman *et al.*, 1982, 1983).It is believed that these agents interact with the DNA in such a way as to alter directly or indirectly the expression of cellular proto-oncogenes (see also Hayward *et al.*, 1981; Neel *et al.*, 1981; Nusse and Varmus, 1982; Marx, 1982). The functions of these genes are thought to be essential in cell differentiation and growth control, and their disturbance results in cellular alterations leading to neoplastic transformation. Carcinogenesis, however, is a multistep process and various agents may affect the process at various stages. Support for the notion that non-virally induced tumors contain oncogene(s) has been presented by Weinberg and co-workers and by Cooper and co-workers, who have shown that fragments of DNA isolated from tumors introduced into nontumor cells transmit the malignant property (Shih *et al.*, 1979, 1981; Cooper *et al.*, 1980; Weinberg, 1981; Krontiris and Cooper, 1981; for reviews see Cooper, 1982; Weinberg, 1982a,b). Only some of the genes so far revealed by this approach, however, correspond to any of the known viral oncogenes.

A great deal of research is being directed along this line of experimentation (for references see Tabin *et al.*, 1982; Reddy *et al.*, 1982). These studies have led so far to the amazing discovery that a single point mutation in the cellular C-Ha-ras 1 gene in T24 bladder carcinoma is sufficient to confer transforming properties on its gene product (Tabin *et al.*, 1982; Reddy *et al.*, 1982). It should be stressed, however, that most tumors studied so far do not yield DNA that is oncogenic in the NIH 3T3 transfection assay. Some of the limitations of this approach have been recently discussed by Logan and Cairns (1982).

An alternative approach for searching for cellular proteins involved in the control of cell proliferation has been the analysis by means of high resolution two-dimensional gel electrophoresis (O'Farrell, 1975; O'Farrell, *et al.*, 1977) of the overall patterns of gene expression in pairs of normal and transformed cells (Strand and August, 1978; Leavitt and Moyzig, 1978; Radke and Martin, 1979; Forchhammer and Klarlund, 1979; Bravo and Celis, 1980a, 1982b; Liebermann *et al.*, 1980; Brzeski and Ege, 1980; Leavitt and Kakunaga, 1980; Bravo *et al.*, 1981d,e, 1982c, 1983b; Laszlo *et al.*, 1981; Lester *et al.*, 1982; Jellum and Thorsrud, 1982; Thorsrud *et al.*, 1982; Tracy *et al.*, 1982; Giometti *et al.*, 1982; Hanash *et al.*, 1982; Gemmel and Anderson, 1982; Bellatin *et al.*, 1982; Forchhammer and Macdonald-Bravo, 1983; see also chapters by Garrels *et al.* and by Tracy and Young). In most cases this approach was stimulated by the fact that most naturally occurring tumors are not of viral etiology. In this laboratory, for example, we hypothesized that neoplastic transformation was due to the abnormal expression of normal genes (Bravo and Celis, 1980a, 1982b; Celis and Bravo, 1981; Bravo *et al.*, 1981d,e, 1983b; Bellatin *et al.*, 1982), and therefore it was expected that a careful and detailed study of the proteins synthesized by normal and transformed cells under a variety of physiological conditions could lead to the identification of cellular proteins that might be involved in the control of cell proliferation. The ultimate goal of these studies is the elucidation of the pathway(s) that control cell proliferation in normal cells and hence the determination of the means by which alterations of this pathway lead to abnormal growth characteristics and/or to neoplastic transformation and cancer.

The primary emphasis of this chapter is to examine the progress made in studying protein synthesis in normal and neoplastic human cultured cells and tumors using two-dimensional gel electrophoresis and to relate it to current thoughts concerning neoplastic transformation. In the interest of a coherent presentation we will concentrate mainly on work carried out in this laboratory, although an effort will be made to correlate this data with that of other laboratories working with human cells as well as those of other species.

II. Transformation-Sensitive Polypeptides in Cultured Epithelial and Fibroblast Human Cells: Work in This Laboratory

A. Transformation-Sensitive Polypeptides: Working with a Small Number of Cells

Given the large number of polypeptides thought to be present in a somatic cell, we turned our attention to the two-dimensional gel electrophoresis technique described by O'Farrell and co-workers (O'Farrell, 1975; O'Farrell *et al.*, 1977; see also chapters by Bravo and by Garrels *et al.*) as this was known to resolve nearly 1000 [^{35}S]methionine-labeled cellular polypeptides based on their charge and molecular weight properties. With improvements in the separation as well as the labeling technique, it has been possible so far to resolve nearly 1400 [^{35}S]methionine-labeled polypeptides from as few as 50 to 100 human cultured cells (Bravo *et al.*, 1981a; Celis and Bravo, 1981; Fey *et al.*, 1981; Bravo and Celis, 1982a; see also chapters by Garrels *et al.* and by Bravo and Celis).

The normal–transformed human cell pair we have studied correspond to early passaged epithelial amnion cells (AF type) and their spontaneously transformed counterparts AMA (Bravo and Celis, 1982b), FL-amnion (Fogh and Lund, 1957) and WISH-amnion (Hayflick, 1961). Other transformed human epithelial cells studied include Detroit 98 (bone marrow; Berman and Stulberg, 1956), Change liver (Chang, 1954), bladder carcinoma EJ19 (a gift from L. M. Franks; Hastings and Franks, 1981), human diploid lung fibroblasts (WI-38, p26; Hayflick, 1965), SV40-transformed lung fibroblasts (WI-38 VA13; Girardi *et al.*, 1965) as well as normal human skin fibroblasts at different stages throughout their lifespan (Bravo *et al.*, 1981e, 1983b; Bravo and Celis, 1982b). Because various human cultured cells examined so far show similar qualitative two-dimensional protein patterns, we have prepared a catalog of human HeLa cell polypeptides that has been used to standardize a numbering system for human proteins (Bravo *et al.*, 1981a; Bravo and Celis, 1982b; see also chapter by Bravo and Celis). In this way it is possible to identify the same protein comigrating in different human cells.

Routinely, between 1000 and 5000 sparsely growing cells are labeled with [^{35}S]methionine (Bravo and Celis, 1980b; Bravo *et al.*, 1981c; Celis and Bravo, 1981; see also chapter by Bravo) and the product of about 1000 cells is analyzed by two-dimensional gel electrophoresis (IEF, O'Farrell, 1975; NEPHGE, O'Farrell *et al.*, 1977; see also chapters by Bravo and by Garrels *et al.*). For quantitation the individual polypeptides are excised from the gel with the aid of a scalpel and their radioactivity determined by scintillation counting (Bravo *et al.*, 1982b). Alternative spots can be quantitated using computerized two-dimensional gel electrophoresis (see chapter by Garrels *et al.*).

Figures 1 and 2, taken from the work of Bravo and Celis (1982b), show representative two-dimensional gels of acidic (IEF) [^{35}S]methionine-labeled polypeptides from normal amnion (Fig. 1a), spontaneously transformed amnion (AMA) (Fig. 1b), normal human diploid lung fibroblasts (WI-38, p26) (Fig. 2a), and SV40-transformed lung fibroblasts (WI-38 VA13) (Fig. 2b). Two-dimensional gels of basic proteins (NEPHGE) are shown in Fig. 3a–d. Essentially the same polypeptide patterns have been obtained for AMA cells labeled at different passages after transformation (over a period of 18 months) and for cells recovered from tumors produced in *nude* mice (results not shown). Very similar polypeptide patterns have been observed in FL-amnion and WISH-amnion (R. Bravo and J. E. Celis, unpublished).

For a given cell type a total of about 1400 acidic and basic polypeptides could be resolved. Visual inspection of many films exposed for different intervals did not reveal new major polypeptides appearing in the trans-

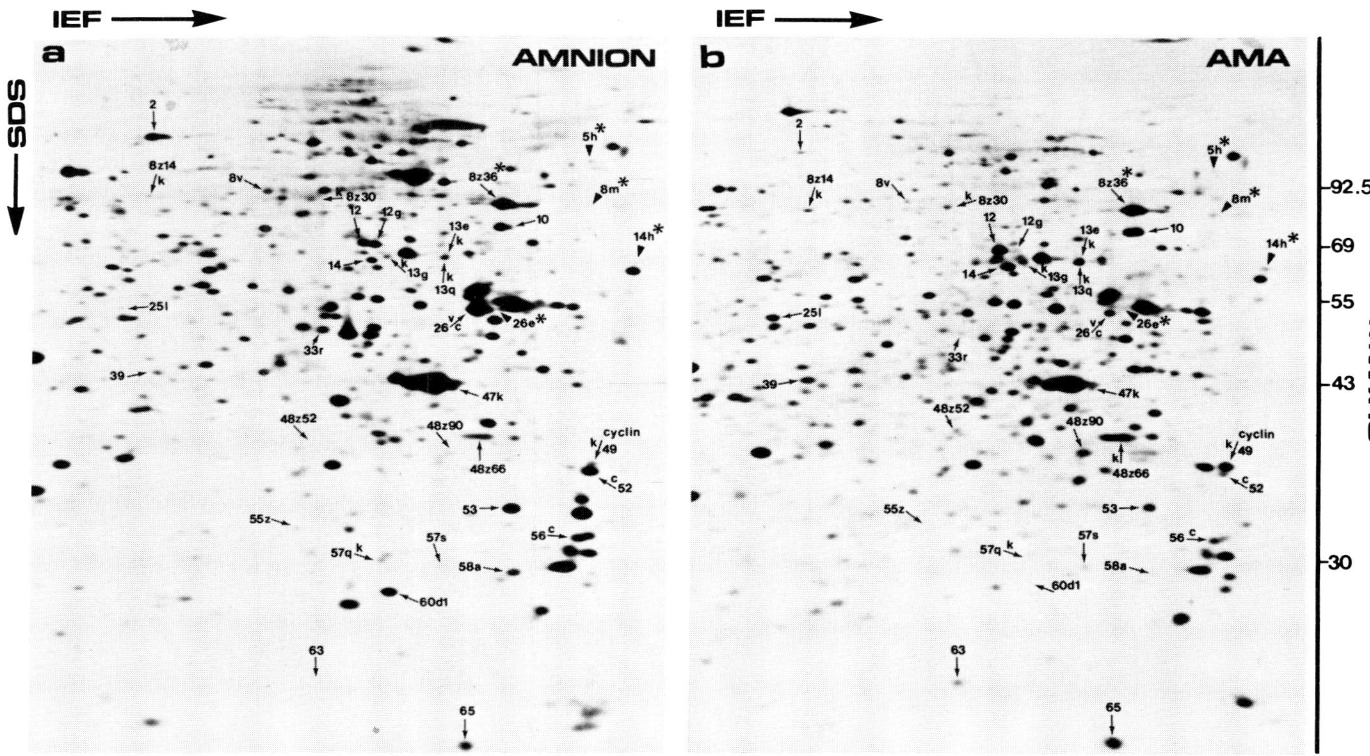

Figure 1. Two-dimensional gel electrophoresis of acidic (IEF) [^{35}S]methionine-labeled polypeptides from (a) normal and (b) spontaneously transformed human amnion cells (AMA). Sparsely growing cells placed in flat-bottomed microtiter wells (0.25 ml, NUNC) were labeled as described (Bravo and Celis, 1980b; Bravo *et al.*, 1981c; Celis and Bravo, 1981; see also chapter by Bravo). v, vimentin; total actin, the large spot located to the left of 47k; c, polypeptides mainly present in cytoplasts; k, polypeptides mainly present in karyoplasts (Bravo *et al.*, 1981b; see chapter by Bravo and Celis); *, phosphoproteins whose relative proportions vary significantly in transformed cells (see also Section V,A). From Bravo and Celis (1982b).

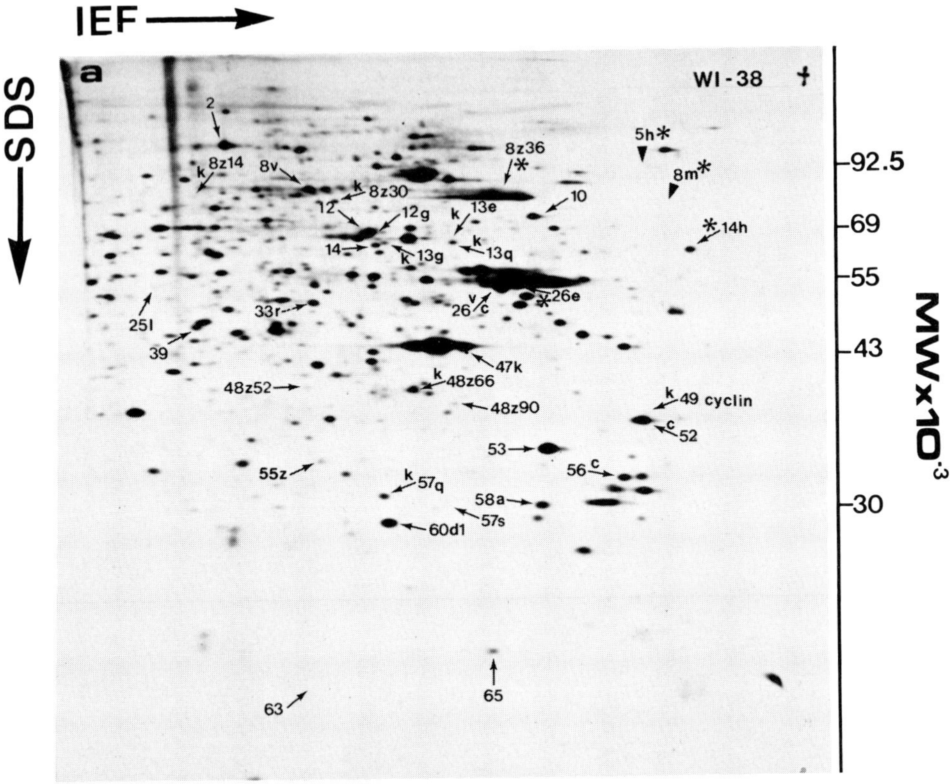

Figure 2. Two-dimensional gel electrophoresis of acidic (IEF) [^{35}S]methionine-labeled polypeptides from (a) normal and (b) SV40-transformed human WI-38 lung fibroblasts. Sparse cultures were labeled for 16 h with [^{35}S]methionine. Only transformation-sensitive polypeptides are indicated. *, phosphoproteins whose relative proportions vary significantly in transformed cells; v, vimentin; c, mainly present in cytoplasts; k, mainly present in karyoplasts. From Bravo and Celis (1982b).

formed cells that could be detected at this level of resolution. The failure to detect T antigen in the SV40-transformed WI-38 cells is most likely due to either comigration with other abundant cellular proteins or to the fact that proteins comprising less than 0.001% of the total labeled protein are difficult to detect (O'Farrell, 1975; Bravo and Celis, 1982b; Duncan and McConkey, 1982). We detected, however, significant changes in the abundance of certain polypeptides that were present both in normal and transformed cells (Figs. 1–3). Even though some proteins were synthesized in relatively small amounts in either the normal or the transformed cells, they could be detected after prolonged exposure of the gels.

Given the large number of polypeptides resolved and considering the

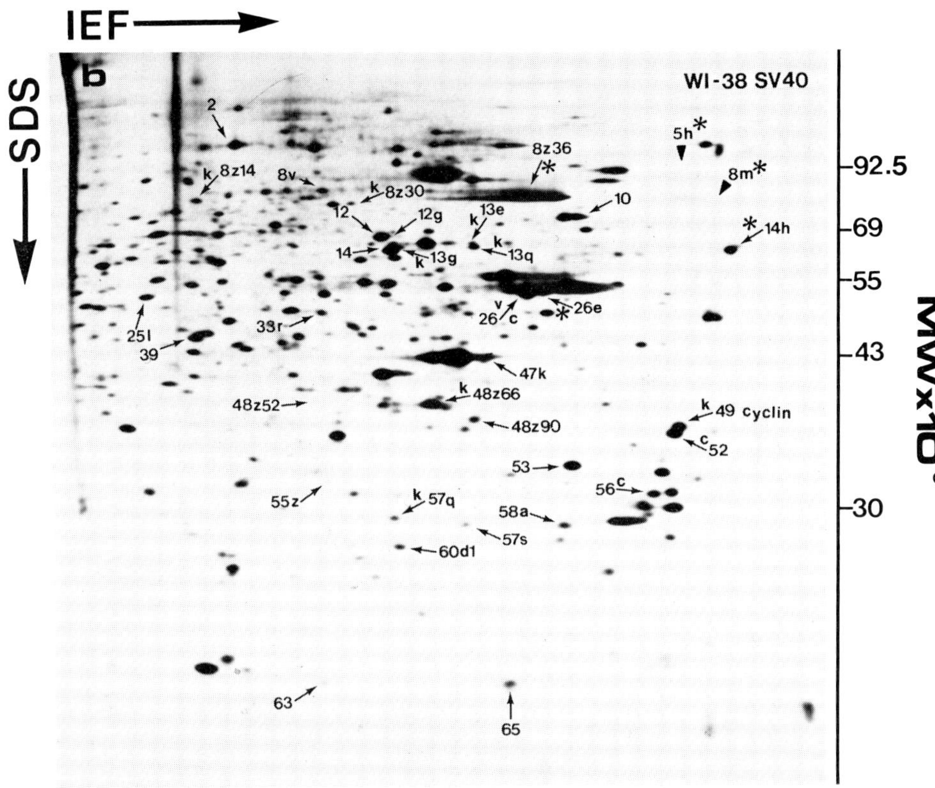

Figure 2. (*continued*)

advantage of identifying transformation-sensitive polypeptides common to various transformed cells, we have focused our attention mainly on those polypeptide changes that are common to various cell types. Of 400 polypeptides quantitated, 53 (22 basic and 31 acidic) were found to vary by 40% or more; these are indicated with arrows and numbers (HeLa numbering system) in Figs. 1–3 and are listed in Table I and II with their coordinates from the HeLa protein catalog (Bravo *et al.*, 1981a; Bravo and Celis, 1982a; see also chapter by Bravo and Celis). These changes have also been observed in gels stained with silver nitrate (results not shown). In cases in which the cellular distribution of these polypeptides is known (cytoplastic or karyoplastic) (Bravo *et al.*, 1981b), this has been indicated with a "c" (cytoplastic) or with a "k" (karyoplastic) in Figs. 1–3. Only a few of these polypeptides have been identified and these correspond to a stress protein (IEF 8z36), vimentin (IEF 26) (Franke *et al.*, 1978), cyclin (IEF 49) (Bravo *et al.*, 1981e), and a tropomyo-

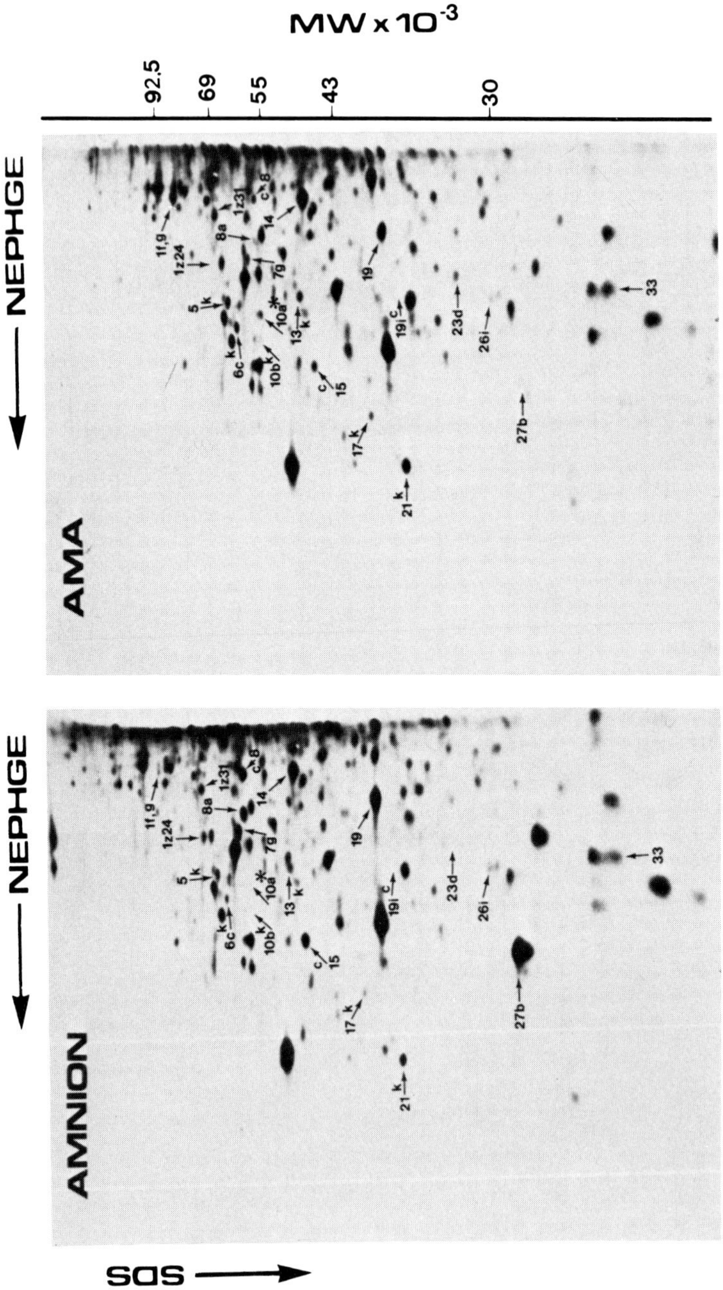
MW x 10-3
92.5
69
55
43
30
NEPHGE
AMA
NEPHGE
AMNION
SDS

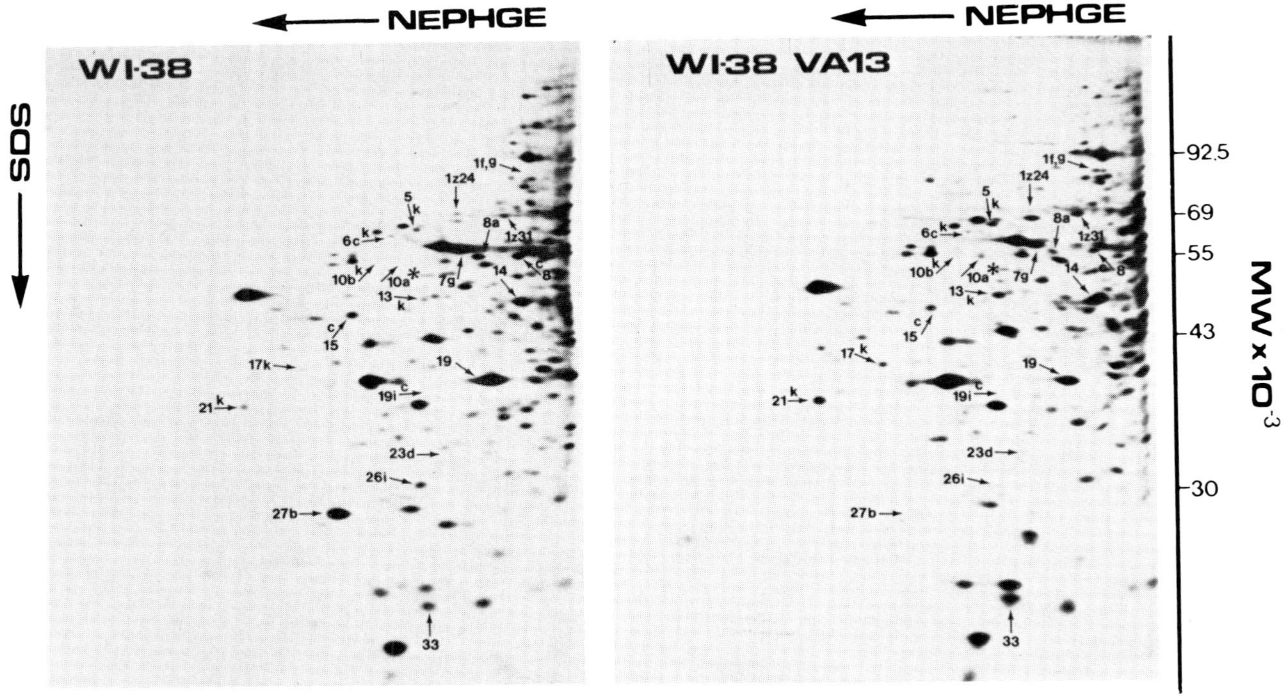

Figure 3. Two-dimensional gel electrophoresis of basic (NEPHGE) [^{35}S]methionine-labeled polypeptides from normal and transformed human cultured cells. Sparse growing cultures were labeled for 16 h with [^{35}S]methionine as described in the legend of Fig. 1. Only variable polypeptides common to both cell pairs are indicated. c, Polypeptides present mainly in cytoplasts; k, polypeptides present mainly in karyoplasts; *, phosphoproteins whose relative proportions vary significantly in transformed cells (see Section V,A). From Bravo and Celis (1982b).

TABLE I

Basic Human Proteins Sensitive to Neoplastic Transformation[a]

Polypeptide (NEPHGE)	Coordinates	cpm-ratio[b] WI-38 VA13 / WI-38	cpm-ratio[b] AMA / AMNION	Relative proportion in transformed cells	Cellular distribution[c]	Other characteristics[d]
1f	92/0.22	14.00	11.90	Increases	—	—
1g	92/0.18	3.66	2.30	Increases	—	—
1z24	71/0.51	0.59	0.50	Decreases	—	Also decreased in human sarcomas
1z31	71/0.30	4.46	1.79	Increases	—	—
5	64/0.70	3.04	2.53	Increases	Mainly in karyoplasts	—
6c	61/0.82	2.10	3.83	Increases	Mainly in karyoplasts	Also increased in human sarcomas
7g	59.5/0.49	0.52	0.37	Decreases	—	—
8	57/0.24	0.54	0.22	Decreases	Cytoplasmic	Also decreased in human sarcomas
8a	57/0.43	0.48	0.14	Decreases	—	Also decreased in human sarcomas
10a	54.5/0.76	7.30	7.00	Increases	Mainly in karyoplasts	—
10b	54.5/0.86	7.00	4.00	Increases	—	—
13	49/0.67	1.82	2.68	Increases	Mainly in karyoplasts	Also increased in human sarcomas
14	48.5/0.24	2.18	1.64	Increases	—	Also increased in human sarcomas
15[e]	47/1.00	0.60	0.59	Decreases	Cytoplasmic	—
17	41/1.22	2.59	1.98	Increases	Mainly in karyoplasts	—
19[e]	39.5/0.40	0.28	0.36	Decreases	—	Also decreased in human sarcomas
19i	38/0.60	0.42	0.54	Decreases	Cytoplasmic	—
21[e]	35/1.45	2.50	2.70	Increases	Mainly in karyoplasts	Also increased in human sarcomas
23d	32.5/0.59	1.90	2.77	Increases	—	—
26i	28.2/0.71	0.55	0.51	Decreases	—	—
27b	27/1.11	0.12	0.02	Decreases	—	—
33	18/0.68	1.84	1.50	Increases	—	—

[a] From Bravo *et al.* (1981a); Bravo and Celis (1982a,b).
[b] The counts were first standardized against total counts applied to the gels. The standard error of these determinations is in the order of 12%.
[c] From Bravo *et al.* (1981b).
[d] From Forchhammer and Macdonald-Bravo (1983).
[e] Also identified in pairs of normal and transformed mouse and hamster fibroblasts (Bravo and Celis, 1980a).

sin-related polypeptide (IEF 52) (Bravo *et al.*, 1981e) (Table II). In the case of the amnion cells the data have been verified by similar quantitative analysis of independently isolated transformed amnion cells such as FL-amnion and WISH-amnion (not shown). The transformation-sensitive polypeptides have also been identified in other transformation cells of epithelial origin such as Detroit 98 (bone marrow), HeLa (cervical carcinoma), Chang liver, and EJ19 bladder carcinoma (results not shown).

TABLE II

Acidic Human Proteins Sensitive to Neoplastic Transformation[a]

Polypeptide (IEF)	Coordinates	cpm-ratio[b] WI-38 VA13 / WI-38	cpm-ratio[b] AMA / AMNION	Relative proportion in transformed cells	Cellular distribution[c]	Other characteristics[d]
2	110/0.44	0.44	0.35	Decreases	—	—
8v	90/0.64	0.34	0.54	Decreases	—	Also decreased in human sarcomas
8z14	84/0.32	1.66	2.59	Increases	Mainly in karyoplasts	Also increased in human sarcomas
8z30	76/0.72	2.18	1.82	Increases	Mainly in karyoplasts	—
8z36 (stress protein)[e]	76/1.27	1.38	1.96	Increases	—	Also increased in human sarcomas
10	72/1.28	2.10	1.60	Increases	—	—
12	68/0.85	1.57	2.41	Increases	—	—
12g	68/0.89	<0.05	<0.05	Decreases	—	—
13e	67/1.03	2.46	2.41	Increases	Mainly in karyoplasts	—
13g	66/0.76	1.87	1.47	Increases	Mainly in karyoplasts	—
13q	66.5/1.03	1.64	3.40	Increases	Mainly in karyoplasts	—
14 (stress protein)	66/0.86	3.08	1.66	Increases	—	Also increases in human sarcomas
251	55/0.25	3.35	2.79	Increases	—	—
26 (vimentin)[e]	54/1.19	0.58	0.35	Decreases	Cytoplasmic	Also decreased in human sarcomas
33r	50/0.70	0.60	0.32	Decreases	—	—
39	45.5/0.91	1.65	1.65	Increases	—	—
47k	43/1.08	0.30	0.09	Decreases	—	—
48z52	35/0.68	0.50	0.63	Decreases	—	Also decreased in human sarcomas
48z66	38.5/0.92	1.76	3.08	Increases	Mainly in karyoplasts	—
48z90	38/1.06	2.32	3.20	Increases	—	—
49 (cyclin)[e]	36/1.48	9.31	1.72	Increases	Mainly in karyoplasts	Also increased in human sarcomas
52 (tropomyosin related)[e]	35/1.43	0.40	0.17	Decreases	Cytoplasmic	—
53	32/1.28	0.21	0.61	Decreases	—	Also decreased in human sarcomas
55z	31.5/0.65	0.46	0.14	Decreases	—	—
56	31/1.41	0.61	0.60	Decreases	Cytoplasmic	Also decreased in human sarcomas
57s	29.5/1.03	2.05	1.87	Increases	—	—
57q	30/0.86	0.60	0.33	Decreases	Mainly in karyoplasts	Also decreased in human sarcomas
58a	29/1.27	0.43	0.20	Decreases	—	—
60d1	26/0.84	0.21	0.07	Decreases	—	—
63	20/0.64	1.89	4.00	Increases	—	—
65	16/1.15	1.86	2.72	Increases	—	Also increased in human sarcomas

[a] From Bravo *et al.* (1981a); Bravo and Celis (1982a,b).

[b] The counts were first standardized against total counts applied to the gels. The standard error of these determinations is in the order of 12%.

[c] From Bravo *et al.* (1981b).

[d] From Forchhammer and Macdonald-Bravo (1983).

[e] Also identified in pairs of normal and transformed mouse and hamster fibroblasts (Bravo and Celis, 1980a).

Of the polypeptides whose relative abundance does not change in transformed cells we have identified total actin (γ, β), α- and β-tubulin, and α-actinin (Bravo and Celis, 1982b).

Quantitative two-dimensional gel electrophoretic analysis by means of computerized microdensitometry of human diploid fibroblasts (KD) transformed with 4-nitroquinoline 1-oxide (Kakunaga, 1978) have been reported by Leavitt and Kakunaga (1980) and by Leavitt *et al.* (1982a,b). These workers estimated that $\leq 2\%$ of the more abundant polypeptides (700 analyzed) were either activated or shut off as a result of neoplastic transformation and that at least 32% were modulated quantitatively (Leavitt *et al.*, 1982a). Interestingly, these workers detected the appearance of mutated β-actins in transformed fibroblasts exhibiting incremental increase in tumorigenicity (discussed in Section IV). Furthermore, these studies revealed a significant decrease in the synthesis of fibronectin in the transformed cells (for references see Vaheri and Alitalo, 1981).

Similar two-dimensional gel electrophoretic analysis of proteins from normal and transformed cells from various species have also indicated that neoplastic transformation is accompanied by changes in the relative abundance of polypeptides present both in normal and transformed cells (Leavitt and Moyzig, 1978; Radke and Martin, 1979; Forchhammer and Klarlund, 1979; Bravo and Celis, 1980a; Brzeski and Ege, 1980; Laszlo *et al.*, 1981) rather than by the appearance of new polypeptides in transformed cells (Strand and August, 1978).

Work is now in progress to prepare antibodies (Mose Larsen *et al.*, 1982; Bravo *et al.*, 1983a; see also legend of Fig. 6 of chapter by Fey *et al.*) against all the transformation-sensitive polypeptides so far identified. These antibodies should be of value for purifying these proteins for functional studies and will provide a rapid probe to screen normal and tumor tissues using indirect immunofluorescence.

B. *The Synthesis of Most Transformation-Sensitive Polypeptides Is Sensitive to Changes in the Growth Properties of the Cells*

1. POLYPEPTIDE SYNTHESIS IN GIANT HELA CELLS

From our previous studies it seemed likely that the changes in the relative proportion of some of the transformation-sensitive polypeptides could reflect variations in the growth properties of the cells (Bravo and Celis, 1980a,b). To approach this question we compared the relative proportion of these polypeptides in HeLa cells and giant HeLa cells produced by irradiation with lethal doses of X rays (1100 rads; 110 rads/

min) (Tolmach and Marcus, 1960; Puck and Marcus, 1965; Okada, 1970; Bellatin *et al.*, 1982). Because giant cells do not proliferate but maintain a high degree of metabolic integrity, they represented a suitable system to search for transformation-sensitive polypeptides whose rate of synthesis may be dependent on cell proliferation. The two-dimensional polypeptide maps (IEF and NEPHGE) of control HeLa cells and giant HeLa cells labeled for 20 h with [^{35}S]methionine are shown in Figs. 4 and 5. The giant cells (pure population) were labeled 6 days after irradiation with 1100 rads of X rays. Only 47 (27 acidic and 20 basic) of the 53 transformation-sensitive polypeptides so far revealed in human cultured cells (Bravo and Celis, 1982b) (Tables I and II) were analyzed and are indicated with arrows and arrowheads in Fig. 4 and 5. With the exception of IEF 12g and 60d1 and NEPHGE 1z31 and 27b the relative proportion of all the other transformation-sensitive polypeptides analysed changed substantially and reproducibly in the giant cells. The polypeptides indicated with arrowheads in Fig. 4 (IEF 8m, 14h and 26e (phosphovimentin)) correspond to transformation-sensitive phosphoproteins (see Section V,A). The relative proportions of IEF 8m and 14h decrease in giant HeLa cells, whereas that of phosphovimentin increases (Fig. 4).

We have carried out similar studies on spontaneously transformed amnion cells (AMA) and giant AMA cells. Again a detailed qualitative analysis showed that with the exception of polypeptides IEF 12g and 60d1 and NEPHGE 1z31 and 27b, the intensities of all other transformation-sensitive polypeptides were similar to that observed in control normal amnion cells (Bravo and Celis, 1982b; R. Bravo and J. E. Celis, unpublished). Because the relative change in radioactivity in NEPHGE 1z31 is much smaller than those observed with the other three polypeptides the importance of this change is difficult to evaluate.

It should be noted that at least at the present level of resolution and sensitivity achieved by two-dimensional gel electrophoresis we failed to detect any new major polypeptide appearing in the giant cells. Thus it would seem likely that the drastic changes in growth properties and cell morphology accompanying giant cell formation may be brought about by changes in the relative proportion (abundance) of proteins synthesized both by control and giant cells.

2. POLYPEPTIDE SYNTHESIS IN GROWING AND SENESCENT HUMAN FIBROBLASTS

Preliminary analyses of the [^{35}S]methionine-labeled polypeptides synthesized by growing and senescent human skin fibroblasts throughout their lifespan (Hayflick, 1965) have shown that the relative proportion of a significant number of the transformation-sensitive polypeptides

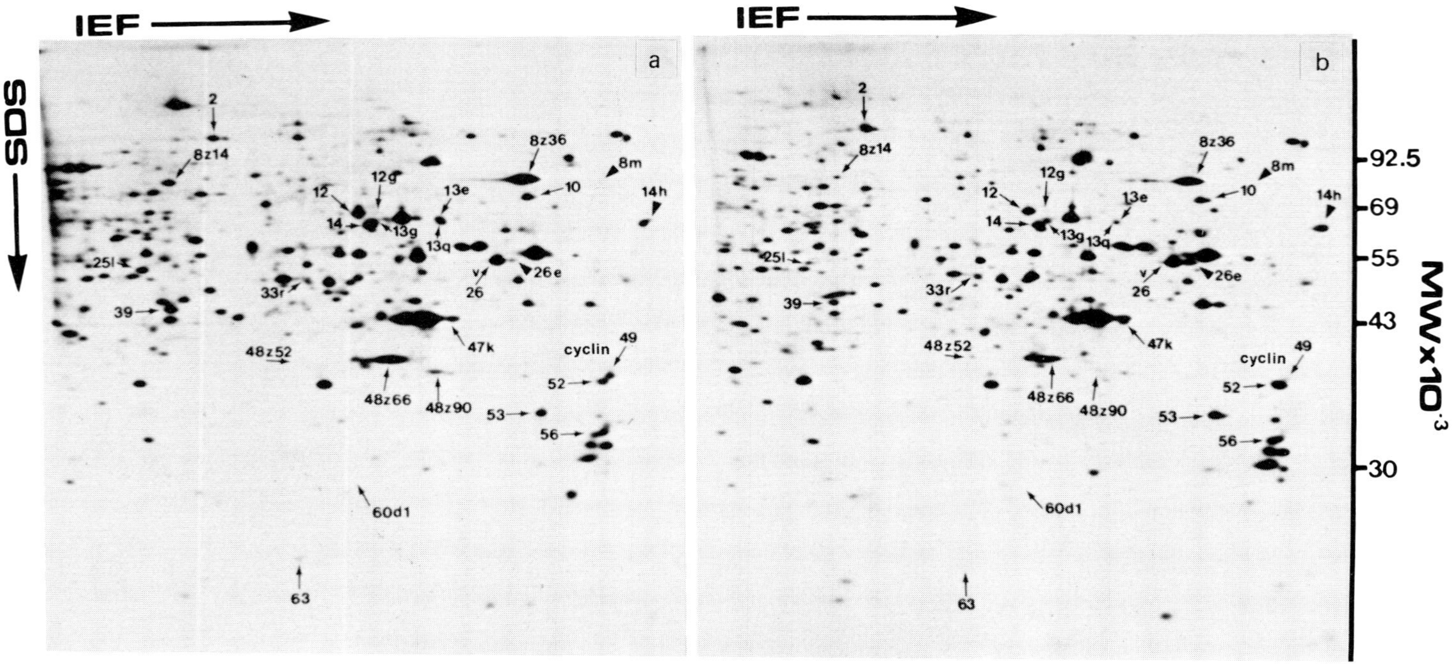

Figure 4. High resolution two-dimensional gel electrophoresis (IEF) of [^{35}S]methionine-labeled polypeptides from HeLa cells (a) and giant HeLa cells (b). Only transformation-sensitive polypeptides are indicated. Arrowheads, phosphorylated proteins; v, vimentin. From Bellatin *et al.* (1982).

changes as the cells cease to divide and become senescent (Bravo *et al.*, 1983b; J. E. Celis and R. Bravo, submitted).

Figures 6 and 7 show representative gels of acidic (IEF, Fig. 6) and basic (NEPHGE, Fig. 7) polypeptides from growing (p3) and senescent (p19) human skin fibroblasts labeled for 16 h with [^{35}S]methionine (Bravo *et al.*, 1983b). Only those polypeptides whose intensity clearly changes are indicated with arrows in Fig. 6 (IEF) and 7 (NEPHGE). There are, however, a few transformation-sensitive polypeptides whose relative intensities do not change in senescent cells and these correspond to IEF 10, 56, and 60d1 (arrowheads in Fig. 6) and to NEPHGE 15 and 27b (arrowheads in Fig. 7). It should be mentioned that the changes in polypeptide synthesis indicated above take place without any significant alteration in the chromosome constitution of the cells (U. Friederich, unpublished). We have obtained evidence indicating that the synthesis of at least some of the transformation-sensitive polypeptides changes following spontaneous growth of senescent human skin fibroblasts. In particular we have observed a significant increase in the synthesis of cyclin following growth stimulation (J. E. Celis and R. Bravo, submitted).

In general our studies of senescent human skin fibroblasts do not lend support to the error catastrophe theory of cellular aging, which predicts that random amino acid substitutions accumulate with aging (Orgel, 1963), as we have failed to detect new proteins (or satellite spots due to amino acid substitutions) in senescent cells (see chapter by J. Pollard). Similar observations have been reported by Wilson *et al.* (1978) in rat superior cervical ganglia, by Engelhardt *et al.* (1978) in human skin fibroblasts, and by Van Keuren *et al.* (1983) in early and late passaged Down's syndrome (trisomy 21) and normal human fibroblasts.

Quantitative changes in protein synthesis following senescence of human fibroblasts have also been reported by Wilson *et al.* (1978) and by Van Keuren *et al.* (1983). These changes, however, are difficult to compare with the ones we have described here because of the different gel running conditions used by various laboratories.

III. Transformation-Sensitive Polypeptides of Known Identity

A. *Vimentin*

Changes in the synthesis of vimentin (Franke *et al.*, 1978) have been reported in SV40- and polyoma-transformed 3T3B and hamster BHK21

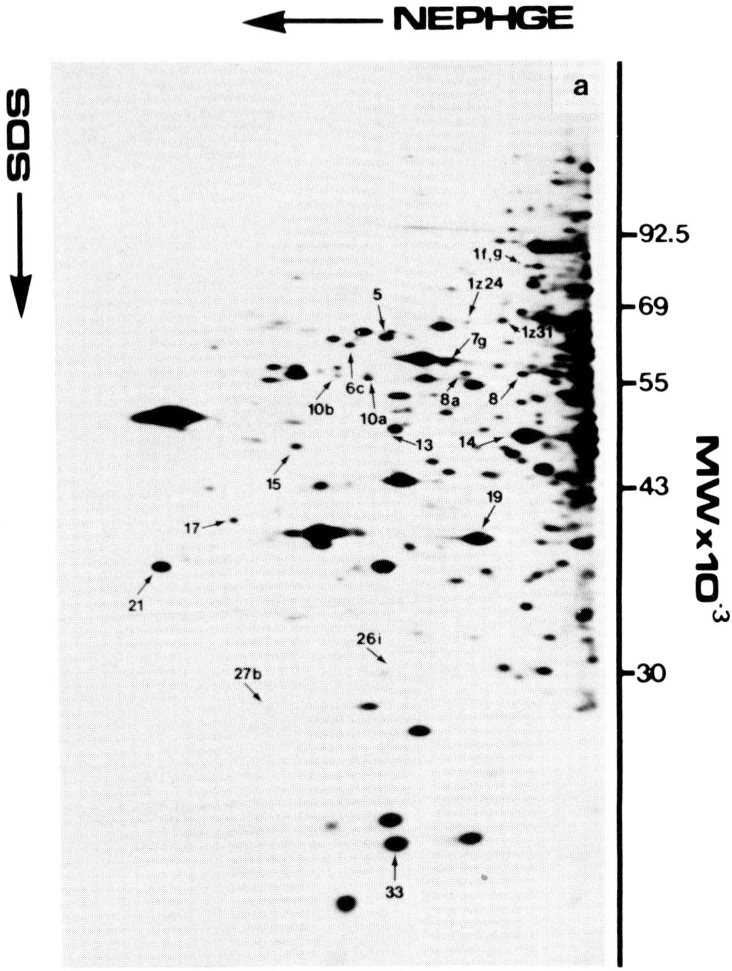

Figure 5. High resolution two-dimensional gel electrophoresis (NEPHGE) of [^{35}S]methionine-labeled polypeptides from control HeLa cells (a) and giant HeLa cells (b). Only transformation-sensitive polypeptides are indicated. From Bellatin *et al.* (1982).

cells (Tuszynski *et al.*, 1979; Bravo and Celis, 1980b) as well as in RSV-transformed chick embryo fibroblasts (Laszlo *et al.*, 1981) and human sarcomas (Forchhammer and Macdonald-Bravo, 1983). Changes in vimentin synthesis have also been observed in embryonal carcinoma cells induced to differentiate by retinoic acid (Linder *et al.*, 1981) and in chick embryo fibroblasts treated with 12-*O*-tetradecanoylphorbol 13-acetate (TPA) (A. Laszlo and M. Bisell, unpublished). The role of vimentin in

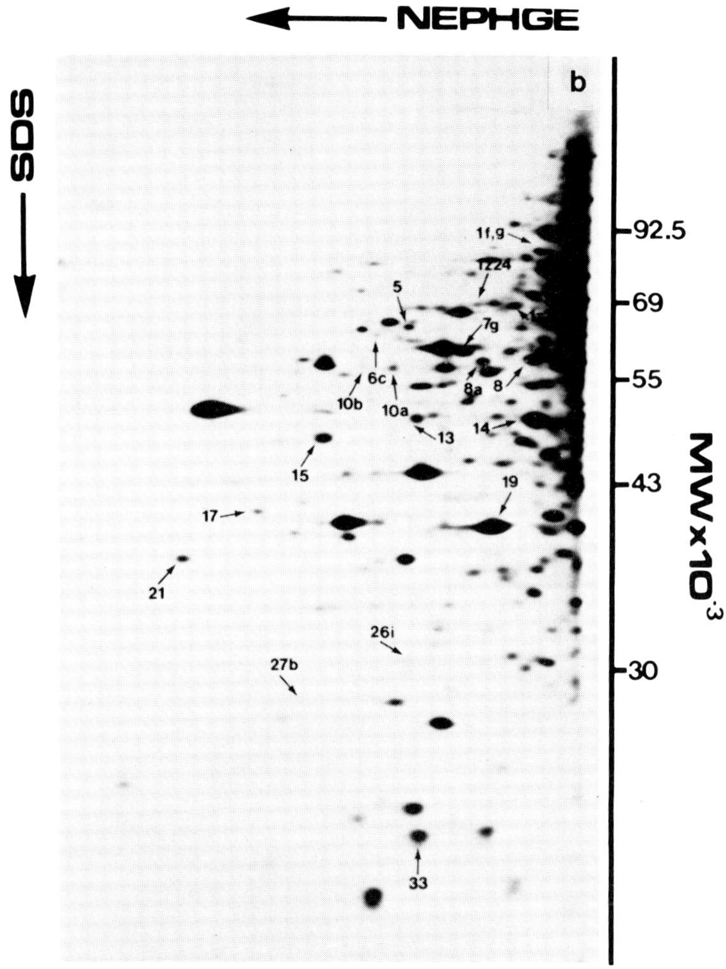

Figure 5. (*continued*)

transformation is, however, unclear as this protein is not present in some cell types *in situ* (Franke *et al.*, 1979c, 1981; Osborn *et al.*, 1981; Osborn and Weber, 1982; Mose Larsen *et al.*, 1983), and it is known to be induced upon *in vitro* culturing of some cell types (Sun and Green, 1978; Franke *et al.*, 1979a,b; Mose Larsen *et al.*, 1983).

Studies of synchronized cultured cells have shown specific degradation (Bravo *et al.*, 1982a,b) as well as phosphorylation (Robinson *et al.*, 1981; Bravo *et al.*, 1982a,b; Evans and Fink, 1982) of vimentin during mitosis, and we have proposed that these modifications may be instrumental in at least part of the process of cell division (Bravo *et al.*, 1982a,b). A rearrangement of the vimentin filament network has been observed in mitosis (Blose, 1979; Aubin *et al.*, 1980) and in asynchronous transformed cells (Hynes and Destree, 1978; A. Laszlo and M. Bisell, unpublished).

Phosphorylation of vimentin (O'Connor *et al.*, 1981) has been found to be modulated by cAMP in various nonmuscle cells (Steinberg and Coffino, 1979; Browning and Sanders, 1981; DePhilip and Kierszenbaum, 1982; Gard and Lazarides, 1982), and O'Connor *et al.* (1981) have demonstrated *in vitro* phosphorylation of vimentin by a cAMP-dependent protein kinase.

At present it is not known whether the changes observed in vimentin synthesis and/or phosphorylation are correlated with the transformed state, proliferation, or adaption to growth in culture; however, despite these caveats this protein may serve as a useful marker for transformation in some cultured cell types (Tuszynski *et al.*, 1979; Bravo and Celis, 1980a, 1982b; Summerhayes *et al.*, 1981; Bellatin *et al.*, 1982) as well as for the characterization of human tumors (Banash *et al.*, 1980; Ramaekers *et al.*, 1981; Osborn and Weber, 1982).

B. Cyclin

Analysis of many transformed cells (immortal or immortal and tumorigenic) of human (Detroit 98, Chang liver HeLa, Fl-amnion, Molt-4, WI-38, WI-38SV40, WISH); hamster (CHO, RJK 39); mouse (Sarcoma 180, CLID, L 929, X63, Friend Leukemia, Ehrlich ascites, Seminoma) and rat origin (Rat 1, NRK, HTC, TCA-A1, BRL 62, L6j) have shown that the nuclear polypeptide cyclin is present in all these cell lines (Bravo *et al.*, 1981d,e; Fey *et al.*, 1981). This has been further confirmed by one-dimensional peptide map analysis (Fey *et al.*, 1981; S. J. Fey, R. Bravo, P. Mose Larsen, and J. Celis unpublished; see also chapter by Fey *et al.*). This protein has also been identified in a few tissues from newborn mouse (Bravo *et al.*, 1981e; Mose Larsen, 1981). We have observed at least two variant forms of cyclin: one acidic variant present in hamster, human, potoroo and rat (Figs. 1 and 2; Bravo and Celis, 1980a; Bravo *et al.*, 1981d, 1982c) and the other, which is more basic, present in mouse (Fig. 8) (also

Figure 6. Two-dimensional gel electrophoresis of acidic proteins (IEF) from growing (p3) and senescent (p19) human skin fibroblasts. Transformation-sensitive polypeptides whose intensities clearly change are indicated with arrows. Polypeptides indicated with arrowheads do not change. From Bravo *et al.* (1983b).

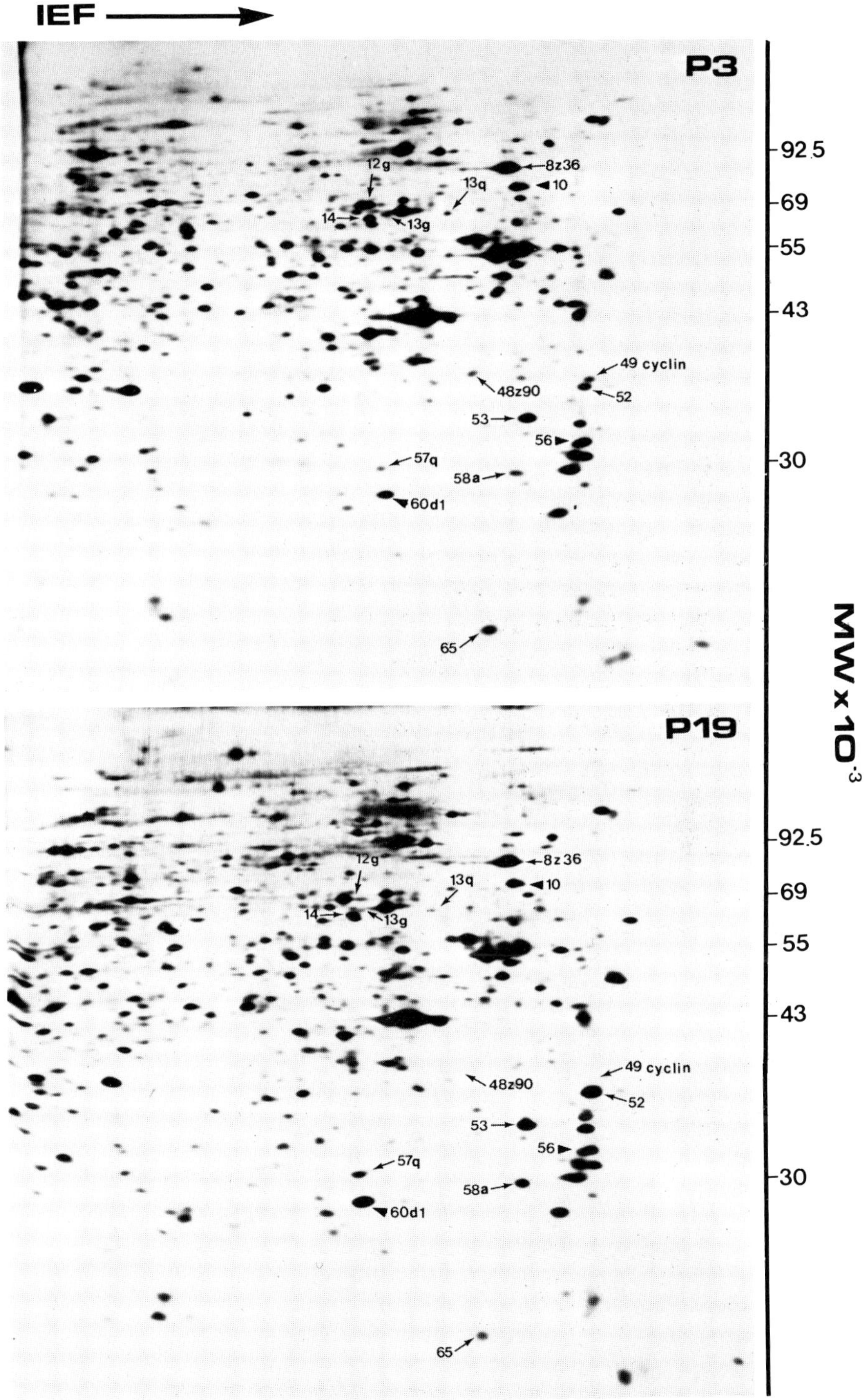

IEF
P3
P19
92.5
69
55
43
30
MW x 10-3
8z36
10
12g
13q
14
13g
49 cyclin
48z90
52
53
56
57q
58a
60d1
65

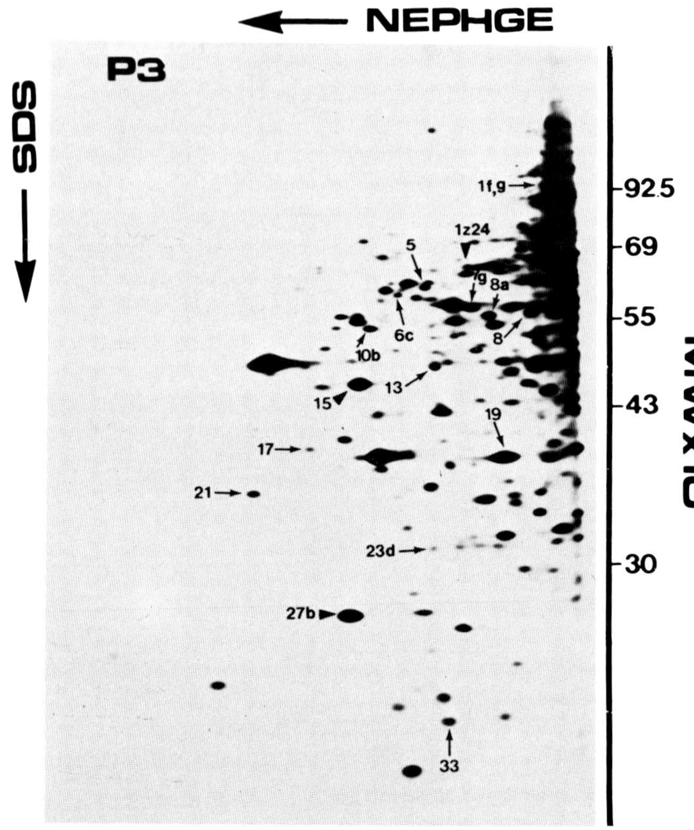

Figure 7. Two-dimensional gel electrophoresis of basic (NEPHGE) proteins from growing (p3) and senescent (p19) human skin fibroblasts. Transformation-sensitive polypeptides whose intensities clearly change are indicated with arrows. Arrowheads indicate polypeptides that do not change. From Bravo *et al.* (1983b).

called IEF 51f; Fey *et al.*, 1981; see also Fig. 3 of chapter by Fey *et al.*) (Bravo *et al.*, 1981d). Both variant types seem to differ by one charge unit but exhibit the same molecular weight (36,000).

There are at least three states of cell proliferation in which there is a variation in the synthesis of cyclin.

1. Noncycling cells synthesize very little or no cyclin.
2. Cycling cultured normal cells (primary and secondary cultures) synthesize this polypeptide at low passages but its synthesis ceases as cell division and DNA synthesis decline.
3. Transformed cells (immortal, or immortal and tumorigenic) synthesize this polypeptide "constitutively" and in varying amounts depending on the degree of transformation.

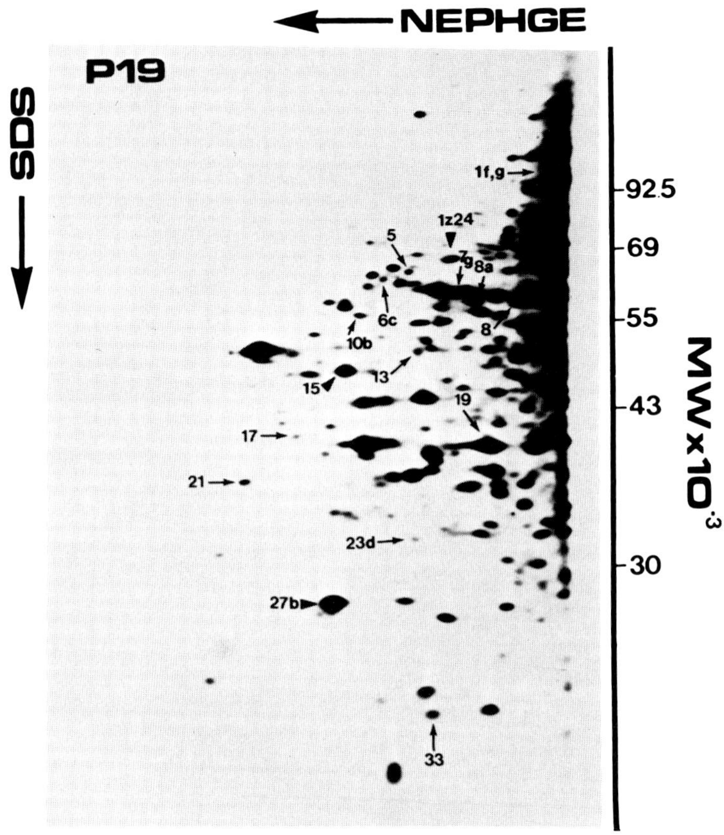

Figure 7. *(continued)*

As an example Fig. 8 shows an enlargement of the appropriate area from gels of slowly growing primary mouse embryonic fibroblasts, immortal mouse 3T3B cells, and tumorigenic SV40-transformed 3T3B cells (Bravo *et al.*, 1981d). Even though we do not know for certain what the parental fibroblast is from which 3T3B cells were derived we have included as control the pattern of an embryonic fibroblast because we have previously demonstrated that this region of the gel is very similar for different mouse fibroblasts (skin, ear, diaphragm, kidney, and lung) (Fey *et al.*, 1981). Clearly, the intensity of cyclin (corresponding to IEF 51f in the mouse) is high in the nontumorigenic transformed 3T3B cells as compared to normal fibroblasts (Fig. 8a,b), and it is higher again in the SV40-transformed tumorigenic cells (Fig. 8c). Giant 3T3BSV40 cells, however, synthesize significantly lower amounts of cyclin (Fig. 9a) as compared to control cells (Fig. 8c), despite the fact that they exhibit normal levels of T antigen (Figs. 9b and c). Both 3T3B and 3T3BSV40

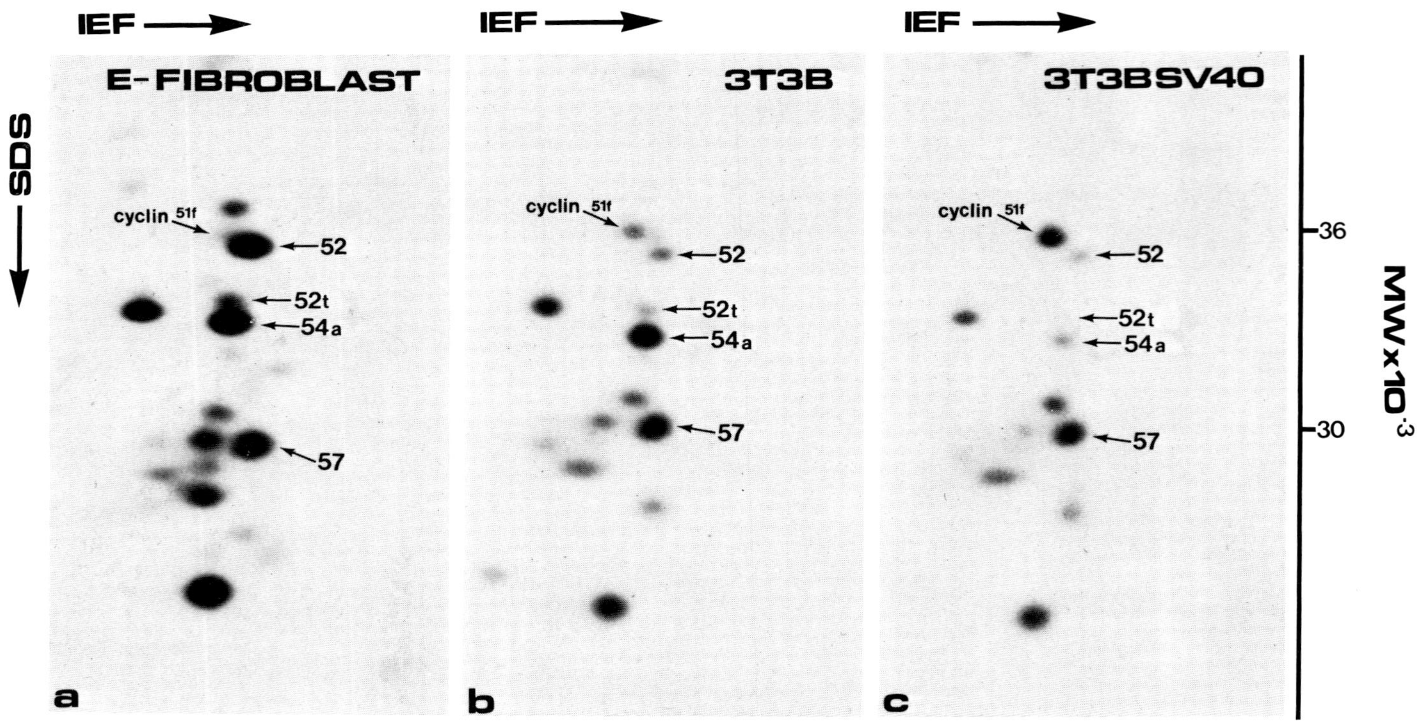

Figure 8. Two-dimensional gel electrophoresis of acidic (IEF) [^{35}S]methionine-labeled proteins from normal and transformed mouse cells. (a) embryonic fibroblasts; (b) 3T3B cells; (c) SV40-transformed cells.

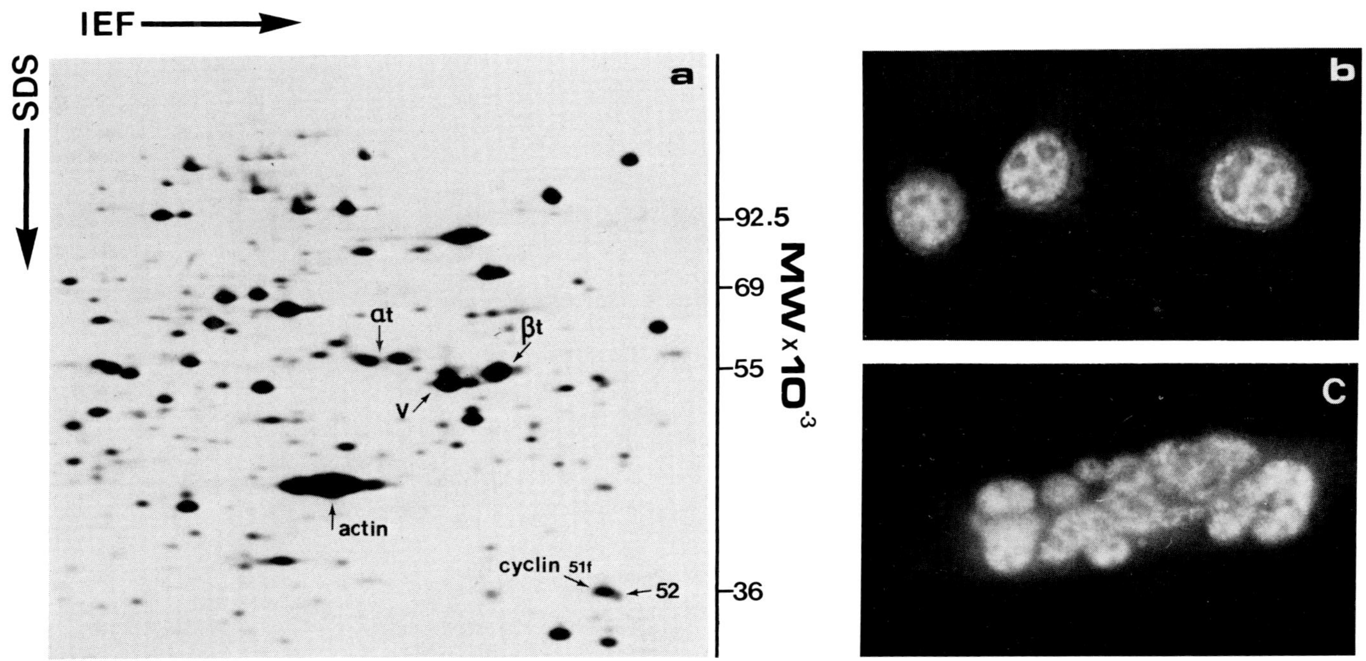

Figure 9. Cyclin synthesis in giant 3T3BSV40 cells produced by irradiation with 1000 rads of X-rays. (a) IEF gel of [^{35}S]methionine-labeled proteins; (b) 3T3BSV40 cells stained for T-antigen; (c) 3T3BSV40 giant cells stained for T-antigen. From J. Bellatin, R. Bravo and J. Celis (unpublished).

cells differ in many of the parameters involved in transformation such as contact inhibition, growth in soft agar, and organization of actin cables (for a review see Franks, 1983), and therefore it is likely that the levels for cyclin reflects the degree of transformation achieved by these cells. The above results support the notion that neoplastic transformation is a multistage process and that different control points may exist along the growth control pathway. Similar conclusions have been reached by Laszlo *et al.* (1981) from studies of the effect of TPA on gene expression and protein phosphorylation in avian cells in culture.

Given the enhanced synthesis of cyclin in growing cultured cells and tumors (Forchhammer and Macdonald-Bravo, 1983) as opposed to non-cycling or slowly cycling cells this polypeptide could be considered as a marker for growing cells. The rate of synthesis of cyclin has been shown to increase in the S phase of the cell cycle in HeLa cells (Bravo and Celis, 1980b). Riddle *et al.* (1979) have detected a polypeptide (M_r = 33,000) whose rate of synthesis increases significantly as quiescent 3T3B cells enter into G_1. At present, we do not know whether cyclin and this protein are related because Riddle *et al.* (1979) reported that the latter was both nuclear and cytoplasmic.

From the studies of normal secondary fibroblast cultures as well as of proliferating tissues, it is obvious that the presence of cyclin by itself is not the primary cause of transformation and that this may be due to other minor protein(s) that in turn regulate(s) its synthesis.

It should be mentioned that in many gel systems it is not possible to resolve cyclin, a fact that may have prevented its identification by other groups.

C. *Tropomyosin*

To date there is evidence indicating that there are several tropomyosins present in nonmuscle cultured cells (Paulin *et al.*, 1979; Bravo *et al.*, 1981d,e, 1983b; Giometti and Anderson, 1981; Hendricks and Weintraub, 1981; Bravo and Celis, 1982a,b; Forchhammer, 1982; Forchhammer and Macdonald-Bravo, 1983; J. I. Garrels, personal communication). Changes in the synthesis of tropomyosin polypeptides have been reported in spontaneously and virally transformed nonmuscle cells (Bravo and Celis, 1980a; Bravo *et al.*, 1981d,e; Hendricks and Weintraub, 1981; Forchhammer, 1982; Bravo *et al.*, 1983b; J. I. Garrels, personal communication) as well as in mouse embryonal carcinoma cells induced to differentiate by hexamethylene bisacetamide and retinoic acid (Paulin *et al.*, 1979; Linder *et al.*, 1981). Although transformation generally leads to a decreased synthesis of tropomyosin polypeptides, the opposite has been observed during differentiation of embryonal carcinoma cells. Furthermore, from studies of chicken embryo fibroblast cells transformed with Rous sarcoma virus, Hendricks and Weintraub (1981) postulated that the decrease in tropomyosin synthesis may play a role in the disruption of the cellular cytoskeleton (actin cables) following transformation (Pollack *et al.*, 1975; Edelman and Yahara, 1976; Ash *et al.*, 1976; Wang and Goldberg, 1976).

At least three tropomyosin-related polypeptides have been described in human nonmuscle cells and these correspond to IEF 52, 52x and 55 (Fig. 10; Bravo *et al.*, 1981b,e; Bravo and Celis, 1982a). A further putative tropomyosin corresponding to polypeptide IEF 56 has been recently identified (indicated with a short arrow in Fig. 10; J. I. Garrels personal communication). Tropomyosins with similar mobility have been described by Giometti and Anderson (1981), Hendricks and Weintraub (1981), Forchhammer (1982) and by J. I. Garrels (personal communication). Of the above mentioned tropomyosins only polypeptide IEF 52 (Figs. 1, 2, and 10) changes consistently (increases) with a decrease in the rate of cell proliferation (Bravo *et al.*, 1981d,e; Bravo and Celis, 1982b; Bellatin *et al.*, 1982; Forchhammer, 1982). For example, the ratio IEF 52/57 (57 being a constant polypeptide) is 1.18 for low passage embryonic mouse fibroblasts (Fig. 8a), 0.40 for 3T3B cells (Fig. 8b), and 0.15 for 3T3BSV40 cells (Fig. 8c). Most mouse and hamster transformed cell lines so far studied (listed in Section III,B) show a IEF 52/57 ratio similar to that of 3T3BSV40. Nontumorigenic rat cell lines such as L6j and Rat 1 exhibit a ratio similar to that of 3T3B cells. Similar results have been observed in transformed human cells (see also Figs. 1, 2, and 10) (Bravo and Celis, 1982b).

Similar gel electrophoretic analysis of the other three putative tropomyosins (IEF 52x (52t in the mouse), 55 (54a in the mouse), and 56), however, have shown that although their synthesis decreases in many transformed cells studied (see also Figs. 8 and 10), it increases in SV40-transformed human lung fibroblasts (Fig. 10d). Because the latter cells exhibit a comparable number of actin cables as the normal counterparts (results not shown), it is concluded that (1) a decreased synthesis of tropomyosin polypeptides is not a universal feature accompanying transformation and (2) that a direct relationship between decreased tropomyosin synthesis and disruption of the actin cables in transformed cells is uncertain.

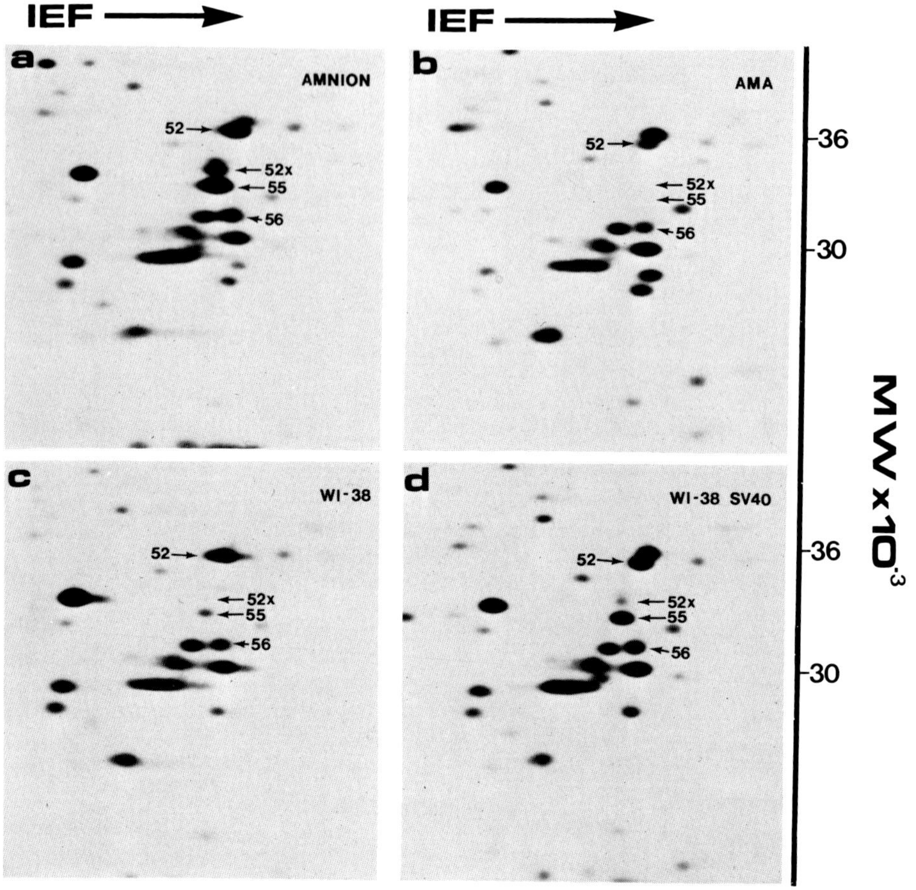

Figure 10. Two-dimensional gel electrophoresis (IEF) of [^{35}S]methionine-labeled polypeptides from normal (a and c) and transformed (b and d) human cells. From Bravo *et al.* (1983a).

IV. Variation in the Expression of Mutant Actins in Chemically Transformed Human Diploid Fibroblasts: Actin in General

Expression of mutant β-actins in transformed human fibroblasts showing incremental increase in tumorigenicity have been described by Leavitt and Kakunaga (1980), Vandekerckhove *et al.* (1980), Hamada *et al.* (1981) and Leavitt *et al.* (1982a,b). One of these mutant actins was

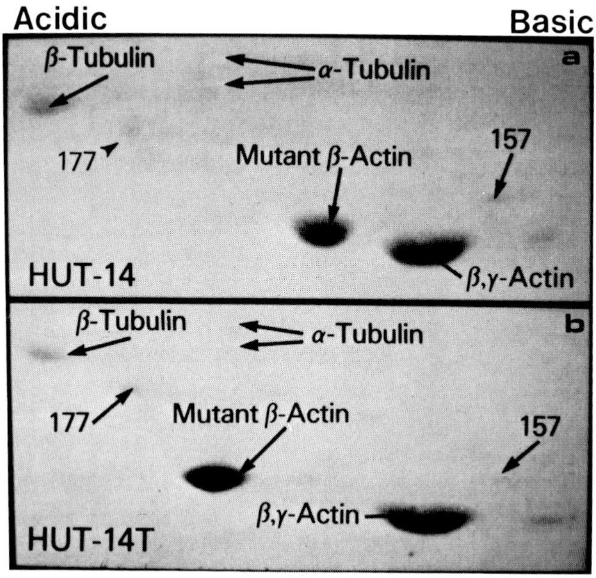

Figure 11. Two-dimensional electrophoretic pattern of [^{35}S]methionine-labeled actin, tubulin, and other neighboring polypeptides in the Triton-soluble proteins of HUT-14 (a) and HUT-14T (b) cells. From Leavitt *et al.* (1982b).

derived from one or two normal β-actin genes by a point mutation (exchange of glycine for aspartic acid at position 244; Vandeckerckhove *et al.*, 1980). This mutant actin, originally termed *A*1 (HUT-14 cells; Leavitt and Kakunaga, 1980), has a pI of 5.2 and is retarded in mobility in the SDS second dimension (Fig. 11a). The second actin mutant was derived from a tumor produced by inoculation of a nude mouse with a highly selective subclone of HUT-14 cells. (Leavitt *et al.*, 1982b). This new mutant actin differs from the original one by an additional negative charge (Fig. 11b). Both actins differ in half-life, rate of synthesis, affinity for DNase I, and cytoskeletal distribution (Leavitt *et al.*, 1982b). These studies provided the first molecular evidence for the occurrence of a genetic mutation in chemically transformed cells.

At present it is not known whether the expression of these β-actin mutants is related to the increased tumorigenicity exhibited by the cells. However, because transformation is frequently but not always accompanied by a change in cytoskeletal properties (actin cables, Pollack *et al.*, 1975; Edelman and Yahara, 1976; Ash *et al.*, 1976; Wang and Goldberg, 1976; see also Celis *et al.*, 1978; Watt *et al.*, 1978), it is possible that the abnormal expression of β-actin may play a role in the process of neoplastic transformation. Cloning as well as microinjection experiments may

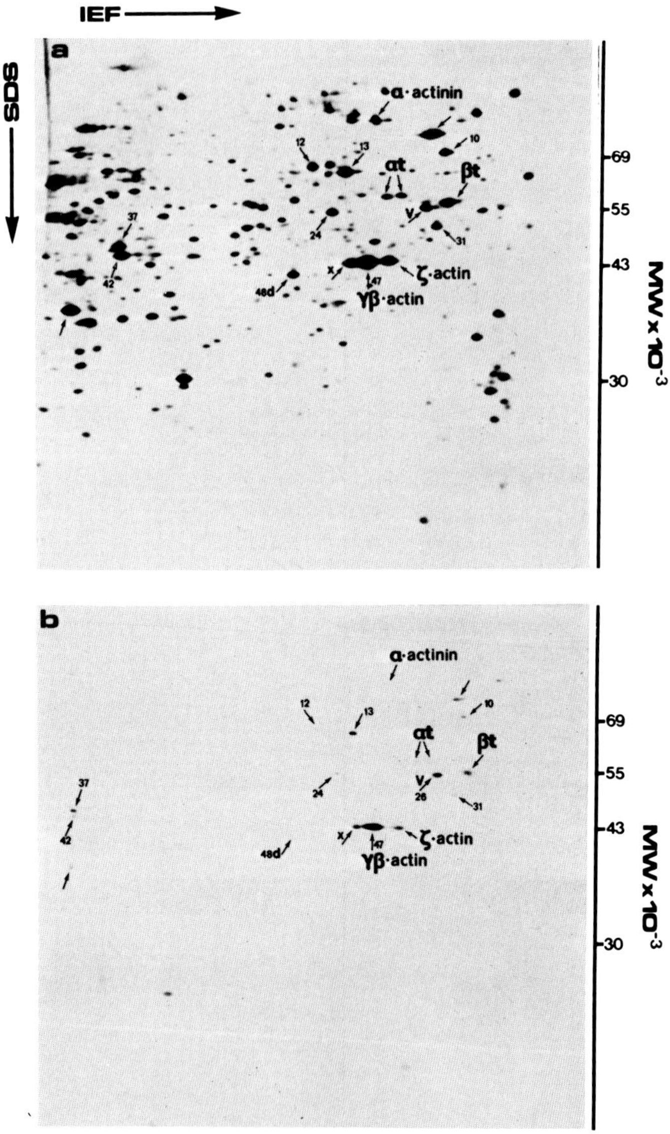

Figure 12. Two-dimensional gel electrophoresis (IEF) of [^{35}S]methionine-labeled Sarcoma 180 cells. (a) Polypeptide pattern of 50 Sarcoma 180 cells labeled for 20 h with [^{35}S]methionine; (b) polypeptide pattern of a single Sarcoma 180 cell labeled for 20 h with [^{35}S]methionine. Labeling of single Sarcoma 180 cells with [^{35}S]methionine was carried out as described in Bravo *et al.* (1981c). The fluorograms have been exposed for 15 days. αt, α-tubulin; βt, β-tubulin; v, vimentin; spot 24 corresponds to a mitochondrial protein. From Bravo *et al.* (1981c).

shed some light on the role of the mutant actins in the transformation process (Kakunaga *et al.*, 1981).

We found that a transformed mouse cell line, Sarcoma 180, synthesized an unusually large amount of an actin polypeptide (termed ζ) that had the same molecular weight as β- and γ-actin but was more acidic than β-actin (Bravo *et al.*, 1981c) (Figure 12a). Two-dimensional gel electrophoresis of the [^{35}S]methionine-labeled polypeptides from a single Sarcoma 180 cell showed that all three major actin species coexisted within the same cell (Fig. 12b). This analysis also demonstrated for the first time the coexistence of α- and β-tubulin, vimentin, and α-actinin. So far ζ-actin has only been observed in Sarcoma 180 cells (Bravo *et al.*, 1981c) and is most likely not a primary cause of transformation.

A change in the relative proportion of total actin (slight decrease) has been observed in SV40-transformed mouse and hamster cells (Bravo and Celis, 1980a), but this is not a general phenomenon. Quantitative analyses of total labeled actin in whole cells as well as in Triton cytoskeletons have shown that the proportion of extractable actin in 3T3B cells increases from 32 to 54% in 3T3BSV40 cells. Similar observations have been reported by Fine and Taylor (1976) on the same pair of cells. At present it is not clear how this increase in extractable actin in transformed cells correlates with the loss of microfilament bundles (Pollack *et al.*, 1975; Edelman and Yahara, 1976; Ash *et al.*, 1976; Wang and Goldberg, 1976) or of any other form of insoluble actin (Goldman *et al.*, 1976; Small and Celis, 1978).

An increased synthesis of total actin has also been reported by Riddle *et al.* (1979) and by Riddle and Pardee (1980) following stimulation of serum-arrested 3T3 cells, and it has been postulated that the enhanced synthesis of this protein may be a critical feature of the transition from quiescence to proliferation. Quantitative analyses of actin during the cell cycle (Milcarek and Zahn, 1978; Bravo and Celis, 1980b; Riddle and Pardee, 1980) have indicated, however, that the synthesis of this protein remains unchanged throughout this process.

V. Phosphoproteins

A. Transformation-Sensitive Phosphoproteins in Cultured Epithelial and Fibroblast Cells

Two-dimensional gel electrophoretic analysis of pairs of normal and transformed human amnion and lung fibroblasts have revealed so far

TABLE III

Transformation-Sensitive Phosphoproteins in Cultured Human Fibroblast and Epithelial Cells[a]

Phosphoprotein	MW	Cellular distribution[b]	Relative proportion in transformed cells	Also varies in other species	Other characteristics
IEF					
5h	96,000		Increases		
8m	91,000		Increases	Ha; M[c]	
8z36 (stress protein)	76,000	Cytoplasmic	Increases		Extractable by Triton X-100
14h	65,000	Present mainly in karyoplasts	Increases	Ha; M[c]	Cytoskeletal
26e (phosphovimentin)	54,000	Cytoplasmic	Decreases	Ha; M[c]	Cytoskeletal
NEPHGE					
10a	54,000	Present mainly in karyoplasts	Increases		Cytoskeletal
unidentified polypeptide	57,000		Decreases		

[a] From Bravo and Celis (1982b).
[b] From Bravo *et al.* (1981b).
[c] Ha, Hamster; M, mouse.

seven phosphoproteins sensitive to transformation among 250 common phosphorylated proteins analyzed by visual inspection (Table III; Bravo and Celis, 1982b; Bravo *et al.*, 1983b; Figs. 13 and 14, only IEF gels are shown). These proteins correspond to NEPHGE 10a (M_r = 54,500) and an unidentified polypeptide (M_r = 57,000) and IEF 5h (M_r = 96,000), 8m (M_r = 91,000), 8z36 (M_r = 76,000), 14h (M_r = 65,000), and phosphovimentin (26e; M_r = 54,000) (Figs. 13 and 14). Except for phosphovimentin and the unidentified basic polypeptide, the relative proportion of all phosphorylated proteins increases considerably in transformed cells (Figs. 13 and 14) (Bravo and Celis, 1982b). Some relevant characteristics of these polypeptides are listed in Table III.

A change in the phosphorylation state of IEF 5h, 8m, 8z36, and 14h has also been observed in FL-amnion; WISH-amnion, EJ19 bladder carcinoma and HeLa cells (Bravo and Celis, 1982b; P. Mose Larsen, S. J. Fey and J. Celis, unpublished) as well as in pairs of normal and transformed hamster (Ha) and mouse (M) cells (Table III) (P. Mose Larsen, S. J. Fey

Figure 13. Two-dimensional gel electrophoresis of [^{32}P]orthophosphate-labeled proteins from normal human amnion (a) and AMA (b) cells. Sparsely growing cells placed in 35-mm dishes were labeled for 8 h at 37°C in 2 ml of phosphate-free medium containing 2 mCi of [^{32}P]orthophosphate (Bravo *et al.*, 1982b). The unnumbered polypeptide has not been identified in the HeLa protein catalog. From Bravo and Celis (1982b).

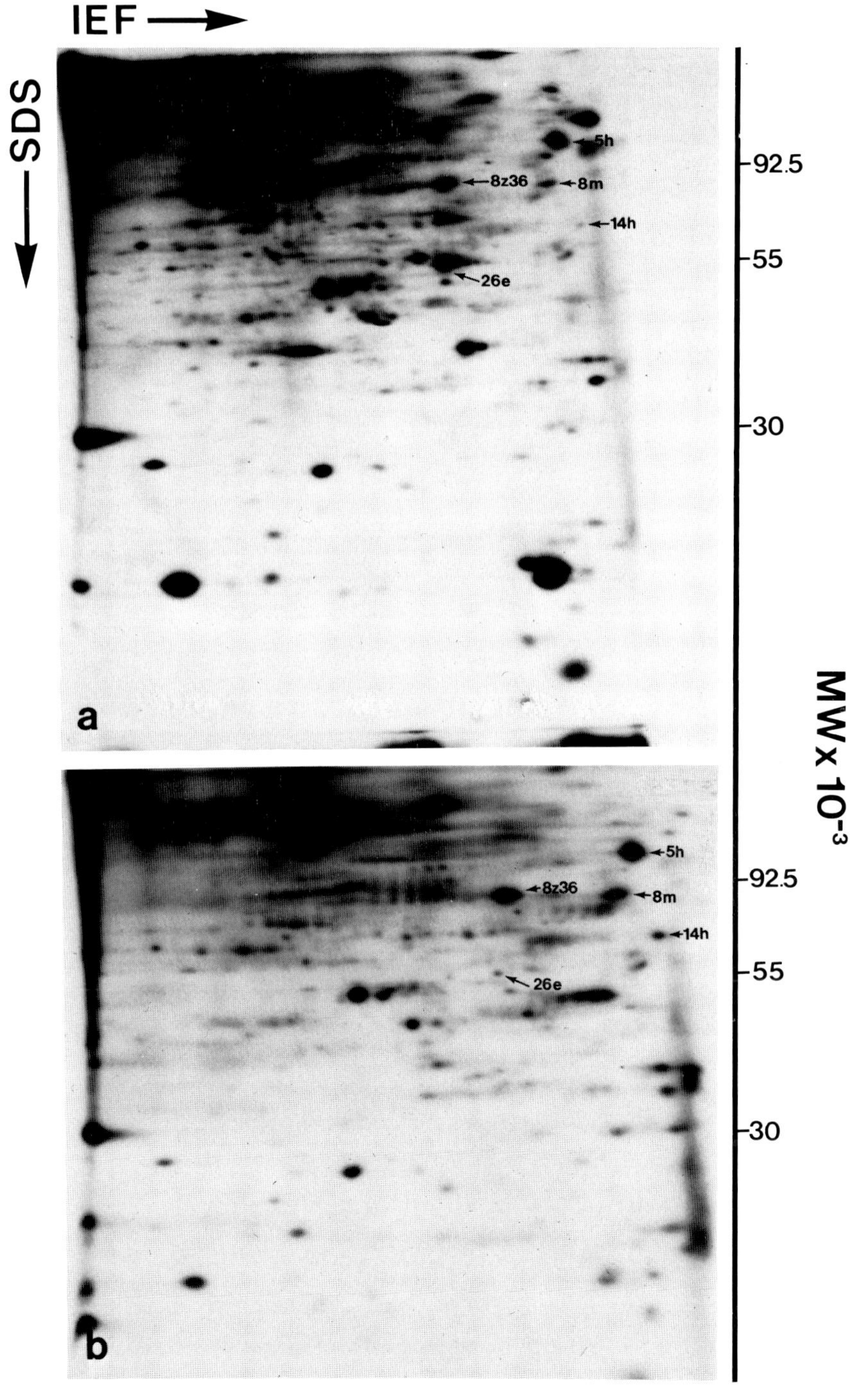

IEF
SDS
5h
8z36
8m
14h
26e
92.5
55
30
a
b
MW x 10^{-3}

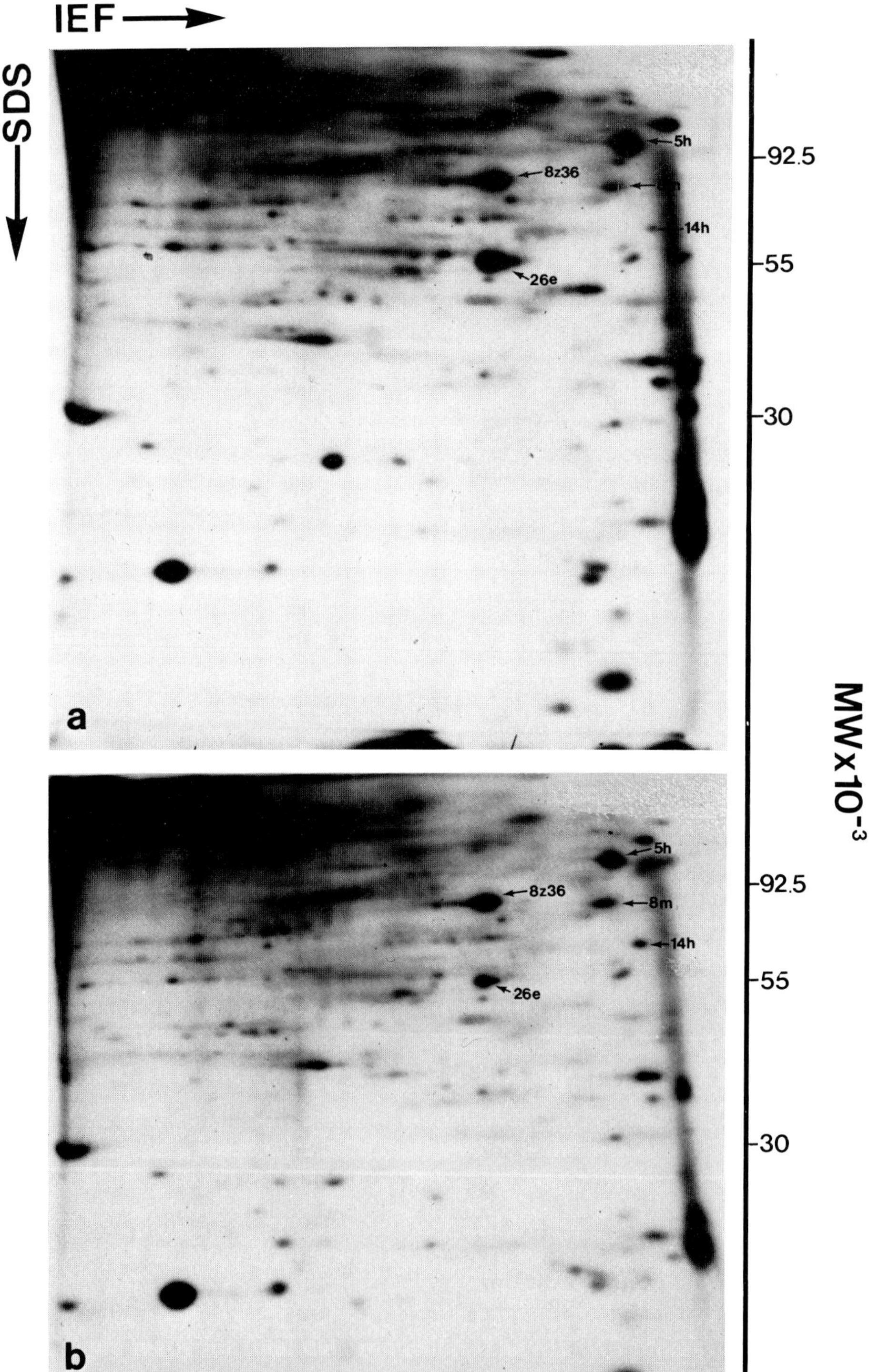
IEF
SDS
5h
8z36
14h
26e
a
5h
8z36
8m
14h
26e
b
92.5
55
30
92.5
55
30
MW x 10⁻³

and J. Celis, unpublished). Some of these phosphoproteins exhibit a M_r and pI very similar to a few of the phosphoproteins that change in chicken embryo fibroblasts treated with TPA (Laszlo *et al.*, 1981).

Interestingly, none of the transformation-sensitive phosphoproteins exhibit a migration in two-dimensional gels similar to the pp60^{src} tyrosine-specific protein kinase (Brugge and Erikson, 1977; Collett and Erikson, 1978; Levinson *et al.*, 1978) or to some of its substrates such as the 36-kdalton protein (Radke and Martin, 1979; Erikson *et al.*, 1979; Radke *et al.*, 1980; Erikson and Erikson, 1980; Cooper and Hunter, 1981b), and vinculin (Sefton *et al.*, 1981a,b). None of these phosphoproteins has a molecular weight or isoelectric point similar to that of the phosphotyrosine-containing phosphoproteins observed in chicken embryo cells transformed by different classes of retroviruses (Cooper and Hunter, 1981a), to the known viral oncogene products (for references see Ghysdael, 1983), or to some of the proteins associated with cellular transforming genes identified by transfection (Cooper, 1982; Tabin *et al.*, 1982; Reddy *et al.*, 1982; Padhy *et al.*, 1982).

B. *Tyrosine Phosphorylation of Proteins in A431 Human Tumor Cells*

Hunter and Cooper (1981) have shown that epidermal growth factor (EGF) induced rapid phosphorylation of two proteins (M_r = 39,000, M_r = 81,000; membrane associated) that are absent or barely detectable in untreated human A431 cells (see Figs. 15a and b). Both phosphoproteins contain phosphotyrosine and phosphoserine and are not induced by fibroblast growth factor or insulin (Hunter and Cooper, 1981). Changes in the phosphorylation state of these proteins, however, have not been detected in HeLa cells (Hunter and Cooper, 1981; Cooper and Hunter, 1981b).

Similar studies of RSV-transformed A431 cells (Cooper and Hunter, 1981b) showed that these cells contained the RSV transforming protein pp60^{src} and the 39-kdalton phosphoprotein (constitutively synthesized; homologous to the 36-kdalton phosphoprotein observed in RSV-transformed mouse cells) (Fig. 15c) but lacked the 81-kdalton phosphoprotein (Fig. 15c). Treatment of RSVA431 cells with EGF, however, resulted in

Figure 14. Two-dimensional gel electrophoresis of [^{32}P]orthophosphate-labeled proteins from (a) WI-38 and (b) SV40-transformed WI-38 cells. The cells were labeled as described in the legend of Fig. 13. The unnumbered polypeptide has not been identified in the HeLa protein catalog. From Bravo and Celis (1982b).

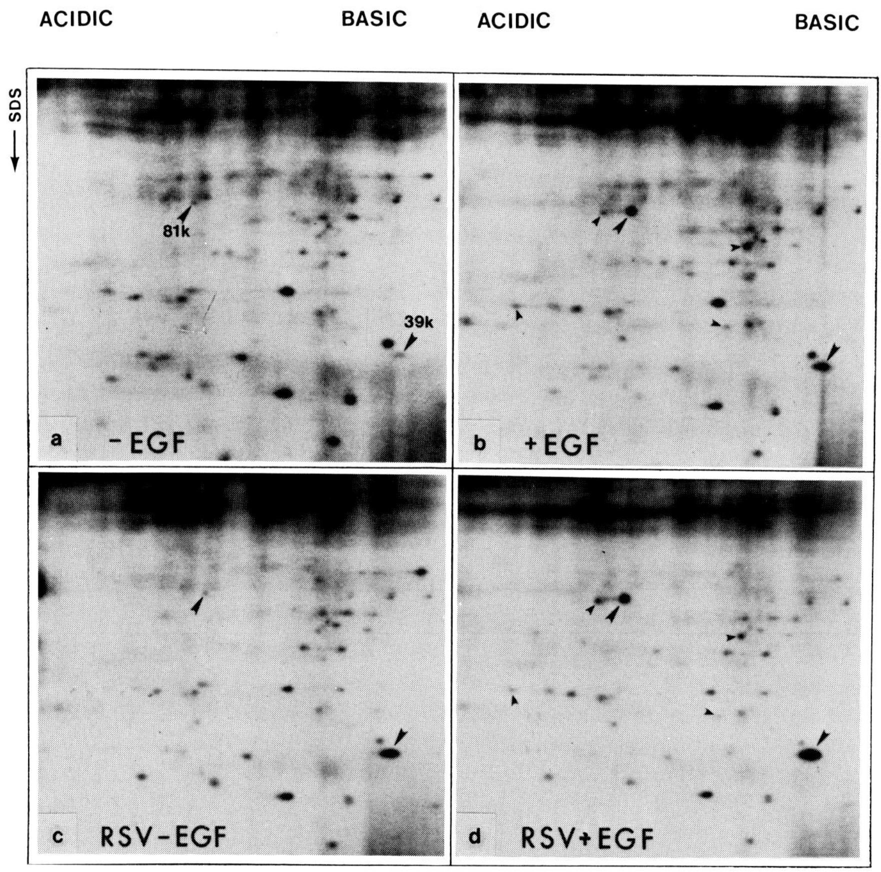

Figure 15. Effect of epidermal growth factor (EGF) on phosphoproteins of control and infected A431 cells. Parallel cultures of A431 (a and b) and RSVA431 clone 6 (c and d) cells were labeled for 16 h with [^{32}P]orthophosphate. EGF (10 m*M* final concentration) was added to one culture of each (b and d) for the final hour of labeling. From Cooper and Hunter (1981b).

an increased phosphorylation of both the 39-kdalton and the 81-kdalton phosphoproteins (Figs. 15c and d). These results clearly showed similarities and differences between the tyrosine-specific kinases induced by RSV and EGF and suggested different routes of action by both effectors. Similar conclusions have been reached by Laszlo *et al.* (1981), who showed that TPA and the pp60^{src} may act at different control points in the pathway that controls cell proliferation. These observations lend further support to the notion that carcinogenesis is a multistep process.

VI. Polypeptide Synthesis in Human Tumors and Normal Tissue

There are few comparative studies of tumors and normal tissue using two-dimensional gel electrophoresis; these include analyses of colon and colorectal cancer (Thorsrud *et al.*, 1982; Tracy *et al.*, 1982; see also chapter by Tracy and Young), renal carcinomas (see chapter by Tracy and Young), and various human sarcomas (Forchhammer and Macdonald-Bravo, 1983). The main problem one encounters in working with tumors is that they are not homogeneous and some may contain areas with more than one tumor grade (for references see Foulds, 1969; Franks, 1979, 1983). In addition normal and tumor tissue contain different proportions of a given cell type, a fact that makes comparisons difficult. Furthermore, if the tumor or normal cells are to be grown in culture prior to analysis (if this is possible at all), one faces the problem of selection, because most differentiated cells die and the cultures are rapidly repopulated from a stem cell population (Hastings and Franks, 1982; see also Franks, 1983).

Bearing these caveats in mind, we will review in this section the work on human sarcomas (Forchhammer and Macdonald-Bravo, 1983). The studies on colon and colorectal cancer as well as of renal carcinomas have already been addressed in the chapter by Tracy and Young. Finally we will discuss emerging technology (microsurgery) that in the future may facilitate the analysis of tumors and normal tissue biopsies.

A. Polypeptide Synthesis in Human Sarcomas and Normal Tissue

Forchhammer and Macdonald-Bravo (1983) have reported a comparative study of the polypeptide patterns of biopsies from three human sarcomas and of corresponding normal differentiated tissue from the same patients. Tissue biopsies (<1 mm^3) were either labeled with [^{35}S]methionine or applied directly to the gels after homogenization and their proteins detected by silver staining (Mose Larsen, 1981; Mose Larsen *et al.*, 1983). Figures 16a and b, taken from the work of these workers, show two-dimensional gel separations (IEF) of [^{35}S]methionine-labeled polypeptides from a malignant mesenchymoma (grade 3) (Fig. 16a) and control subcutaneous fat tissue (Fig. 16b). Arrows and arrowheads pointing upward or downward indicate polypeptides whose rela-

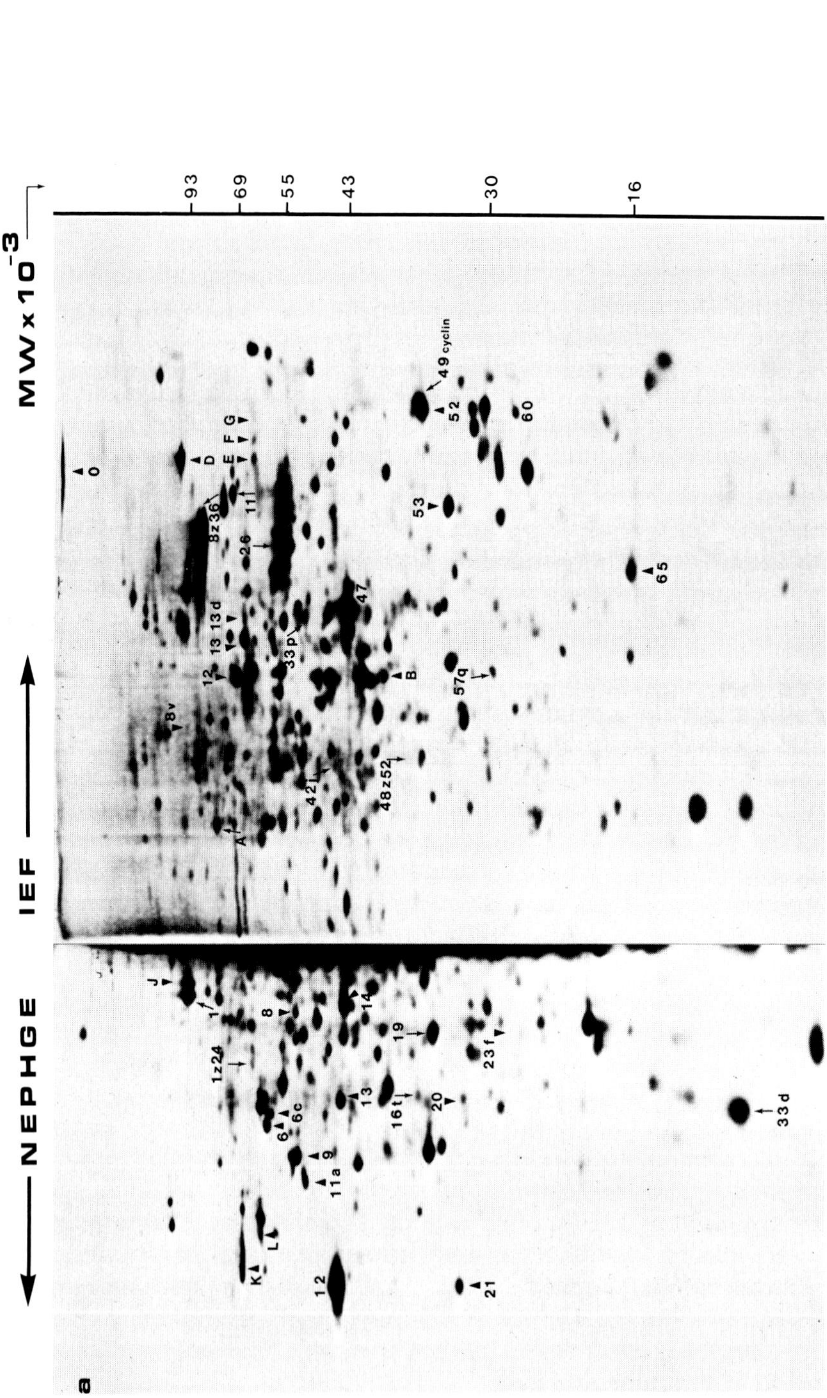
MW x 10^-3
IEF
NEPHGE
93
69
55
43
30
16
a

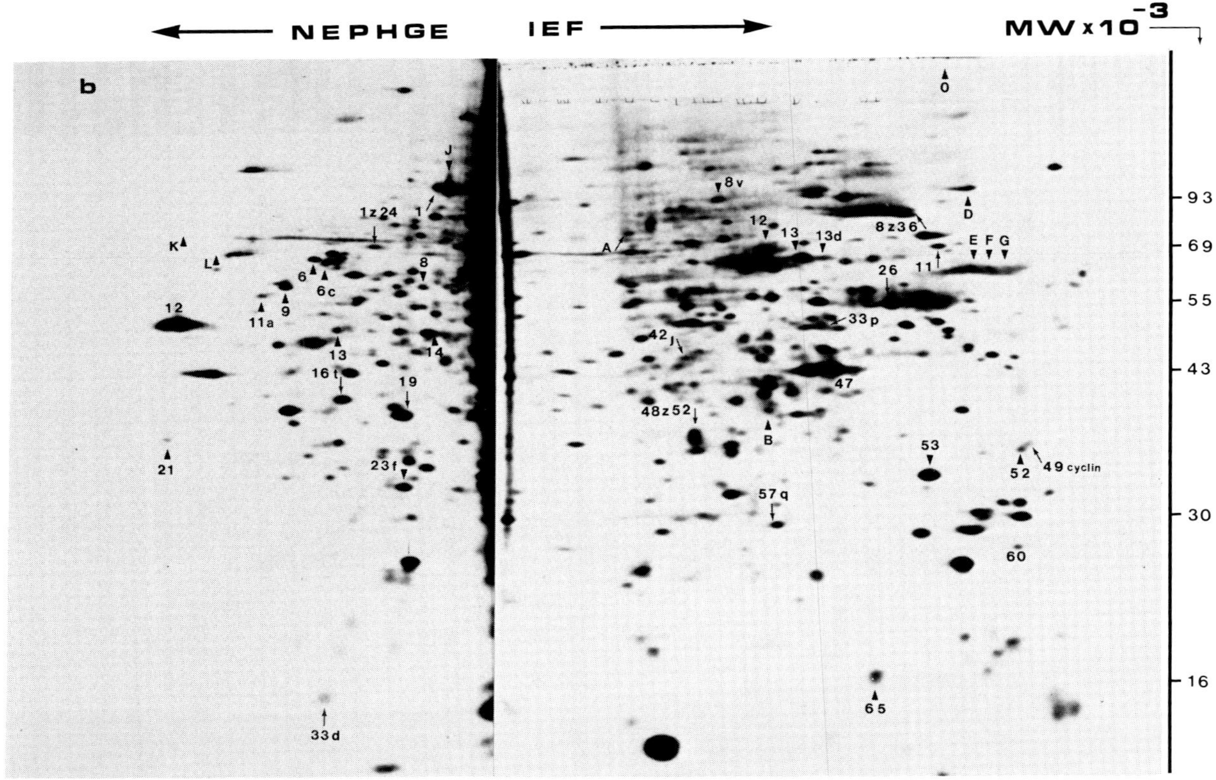

Figure 16. [^{35}S]methionine polypeptides labeled from (a) a patient with malignant mesenchymoma (grade 3); (b) from control subcutaneous fat tissue. Pieces of tumors and control tissue (less than 1 mm^3) were grown in short-term cultures in a microtiter plate pressed under a 3 × 3 mm coverslip. The pieces were labeled with [^{35}S]methionine as previously described (see chapter by Bravo). Arrows and arrowheads pointing upward or downward indicate polypeptides whose relative proportion increased or decreased, respectively, by 30% or more in tumors as compared to controls. Arrows indicate changes observed in three tumor types (mesenchymoma, histiocytoma, and chondrosarcoma). Arrowheads indicate changes only seen in two of the three tumor types. From Forchhammer and Macdonald-Bravo (1983).

tive proportions increased or decreased more than 30% in the tumor as compared to the control tissue. Polypeptides are numbered according to the HeLa protein numbering system (Bravo *et al.*, 1981a; Bravo and Celis, 1982a). Similar analyses of other sarcomas (histiocytoma, chondrosarcoma) revealed 49 polypeptide differences (30% or more variation) among 250 polypeptides quantitated (Forchhammer and Macdonald-Bravo, 1983). Eighteen of these polypeptide changes have also been observed in cultured cells (see Tables I and II), whereas 31 corresponded to changes not previously observed (see Figs. 16a and b). As suggested by these authors this apparent discrepancy may reflect differences in the control mechanisms present in normal tissue that are lost during the establishment of the cell lines. It is also possible, however, that this discrepancy may reflect slight differences in the cell population present in the tumor and in the normal biopsy.

B. Emerging Technology

1. MICROSURGERY

As pointed out at the beginning of this section, there are many cases in which the heterogeneity of the tumor and normal tissue biopsies is such that a meaningful comparison is not possible. One way to circumvent this problem would be to work with a small number of purified cells. In this laboratory we have pursued the possibility of dissecting groups of cells from cryostat sections (10 μm in thickness) using microsurgery. As an example, Fig. 17 shows the microdissection of a group of about 45 cells from the dermis of human skin. Figure 17a shows hematoxylin-eosin staining while Figs. 17b–f show phase-contrast microscopy of a cryostat section similar to that used for the microdissection. Both the apparatus for microsurgery and the microneedles are the same as those used for microinjection of somatic cells. The procedures and manipulations have been described in detail elsewhere (Celis, 1977; Celis *et al.*, 1980).

The number of cells required in order to be able to detect their proteins in a two-dimensional gel by silver staining is, however, much higher than the 45 cells dissected. For comparison Fig. 18 shows a selected area of a silver-stained IEF gel made from a total protein extract from a single cryostat section of human skin similar to that used for microsurgery (about 490 times larger in surface area) (Giometti and Anderson, 1981).

Under these conditions only major proteins are revealed. If microdissection is to be used on unlabeled specimens, it will be necessary to pool the cells from several dissections. Alternatively, if the samples are labeled prior to sectioning ([^{35}S]methionine) or to electrophoresis ([^{14}C]iodoacetamide, Giometti and Anderson, 1981; [^{14}C]formaldehyde; Jentoft and Dearborn, 1979; Billeter *et al.*, 1981), this procedure could well be applied to a very small number of cells.

2. SPECIFIC ANTIBODIES

An alternative approach for isolating a given cell type is to use specific antibodies (e.g., against membrane markers) in conjunction with the cell sorter. Here again two-dimensional gel electrophoresis promises to be a valuable tool. Specific antibodies have been raised in mouse using very small amounts of pure proteins extracted from two-dimensional gels (Mose Larsen *et al.*, 1982; Bravo *et al.*, 1983a; for preparation of antibodies see Fig. 6 in the chapter by Fey *et al.*). Thus if specific membrane protein markers are identified in gels it should be possible in the future to raise antibodies that may be instrumental in distinguishing among groups of related cells. In this laboratory a library of polyclonal antibodies is being produced against membrane, cytoplasmic, and nuclear proteins to study differentiation and transformation of human epidermal basal cells.

VII. Conclusions

The most striking observation in the foregoing studies is the small number of quantitative changes in polypeptide synthesis that occur following transformation (immortality and tumorigenicity). This is most remarkable considering the major changes in cell morphology, architecture, and behavior that accompany this process.

In general neoplastic transformation (virally, chemically, or spontaneously induced) results in changes in the relative proportions (rate of synthesis, modification, degradation) of proteins normally synthesized by somatic cells rather than in the appearance of new proteins. Induction of new protein(s) has been reported in virally transformed cells but with few exceptions is not commonly observed in chemically or spontaneously transformed cells. Failure to detect new proteins, however, may

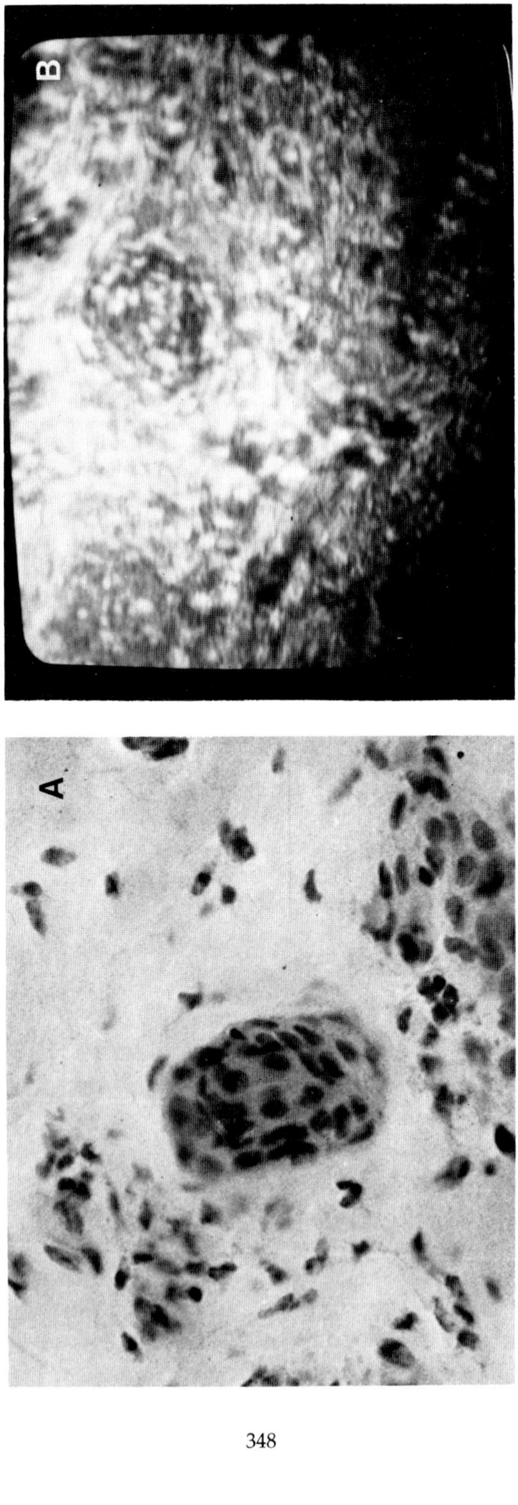
B
A

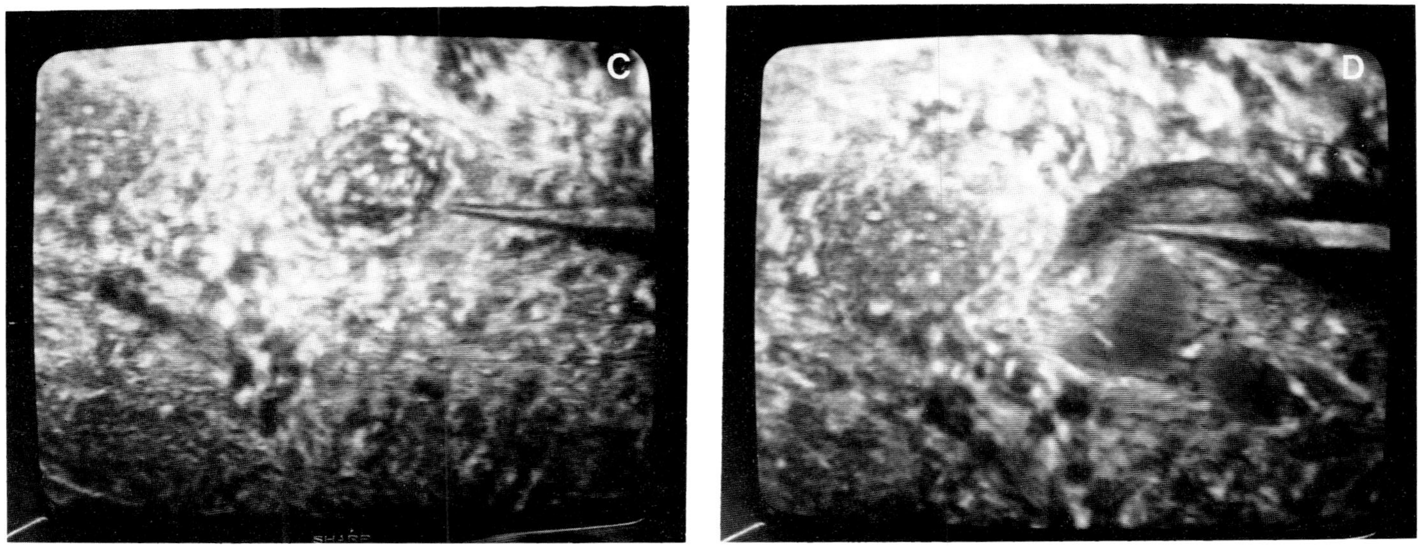

Figure 17. Microdissection of a small group of cells from a cryostat section (10 μm in thickness) of human skin. (A) Hematoxylin–eosin–stained section; (B–F) microsurgery: phase contrast photographs (as seen in the television minitor) of unfixed cryostat section similar to that shown in (a). Various stages of the microsurgery procedure. The section bound to 12 × 12 glass coverslips was dissected using needles of 1–2 μm at the tip. The procedure has been described in detail elsewhere (Celis, 1977; Celis *et al.*, 1978).

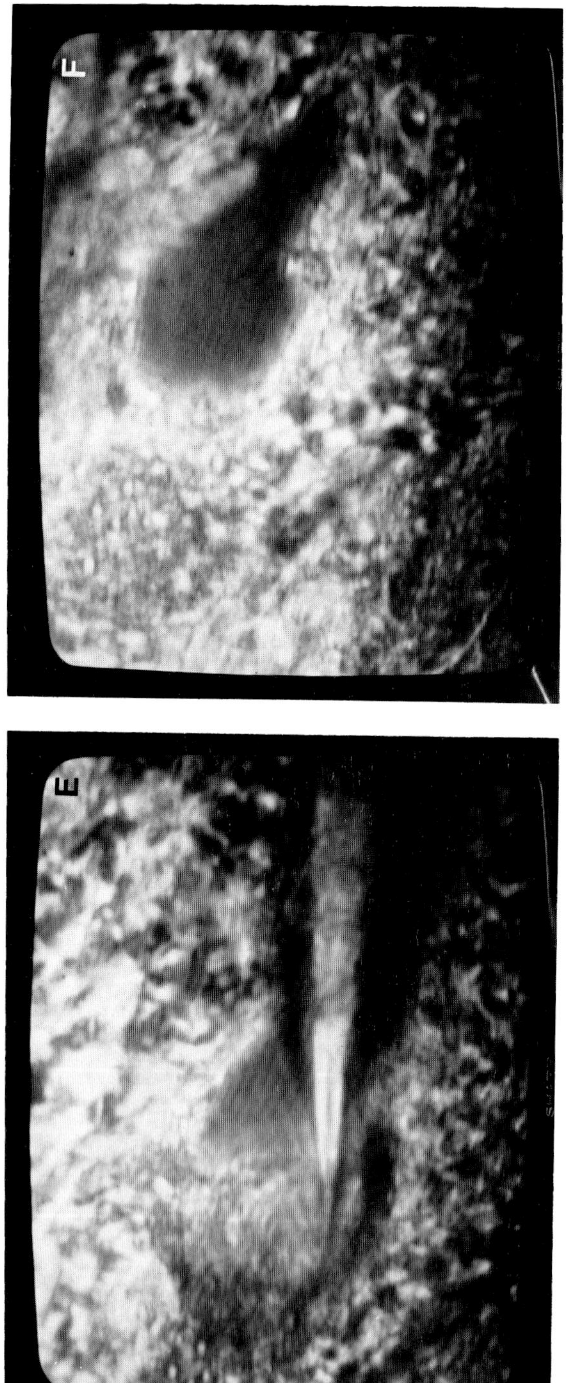

Figure 17. *(continued)*

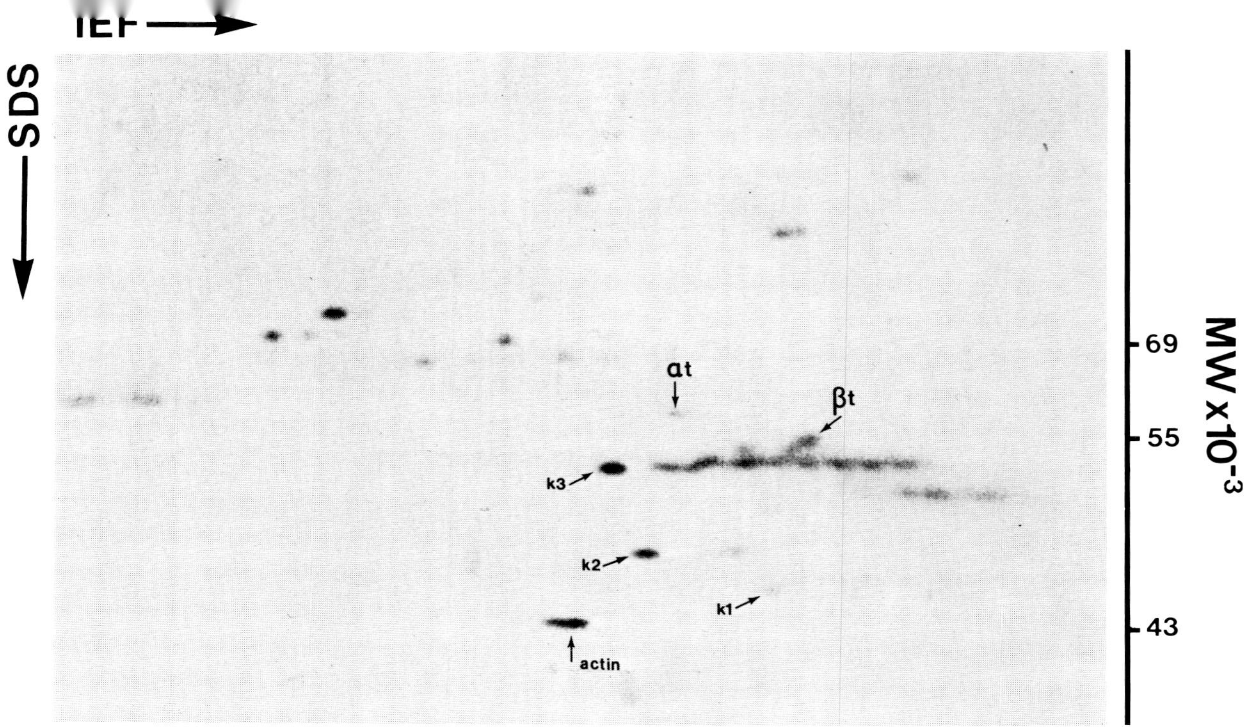

Figure 18. Two-dimensional gel electrophoresis (IEF) of proteins from a single cryostat section of split skin (10 μm in thickness) stained with silver nitrate. Protein gels were stained with silver nitrate as previously described (Mose Larsen, 1981; Mose Larsen *et al.*, 1983). Following electrophoresis the gels (0.8 mm) were shaken in the following solutions: (a) 50% methanol containing 12% trichloroacetic acid (TCA) (30 min); (b) 10% ethanol containing 5% acetic acid (20 min); (c) 0.06% $KMnO_4$ containing 0.02% $CuSO_4$ (12 min), (d) 10% ethanol (10, 20 and 30 min, respectively); (e) triple distilled H_2O (20 min); (f) 0.1% silver nitrate (60 min) and 0.015% formaldehyde containing 12% K_2CO_3 (until spots become dark brown or black and the background is light). Finally, the gels were shaken in 0.1% acetic acid until the background turned yellow and in 5% Kodak FX-40 until they turned transparent. K1, K2, and K3 correspond to keratins, αt, α-tubulin; βt, β-tubulin.

be due to their comigration with other cellular proteins, low abundance, or both.

Interestingly, two-dimensional gel electrophoretic analysis of many other biological processes as well as of normal and altered physiological conditions have given similar conclusions. These studies have included analyses of the following: the cell cycle (Milcarek and Zahn, 1978; Elliot and McLaughlin, 1978; Bravo and Celis, 1980b), differentiation (Peterson and McConkey, 1976; Devlin and Emerson, 1978; Paulin *et al.*, 1979; Liebermann *et al.*, 1980; Evans *et al.*, 1983; R. Bravo and J. E. Celis, unpublished; see also chapter by Garrels *et al.*), *in vitro* aging of fibroblasts (Wilson *et al.*, 1978; Engelhardt *et al.*, 1978; Bravo *et al.*, 1983b; Van Keuren *et al.*, 1983), *in vivo* aging of *nude* mice (skin; R. Bravo and J. E. Celis, unpublished observations), giant cell formation (Bellatin *et al.*, 1982), dividing versus stationary cells (Garrels, 1979; Riddle *et al.*, 1979; Lester *et al.*, 1981), the effect of tumor promoters (Laszlo *et al.*, 1981; Cabral *et al.*, 1981), hormones (Ivarie and O'Farrell, 1978; Ivarie *et al.*, 1981; Voris and Young, 1981; DePhilip and Kierszenbaum, 1982), interferon (Knight and Korant, 1979), growth factors (Garrels and Schubert, 1979; Cooper and Hunter, 1981b; R. Bravo, in press), and cAMP (Steinberg and Coffino, 1979; Rapoport *et al.*, 1982).

A number of transformation-sensitive polypeptides common to various cell types have been identified in cultured human cells and tumors (Bravo and Celis, 1982b; Forchhammer and Macdonald-Bravo, 1983). The synthesis of many of these polypeptides is also sensitive to changes in growth rate (Bellatin *et al.*, 1982) and a significant number have been identified in other species (Bravo and Celis, 1980a). Even though the normal role of these proteins is unknown, it is possible that some may be components of the cellular pathway that controls cell proliferation. Such a fundamental pathway would be expected to be strongly conserved during evolution, showing similar changes upon transformation irrespective of the species origin of the cell. None of these proteins, however, is similar to the products of the known viral oncogenes (for references see Ghysdael, 1983), the cellular oncogenes identified by transfection experiments (Cooper, 1982; Tabin *et al.*, 1982; Reddy *et al.*, 1982; Padhy *et al.*, 1982), or to some of the phosphoproteins induced by epidermal growth factors (Cooper and Hunter, 1981b).

From the various studies considered in this chapter it is clear that the cell proliferation pathway is intricate and must contain many control points. This finding supports the theory that carcinogenesis is a multistep process. Further elucidation of the functions of the transformation-sensitive proteins as well as of the proteins coded by the oncogenes should eventually lead us to a better understanding of the mechanisms underlying normal control of cell proliferation and hence neoplastic transformation and cancer.

Acknowledgments

We would like to thank H. Macdonald-Bravo and J. Forchhammer for helpful discussions and to B. Thomsen and O. Jensen for photography. R. B. was recipient of a fellowship from the Danish Medical and Natural Science Research Councils. P. M. L. is a recipient of a fellowship from the Aarhus University. S. J. F. is a recipient of a fellowship from the Danish Cancer Society. J. B. was a recipient of a DANIDA fellowship. This work was supported by grants from Euratom, the Danish Medical and Natural Science Research Councils, the Danish Cancer Society, Carlsberg, and Novo.

References

Ash, J. F., Vogt, P. K., and Singer, S. J. (1976). Reversion from transformed to normal phenotype by inhibition of protein synthesis in rat kidney cells infected with a temperature-sensitive mutant of Rous sarcoma virus. *Proc. Natl. Acad. Sci. USA* **73**, 3603–3607.

Aubin, J. E., Osborn, M., Franke, W. W., and Weber, K. (1980). Intermediate filaments of the vimentin and cytokeratin-type are distributed differently during mitosis. *Exp. Cell Res.* **129**, 149–165.

Banash, P., Zerban, H., Schmid, E., and Franke, W. W. (1980). Liver tumours distinguished by immunofluorescence microscopy with antibodies to proteins of intermediate-sized filaments. *Proc. Natl. Acad. Sci. USA* **77**, 4948–4952.

Beemon, K. (1981). Transforming proteins of some feline and avian sarcoma viruses are related structurally and functionally. *Cell* **24**, 145–153.

Bellatin, J., Bravo, R., and Celis, J. E. (1982). Changes in the relative proportion of transformation sensitive polypeptides in giant HeLa cells produced by irradiation with lethal doses of X-rays. *Proc. Natl. Acad. Sci. USA* **79**, 4367–4370.

Benchimol, S., Pim, D., and Crawford, L. (1982). Radioimmunoassay of the cellular protein p53 in mouse and human cell lines. *EMBO J.* **1**, 1055–1062.

Berman, L., and Stulberg, C. S. (1956). Eight culture strains (Detroit) of human epithelial-like cells. *Proc. Soc. Exp. Biol. Med.* **92**, 730–735.

Billeter, R., Heizmann, C. W., Howald, H., and Jenny, E. (1981). Analysis of myosin light and heavy chain types in single human skeletal muscle fibers. *J. Biochem. (Tokyo)* **116**, 389–395.

Bishop, J. M. (1981). Enemies within: The genesis of retrovirus oncogenes. *Cell* **23**, 5–6.

Bishop, J. M. (1982). Oncogenes, *Sci. Am.* **246,** 69–78.

Bishop, J. M., and Varmus, H. E. (1982). Functions and origins of retroviral transforming genes. *In* "The Molecular Biology of Tumour Viruses" (R. A. Weiss, N. Teich, H. E. Varmus, and J. M. Coffin, eds.), Part 3, pp. 999. Cold Spring Harbor Lab., Cold Spring Harbor, New York.

Blose, S. H. (1979). Ten-nanometer filaments and mitosis: Maintenance of structural continuity in dividing endothelial cells. *Proc. Natl. Acad. Sci. USA* **76**, 3372–3376.
Bravo, R. (1983). Epidermal growth factor induced changes in the polypeptide synthesis of A 431 cells. *In* "The Cancer Cell" (A. J. Levine, W. C. Topp, G. Van de Woude, and J. D. Watson, eds.). Cold Spring Harbor.
Bravo, R., and Celis, J. E. (1980a). Gene expression in normal and virally transformed mouse 3T3B and hamster BHK21 cells. *Exp. Cell Res.* **127**, 249–260.
Bravo, R., and Celis, J. E. (1980b). A search for differential polypeptide synthesis throughout the cell cycle of HeLa cells. *J. Cell Biol.* **48**, 795–802.
Bravo, R., and Celis, J. E. (1982a). Updated catalogue of HeLa cell proteins: percentages and characteristics of the major cell polypeptides labelled with a mixture of 16[^{14}C]amino acids. *Clin. Chem.* **28**, 766–781.
Bravo, R., and Celis, J. E. (1982b). Human proteins sensitive to neoplastic transformation in cultured epithelial and fibroblast cells. *Clin. Chem.* **28**, 949–954.
Bravo, R., Bellatin, J., and Celis, J. E. (1981a). [^{35}S]-Methionine labeled polypeptides from HeLa cells. Coordinates and percentage of some major polypeptides. *Cell Biol. Int. Rep.* **5**, 93–96.
Bravo, R., Celis, A., Mosses, D., and Celis, J. E. (1981b). Distribution of HeLa cells polypeptides in cytoplasts and karyoplasts. *Cell Biol. Int. Rep.* **5**, 479–489.
Bravo, R., Fey, S. J., Small, J. V., Mose Larsen, P., and Celis, J. E. (1981c). Coexistence of three major isoactins in a single Sarcoma 180 cell. *Cell* **25**, 195–202.
Bravo, R., Fey, S. J., and Celis, J. E. (1981d). Gene expression in murine hybrids exhibiting different morphologies and tumorigenic properties. *Carcinogenesis (N.Y.)* **2**, 769–782.
Bravo, R., Fey, S. J., Bellatin, J., Mose-Larsen, P., Arevalo, J., and Celis, J. E. (1981e). Identification of a nuclear and of a cytoplasmic polypeptide whose relative proportions are sensitive to changes in the rate of cell proliferation. *Exp. Cell Res.* **136**, 311–319.
Bravo, R., Fey, S. J., Mose-Larsen, P., and Celis, J. E. (1982a). Modification of vimentin polypeptides during mitosis. *Cold Spring Harbor Symp. Quant. Biol.* **46**, 379–385.
Bravo, R., Small, J. V., Fey, S. J., Mose Larsen, P., and Celis, J. E. (1982b). Architecture and polypeptide composition of HeLa cell cytoskeletons. Modification of cytoarchitectural proteins during mitosis. *J. Mol. Biol.* **159**, 121–143.
Bravo, R., Schäfer, R., Willecke, K., Macdonald-Bravo, H., Fey, S. J., and Celis, J. E. (1982c). More than one third of the discernible mouse polypeptides are not expressed in a Chinese hamster—mouse embryo fibroblast hybrid that retains all mouse chromosomes. *Proc. Natl. Acad. Sci. USA* **79**, 2281–2285.
Bravo, R., Fey, S. J., Mose Larsen, P., Coppard, N., and Celis, J. E. (1983a). Proteins IEF (Isoelectric focussing) 31 and IEF 46 are keratin-type components of the intermediate-sized filaments: keratins of various human cultured epithelial cells. *J. Cell Biol.* **96**, 416–423.
Bravo, R., Bellatin, J., Fey, S. J., Mose Larsen, P., and Celis, J. E. (1983b). Expression of cellular proteins in normal and transformed human cultured cells. *In* "Gene Expression in Normal and Transformed Cells" (J. E. Celis and R. Bravo, eds.), pp. 263–290. Plenum, New York.
Browning, E. T., and Sanders, M. M. (1981). Vimentin: a protein under hormonal regulation. *J. Cell Biol.* **90**, 803–808.
Brugge, J. S., and Erikson, R. L. (1977). Identification of a transformation-specific antigen induced by an avian sarcoma virus. *Nature (London)* **269**, 346–348.
Brzeski, J., and Ege, T. (1980). Changes in polypeptide pattern in ASV-transformed rat cells are correlated with the degree of morphological transformation. *Cell* **22**, 513–522.
Cabral, F., Gottesman, M. M., and Yuspa, S. H. (1981). Induction of specific protein synthesis by phorbol esters in mouse epidermal cell culture. *Cancer Res.* **41**, 2025–2031.

Cairns, J. (1975). Mutation selection and natural history of cancer. *Nature (London)* **255**, 197–200.

Cairns, J. (1978). "Cancer, Science and Society." Freeman, San Francisco, California.

Cairns, J. (1981). The origin of human cancers. *Nature (London)* **289**, 353–357.

Celis, J. E. (1977). Injection of tRNAs into somatic cells. Search for *in vivo* systems to assay potential nonsense mutations in somatic cells. *Brookhaven Symp. Biol.* **29**, 178–196.

Celis, J. E., and Bravo, R. (1981). Cataloguing human and mouse proteins. *Trends Biochem. Sci.* **6,** 197–201.

Celis, J. E., Small, J. V., Anderson, P., and Celis, A. (1978). Microfilament bundles in cultured cells. Correlation with anchorage independence and tumorigenicity in nude mice. *Exp. Cell Res.* **114,** 335–348.

Celis, J. E., Kaltoft, K., and Bravo, R. (1980). Microinjection of somatic cells with micropipettes and PEG-erythrocyte ghost mediated microinjection. *In* "Transfer of Cell Constituents into Eukaryotic Cells" (J. E. Celis, A. Graessmann, and A. Loyter, eds), pp. 1–28. Plenum, New York and London.

Chang, R. S. (1954). Continuous subcultivation of epithelial cells from normal human tissues. *Proc. Soc. Exp. Biol. Med.* **87**, 440–443.

Coffin, J. M., Varmus, H. E., Bishop, J. M., Essex, M., Hardy, W. D., Martin, G. S., Rosenberg, N. E., Scolnick, E. M., Weinberg, R. A., and Vogt, P. K. (1981). Proposal for naming host-derived inserts in retrovirus genomes. *J. Virol.* **40**, 953–957.

Collett, M. S., and Erikson, R. L. (1978). Protein kinase activity associated with the avian sarcoma virus *src* gene product. *Proc. Natl. Acad. Sci. USA* **75**, 2021–2024.

Collett, M. S., Purchio, A. F., and Erikson, R. L. (1980). Avian sarcoma virus transforming protein $pp60^{src}$ shows protein kinase activity specific for tyrosine. *Nature (London)* **285**, 167–169.

Cooper, G. M. (1982). Cellular transforming genes. *Science (Washington, D.C.)* **217**, 801–806.

Cooper, G. M., Okenquist, S., and Silverman, L. (1980). Transforming activity of DNA of chemically transformed and normal cells. *Nature (London)* **284**, 418–421.

Cooper, J. A., and Hunter, T. (1981a). Four different classes of retrovirus induce phosphorylation of tyrosines present in similar cellular proteins. *Mol. Cell. Biol.* **1**, 394–407.

Cooper, J. A., and Hunter, T. (1981b). Similarities and differences between the effects of epidermal growth factor and Rous sarcoma virus. *J. Cell Biol.* **91**, 878–883.

Cooper, J. A., Bowen-Pope, D. F., Raines, E., Ross, R., and Hunter, T. (1982). Similar effects of platelet-derived growth factor and epidermal growth factor on the phosphorylation of tyrosine in cellular proteins. *Cell* **31**, 263–273.

DePhilip, R. M., and Kierszenbaum, A. L. (1982). Hormonal regulation of protein synthesis secretion, and phosphorylation in cultured rat Sertoli cells. *Proc. Natl. Acad. Sci. USA* **79**, 6551–6555.

Devlin, R. B., and Emerson, C. J. (1978). Coordinate regulation of contractile protein synthesis during myeloblast differentiation. *Cell* **14**, 599–611.

Duncan, R., and McConkey, E. H. (1982). How many proteins are there in a typical mammalian cell? *Clin. Chem.* **28**, 749–755.

Edelman, G. M., and Yahara, I. (1976). Temperature-sensitive changes in surface modulating assemblies of fibroblasts transformed by mutants of Rous sarcoma virus. *Proc. Natl. Acad. Sci. USA* **73**, 2047–2051.

Ek, B., Westermark, B., Wasteson, A., and Heldin, C. H. (1982). Stimulation of tyrosine-specific phosphorylation by platelet-derived growth factor (PDGF). *Nature (London)* **295**, 419–420.

Elliot, S. G., and McLaughlin, S. (1978). Rate of macromolecular synthesis through the cell cycle of the yeast *Saccharomyces cerevisae, Proc. Natl. Acad. Sci. USA* **75**, 4384–4388.

Engelhardt, D. L., Lee, G. T. Y., and Moleym, S. F. (1978). Patterns of peptide synthesis in senescent and presenescent human fibroblasts. *J. Cell. Physiol.* **98**, 193–198.

Erikson, E., and Erikson, R. L. (1980). Identification of a cellular protein substrate phosphorylated by the avian sarcoma virus transforming gene product. *Cell* **21,** 829–836.

Erikson, R. L., Collett, M. S., Erikson, E. L., and Purchio, A. F. (1979). Evidence that the avian sarcoma virus transforming gene product is a cyclic AMP–independent protein kinase. *Proc. Natl. Acad. Sci. USA* **76**, 6260–6264.

Evans, M. J., Lovell-Badge, R. H., Latchman, D., Stacey, A., and Brzeski, H. (1983). Approaches to the biochemistry of differentiation of mouse embryonal carcinoma cells. *In* "Gene Expression in Normal and Transformed Cells" (J. E. Celis and R. Bravo, eds.) pp. 87–99. Plenum, New York.

Evans, R. M., and Fink, L. M. (1982). An alteration in the phosphorylation of vimentin type intermediate filaments is associated with mitosis in cultured mammalian cells. *Cell* **29**, 43–52.

Fey, S. J., Bravo, R., Mose Larsen, P., Bellatin, J., and Celis, J. E. (1981). [^{35}S]-methionine labelled polypeptides from secondary mouse kidney fibroblasts. Coordinates and one-dimensional peptide maps of some major polypeptides. *Cell Biol. Int. Rep.* **5**, 491–500.

Fine, R. E., and Taylor, L. (1976). Decreased actin and tubulin synthesis in 3T3 cells after transformation by SV40 virus. *Exp. Cell Res.* **102**, 162–168.

Fogh, J., and Lund, R. O. (1957). Continuous cultivation of epithelial cell strain (FL) from human amniotic membrane. *Proc. Soc. Exp. Biol. Med.* **94,** 532–537.

Forchhammer, J. (1982). Quantitative changes of some cellular polypeptides in C3H mouse cells following transformation by moloney sarcoma virus. *In* "Biological Markers of Neoplastic Transformation" (P. Chandra, ed.). Plenum, New York.

Forchhammer, J., and Klarlund, J. (1979). Changes in proteins from transformed cultures and tumours induced by sarcoma virus. *Adv. Med. Oncol. Res. Educ.* **1**, 51–60.

Forchhammer, J., and Macdonald-Bravo, H. (1983). Polypeptide synthesis in human sarcoma and normal tissue. *In* "Gene Expression in Normal and Transformed Cells" (J. E. Celis and R. Bravo, eds.), pp. 291–314. Plenum, New York.

Foulds, L. (1969). "Neoplastic Development," Vol. 1. Academic Press, London and New York.

Franke, W. W., Schmid, E., Osborn, M., and Weber, K. (1978). Different intermediate-sized filaments distinguished by immunofluorescence microscopy. *Proc. Natl. Acad. Sci. USA* **75**, 5034–5038.

Franke, W. W., Schmid, E., Breitkreutz, D., Luder, M., Boukamp, P., Fusenig, N. E., Osborn, M., and Weber, K. (1979a). Simultaneous expression of two different types of intermediate sized filaments in mouse keratinocytes proliferating *in vitro*. *Differentiation (Berlin)* **14**, 35–50.

Franke, W. W., Schmid, E., Weber, K., and Osborn, M. (1979b). HeLa cells contain intermediate-sized filaments of the prekeratin type. *Exp. Cell Res.* **118**, 95–109.

Franke, W. W., Schmid, E., Kartenbeck, J., Haker, H. J., Bannash, P., Osborn, M., Weber, K., Denk, H., Wanson, J. C., and Drachmans, P. (1979c). Characterization of the intermediate-sized filaments in liver cells by immunofluorescence and electron microscopy. *Biol. Cell.* **34**, 99–110.

Franke, W. W., Schmid, E., Schiller, D. L., Winter, S., Jarasch, E. E., Moll, R., Denk, H., Jackson, B. W., and Illmensee, K. (1981). Differentiation-related patterns of expression of proteins of intermediate-sized filaments in tissues and cultured cells. *Cold Spring Harbor Symp. Quant. Biol.* **46**, 431–453.

Franks, L. M. (1979). What is a cancer cell? Phenotypic markers in *in vitro* carcinogenesis in epithelial systems. *In* "Neoplastic Transformation in Differentiated Epithelial Cell

Systems *in vitro*." (L. M. Franks and C. B. Wigley, eds.), pp. 287–300. Academic Press, London and New York.

Franks, L. M. (1983). The neoplastic cell and its analysis by cell hybridization (1) The nature of the transformation process and its markers (2) Analysis of transformation by cell hybridization. *In* "Gene Expression in Normal and Transformed Cells" (J. E. Celis and R. Bravo, eds.), pp. 141–176. Plenum Press, New York.

Gard, D. L., and Lazarides, E. (1982). Analysis of desmin and vimentin phosphopeptides in cultured avian myogenic cells and their modulation by 8-bromoadenosine 3',5'-cyclic monophosphate. *Proc. Natl. Acad. Sci. USA* **79**, 6912–6916.

Garrels, J. I. (1979). Two dimensional gel electrophoresis and computer analysis of proteins synthesized by clonal cell lines. *J. Biol. Chem.* **254**, 7961–7977.

Garrels, J. I., and Schubert, D. (1979). Modulation of protein synthesis by nerve growth factor. *J. Biol. Chem.* **254**, 7978–7985.

Gemmel, M. A., and Anderson, N. L. (1982). Lymphocyte, monocyte and granulocyte proteins compared by use of two dimensional electrophoresis. *Clin. Chem.* **28**, 1062–1066.

Ghysdael, J. (1983). Protein kinases specific for tyrosine residues and the role of tyrosine phosphorylation of proteins in cell transformation. *In* "Gene Expression in Normal and Transformed Cells" (J. E. Celis and R. Bravo, eds.), pp. 209–230. Plenum, New York.

Giometti, C. S., and Anderson, N. G. (1981). Muscle protein analysis. III. Analysis of solubilized frozen-tissue sections by two dimensional electrophoresis. *Clin. Chem.* **27**, 1918–1921.

Giometti, C. S., Willard, K. E., and Anderson, N. L. (1982). Cytoskeletal proteins from human skin fibroblasts, peripheral blood leukocytes, and a lymphoblastoid cell line compared by two dimensional gel electrophoresis. *Clin. Chem.* **28**, 955–961.

Girardi, A. J., Jensen, F. C., and Koprowski, H. (1965). SV40 induced transformation of human diploid cells: crisis and recovery. *J. Cell. Comp. Physiol.* **65**, 69–83.

Goldman, R. D., Yerna, M. J., and Schloss, J. A. (1976). Localization and organization of microfilaments and related proteins in normal and virus transformed cells. *J. Supramol. Struct.* **5**, 155–183.

Hamada, H., Leavitt, J., and Kakunaga, T. (1981). Mutated *beta*-actin gene: Coexpression with an unmutated allele in a chemically transformed human fibroblast cell line. *Proc. Natl. Acad. Sci. USA* **78**, 3634–3638.

Hanafusa, H. (1977). Cell transformation by RNA tumour viruses. *Compr. Virol.* **10**, 401–483.

Hanash, S. M., Tubergen, D. G., Heyn, R. M., Neel, J. V., Sandy L., Stevens, G. S., Rosenblum, B. B., and Krzesicki, R. F. (1982). Two dimensional gel electrophoresis of cell proteins in childhood leukemia with silver staining. A preliminary report. *Clin. Chem.* **28**, 1026–1030.

Hastings, R. J., and Franks, L. M. (1981). Chromosome pattern, growth in agar and tumorigenicity in *nude* mice of four human bladder carcinoma cell lines. *Int. J. Cancer* **27**, 15–21.

Hastings, R. J., and Franks, L. M. (1982). Cellular heterogeneity in a tissue culture cell line derived from a human bladder carcinoma. *Int. J. Cancer.*

Hayflick, L. (1961). The establishment of a line (WISH) of human amnion cells in continuous cultivation. *Exp. Cell Res.* **23**, 14–20.

Hayflick, L. (1965). The limited *in vitro* lifetime of human diploid cell strains. *Exp. Cell Res.* **37**, 614–636.

Hayward, W. G., Neel, B. E., and Astrin, S. M. (1981). Activation of a cellular oncogene by promoter insertion in ALV induced lymphoid leukosis. *Nature (London)* **290**, 475–480.

Hendricks, M., and Weintraub, H. (1981). Tropomyosin is decreased in transformed cells. *Proc. Natl. Acad. Sci. USA* **78**, 5633–5637.
Howley, P. M. (1980). Molecular biology of SV40 and the human polyomaviruses BK and JC. *In* "Viral Oncology" (G. Klein, ed.), pp. 489–550. Raven, New York.
Hunter, T. (1982). Phosphotyrosine—a new protein modification. *Trends Biochem. Sci.* **7**, 246–249.
Hunter, T., and Cooper, J. A. (1981). Epidermal growth factor induces rapid tyrosine phosphorylation of proteins in A431 human tumour cells. *Cell* **24**, 741–752.
Hunter, T., and Sefton, B. M. (1980). The transforming gene product of Rous sarcoma virus phosphorylates tyrosine. *Proc. Natl. Acad. Sci. USA* **77**, 1311–1315.
Hynes, R. O., and Destree, A. T. (1978). 10-nm filaments in normal and transformed cells. *Cell* **13**, 151–163.
Ito, Y. (1980). Organization and expression of the genome of polyoma virus. *In* "Viral Oncology" (G. Klein, ed.), pp. 447–480. Raven, New York.
Ivarie, R. D., and O'Farrell, P. H. (1978). The glucocorticoid domain: Steroid-mediates changes in the rate of synthesis of rat hepatoma proteins. *Cell* **13**, 41–55.
Ivarie, R. D., Baxter, D., and Morris, J. A. (1981). Interaction of thyroid and glucocorticoid hormones in rat pituitary tumour cells. *J. Biol. Chem.* **256**, 4520–4528.
Jellum, E., and Thorsrud, A. K. (1982). Clinical applications of two dimensional gel electrophoresis. *Clin. Chem.* **28**, 876–883.
Jentoft, N., and Dearborn, D. G. (1979). Labelling of proteins by reductive methylation using sodium cyanoborohydride. *J. Biol. Chem.* **254**, 4349–4355.
Kakunaga, T. (1978). Neoplastic transformation of human diploid fibroblast cells by chemical carcinogenesis. *Proc. Natl. Acad. Sci. USA* **75**, 1334–1338.
Kakunaga, T., Crow, J., Lo, K. Y. L., Hamada, H., and Hirakawa, T. (1981). Cell transformation as a system for studying mechanisms of carcinogenesis. *Gann Monogr. Cancer Res.* **27**, 231–242.
Klein, G. (1981). The role of gene dosage and genetic transpositions in carcinogenesis. *Nature (London)* **294**, 313–318.
Klein, G., Luca, J., and Zeuthen, J. (1979). Transformation induced by Epstein–Barr virus and the role of the nuclear antigen. *Cold Spring Harbor Symp.* **44**, 253–261.
Knight, E., and Korant, B. (1979). Fibroblast interferon induces synthesis of four proteins in human fibroblast cells. *Proc. Natl. Acad. Sci. USA* **76**, 1824–1827.
Krontiris, T. G., and Cooper, G. M. (1981). Transforming activity of human tumour DNA. *Proc. Natl. Acad. Sci. USA* **78**, 1181–1184.
Lane, D. P., and Crawford, L. V. (1979). T-antigen is bound to a host protein in SV40-transformed cells. *Nature (London)* **278**, 261–263.
Laszlo, A., Radke, K., Chin, S., and Bissell, M. (1981). Tumour promoters alter gene expression and protein phosphorylation in avian cells in culture. *Proc. Natl. Acad. Sci. USA* **78**, 6241–6245.
Leavitt, J., and Kakunaga, T. (1980). Expression of a variant form of actin and additional polypeptide changes following chemical induced *in vitro* neoplastic transformation of human fibroblasts. *J. Biol. Chem.* **255**, 1650–1661.
Leavitt, J., and Moyzig, R. (1978). Changes in gene expression accompanying neoplastic transformation of Syrian hamster cells. *J. Biol. Chem.* **253**, 2497–2500.
Leavitt, J., Goldman, D., Merril, C., and Kakunaga, T. (1982a). Changes in gene expression accompanying chemically induced malignant transformation of human fibroblasts. *Carcinogenesis (N.Y.)* **3**,. 61–70.
Leavitt, J., Bushar, G., Kakunaga, T., Hamada, H., Hirakawa, T., Goldman, D., and Merril, C. (1982b). Variations in expression of mutant β-actin accompanying incremental increases in human fibroblast tumorigenicity. *Cell* **28**, 259–268.

Lester, E. P., Lemkin, P., Lipkin, L., and Cooper, H. L. (1981). A two-dimensional electrophoretic analysis of protein synthesis in resting and growing lymphocytes *in vitro*. *J. Immunol.* **126**, 1428–1434.

Lester, E. P., Lemkin, P., and Lipkin, L. (1982). A two-dimensional gel analysis of autologous T and B lymphoblastoid cell lines. *Clin. Chem.* **28**, 829–839.

Levinson, A. D., Oppermann, H., Levintow, L., Varmus, H. E., and Bishop, J. M. (1978). Evidence that the transforming gene of avian sarcoma virus encodes a protein kinase associated with a phosphoprotein. *Cell* **15**, 561–572.

Levinson, A. D., Oppermann, H., Varmus, H. E., and Bishop, J. M. (1980). The purified product of the transforming gene of avian sarcoma virus phosphorylates tyrosine. *J. Biol. Chem.* **255**, 11973–11980.

Liebermann, D., Hoffman-Liebermann, B., and Sachs, L. (1980). Molecular dissection of differentiation in normal and leukemic myeloblasts: Separately programmed pathways of gene expression. *Dev. Biol.* **79**, 46–63.

Linder, S., Krondahl, U., Sennerstam, R., and Ringertz, N. (1981). Retinoic acid-induced differentiation of F9 embryonal carcinoma cells. *Exp. Cell Res.* **132**, 453–460.

Linzer, D. I., and Levine, A. J. (1979). Characterization of a 54Kdalton cellular SV40 tumour antigen present in SV40-transformed cells and uninfected embryonal carcinoma cells. *Cell* **17**, 43–52.

Logan, J., and Cairns, J. (1982). The secrets of cancer. *Nature (London)* **300**, 104–105.

Martin, R. G. (1981). The transformation, cell growth and transmodification of DNA synthesis by simian virus 40. *Adv. Cancer Res.* **34**, 1–68.

Marx, J. L. (1982). The case of the misplaced gene. *Science (Washington, D.C.)* **218**, 983–985.

Milcarek, C., and Zahn, K. (1978). The synthesis of ninety proteins including actin throughout the HeLa cell cycle. *J. Cell Biol.* **79**, 833–838.

Mose Larsen, P. (1981). An assessment of the potential offered by two-dimensional gel electrophoresis and silver staining for developmental biology. Cand. Sci. Dissertation, Aarhus Univ., Aarhus, Denmark.

Mose Larsen, P., Bravo, R., Fey, S. J., Small, J. V., and Celis, J. E. (1982). Putative association of mitochondria with a subpopulation of intermediate-sized filaments in cultured human skin fibroblasts. *Cell* **31**, 681–692.

Mose Larsen, P., Fey, S. J., Bravo, R., and Celis, J. E. (1983). Mouse mitochrondrial protein IEF 24: Identification and immunohistochemical localization of mitochondria in various tissues. *Electrophoresis* **4**, 247–256.

Neel, B. G., Hayward, W. S., Robinson, H. L., Fang, J., and Astrin, S. M. (1981). Avian leukosis virus-induced tumours have common proviral integration sites and synthesize discrete new RNA's: oncogenesis by promoter insertion. *Cell* **23**, 323–334.

Nusse, R., and Varmus, H. E. (1982). Many tumours induced by the mouse mammary tumour virus contain a provirus integrated in the same region of the host genome. *Cell* **31**, 99–109.

O'Connor, C. M., Gard, D. L., and Lazarides, E. (1981). Phosphorylation of intermediate filament proteins by cAMP-dependent protein kinases. *Cell* **23**, 135–143.

O'Farrell, P. H. (1975). High-resolution two dimensional electrophoresis of proteins. *J. Biol. Chem.* **250**, 4007–4021.

O'Farrell, P. Z., Goodman, H. M., and O'Farrell, P. H. (1977). High resolution two-dimensional electrophoresis of basic as well as acidic proteins. *Cell* **12**, 113–1142.

Okada, S. (ed.) (1970). "Radiation Biochemistry," Vol. 1. Academic Press, New York and London.

Orgel, L. (1963). The maintenance of the accuracy of protein synthesis and its relevance to ageing. *Proc. Natl. Acad. Sci. USA* **49**, 517–521.

Osborn, M., and Weber, K. (1982). Intermediate filaments: cell type specific markers in differentiation and pathology. *Cell* **31**, 303–306.
Osborn, M., Geisler, N., Shaw, G., Sharp, G., and Weber, K. (1981). Intermediate filaments. *Cold Spring Harbor Symp. Quant. Biol.* **46**, 413–429.
Padhy, L. C., Shih, C., Cowing, D., Finkelstein, R., and Weinberg, R. A. (1982). Identification of a phosphoprotein specifically induced by the transforming DNA of rat neuroblastomas. *Cell* **28**, 865–871.
Paulin, D., Perreu, J., Jakob, H., Jacob, F., and Yaniv, M. (1979). Tropomyosin synthesis accompanies formation of actin filaments in embryonal carcinoma cells induced to differentiate by hexamethylene bisacetamide. *Proc. Natl. Acad. Sci. USA* **76**, 1891–1895.
Peterson, J. L., and McConkey, E. H. (1976). Proteins of Friend leukemia cells. Comparison of hemoglobin-synthesizing and non-induced populations. *J. Biol. Chem.* **251**, 555–558.
Peto, R. (1977). Epidemiology, multistage models and short-term mutagenicity tests. *In* "Origin of Human Cancer" (H. H. Hiatt, J. D. Watson, and J. A. Winsten, eds.), p. 403. Cold Spring Harbor Lab., Cold Spring Harbor, New York.
Pollack, R., Osborn, M., and Weber, K. (1975). Patterns of organization of actin and myosin in normal and transformed non-muscle cells. *Proc. Natl. Acad. Sci. USA* **72**, 994–998.
Puck, T. T., and Marcus, P. I. (1965). Action of X-rays on mammalian cells. *J. Exp. Med.* **103**, 653–666.
Purchio, A. F., Erikson, E., Brugge, J. S., and Erikson, R. L. (1978). Identification of a polypeptide encoded by the avian sarcoma virus *src* gene. *Proc. Natl. Acad. Sci. USA* **75**, 1567–1571.
Radke, K., and Martin, G. S. (1979). Transformation by Rous sarcoma virus: Effects of *src* gene expression on the synthesis and phosphorylation of cellular polypeptides. *Proc. Natl. Acad. Sci. USA* **76**, 5212–5216.
Radke, K., Gilmore, T., and Martin, G. S. (1980). Transformation by Rous sarcoma virus: A cellular substrate for transformation-specific protein phosphorylation contains phosphotyrosine. *Cell* **21**, 821–828.
Radman, M., Jeggo, P., and Wagner, R. (1982). Chromosomal rearrangements and carcinogenesis. *Mutat. Res.* **98**, 249–264.
Radman, M., Wagner, R., and Jeggo, P. (1983). DNA and time in carcinogenesis. *In* "Gene Expression in Normal and Transformed Cells" (J. E. Celis and R. Bravo, eds.), pp. 177–191. Plenum, New York.
Ramaekers, F. C. S., Puts, J. J. G., Kant, A., Moesker, O., Jap, P. H. K., and Vooijs, G. P. (1981). Use of antibodies to intermediate filaments in the characterization of human tumours. *Cold Spring Harbor Symp. Quant. Biol.* **46**, 331–339.
Rapoport, R. M., Draznin, M. B., and Murad, F. (1982). Sodium nitroprusside-induced protein phosphorylation in intact rat aorta is mimicked by 8-bromo cyclic AMP. *Proc. Natl. Acad. Sci. USA* **79**, 6470–6474.
Reddy, E. P., Reynolds, R. K., Santos, E., and Barbacid, M. (1982). A point mutation is responsible for the acquisition of transforming properties by the T24 human bladder carcinoma oncogene. *Nature (London)* **300**, 149–152.
Riddle, V. H. G., and Pardee, A. B. (1980). Quiescent cells but not cycling cells exhibit enhanced actin synthesis before they synthesize DNA. *J. Cell. Physiol.* **103**, 11–15.
Riddle, V. G. H., Dubrow, R., and Pardee, A. (1979). Changes in the synthesis of actin and other cell proteins after stimulation of serum-arrested cells. *Proc. Natl. Acad. Sci. USA* **76**, 1298–1302.
Robinson, S. I., Nelkin, B., Kaufmann, S., and Vogelstein, B. (1981). Increased phosphorylation rate of intermediate filaments during mitotic arrest. *Exp. Cell Res.* **133**, 445–449.

Schaffhausen, B. S., and Benjamin, T. L. (1979). Phosphorylation of polyoma T antigens. *Cell* **18**, 935–946.

Sefton, B. M., Hunter, T., Beemon, K., and Eckhart, W. (1980). Evidence that the phosphorylation of tyrosine is essential for cellular transformation by Rous sarcoma virus. *Cell* **20**, 807–816.

Sefton, B. M., Hunter, T., and Raschke, W. C. (1981a). Evidence that the Abelson virus protein functions *in vivo* as a protein kinase that phosphorylates tyrosine. *Proc. Natl. Acad. Sci. USA* **78,** 1552–1556.

Sefton, B. M., Hunter, T., Nigg, E. A., Singer, S. J., and Walter, G. (1981b). Cytoskeletal targets for viral transforming proteins with tyrosine protein kinase activity. *Cold Spring Harbor Symp. Quant. Biol.* **46**, 939–951.

Segawa, K., and Ito, Y. (1982). Differential subcellular localization of *in vivo* phosphorylated and nonphosphorylated middle-sized tumour antigen of polyoma virus and its relationship to middle-sized tumour antigen phosphorylating activity *in vitro*. *Proc. Natl. Acad. Sci. USA* **79**, 6812–6816.

Shih, C., Shilo, B., Goldfarb, M. P., Dannenberg, A., and Weinberg, R. A. (1979). Passage of phenotype of transformation via transfection of DNA and chromatin. *Proc. Natl. Acad. Sci. USA* **76**, 5714–5718.

Shih, C., Padhy, L. C., Murray, M. J., and Weinberg, R. (1981). Transforming genes of carcinomas and neuroblastomas introduced into mouse fibroblasts. *Nature (London)* **290,** 261–264.

Small, J. V., and Celis, J. E. (1978). Filament arrangements in negatively stained cultured cells: The organization of actin. *Cytobiologie* **16**, 308–325.

Steinberg, R. A., and Coffino, P. (1979). Two dimensional gel analysis of cyclic AMP effects in cultured S49 mouse lymphoma cells: protein modifications, induction and repressions. *Cell* **18**, 719–733.

Strand, M., and August, J. T. (1978). Polypeptide maps of cells infected with murine type C leukemia or sarcoma oncovirus. *Cell* **13**, 339–408.

Summerhayes, J. C., Cheng, Y. E., Sun, T. T., and Chen, L. B. (1981). Expression of keratin and vimentin intermediate filaments in rabbit bladder epithelial cells at different stages of benzo[a]pyrene-induced neoplastic progression. *J. Cell. Biol.* **90,** 63–69.

Sun, T. T., and Green, H. (1978). Keratin filaments of cultured epidermal cells. *J. Cell Biol.* **253**, 2053–2060.

Thorsrud, A. K., Vatn, M. H., and Jellum, E. (1982). Two dimensional electrophoretic patterns of protein of normal mucosa, polyps and carcinomas of the large intestine. *Clin. Chem.* **28**, 884–889.

Tabin, C. J., Bradley, S. M., Bargmann, C. I., Weinberg, R., Papageorge, A. G., Scolnick, E. M., Dhar, R., Lowy, D. R., and Chang, E. A. (1982). Mechanisms of activation of a human oncogene. *Nature (London)* **300**, 143–149.

Tolmach, L. J., and Marcus, P. I. (1960). Development of X-ray induced giant HeLa cells. *Exp. Cell Res.* **20,** 350–360.

Tooze, J. (1981). "DNA tumor viruses—Molecular biology of tumor viruses," (2nd ed.). Harbor Lab., Cold Spring Harbor, New York.

Tracy, R. P., Wold, L. E., Currie, R. M., and Young, D. S. (1982). Patterns for normal colon mucosa and colon adenocarcinoma compared by two dimensional gel electrophoresis. *Clin. Chem.* **28**, 915–919.

Tuszynski, G. P., Frank, E. D., Damsky, C. H., Buck, C. A., Warren, L. (1979). The detection of smooth muscle desmin like protein in BHK21/C_{13} fibroblasts. *J. Biol. Chem.* **254**, 6138–6143.

Ushiro, H., and Cohen, S. (1980). Identification of phosphotyrosine as a product of epidermal growth factor-activated protein kinase in A431-cell membranes. *J. Biol. Chem.* **255**, 8363–8365.

Vaheri, A., and Alitalo, K. (1981). Pericellular matrix glycoproteins in cell differentiation and in malignant transformation. *In* "Cellular Controls in Differentiation" (C. W. Lloyd and D. A. Rees, eds.), pp. 29–54. Academic Press, New York.

Vandekerckhove, J., Leavitt, J., Kakunaga, T., and Weber, K. (1980). Coexpression of a mutant *beta*-actin and the two normal *beta* and *gamma* cytoplasmic actin in a stably transformed human cell line. *Cell* **22**, 893–899.

Van Keuren, M. L., Merril, C. R., and Goldman, D. (1983). Proteins affected by chromosome 21 and ageing *in vitro*. *In* "Gene Expression in Normal and Transformed Cells" (J. E. Celis and R. Bravo, eds.), pp. 349–378. Plenum, New York.

Vogt, P. K. (1977). Genetics of RNA tumour viruses. *Compr. Virol.* **9**, 341–455.

Voris, B. P., and Young, D. A. (1981). Glucocorticoid-induced proteins in rat thymus cells. *J. Biol. Chem.* **256**, 11319–11329.

Wang, E., and Goldberg, A. R. (1976). Changed in microfilament organization and surface topography upon transformation of chick embryo fibroblasts with Rous sarcoma virus. *Proc. Natl. Acad. Sci. USA* **73**, 4065–4069.

Watt, F. M., Harris, H., Weber, K., and Osborn, M. (1978). The distribution of actin cables and microtubules in hybrids between malignant and non-malignant cells and in human tumours derived from them. *J. Cell Sci.* **32**, 419–432.

Weinberg, R. A. (1981). Use of transfection to analyze genetic information and malignant transformation. *Biochim. Biophys. Acta* **651**, 25–35.

Weinberg, R. A. (1982a). Fewer and fewer oncogenes. *Cell* **30**, 3–7.

Weinberg, R. A. (1982b). Oncogenes of human tumour genes. *Trends Biochem. Sci.* **7**, 135–136.

Weiss, R., Teich, N., Varmus, H., and Coffin, J. (eds.) (1982) *In* "Molecular biology of tumour viruses: RNA tumour viruses. Cold Spring Harbor Lab., Cold Spring Harbor, New York.

Wilson, D. L., Hall, M. E., and Stone, G. C. (1978). Test of some aging hypothesis using two-dimensional protein mapping. *Gerontology (Basel)* **24**, 426–433.

CHAPTER 11

Application of Two-Dimensional Polyacrylamide Gel Electrophoresis to Studies of Mistranslation in Animal and Bacterial Cells

JEFFREY W. POLLARD
MRC Human Genetic Diseases Research Group,
Department of Biochemistry
Queen Elizabeth College
University of London
London, England

I. Introduction 363
II. Detection of Error-Containing Proteins by Two-Dimensional Polyacrylamide Gel Electrophoresis 364
III. Quantification of the Error Frequency of Protein Synthesis 373
IV. Assessment of the Error Frequencies of Protein Synthesis under Varying Physiological Conditions 379
A. The Error Catastrophe Theory of Cellular Aging 379
B. The Stringent Control System in Bacteria and Its Effect on Protein Synthetic Accuracy 386
V. Other Uses of the Two-Dimensional Gel Assay 388
VI. Two out of Three Reading 390
References 392

I. Introduction

A minimum level of accuracy is necessary for successful information transfer. In cascade processes and most particularly in cyclic cascade processes, a high degree of fidelity of information transfer is crucial because under these circumstances errors can be autocatalytic, leading

TWO-DIMENSIONAL GEL
ELECTROPHORESIS
OF PROTEINS

ISBN 0-12-164720-X

to the eventual collapse of the system. This problem was first discussed for protein synthesis by Orgel (1963), who went so far as to suggest that such a logarithmic increase in protein synthetic errors may be responsible for cellular aging (Orgel, 1963, 1970). Similar considerations also led to the conclusion that in order to maintain cellular life the genetic code must have evolved from a system of low precision to one of greater accuracy (Hoffman, 1974). These reasons, as well as the desire to more fully understand the fundamental processes of protein synthesis, led to many attempts to define the frequency and pattern of protein synthetic errors under differing physiological states.

Most assays of mistranslation have exploited the illegitimate incorporation of amino acids into polypeptides either in *in vitro* protein-synthesizing systems programmed with synthetic polynucleotides or into peptide fragments of selected proteins synthesized *in vivo.* Neither of these assays have been completely satisfactory. The former is prone to artifact and until recently exhibited higher error levels than observed *in vivo,* whereas the latter procedure involves time-consuming purification of selected abundant proteins coupled with highly sensitive assays for mistranslation and thus does not provide a readily available assay for protein synthetic accuracy. However, mistranslated forms of proteins generated by amino acid starvation of both bacterial and animal cells have been detected using two-dimensional polyacrylamide gel electrophoresis (2-D PAGE) (O'Farrell, 1978; Parker *et al.*, 1978). This approach has led to a rapid and quantitative *in vivo* assay for mistranslation, which has enabled error frequencies to be determined in a wide range of cells under different growth conditions (Harley *et al.*, 1980, 1981).

II. Detection of Error-Containing Proteins by Two-Dimensional Polyacrylamide Gel Electrophoresis

Acute starvation of both bacterial and animal cells for certain amino acids leads to the synthesis of aberrant proteins detected on autoradiograms of 2-D gels as a series of satellite spots with similar molecular weights to the authentic protein detected by Coomassie Blue staining but separated from it in the isoelectric focusing dimension (O'Farrell, 1978; Parker *et al.*, 1978), a phenomenon termed *stuttering* (Figs. 1 and 2). This charge heterogeneity has been interpreted as a consequence of the decoding of the codon for the limiting amino acid by the incorrect charged tRNA most closely related to that codon at its third position.

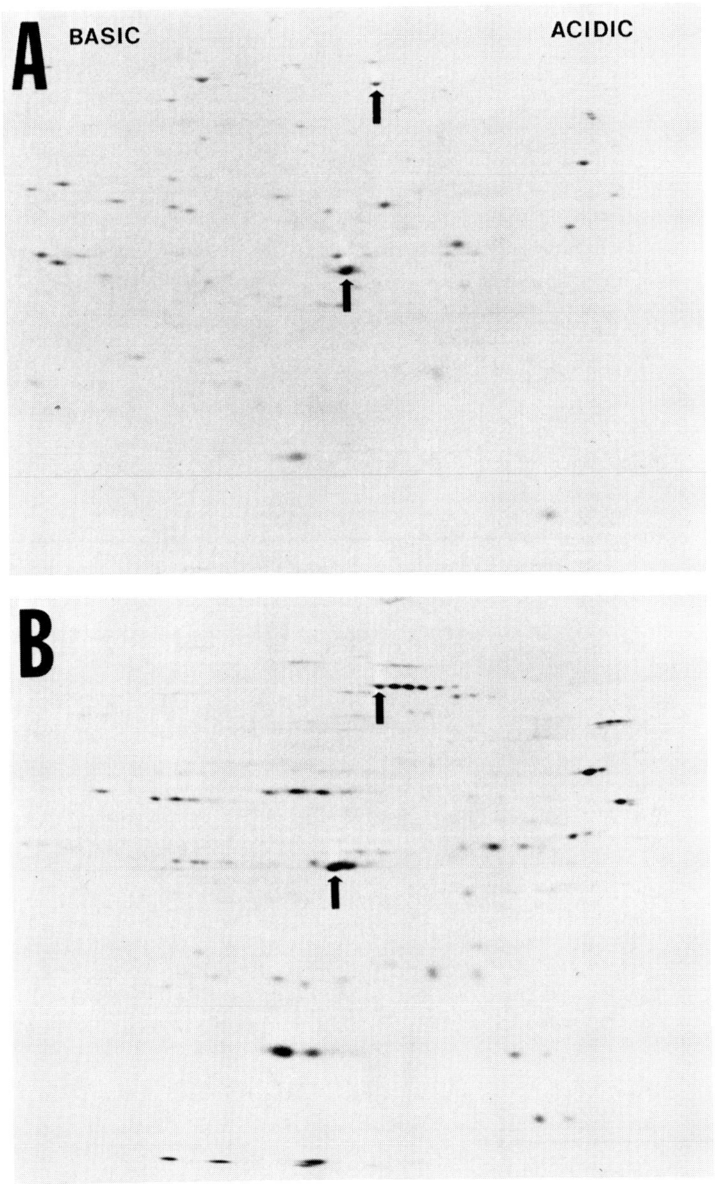

Figure 1. Autoradiograms of 2-D gels of [^{35}S]methionine-labeled histidine auxotroph *E. coli* (strain Nf931) proteins synthesized by exponentially growing cells labeled for 5 min (A) and cells starved for histidine for 20 min followed by 5 min labeling under starvation conditions (B). In each panel the upper arrow indicates the position of authentic EF-G and the lower arrow EF-Tu. The gels in this figure, and all other figures, are oriented with the pH gradient running from the right (acidic) to the left (basic). From Parker *et al.* (1978) with permission.

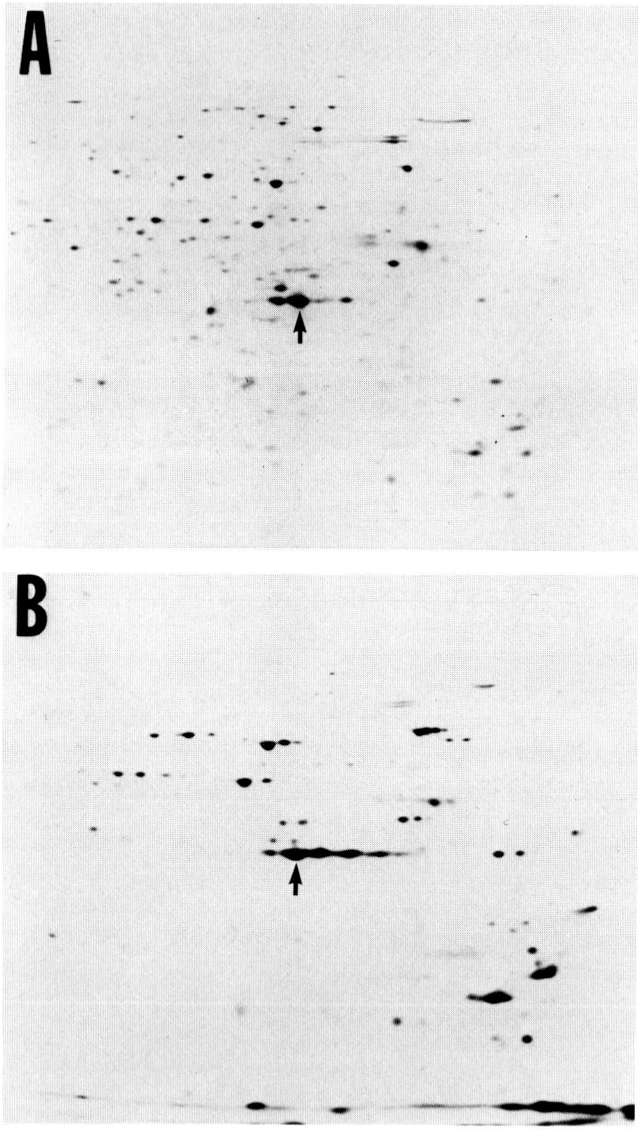

Figure 2. Fluorograms of 2-D gels of [^{35}S]methionine-labeled cytoplasmic proteins of L-cells synthesized by exponentially growing cells in α-minimal essential medium plus 10% fetal calf serum (FCS) lacking methionine (A) and by cells starved for histidine (B) grown in α-minimal essential medium lacking histidine and methionine but containing 10% FCS and 2 m*M* histidinol. All cells were preincubated for 30 min and then labeled with [^{35}S]methionine for 1 h under the above conditions. The arrow indicates the position of β-actin. From Parker *et al.* (1978) with permission.

The most persuasive evidence in favour of this interpretation is that both the existence and direction of the mistranslated forms generated by amino acid starvation can be predicted from the genetic code by applying rules derived from the patterns and hierarchy of errors found using synthetic polynucleotides in cell-free systems (Woese, 1967). These rules indicate the following:

1. Misreading affects only one base at a time and is most common at the third base of the codon (wobble base).
2. Pyrimidines are mostly misread as the alternative pyrimidine, then less frequently as adenine and rarely as guanine.
3. Purines are misread as pyrimidines at much lower frequencies than pyrimidines as purines.

Thus cells were deprived of individual amino acids in one of three ways followed by the detection of error-containing newly synthesized proteins on 2-D gels. The three methods of amino acid starvation were as follows.

1. In addition to being substrate restricted, available temperature-sensitive (*ts*) amino acyl tRNA synthetase mutants were starved for a single amino acid (charged to its tRNA) by expression of the *ts* mutation at a nonpermissive temperature;
2. In addition to being histidine restricted, mammalian cells were exposed to the drug histidinol, which competes with histidine at the active site of the histidyl-tRNA synthetase but does not get charged to the tRNA, resulting in effective histidine starvation (Hansen *et al.*, 1972);
3. An amino acid was removed from the medium corresponding to the requirement of the appropriate bacterial auxotroph.

It is worth pointing out that simple removal of amino acids from the medium of mammalian cells does not result in effective starvation because of the high level of protein degradation, which gives large intracellular pools of amino acids and thus restricts the choice of starvation to one of the first two methods described.

Thus by consideration of the genetic code it would be predicted that following histidine (a basic amino acid) starvation of either bacterial or mammalian cells, the codons (CAU, CAC) would first be misread at the wobble base, then as a glutamine codon (CAA, CAG) resulting in the incorporation of glutamine, a neutral amino acid, and a net reduction in the protein's isoelectric point. Figures 1, 2, and 3B show the effect of histidine starvation on proteins synthesized in both bacterial and mammalian cells. It can be seen that the displacement of the satellite spots is

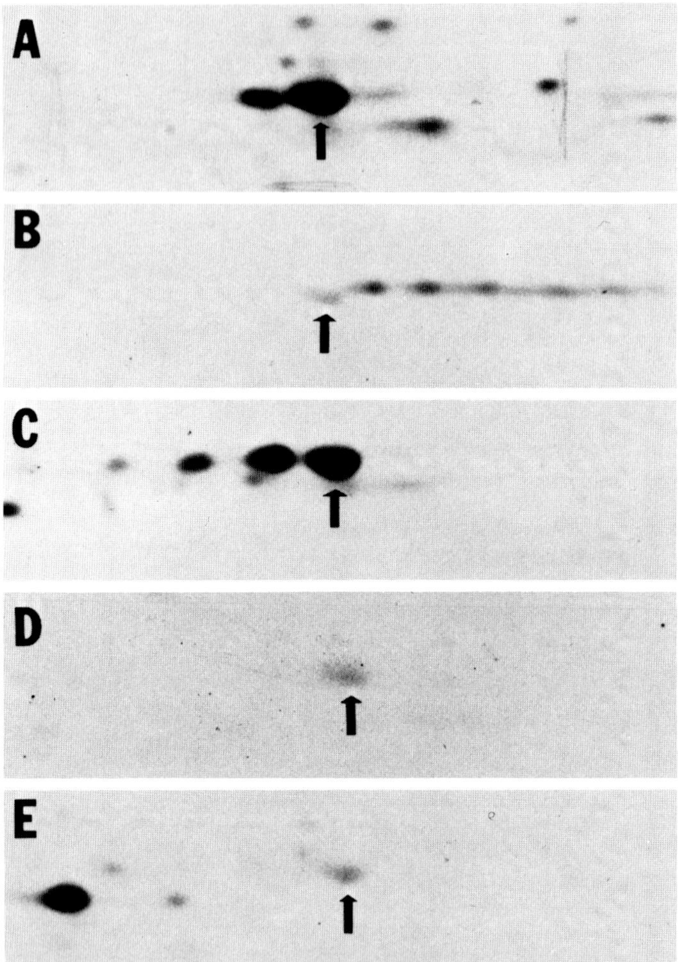

Figure 3. Autoradiograms of the central portions of 2-D gels of [^{35}S]methionine-labeled Chinese hamster ovary cells (CHO) cytoplasmic proteins synthesized under various conditions. All cells were preincubated in α-minimal essential medium lacking methionine and supplemented with 10% FCS and then incubated in the presence of [^{35}S]methionine for 1 h. The position of β-actin is shown by the arrows. (A) Exponentially growing *ts* asparaginyl-tRNA synthetase mutant (Asn-7) cells at 34°C. (B) Asn-7 cells starved for histidine by 5 m*M* histidinol in medium lacking histidine. (C) Asn-7 cells starved for asparagine by exposure to a nonpermissive temperature, 39.5°C, in medium lacking asparagine. (D) A *ts* methionyl-tRNA synthetase mutant (Met-1) of CHO cells starved for methionine by exposure to a nonpermissive temperature, 38.5°C, in medium lacking methionine. (E) CHO (wild type) cells treated with cycloheximide (20 μg/ml) in complete medium at 34°C. From Parker *et al.* (1978).

in the predicted direction, toward the acidic end of the gel. Similarly, starvation for asparagine (AAU, AAC, neutral) is predicted to result in the misincorporation of the basic amino acid lysine (AAA, AAG) with a resultant increase in isoelectric point of the protein. This shift of the satellite forms toward the basic end of the 2-D gel following asparagine starvation can be seen in both Figs. 3C and 8.

Table I shows the results of similar experiments exploiting the amino acid deprivation produced by expression of the eight available *ts* aminoacyl tRNA synthetase mutations of Chinese hamster ovary (CHO) cells. In all cases the experimental observations were consistent with the predictions derived from the genetic code. These include starvation for amino acids whose codons end in pyrimidines (histidine, asparagine, leucine, valine, and arginine) and those ending in purines (methionine, lysine, glutamine, valine, leucine, and arginine). In the latter case, because misreading of a purine as a pyrimidine would occur at a low frequency, it was predicted that if satellite forms would be expected (e.g., following lysine starvation), their presence would be at or below the level of detection by autoradiography of 2-D gels. In mammalian cells this appeared to be the case, because deprivation for any amino acid whose codon ended in a purine did not result in detectable satellite forms. For technical reasons, however, *E. coli* proteins may be radioactively labeled to higher specific activities than mammalian proteins and under these circumstances a shift towards the acidic end of the gel following lysine starvation was detected (Parker and Friesen, 1980). This charge shift may be interpreted as a misreading of the lysine codon ending in a purine (AAA, AAG) for asparagine (AAU, AAC). This misreading occurs at about 6% of the rate of misreading asparagine to lysine and is entirely in accord with the patterns described above.

The data are therefore consistent with misreading at the third base of codon. This conclusion is further strengthened by consideration of the glutamine codons (CAA, CAG) in which third position misreading would be restricted by the presence of the purine, resulting in the lack of detectable satellite forms, but significant first position misreading would result in the misincorporation of lysine (AAA, AAG) and shifts towards the basic end of the gel. There is also the possibility of misreading as glutamic acid (GAA, GAG) and shift to the acidic end of the gel or for premature termination (UAA, UAG). Experimentally in CHO cells, however, following expression of the *ts* glutamyl-tRNA synthetase mutation no satellite proteins were observed on the 2-D gel (Parker *et al.*, 1978). By similar considerations of the genetic code second position misreading (considered to be a rare event in any case) would result in a completely different pattern of movement in the isoelectric point of the

TABLE I

The Predicted Movement on 2-D Gels of Mistranslated Proteins Synthesized following Amino Acid Starvation Compared to the Experimental Results Obtained by Starving CHO Cells for the Relevant Amino Acids[a,b]

Amino acid	Charge[c]	Code	Misread as	Substituted amino acid	Charge	Predicted direction of movement of satellite spots	Observed direction of movement of satellite spots
His	Basic	CAU	CAA	Gln	Neutral	Acidic	Acidic
			CAG	Gln			
		CAC	CAA	Gln			
Asn	Neutral	AAU	AAA	Lys	Basic	Basic	Basic
			AAG	Lys			
		AAC	AAA	Lys			
Leu	Neutral	CUU	CUA	Leu	Neutral	None	None
			CUG	Leu			
		CUC	CUA	Leu			
Val	Neutral	GUU	GUA	Val	Neutral	None	None
			GUG	Val			
		GUC	GUA	Val			
Arg	Basic	CGU	CGA	Arg	Basic	None	None
			CGG	Arg			
		CGC	CGA	Arg			

[a] Predictions of misreading after Woese (1967). U misread as C > A ≫ G; C misread as U > A. Misreading at purines was considered to have a much lower frequency than misreading at pyrimidines. Therefore, as predicted starvation for methionine, glutamine, lysine, leucine, arginine, and valine, all amino acids having codons terminating in purines, did not produce any detectable satellite proteins on the 2-D gels. Data from Parker *et al.* (1978); J. W. Pollard, J. W. Chamberlain, and C. P. Stanners (unpublished).

[b] All experiments to determine the direction of movement of satellite spots were performed by shifting the relevant *ts* aminoacyl tRNA synthetase mutant of CHO cells to its nonpermissive temperature in medium lacking the appropriate amino acid. Following the expression of the mutation, newly synthesized proteins were labeled with [^{35}S]methionine and displayed on 2-D gels. Gels were dried and exposed to Kodak XR-1 film and the direction of misreading on the autoradiograms estimating visually.

[c] The charge of the amino acid within the protein was deduced from consideration of the p*K* of the exposed side group.

satellite proteins than was observed, and therefore the hypothesis can be effectively rejected (see Parker *et al.*, 1978; Parker and Friesen, 1980).

Further strengthening the above interpretation, Parker *et al.* (1980) examined the effect of amino acid starvation on bacteriophage MS2 coat protein synthesis following infection of *E. coli* under conditions in which host protein synthesis was inhibited by rifampicin. Similar patterns of protein synthesis to those described above were detected following amino acid starvation with the exception of histidine starvation, which produced no detectable charge heterogeneity. This would be expected because of the complete absence of histidine residues in MS2 coat protein. It was also shown that following asparagine starvation there was a relative increase in the incorporation of lysine as compared to leucine in the satellite spots and this ratio increased the further the satellite forms were from the authentic protein in a fashion predicted from the nucleotide sequence of the coat protein RNA, given that asparagine was being misread as lysine. In mammalian cells it has so far been difficult to incorporate enough radioactive isotope to carry out similar experiments.

Four other strands of evidence favor the interpretation that the heterogeneity of the 2-D gel patterns following amino acid starvation is due to mistranslation rather than to postsynthetic modifications or to the presence of unique proteins synthesized as a response to amino acid starvation conditions.

1. A ribosomal mutation of *E. coli* confering resistance to streptomycin (rpsL), which is known to increase ribosomal fidelity, also decreased the level of mistranslation detected on the 2-D gel following asparagine starvation (Parker and Friesen, 1980). Streptomycin itself in wild-type *E. coli* cells decreases ribosomal fidelity and a corresponding increase in stuttering was detected on the 2-D gels (Parker *et al.*, 1981). Because streptomycin treatment does not lead to any apparent posttranslational alteration of the proteins synthesized and it is known to act specifically at the ribosomal level, it would seem that it is not misacylation that leads to the observed mistranslation but rather erroneous codon–anticodon interactions within the ribosome. Similar arguments may be made for ribosomal involvement from the observation that relaxed mutants (relA) of *E. coli* have a significantly higher error level after amino acid starvation than stringent strains ($relA^+$) (Hall and Gallant, 1972). This is also detectable on 2-D gels (Section IV,B, O'Farrell, 1978).

2. It has been shown that satellite spots that had been detected on the 2-D gels were related to the native protein by peptide mapping in the case of MS2 coat protein (Parker *et al.*, 1980), and by immunoprecipitation of actin satellites in mammalian cells (J. W. Chamberlain, J. W. Pollard, and C. P. Stanners, unpublished). In the latter case, DNase I, which specifically interacts with actin on a molar basis, was reacted with [^{35}S]methionine-labeled cytoplasmic extracts of CHO cells and the actin–DNase I complex immunoprecipitated by anti-DNase I. Following histidine starvation, only actin and its satellite forms were preferentially

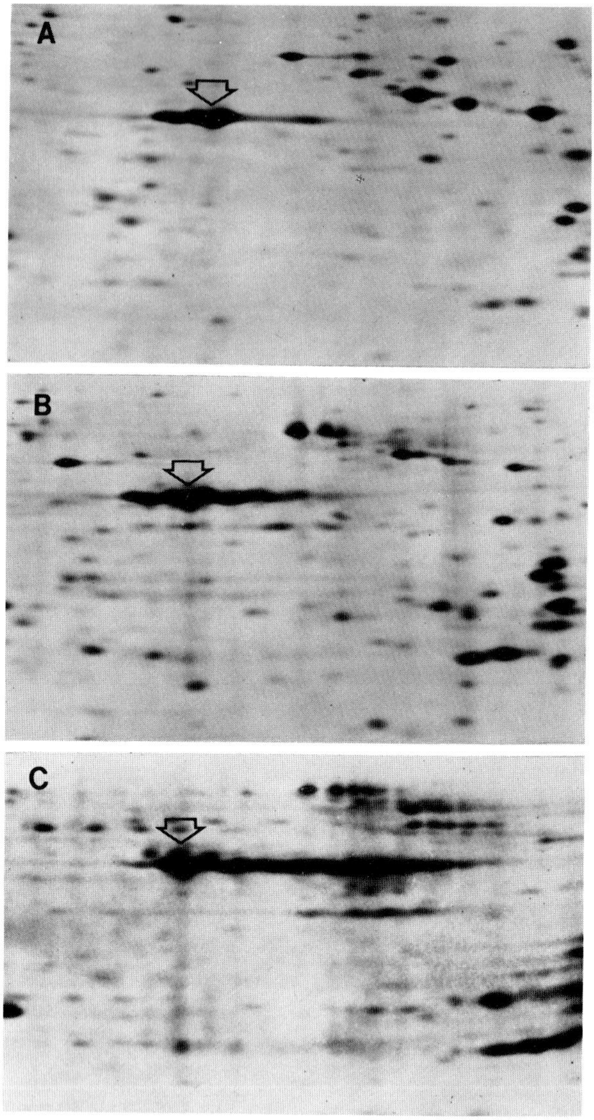

Figure 4. Autoradiogram of the central portions of 2-D gels of [^{35}S]methionine labeled proteins of chick embryo fibroblasts starved for histidine to increasing levels by exposure to increasing concentrations of histidinol in histidine-free α-minimal essential medium supplemented with 10% FCS and lacking methionine. The autoradiograms presented are (A) control, unstarved cells; (B) cells starved for histidine with 0.1 m*M* histidinol; and (C) 0.5 m*M* histidinol. The protein synthetic rates were 34.2 and 14.5%, respectively, of the control rate. Mistranslation increases with the severity of the histidine starvation. Note the extensive amount of mistranslation in chick cells, which had previously been shown to have the highest error level of any eukaryotic cell tested (Harley, 1979). This extensive mistranslation is particularly noticeable for actin (β-actin is indicated by the arrow), which in chick is present in multiple forms and whose stuttered forms merge to give a smeared pattern toward the acidic end of the gel.

enriched by the immunoprecipitation, implying that the satellite forms were related to actin and were not novel proteins synthesized in response to amino acid starvation. Similarly, it was shown that the charge heterogeneity was not simply the results of an inhibition of protein synthesis, both by consideration of the data on amino acid specificity and also because if protein synthesis in CHO cells was inhibited with cycloheximide to the same extent as found during amino acid starvation, charge heterogeneity was not induced (Fig. 3E) (Parker *et al.*, 1978).

3. In pulse-chase experiments a sequential incorporation into the satellite forms in relation to the native form was not observed as would have been predicted for a postsynthetic modification.

4. No simple observed postsynthetic modification will increase the isoelectric point of a protein. The observed change in the isoelectric point of the satellite forms of actin following both starvation for asparagine and histidine closely paralleled that predicted from the amino acid sequence of actin for substitution of lysine and glutamine, respectively (Harley, 1979).

It can therefore be safely concluded that the phenomenon referred to as stuttering is the consequence of missense substitution caused by third position misreading. The simplest model to describe such a phenomenon is one of substrate limitation, that is, the error rate is directly related to the increased step time of the ribosome at the hungry codon (Caplan and Menninger, 1979), which increases the probability of the noncognate tRNA interacting with the codon sufficiently often or for long enough for a peptide bond to be formed and consequently an incorrect amino acid being incorporated into the growing peptide chain. From this model it would be predicted that the more severe the amino acid starvation, the more severe the degree of mistranslation. This has been shown to be so in both bacterial and mammalian cells (Fig. 4; Parker *et al.*, 1978; Parker and Friesen, 1980; Harley *et al.*, 1980, 1981) and, in fact, under situations of extreme starvation over 90% of a specific protein synthesized may be found as aberrant forms (Parker *et al.*, 1978).

III. Quantification of the Error Frequency of Protein Synthesis

Determination of error frequencies in protein synthesis have been fraught with difficulties. The most reliable measurements have been by analysis of illegitimate incorporation of isoleucine into a peptide fragment of hemoglobin (Loftfield, 1963; Loftfield and Vanderjagt, 1972) or cysteine into *E. coli* flagellin (Edelmann and Gallant, 1977a). These estimates were of first position misreading of the codon and gave frequencies of about 1 to 2 $\times$ 10^{-4}. Both involved the purification of selected proteins coupled with highly sensitive assays for mistranslation and therefore do not provide readily available assays for mistranslation. They are also open to a number of objections because the values obtained would lead to underestimates if the error-containing proteins were selectively degraded in the cell or if as a consequence of the amino acid substitution, the method of isolation preferentially excluded the error-containing form of a particular protein. They can also lead to an overestimation because one cannot be certain that the protein is absolutely pure nor can one be certain that the illegitimate incorporation of an amino acid is not due to translation of an altered mRNA arising from aberrant transcription or from the expression of multiple polymorphic genes coding for the same protein.

The detection of error-containing proteins on 2-D gels provides an alternative means of estimating error frequencies that avoids the above problems. Although it has its own set of constraints, it is in principle a general method for estimating error frequencies in any cell (Harley *et al.*, 1981).

In order to obtain this estimate of error frequency, the fraction of substituted sites in a protein of known sequences has to be determined following amino acid starvation. These error frequencies of course are several orders of magnitude above those normally observed (Parker *et al.*, 1978; Harley *et al.*, 1981), are related to the degree of inhibition of the protein synthetic rate, and therefore need to be converted to a normalized error frequency by analytical means. In the initial analysis actin was chosen as a reference protein because its sequence is known, it is the major protein synthesized under starvation conditions in all mammalian cells tested, its mistranslated forms are readily resolved and quantitated, and they are not preferentially degraded during the time course of the experiments (Harley *et al.*, 1981). However, any protein fulfilling the above criteria could be chosen; for example, bacteria EF-G or EF-Tu would be good candidates. After starvation for an appropriate amino acid, [^{35}S]methionine-labeled proteins were displayed on a 2-D gel and detected by autoradiography. The degree of mistranslation was estimated by quantitative scanning densitometry of actin and its satellite spots. Weighting the fractional areas for each mistranslated form (S_1, S_2, ..., S_m), a fraction of substituted sites, or error frequencies (p'), may be determined by

$$p'_x = \sum_{i=1}^{m} iS_i/m, \tag{1}$$

where m is the number of substituted sites and x the limiting amino acid. Because both β- and γ-actin contain nine histidine and eight asparagine residues, respectively, then

$$p'_H = \sum_{i=1}^{9} iS_i/9, \tag{2}$$

$$p'_N = \sum_{i=1}^{8} iS_i/8. \tag{3}$$

Given that the rate-limiting step during the inhibition of protein synthesis by amino acid deprivation is the time spent at the hungry codon, a linear relationship between the error frequency due to codon-anticodon mispairing and the time spent on the hungry codon may be assumed. With this assumption, the observed high level of mistranslation (p') is related to the error frequency (p) during normal protein synthesis by

$$p = p'/t'_{s,x} \tag{4}$$

where $t'_{s,x}$ is the ratio of the ribosome step time at the hungry codon x during starvation to the step time under normal conditions. Values for $t'_{s,x}$ were derived both analytically and using computer models of mRNA translation (Harley *et al.*, 1979, 1981) relating step time to the protein synthetic rate. Because for technical reasons in these experiments relative protein synthetic rate was determined by incorporation of a mixture of radioactively labeled amino acids into total protein, the relationship between relative step time of the ribosome and the relative protein synthetic rate (R') for proteins of average amino acid composition was derived. These were $t'_{s,H} = 17/R'$ for extreme histidine starvation, and $t'_{s,N} = 12/R'$ for extreme asparagine starvation. The validity of these solutions was confirmed by Monte Carlo simulations of protein synthesis using hundreds of hypothetical proteins.

When these equations were applied to actin, the normalized average error frequency for the histidine codons in actin of WI-38 and CHO cells was calculated to be 0.4×10^{-4} and 1.3×10^{-4} respectively; the error frequency for the asparagine codons in actin of CHO cells was 0.7×10^{-4}. There is less precision in the calculation of error frequencies for asparagine codons because of the presence of proteins on the basic side of native β-actin that interfere with the densitometric determinations.

The error frequency derived using the equations above varies substantially among cell types and among species (see Tables II and III and

TABLE II

Error Frequencies of Cultured Human Cells[a]

Cell type	Replicative life span[b]	Normalized error frequency × 10^{-4} ± SEM (N)	
		Early passage	Late passage
Young donors			
WI-38	55	0.6 ± 0.1 (7)	0.3 ± 0.1 (4)
MRC5	65	1.1 ± 0.2 (2)	
A2	65	0.8 ± 0.1 (10)	0.6 ± 0.1 (8)
GM37	50	1.1 ± 0.1 (3)	
Mean		0.9 ± 0.1 (22)	0.5 ± 0.1 (12)
Old donors			
JO69	50	0.6 ± 0.1 (3)	0.6 ± 0.1 (3)
JO88	48	0.7 ± 0.2 (4)	
Progeria			
P5	42	0.8 ± 0.1 (6)	ND
P18	53	0.9 ± 0.2 (4)	
Werner's syndrome			
WS2	37	0.9 ± 0.2 (2)	ND
WS4	ND	0.4 ± 0.1 (3)	
Mean of old, progeria, and Werner's syndrome donors	ND	0.7 ± 0.1 (22)	ND
Mean of diploid cells	ND	0.8 ± 0.1 (44)	

[a] Data represent the mean of the normalized error frequency ± the standard error of the mean (SEM) of N separately analyzed extracts of proteins labeled with [^{35}S]methionine during histidine starvation and corresponds to the data of Harley *et al.* (1980) modified for the slightly altered equation produced by analytical means (Harley *et al.*, 1981) relating error frequencies under amino acid starvation to those during normal protein synthesis.

[b] Maximal number of population doublings. ND, Not determined.

Section IV,A for description of cell lines). It also varies among different proteins within cell types. An impression of this can be formed in mammalian cells by observing that actin is mistranslated to a greater extent than other higher molecular weight proteins with potentially more histidine residues (Fig. 2). In *E. coli*, however, EF-G and EF-Tu can be compared directly because they have equivalent histidine contents. Nevertheless EF-G is mistranslated to a higher extent (Parker *et al.*, 1978). Similarly, the coat protein of MS2 is mistranslated to a higher level than equivalent *E. coli* proteins (Parker *et al.*, 1980). This has been interpreted as reflecting context effects within the messenger RNA, but it could also relate to codon usage between different proteins (see Grosjean *et al.*,

TABLE III

Calculated Error Frequencies for Cultured Mammalian Cells[a]

Tissue of origin	Cell strain	Error frequency $\times 10^{-4} \pm$ SEM at the histidine codons in actin	*n*
Human lung	WI-38	0.6 ± 0.1	7
	WI-38SV40	1.9 ± 0.3	10
	MRC-5	1.3 ± 0.1	2
	MRC-5SV40	1.7 ± 0.1	3
Human skin	GM37	0.3 ± 0.1	3
	GM37SV40	0.9 ± 0.1	2
	C5-IOW1	1.7 ± 0.5	2
	Adeno-C5IOW1	1.5 ± 0.7	2
Mouse embryo	3T3	1.1 ± 0.1	2
	3T3SV40	3.5 ± 0.7	2
Rat kidney	NRK	1.5 ± 0.4	2
	Ki-MSV-NRK	1.5 ± 0.2	2
Rat embryo	RE	0.5 ± 0.1	2
Rat kidney	RE-Adeno-709-C5	0.4 ± 0.1	2
Syrian hamster embryo	Nil	0.6 ± 0.1	2
	Nil-HSV	0.6 ± 0.1	2
	HA	0.5	1
	HA-Py(T)	0.5	1
Chinese hamster ovary	LR-73	1.2 ± 0.2	5
	TR-3	0.8 ± 0.2	4

[a] Data for normalized error frequencies of [^{35}S]methionine-labeled proteins following histidine starvation are taken from Pollard *et al.* (1982). In all cases the normal cell is placed above its transformed derivative. The origin of the cell strains and the protocol for histidine starvation may be found in Pollard *et al.* (1982), from which the data is reproduced with permission.

1978; Parker *et al.*, 1980). Context effects influencing the degree of misreading mRNA are well documented in many bacterial systems by the degree of supression of both nonsense (Davies *et al.*, 1966; Salser *et al.*, 1969; Bossi and Roth, 1980) and missense codons (Gallant and Foley, 1980).

The theoretical model for estimated error frequencies just described has substantial advantages over other models because it does not involve protein purification or extensive experimental manipulations to extracts and it may, in principle, be applied to any cell in culture. Because the 2-D gel technique also has great flexibility both in the isoelectric focusing and molecular weight dimensions, it may be applied to any polypeptide. However, it has a number of limitations: it cannot be applied in the organism it is restricted to certain types of error that cause a charge shift on the 2-D gel, and it requires acute starvation for amino acids to increase errors to a detectable level. This latter point requires that there is a direct deducible relationship between the observed errors and those occurring in normal protein synthesis. This assumption is subject to two major criticisms. The first is that the system must be treated as responding in a passive fashion to amino acid starvation, which is clearly not the case in stringent bacteria (Gallant and Foley, 1980) but seems more appropriate in mammalian cells where a stringent-like response has not been demonstrated (Stanners and Thompson, 1974; Pollard and Parker, 1977; Pollard *et al.*, 1980). To support this assumption of a direct relationship, the normalized error frequency for actin under various degrees of histidine starvation was found to be independent of the protein synthetic rate. This is contrary to the prediction for a system intrinsically increasing in fidelity. The second objection is that it does not take into account premature termination, either due to a type of error correction proposed by Menninger (e.g., Caplan and Menninger, 1979) or as the result of frameshift errors that also appear to be induced by amino acid starvation (Gallant and Foley, 1980). It was shown, using MS2-infected *E. coli* starved for tryptophan that a small amount of premature termination may occur at the MS2 tryptophan codons (Goldman, 1982). But these data are consistent with the model discussed in Section III, because the tryptophan codon (UGG) is related to the termination codon (UGA) in the third position. In the same series of experiments, Goldman (1982) also demonstrated that during leucine starvation premature termination was substantially lower at the leucine codons where misreading readily occurs. It is also noteworthy that Menninger *et al.* (1983) have observed that premature termination, proposed to occur as a consequence of ribosome editing, is substantially lower in relaxed bacteria than in stringent bacteria. Similarly, ribosome stalling, rather than extensive premature termination, at hungry codons, must occur in the attenuator control system of bacterial amino acid biosynthetic operons because the change in secondary structure of mRNA, associated with increased transcription rates of the operon, is activated by the presence of a ribosome stalled at the tandem hungry codons in the leader peptide (Yanofsky, 1981).

Thus, at least for histidine and asparagine codons in higher eukaryotic cells and *relA* bacteria that do not have a stringent response, premature termination appears to be a minor factor in calculation of error rates, which is consistent with the data, as stated above, that error rates calculated at 10% protein synthesis where little premature termination would occur, are very similar to those calculated at much lower levels of protein synthesis, where more premature termination might be expected (Harley *et al.*, 1981). Similarly, polysome size increases with more extreme amino acid starvation (Harley *et al.*, 1981); the converse is predicted if extensive premature termination were occurring.

This theoretical analysis provides information about error levels as well as information on the response of the protein synthesizing machinery to amino acid starvation. In particular, it demonstrates that the rate-limiting step in protein synthesis during extreme amino acid starvation is the time required to translate the largest cluster of hungry codons (i.e., the greatest number of hungry codons separated by less than the minimum interribosome distance) in the mRNA. Consequently, because of the tail-back effect of ribosomes, changes in initiation rate observed after mild amino acid starvation (Stanners and Thompson, 1974) have no effect on protein synthetic rate under extreme starvation conditions, because the initiation codons are covered by queueing ribosomes. Substantial experimental support for this model has been provided by Goldman (1982), who demonstrated that the effect of leucine starvation on MS-2 protein synthesis was to prevent effective protein synthesis initiation resulting from ribosomes stalling at an early leucine codon in MS-2 mRNA.

Because of this effect on proteins synthetic rate of clustering of hungry codons, large mRNAs having values for the expected minimum cluster size that are larger than for small mRNAs are translated at a slower rate during amino acid starvation. Analytically this effect was shown to explain quantitatively the decrease observed experimentally in the mean molecular weight of proteins synthesized following amino acid starvation (Harley *et al.*, 1981).

IV. Assessment of the Error Frequencies of Protein Synthesis under Varying Physiological Conditions

A. The Error Catastrophe Theory of Cellular Aging

Perhaps the greatest stimulus to research on mistranslation was the proposal by Orgel (1963, 1970) that errors in the translational machinery would lead to a protein-synthesizing apparatus with progressively lower fidelity yielding an exponential accumulation of aberrant proteins until a lethal level was reached and that this error feedback might be the cause of cellular aging. In the original model it was assumed for simplicity that each generation of proteins was discrete and distinguishable and that each generation of proteins produced the next generation of proteins. The positive feedback loop can therefore be expressed in the formulation of Gallant and Prothero (1980) as a recursion relationship between the error frequency e_i in a given generation and the error frequency, $e_i - 1$, in its predecessor. The simplest relationship is therefore

$$e_i = E + \alpha(e_i - 1),$$

where E is the error frequency characteristic of a "perfect" protein synthetic apparatus with the correct amino acid at every position and α is a proportionality constant relating new and old errors. For $\alpha < 1$ the recursion relationship converges to a stable value of $E/(1 - \alpha)$, as must be the case in vegetatively growing microorganisms. For $\alpha > 1$, however, there is no stable solution and the error frequency must rise without limit until the cell dies. If this were the case in eukaryotic somatic cells it could be the cause of cellular aging (Orgel, 1970).

This prediction that senescent cells exhibit a higher error frequency than young cells is one of the few testable hypotheses in cellular aging. Given the sensitivity of 2-D gels, several groups therefore have endeavored to detect higher error frequencies in aging cells with uniformly negative results. Initially, however, a substantial research effort was directed into attempting to detect aberrant proteins in aging cells, usually by indirect means such as thermolability of enzymes, intrinsic enzyme activity, or the production of aberrant virus particles. The results were equivocal to say the least, and Gershon (1979) showed that for every positive result published showing an increase in aberrant proteins, there is a negative report published showing the lack of detectable change. Similarly, a given enzyme taken from two different tissues of the same animal can show either aberrant or normal thermolability (Rothstein *et al.*, 1980) during aging. It has been suggested that the most likely explanation of such results is that proteins are posttranslationally modified in aging cells (Gershon, 1979; Rothstein, 1979). Given these data, it is clear, however, that the debate over mistranslation and aging will only be resolved by direct measurements of error frequencies.

Using the ability to detect error-containing proteins on 2-D gels, Harley *et al.* (1980) determined the error frequency in a variety of different cell types in culture. These data included early and late passage diploid fibroblasts, fibroblasts derived from young and old individuals and from patients with the so-called early aging diseases, Werner's syndrome and progeria (Goldstein and Moerman, 1975; Yatscoff *et al.*, 1978). In these experiments histidine starvation was induced by treating the cells in

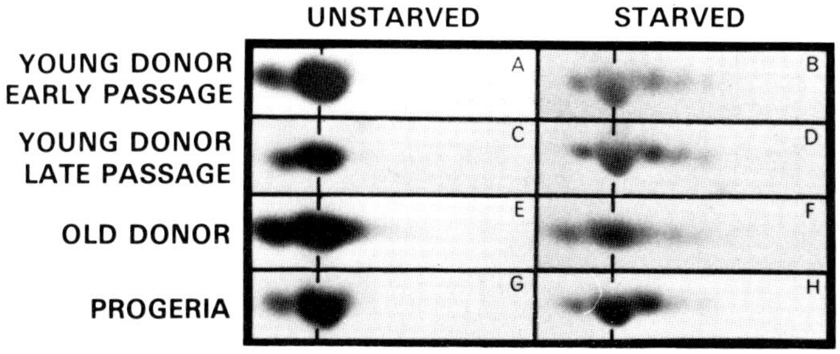

Figure 5. Autoradiograms of [^{35}S]methionine-labeled proteins of the actin regions of 2-D gels. Cells were grown in complete α-minimal essential medium supplemented with 15% FCS and lacking methionine (A, C, E, and G), or starved to equivalent levels of protein synthesis inhibition by incubation in medium lacking histidine and containing histidinol at 20 m*M* (B) or 10 m*M* (D, F and H). Cell strains include early passage (A and B) and late passage (C and D) A_2 skin fibroblasts, old donor skin fibroblasts (JO69) (E and F) and cells from a patient with the premature aging syndrome progeria P_5 (G and H). The derivation of the cell strains may be found in Harley *et al.* (1980). Lines indicate the position of β-actin. From Harley *et al.* (1980) with permission.

histidine-free medium with histidinol, a treatment that produced effective histidine starvation in all mammalian cells tested. Following the histidine starvation, [^{35}S]methionine-labeled newly synthesized proteins were displayed on 2-D gels and the degree of mistranslation compared between cell types by scanning densitometry of actin and its mistranslated forms on the resultant autoradiograms. In all cell types tested there did not appear to be an increase in error frequency in the senescent cells if compared at equivalent degrees of protein synthesis inhibition (Figs. 5 and 6). This statement is also valid if proteins other than actin are compared, but under these circumstances an error frequency cannot be determined because the relative histidine content is unknown. Using the model discussed in Section III, normalized error frequencies can be derived, and similarly no increase in error frequency could be detected in aging cells (Table II), or in cells derived from patients with the early aging diseases progeria and Werner's syndrome. If anything there appears to be a statistically significant ($p < .05$) increase in protein synthetic fidelity in late passage cells compared to early passage cells. There is a substantial variance in error frequencies between cell strains, but there was no correlation between this and their *in vitro* life span (Harley *et al.*, 1980). These data may be criticized because the experimental protocol requires acute starvation for histidine and if the relationship between step time and the normal error frequency is more complex than supposed and different between young and old cells, then the conclusions would be invalid. However, this argument would suggest that old

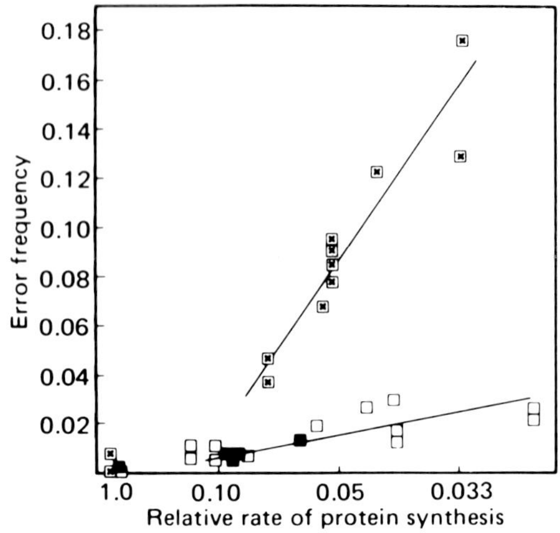

Figure 6. Error frequencies are plotted against the reciprocal protein synthetic rate for early passage WI-38 (□), late passage WI-38 (■), and WI-38 SV40 (☒); Error frequencies represent the fraction of substituted histidine sites (p'). Each point represents the error frequency determined by quantifying the densitometric scan from a single gel of actin and its substituted species following histidine starvation. From Harley *et al.* (1980) with permission.

cells were better able to modulate error levels under amino acid starvation than young cells, a notion that is unlikely because most other control systems are more relaxed in old cells. The close similarity between the derived error frequencies in pairs of young and old cells also lends strong support to the hypothesis of an equivalent relationship between step time and error level in the young and old cells. The conclusions with respect to senescence are valid, even if the relationship between the starved state and normal protein synthesis is more complicated than proposed, providing that there is a relationship and it is the same between the young and old cell pairs.

Popp *et al.* (1976) measured the *in vivo* error frequency of isoleucine substitution into hemoglobin in young and old human subjects and did not find an increase with age. However, in this study the sample of aging subjects was small. Several other groups have attempted to measure error frequencies in cell-free systems derived from early and late passage cells in culture. Although intrinsically the error levels are much higher in cell-free systems than in the cell, the majority of estimates do not show an increase in error level with age (Wojtyk and Goldstein, 1980, 1982; Goldstein *et al.*, 1980). In fact, in a recent study, a decline in *in vitro* protein synthetic error frequency was observed in mass cultures of human diploid fibroblasts as they traversed their replicative lifespan. This was interpreted as selection during growth for clones with higher protein synthetic accuracy (Wojtyk and Goldstein, 1982) and is in accord with (although greater in magnitude than) the decline seen *in vivo* by Harley *et al.* (1980). This clonal selection hypothesis proposes that loss of cells results from a growth disadvantage (not an error catastrophe) of cells with elevated error frequencies.

Total proteins synthesized in young and old cells have also been analyzed on 2-D gels for charge heterogeneity, which if present in aging cells, would be indicative of an increased error level. In all cases reported to date there has not been a significant increase in charge heterogeneity detected (Wilson *et al.*, 1978; Engelhardt *et al.*, 1979; Harley *et al.*, 1980; Parker *et al.*, 1981; see chapter by Celis *et al.*). In one study (Parker *et al.*, 1981) the proteins synthesized by aging *Drosophila melanogaster* were compared with those synthesized by juvenile flies and shown not to exhibit increased charge heterogeneity. In this case a quantitative estimate of the level of detection of errors was determined as 4 in 10^4. Consequently if aging does involve an error catastrophe, then lethal levels of protein synthetic errors would have to be less than 4 mistakes in 10,000. There would, however, be some errors compensating for charge shifts on the 2-D gel, thus the true error frequency may be higher; nevertheless in bacteria error levels of 20 times the normal level of 1 in 10^4 can be induced without loss in cell viability (Edelmann and Gallant, 1977b; Gallant and Palmer, 1979). Similarly, cultured human or *Drosophila* cells can tolerate substantial incorporation of amino acid analogs into protein without a significant reduction in lifespan (Ryan *et al.*, 1974; Shmookler-Reis, 1976). Thus if errors can be detected on 2-D gels at about 4 in 10^4 and the normal error frequency is about 1 in 10^4, then a fourfold increase in error frequency would seem to be insufficient to tip the cell into an error catastrophe unless the cells were poised on the brink of the transition between a stable state and an unstable state. Such a prediction has been made by Kirkwood and Holliday (1975, 1979) who extended the Hoffman (1974) formulation of Orgel's error catastrophe model to show that cells could exist in a metastable state wherein a stable error frequency is usually assured but statistical variation in the error frequency could push the system into the unstable mode, resulting in error catastrophe. If this metastable state is characteristic of somatic cells then senescence is due to a random drift of cells into an error catastrophe. Both theoretical (Gallant and Prothero, 1980) and empirical (Harley *et al.*, 1980; Parker *et al.*, 1981) estimates, however, argue against this model, because error levels in somatic cells are not very different from those observed in bacteria, whose protein-synthesizing machinery is well removed from the hypothetical metastable level (Edelmann and Gallant, 1977b; Gallant and Palmer, 1979), and during aging of somatic cells the error level is not sufficiently elevated to push the translational apparatus into error catastrophe (Harley *et al.*, 1980; Parker *et al.*, 1981). It was also observed that SV40 tansformation of mortal diploid fibroblasts resulted in an increased error level (Pollard *et al.*, 1982) but nevertheless the transformed cells were immortal.

Therefore the data, particularly those derived from 2-D gel analysis, strongly argue against the error catastrophe theory of cellular aging. This refutation of the error catastrophe hypothesis still leaves open the question of the cause of aging. It is clearly not appropriate to go into any detail in this chapter, but it is worth pointing out that many aspects of aging appear to reflect a random intrinsic process. This randomness was one of the reasons that the error catastrophe theory was so appealing, and any theory to explain aging needs to take this apparent randomness into account. I feel that the best view is that aging is due to a random but increasing accumulation of genetic damage, particularly of the type that produces chromosomal rearrangements. This usually results in aberrant cellular functions and cell death but occasionally results in neoplastic transformation. Evidence for the accumulation of genetic damage during aging has recently been reviewed by Gensler and Bernstein (1981).

An alternative error accumulation model may be considered for neoplastically transformed cells. Thus it may be proposed that during transformation protein synthetic fidelity is reduced, which results in an increased synthesis of aberrant proteins, including enzymes involved in DNA replication and repair. These aberrant proteins could produce both an increased mutation rate and a decreased efficiency of regulatory controls. Mutant cells unable to respond to environmental controls and with an increasing ability to proliferate would be selected, a phenomenon that could be the basis of tumor progression. This hypothesis was tested by comparing the error frequencies of a variety of transformed-normal cell pairs using the 2-D gel assay, either by comparing the error level at comparable degrees of protein synthesis inhibition or by normalizing the error frequency to normal protein synthesis using the model of Harley *et al.* (1981). It was found that transformation is not generally associated with an increased error level but that SV40 resulted in an increased error level in all cell lines tested (Table III and Fig. 7) (Pollard *et al.*, 1982). The cause of the SV40 effect is unknown, but it may explain the reported mutagenic action of this virus in mammalian cells (Geissler *et al.*, 1980). The use of *ts* mutants of SV40 may help to identify components of mammalian cells involved in protein synthetic fidelity.

It is interesting to note that there is a wide variation of error levels

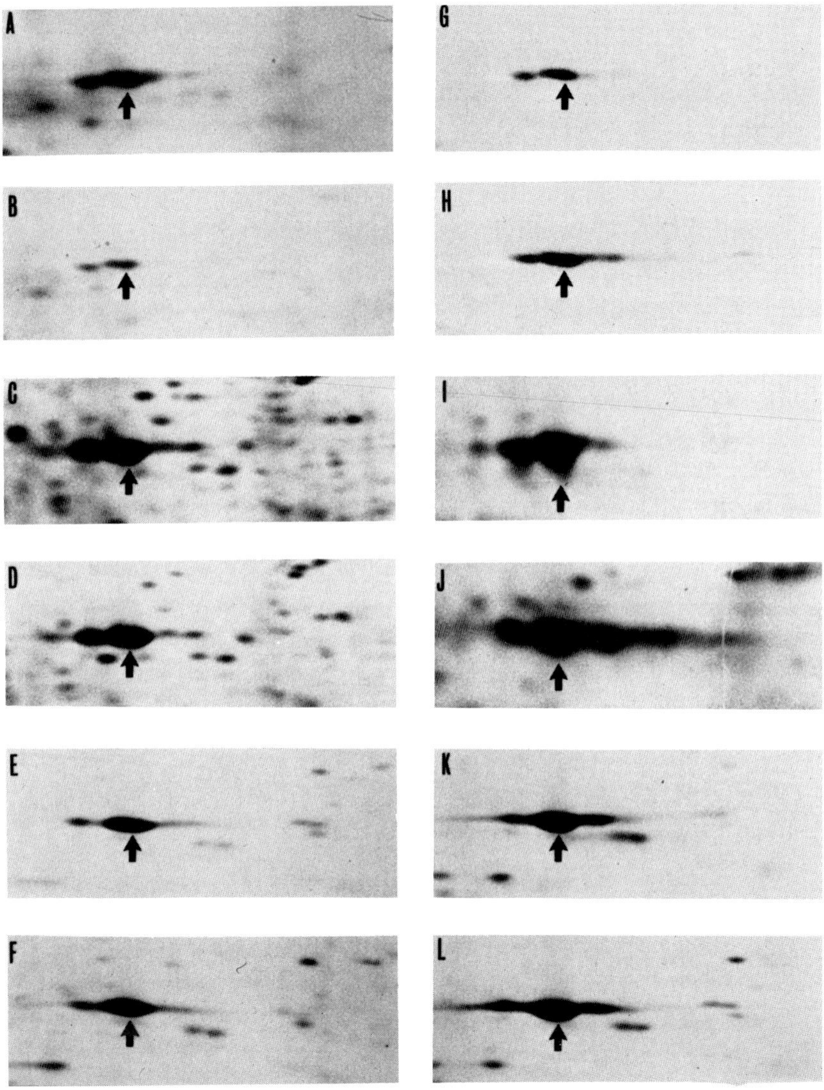

Figure 7. A comparison of the degree of mistranslation in normal and transformed cell strains. The figure shows autoradiograms of the actin region of [^{35}S]methionine-labeled proteins displayed on 2-D gels. Cell strains are described in Pollard *et al.* (1982) and are GM37 (normal A,G), SV40-GM37 (transformed B, H), 3T3 (normal C, I), SV40-3T3 (transformed D, J), Nil (normal E, K_1) and HSV-Nil (transformed F, L). The left panel represents the patterns of proteins synthesized by exponentially growing control cells, and the right panels show the proteins synthesized following histidine starvation to approximately equivalent degrees of protein synthesis inhibition for each cell pair. The arrow marks the position of β-actin. From Pollard *et al.* (1982) with permission.

among cell lines (Tables II and III) and also within primary cultures of human cells (Harley *et al.*, 1980). The explanation for this is unclear, but similar variance is observed using *in vitro* protein synthesis assays derived from sibling clones of human fibroblasts (Wojtyk and Goldstein, 1982) or from different human subjects (Goldstein *et al.*, 1980). This is probably a reflection of biochemical individuality and clonal selection, because it does not appear to exhibit any systematic correlation with any physiological state.

B. *The Stringent Control System in Bacteria and its Effect on Protein Synthetic Accuracy*

Bacteria starved for amino acids undergo a profound and rapid change in many cellular functions known as the stringent response (Cashel and Gallant, 1974). In stringent bacteria ($relA^+$) this response is mediated by two unusual guanosine nucleotides, guanosine 5′-diphosphate-3′-diphosphate (ppGpp) and guanosine 5′-triphosphate-3′-diphosphate (pppGpp). Relaxed mutants (relA) unable to respond to amino acid starvation fail to show this complex pleiotropic response and also fail to synthesize to any appreciable level either ppGpp or pppGpp. Hall and Gallant suggested (1972) that one aspect of the stringent response was to modulate protein synthetic accuracy because in a cell unable to regulate its fidelity the inevitable consequence of amino acid deprivation would be a massive increase in errors that would not be tolerated by the cell. A substantial body of evidence including data derived from 2-D gel analysis has now been accumulated to support this contention. In his original paper showing that protein synthetic errors were detectable on 2-D gels following amino acid starvation of relaxed bacteria, O'Farrell (1978) was unable to detect errors in an isogenic stringent strain. Upon more thorough analysis, however, errors following amino acid starvation of stringent strains were detected on 2-D gels, but these were at a lower level than in the corresponding relaxed strain (Parker and Friesen, 1980) (Fig. 8). These data, as well as data on the increased phenotypic supression of nonsense and frameshift mutants in relaxed, compared to stringent, bacteria, have given strong support to the hypothesis that one of the functions of the stringent response is the modulation of translation fidelity (Edelman and Gallant, 1977a; Gallant, 1979; Gallant and Foley, 1980).

Relevant to this topic is the "kinetic amplification" model of Ninio (1975) derived to explain the specificity of substrate selection. Ninio's

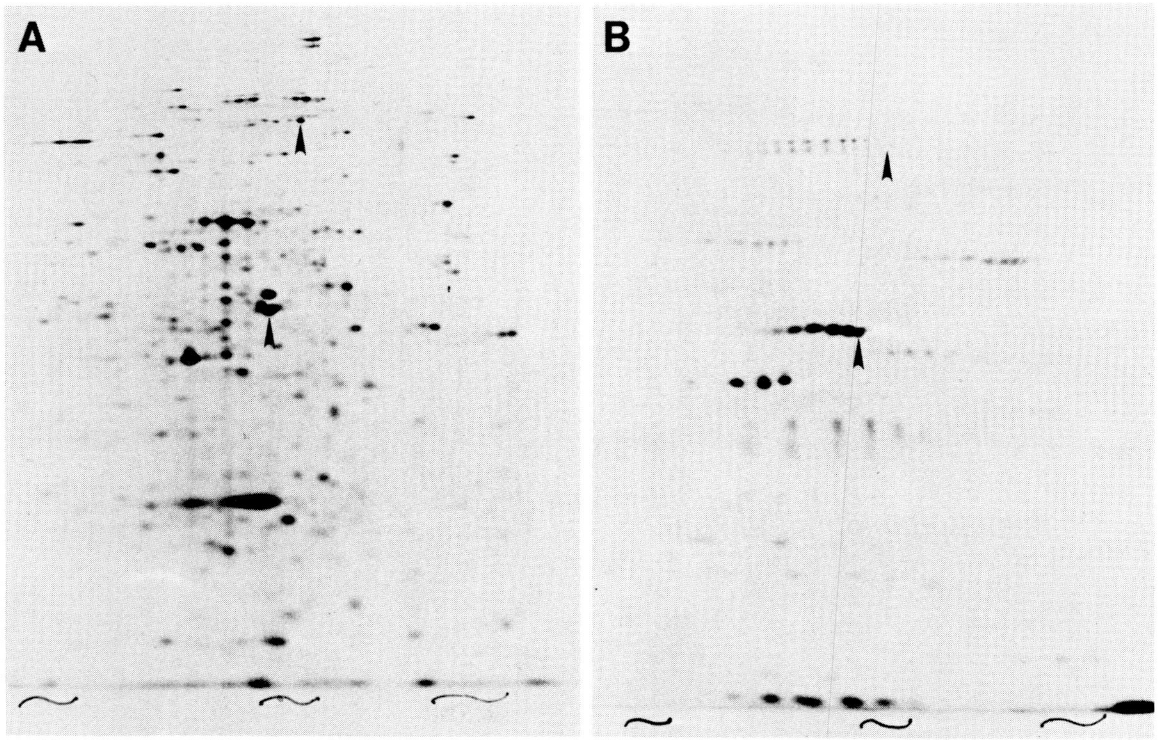

Figure 8. Autoradiograms of 2-D gels of [^{35}S]methionine-labeled proteins synthesized after asparagine starvation in stringent and relaxed *E. coli.* Asparagine auxotrophic stringent strain (A) JK120 (rel A^+) and relaxed isogenic strain (B) JK100 (rel A) were starved for asparagine by removal of asparagine from the medium and labeled 15–20 min after starvation with [^{35}S]methionine. The pattern of proteins displayed on the 2-D gel is different in the relaxed strain than in the stringent strain, as expected (Parker and Friesen, 1980), but the degree of mistranslation of the remaining proteins synthesized in the relaxed strain is clearly greater than in the stringent strain. The upper arrow in each panel represents EF-G, and the lower arrow represents EF-Tu. Reproduced with permission from Parker and Friesen (1980).

model in essence states that when correct and incorrect substrates compete for a site they both bind reversibly but the incorrect one dissociates more rapidly. This binding is followed by an irreversible catalytic step that alters the state of the complex, removing it from equilibrium with the free substrate. Selectivity therefore depends upon the relationship between the catalytic step and the rate of dissociation of the two substrates. Thus if the catalytic step is fast in comparison with the dissociation of the incorrect substrate, then the latter will frequently be trapped in the product. However, if there is more time for the dissociation of the incorrect substrate, the system's specificity is improved. Ninio proposed that this trade-off between speed and accuracy may be modulated by a secondary element. Gallant (Gallant, 1979; Gallant and Foley, 1980) suggested that during the stringent response this secondary element may be ppGpp acting by inhibiting GTP hydrolysis in reactions involving EF-Tu and EF-G. These workers presented evidence to support this view by showing that *in vitro* ppGpp increased protein synthetic fidelity (Gallant and Foley, 1980; Wagner *et al.*, 1982).

In conclusion, therefore, mistranslation induced by amino acid starvation is best explained by models of single substrate limitation (Harley *et al.*, 1981). Thus deprivation of a single amino acid results in misincorporation because the incorrect amino acyl tRNA is present for increased times at the hungry codon, resulting in the formation of a peptide bond with the incorrect amino acid. Because in phase translocation also appears to involve correct codon–anticodon recognition, misrecognition also occasionally shifts the reading frame, resulting in phenotypic suppression of frameshift mutants, particularly in relA mutants (Gallant and Foley, 1980; Kurland and Gallant, 1982). $RelA^+$ bacteria modulate translational accuracy by altering the rate of peptide bond formation at key steps, thus trading off speed against accuracy and consequently reducing the degree of missense misreading and limiting phenotypic suppression.

V. Other Uses of the Two-Dimensional Gel Assay

The 2-D gel assay provides a rapid and effective means both for assessing mistranslation *in vivo* (or presumably *in vitro*), and for assaying the rate of turnover and fate of defective protein *in vivo*.

Measurements in both mammalian and bacterial cells indicate that some aberrant proteins induced by amino acid starvation are not preferentially degraded (Harley *et al.*, 1981; Parker, 1981). It is important to note, however, that some proteins, particularly in mammalian cells, exhibit very little or no mistranslation. This is normally interpreted as the consequence of context effects of codons within the mRNA (see Section III) coupled with an increased fidelity of mammalian ribosomes compared to bacterial ribosomes (Burrans and Kurtz, 1977), but the alternative explanation cannot be excluded that the mistranslated form of these proteins are very rapidly degraded and thus avoid detection. If proteins involved in information transfer were of this type, this might be one method that mammalian cells exploit to avoid an error catastrophe. It is also possible that a subfraction of mistranslated forms may undergo very rapid degradation and never contribute to the 2-D gel patterns. If such a class exists, then error frequency measurements by the 2-D gel method would underestimate the true error frequency. Nevertheless, the error-containing forms for several proteins have been analyzed and are as stable as the native form, which is contrary to data for other types of error-containing proteins (Goldberg and St. John, 1976). Parker (1981) has provided an explanation for this discrepancy by suggesting that third position errors are the most common ones, even during normal growth, but that despite the fact that the substitution causes a different isoelectric point, these errors are least likely to cause damage to the cell because the replacement amino acid is in many cases similar to the correct amino acid in physicochemical properties (Wolfenden *et al.*, 1979). In support of this view, Parker (1981) showed that error-containing proteins are found in fully assembled ribosomes, suggesting that the aberrant proteins are not sufficiently different to be excluded from such structures and may not therefore be preferentially recognized by the degradative machinery. Experiments of this type should play a part in defining the constraints on the degradative machinery both in prokaryotes and eukaryotes.

The 2-D gel assay also provides a means of determining cellular components or external factors that affect protein synthetic fidelity. Thus the presence of the rpsL mutation, which confers resistance to the antibiotic streptomycin, also increases ribosomal fidelity (Gorini, 1974) and this effect may be detected on 2-D gels (Parker and Friesen, 1980). Similarly by comparing error levels on 2-D gel mutations in pseudouridylate synthetase (hisT) were shown to inhibit mistranslation following histidine but not asparagine starvation (Parker, 1982). This was interpreted as the consequence of the replacement of pseudouridine in the anticodon of the glutamyl-tRNA leading to a less stable binding with the histidine codons and consequently as increased translational fidelity. Interestingly, Parker *et al.* (1978) reported a high degree of mistranslation in the stringent *E. coli* strain CP78. This high level of errors was almost certainly due to the fortuitous presence of an additional ribosomal mutation in this strain (Parker and Friesen, 1980). Such experiments indicate the potential power of this assay for detecting ribosomal mutants that affect fidelity. We have investigated the emetine resistant mutants of CHO cells (Gupta and Siminovitch, 1977), but to date we have been unable to detect any effects on translational accuracy (L. M. Wong and J. W. Pollard, unpublished).

VI. Two out of Three Reading

This chapter is not the place to discuss mistranslation in general, so I will restrict myself to a few comments. It appears that the major source of normal errors is ribosomal decoding and not misacylation of tRNA (Yarus, 1979; Kurland and Gallant, 1982). The problem of accuracy is therefore to explain how errors occur at only about a rate of 1 in 10^4, because the degree of codon–anticodon discrimination is at best only 1 in 100 and yet this discrimination is the sole determinant of the amino acid sequence of the peptide chain. The increase in codon–anticodon discrimination is due to ribosomal functions (Kurland, 1980; Kurland and Gallant, 1982) and several models have emerged to explain this ribosomal effect coherently. These include the "kinetic amplification" models described in Section IV,B (Ninio, 1975; Kurland, 1980) and proofreading models (Hopfield and Yamane, 1980); I favor the former. However, I think the data are not entirely persuasive for either model, with a little ingenuity compromise models may be constructed (see Kurland, 1980).

Errors appear to occur at all positions of the codon (Davies *et al.*, 1965; Gorini, 1974; Parker *et al.*, 1978), but for the first time, third position errors have been unequivocably demonstrated using the 2-D gels, and at least during amino acid starvation these errors predominate. This implies that tRNAs are able to read the third position with a lower degree of specificity than predicted by the wobble hypothesis (Crick, 1966). Lagerkvist (1978) has provided evidence that within a codon group (a set of four codons differing only at the third position) that codes for a single amino acid all four codons may be read by a single tRNA species. He termed this *two-out-of-three* reading. It is clear that in codon groups that code for more than one amino acid, two-out-of-three reading would result in a high level of mistranslation and could not be tolerated. Evidence suggests that in these groups third position specificity is greater than in those which code for a single amino acid (Goldman *et al.*, 1979; Elias *et al.*, 1979). During amino acid starvation for the appropriate amino acid, however, these errors are amplified to a readily detected level. Nevertheless, even during normal protein synthesis, these errors, at least for the histidine and asparagine codons, appear to be among the most common (Harley *et al.*, 1981; Parker, 1981), which is in accordance with the predictions made by Woese (1967) during the period of the elucidation of the code. Error rates, however, do appear to vary substantially between different codons such that first position errors may be more frequent for certain codons (obviously so for codon groups) and even among third position errors not all types are equally likely, apparently following a strict order with substantial restrictions being placed on misreading codons ending in purines. Thus the lysine to asparagine misincorporation is much less frequent than the reverse and arginine starvation leads to a high level of mistakes in bacteria, whereas serine starvation does not (Parker, 1981). These restrictions appear to follow the strength of codon–anticodon binding as predicted from studies using tRNA anticodon–anticodon binding (Grosjean *et al.*, 1978), presumably because these binding constants determine the length of stay of the incorrect aminoacyl tRNA at hungry codons.

The 2-D gel assay has been shown to detect other than third position errors produced by treating *E. coli* with streptomycin (Parker *et al.*, 1980). This observation has allowed an estimate for the total level of misincorporation per codon of about 2 in 10^{-4} (Ellis and Gallant, 1982), which is in good accord with previous estimates (Loftfield and Vanderjagt, 1972; Edelman and Gallant, 1977a; Harley *et al.*, 1981).

In conclusion, therefore, with judicious selection of either mutants or drugs coupled with the detection of errors on 2-D gels, a greater understanding of ribosomal decoding has been obtained. In the future, given the versatility of the 2-D gel technique, the increasing availability of sequence data, and the ability to manipulate nucleotide sequences, the methods described in this paper should profoundly increase our understanding of the constraints placed upon codon–anticodon interactions.

Acknowledgments

Both the work and ideas presented in this paper could not have occurred without my collaborators John Chamberlain, Jim Friesen, Sam Goldstein, Calvin Harley, Jack Parker, and Cliff Stanners. I would also like to thank N. Dance, C. Harley, and C. Thurston for their critical comments on the manuscript. Many of the experiments were performed while I was a National Cancer Institute of Canada Fellow and were carried out under grants C. P. Stanners, from the MRC of Canada (MT1877), NIH, USA (NOI-CP-4331), and the National Cancer Institute of Canada Institute of Canada; to S. Goldstein from the MRC of Canada (MT3515) and to J. Friesen, from the National Research Council and the National Cancer Institute of Canada. I would also like to thank the Wellcome Foundation and the Central Research Fund of the University of London for grant support of my work in England.

References

Bossi, L., and Roth, J. R. (1980). The influence of codon context on genetic code translation. *Nature (London)* **286**, 123–127.

Burrans, L., and Kurtz, D. I. (1977). Should animal ribosomes be expected to participate in an "error catastrophe"? *J. Mol. Biol.* **112**, 349–352.

Caplan, A. B., and Menninger, J. R. (1979). Tests of the ribosomal editing hypothesis. Amino-acid starvation differentially enhances the dissociation of peptidyl-transfer RNA from the ribosome. *J. Mol. Biol.* **134**, 621–637.

Cashel, M., and Gallant, J. (1974). Cellular regulation of guanosine tetra-phosphate and guanosine penta-phosphate. *In* "Ribosomes" (M. Nomura, A. Tissières, and P. Lengyel, eds.), pp. 733–745. Cold Spring Harbor Lab., Cold Spring Harbor, New York.

Crick, F. H. C. (1966). Codon-anticodon pairing: The wobble hypothesis. *J. Mol. Biol.* **19**, 548–555.

Davies, J., Gorini, L., and Davis, B. D. (1965). Misreading of RNA codewords induced by aminoglycoside antibiotics. *Mol. Pharmacol.* **1**, 93–106.

Davies, J., Jones, D. S., and Khorana, H. G. (1966). A further study of misreading of codons induced by streptomycin and neomycin using ribopolynucleotides containing two nucleotides in alternating sequence as templates. *J. Mol. Biol.* **18**, 48–57.

Edelmann, P., and Gallant, J. (1977a). Mistranslation in *E. coli. Cell* **10**, 131–137.

Edelmann, P., and Gallant, J. (1977b). On the translational error theory of aging. *Proc. Natl. Acad. Sci. USA* **74**, 3396–3398.

Elias, P., Lustig, F., Axberg, T., Akesson, B., and Lagerkvist, U. (1979). Reading of the lysine codons in MS2 coat protein cistron during protein synthesis in vitro. *FEBS Lett.* **98**, 145–151.

Ellis, N., and Gallant, J. (1982). An estimate of global error frequency in translation. *Mol. Gen. Genet.* **188**, 169–172.

Engelhardt, D. L., Lee, G. T. Y., and Moley, J. F. (1979). Patterns of peptide synthesis in senescent and pre-senescent human fibroblasts. *J. Cell. Physiol.* **98**, 193–198.

Gallant, J. A. (1979). Stringent control in *E. coli. Annu. Rev. Genet.* **13**, 393–415.

Gallant, J., and Foley, D. (1980). On the causes and prevention of mistranslation. *In* "Ribosomes: Structure, Function and Genetics" (G. Chambliss, G. R. Craven, J. Davies, K. Davis, L. Kahan, and N. Nomura, eds.), pp. 615–638. Univ. Park Press, Baltimore, Maryland.

Gallant, J., and Palmer, L. (1979). Error propagation in viable cells. *Mech. Ageing Dev.* **10**, 27–38.

Gallant, J. A., and Prothero, J. (1980). Testing models of error propagation. *J. Theor. Biol.* **83**, 561–578.

Geissler, E., Schernick, S., Theile, M., Herold, H. J., Staneczek, W., Zimmermann, W.,

Krause, H., Prokoph, H., Vogel, F., and Platzer, H. (1980). Studies on the mutagenic and transforming activities of SV-40 like viruses. *In* "Leukaemias, Lymphomas and Papillomas: Comparative Aspects" (P. A. Bachmann, ed.), pp. 43–55. Taylor & Francis, London.

Gensler, H. L., and Bernstein, H. (1981). DNA damage as the primary cause of aging. *Q. Rev. Biol.* **56**, 279–303.

Gershon, D. (1979). Current status of age-altered enzymes: Alternative mechanisms. *Mech. Ageing Dev.* **9**, 197–202.

Goldberg, A. L., and St. John, A. C. (1976). Intracellular degradation in mammalian and bacterial cells. *Annu. Rev. Biochem.* **45**, 747–803.

Goldman, E. (1982). Effect of rate-limiting elongation on bacteriophage MS2 RNA-directed protein synthesis in extracts of *Eschericia coli*. *J. Mol. Biol.* **158**, 619–636.

Goldman, E., Holmes, W. M., and Hatfield, G. W. (1979). Specificity of codon recognition by *Escherichia coli* $tRNA^{Leu}$ isoaccepting species determined by protein synthesis *in vitro* directed by phage RNA. *J. Mol. Biol.* **129**, 567–585.

Goldstein, S., and Moerman, E. (1975). Heat-labile enzymes in cultured progeria fibroblasts. *N. Engl. J. Med.* **292**, 1305–1309.

Goldstein, S., Wojtyk, R. I., Harley, C. B., Pollard, J. W., Chamberlain, J. W., and Stanners, C. P. (1980). Fidelity of protein synthesis in aging human fibroblasts. *In* "Aging: Its Chemistry" (A. A. Dietz, ed.), pp. 248–258. Amer. Assoc. Clin. Chem., Washington, D.C.

Gorini, L. (1974). Streptomycin and misreading of the genetic code. *In* "Ribosomes" (M. Nomura, A. Tissières, and P. Lengyel, eds.), pp. 791–803. Cold Springs Harbor Lab., Cold Spring Harbor, New York.

Grosjean, H. J., de Henau, S., and Crothers, D. M. (1978). On the physical basis for ambiguity in genetic coding interaction. *Proc. Natl. Acad. Sci. USA* **75**, 610–614.

Gupta, R. S., and Siminovitch, L. (1977). The molecular basis of emetine resistance in Chinese hamster ovary cells: Alteration in the 40S ribosomal sub-unit. *Cell* **10**, 61–66.

Hall, B., and Gallant, J. (1972). Defective translation in RC^- cells. *Nature New Biol.* **237**, 131–135.

Hansen, B. S., Vaughan, M. H., and Wang, L. (1972). Reversible inhibition by histidinol of protein synthesis in human cells at the activation of histidine. *J. Biol. Chem.* **247**, 3854–3857.

Harley, C. B. (1979). "Ageing, Protein Synthesis and Mistranslation in Cultured Human Cells." Ph.D. Thesis, McMaster Univ., Hamilton, Canada.

Harley, C. B., Goldstein, S., Pollard, J. W., and Stanners, C. P. (1979). A novel method of determining error rates in protein synthesis. *Fed. Proc. Fed. Am. Soc. Exp. Biol.* **38**, 328.

Harley, C. B., Pollard, J. W., Chamberlain, J. W., Stanners, C. P., and Goldstein, S. (1980). Protein synthetic errors do not increase during aging of cultured human fibroblasts. *Proc. Natl. Acad. Sci. USA* **77**, 1885–1889.

Harley, C. B., Pollard, J. W., Stanners, C. P., and Goldstein, S. (1981). Model for messenger RNA translation during amino acid starvation applied to the calculation of protein synthetic error rates. *J. Biol. Chem.* **256**, 10786–10794.

Hoffman, G. W. (1974). On the origin of the genetic code and the stability of the translation apparatus. *J. Mol. Biol.* **86**, 349–362.

Hopfield, J. J., and Yamane, T. (1980). The fidelity of protein synthesis. *In* "Ribosomes, Structure, Function and Genetics" (G. Chambliss, G. R. Craven, J. Davies, K. Davis, L. Kahan, and M. Nomura, eds.), pp. 585–596. Univ. Park Press, Baltimore, Maryland.

Kirkwood, T. B., and Holliday, R. (1975). The stability of the translation apparatus. *J. Mol. Biol.* **97**, 257–265.

Kirkwood, T. B., and Holliday, R. (1979). Evolution of aging and longevity. *Proc. Soc. London Ser. B* **205**, 531–546.

Kurland, C. G. (1980). On the accuracy of elongation. *In* "Ribosomes, Structure, Function and Genetics" (G. Chambliss, G. R. Craven, J. Davies, K. Davis, L. Kahan, and M. Nomura, eds.), pp. 597–614. Univ. Park Press, Baltimore, Maryland.

Kurland, C. G., and Gallant, J. (1983). The secret life of the ribosome. *In* "Accuracy in Biology" (D. Galas, ed.), Dekker, New York, in press.

Lagerkvist, U. (1978). "Two out of three:" An alternative method for codon reading. *Proc. Natl. Acad. Sci. USA* **75**, 1759–1762.

Loftfield, R. B. (1963). The frequency of errors in protein biosynthesis. *Biochem. J.* **89**, 82–92.

Loftfield, R. B., and Vanderjagt, D. (1972). The frequency of errors in protein synthesis. *Biochem. J.* **128**, 1353–1356.

Menninger, J. R., Caplan, A. B., Gingrich, P. K. E., and Atherly, A. G. (1983). Tests of the ribosome editor hypothesis. II. Relaxed (relA) and Stringent ($relA^+$) *E. coli* differ in rates of dissociation of peptidyl-tRNA from ribosomes. *Mol. Gen. Genet.* **190**, 215–221.

Ninio, J. (1975). Kinetic amplification of enzyme discrimination. *Biochimie* **57**, 487–595.

O'Farrell, P. H. (1978). The suppression of defective translation by ppGpp and its role in the stringent response. *Cell* **14**, 545–557.

Orgel, L. E. (1963). The maintenance of the accuracy of protein synthesis and its relevance to ageing. *Proc. Natl. Acad. Sci. USA* **49**, 517–521.

Orgel, L. E. (1970). The maintenance of the accuracy of protein synthesis and its relevance to ageing. A correction. *Proc. Natl. Acad. Sci. USA* **67**, 1476.

Parker, J. (1981). Mistranslated protein in *Escherichia coli*. *J. Biol. Chem.* **256**, 9770–9773.

Parker, J. (1982). Specific mistranslation in hisT mutants of *Escherichia coli*. *Mol. Gen. Genet.* **187**, 405–409.

Parker, J., and Friesen, J. D. (1980). "Two out of three" codon reading leading to mistranslation *in vivo*. *Mol. Gen. Genet.* **177**, 439–446.

Parker, J., Pollard, J. W., Friesen, J. D., and Stanners, C. P. (1978). Stuttering: High-level mistranslation in animal and bacterial cells. *Proc. Natl. Acad. Sci. USA* **75**, 1091–1095.

Parker, J., Johnston, T. C., and Borgia, P. T. (1980). Mistranslation in cells infected with the bacteriophage MS2: Direct evidence of Lys for Asn substitution. *Mol. Gen. Genet.* **180**, 275–281.

Parker, J., Flanagan, J., Murphy, J., and Gallant, J. (1981). On the accuracy of protein synthesis in *Drosophila melanogaster*. *Mech. Ageing Dev.* **16**, 127–139.

Pollard, J. W., and Parker, J. (1977). Guanosine tetra- and penta-phosphate synthesis by bacteria stringent factor and eukaryotic ribosomes. *Nature (London)* **267**, 371–373.

Pollard, J. W., Lam, T., and Stanners, C. P. (1980). Mammalian cells do not have a stringent response. *J. Cell. Physiol.* **105**, 313–325.

Pollard, J. W., Harley, C. B., Chamberlain, J. W., Goldstein, S., and Stanners, C. P. (1982). Is transformation associated with an increased error frequency? *J. Biol. Chem.* **257**, 5977–5979.

Popp, R. A., Bailiff, E. G., Hirsch, G. P., and Conrad, R. A. (1976). Errors in human haemoglobin as a function of age. *Interdiscip. Top. Gerontol.* **9**, 209–218.

Rothstein, M. (1979). The formation of altered enzymes in ageing mammals. *Mech. Ageing Dev.* **9**, 197–202.

Rothstein, M., Coppens, M., and Sharma, H. K. (1980). Effect of aging on enolase from rat muscle, liver and heart. *Biochem. Biophys. Acta* **614**, 591–600.

Ryan, J. M., Duda, G., and Cristofalo, V. J. (1974). Error accumulation and aging in human diploid cells. *J. Gerontol.* **29**, 616–662.

Salser, W., Fluck, M., and Epstein, R. (1969). The influence of the reading context upon the suppression of nonsense codons, III. *Cold Spring Harbor Symp. Quant. Biol.* **34**, 513–520.

Shmookler-Reis, R. J. (1976). Enzyme fidelity and metazoan ageing. *Interdiscip. Top. Gerontol.* **10**, 11–23.

Stanners, C. P., and Thompson, L. H. (1974). Studies on a mammalian cell mutant with a temperature-sensitive leucyl-tRNA synthetase. *In* "Control of Proliferation in Animal Cells" (B. Clarkson and R. Baserga, eds.), pp. 191–203. Cold Spring Harbor Lab., Cold Spring Harbor, New York.

Wagner, E., Gerhart, H., Ehrenberg, M., and Kurland, C. G. (1982). Kinetic suppression of translational errors by guanosine 5′-diphosphate 3′-diphosphate or guanosine 5′-triphosphate 3′-diphosphate. *Mol. Gen. Genet.* **185**, 269–274.

Wilson, D. L., Hall, M. E., and Stone, G. C. (1978). Test of some aging hypotheses using 2-dimensional protein mapping. *Gerontology (Basel)* **24**, 426–433.

Woese, C. R. (1967). The present status of the genetic code. *Prog. Nucleic Acids Res. Mol. Biol.* **7**, 107–172.

Wojtyk, R. I., and Goldstein, S. (1980). Fidelity of protein synthesis does not decline during aging of cultured human fibroblasts. *J. Cell. Physiol.* **103**, 299–303.

Wojtyk, R. I., and Goldstein, S. (1982). Clonal selection in cultured human fibroblasts: Role of protein synthetic errors. *J. Cell Biol.* **95**, 704–710.

Wolfenden, R. V., Cullis, P. M., and Southgate, C. C. F. (1979). Water, protein folding and the genetic code. *Science (Washington, D.C.)* **206**, 575–577.

Yanofsky, C. (1981). Attenuation in the control of expression of bacterial operons. *Nature (London)* **289**, 751–758.

Yarus, M. (1979). The accuracy of translation. *Prog. Nucleic Acids Res. Mol. Biol.* **23**, 195–225.

Yatscoff, R. W., Goldstein, S., and Freeman, K. B. (1978). Conservation of genes coding for proteins synthesised in human mitrochondria. *Somatic Cell Genet.* **4**, 633–645.

CHAPTER 12

Application of Two-Dimensional Gel Electrophoresis in Studies of Gene Expression during Early Plant Development

Z. R. SUNG
Department of Plant Pathology and Department of Genetics
University of California
Berkeley, California

I.	Introduction	397
II.	Gene Expression during Early Embryogenesis	398
III.	Somatic Embryogenesis	399
IV.	Gene Expression during Somatic Embryogenesis	399
V.	Coordinate Regulation of Sets of Biochemical Phenotypes	403
VI.	Gene Expression in Temperature-Sensitive Cell Lines Impaired in Embryogeny	405
VII.	Gene Expression during Late Embryogeny	406
VIII.	Gene Expression in Cotton Cotyledons	406
IX.	Gene Expression in Germinating Seeds	407
X.	Conclusion	411
	References	412

I. Introduction

Two-dimensional polyacrylamide gel electrophoresis (2-D PAGE) has been employed in several plant systems to study gene expression. Several different types of 2-D PAGE have been employed: (1) isoelectric focusing (IEF) and sodium dodecyl sulfate (SDS) gels (O'Farrell and O'Farrell, 1977), (2) acetic acid–urea and Triton gels, in which the mobility of the protein depends on its hydrophobic properties in addition

TWO-DIMENSIONAL GEL
ELECTROPHORESIS
OF PROTEINS

ISBN 0-12-164720-X

to the molecular size and charge (Panyim and Chalkley, 1969), and (3) nondenaturing(native)–SDS 2-D gels, in which the first dimension separates proteins by molecular weight and by charge (Sachs *et al.*, 1980).

Modifications have been introduced to the original IEF–SDS gels. In addition to variations in total volt-hr, acrylamide concentration, etc, Baulcombe *et al.* (1980) have made the following modifications: (1) reversing the polarity of the system such that samples are loaded on the acidic end of the IEF gel and (2) using sulfuric acid and ethanolamine as the upper and lower buffers, respectively (Zurfluh and Guilfoyle, 1982a).

II. Gene Expression during Early Embryogenesis

Embryogenesis begins with the division of the zygote and the generation of two daughter cells. The cell situated on top develops into the embryo proper, whereas the one below develops into a suspensor tissue that serves as a support for the embryo. It is believed that the suspensor not only provides physical support but also nutrition, energy, and growth hormones, because of concentration differences of malate, $NAPH^+$ (D. P. Malik, personal communication), gibberelic acid (Cionini *et al.*, 1976), and auxins (Przybyllok and Nagl, 1977) found in the two tissues. Limited information is available on differential gene expression during early embryogenesis because the zygote and the embryos are small cells embedded in the maternal tissue, that is, the ovary. In order to obtain large amounts of material for biochemical work, one needs to dissect the embryos out of the ovary individually.

The biochemistry of early embryogenesis can be better studied *in vitro* using cultures of somatic carrot cells. Carrot cells proliferate as undifferentiated tissues (*callus*) in a chemically defined medium containing an auxin such as 2,4-dichlorophenoxyacetic acid (2,4-D). Upon removal of 2,4-D these cells grow in an organized fashion and form somatic embryos. This *in vitro* system produces 10^5 free embryos per milliliter in liquid culture without the contamination of callus tissue (Sung *et al.*, 1979). However, the embryonic culture contains a population of terminally differentiated, highly enlarged, vacuolated single cells. The free-floating embryos sometimes have a suspensor-like structure attached to them, but apparently do not depend on suspensors for nutrition. Synchrony can be achieved during early embryogenesis (Fujimura and Komamine, 1979).

III. Somatic Embryogenesis

The initial stage of plant embryogeny is similar to animal embryo development in that the process involves rapid cell proliferation resulting in the formation of a spherical structure termed the *globular stage* embryo. When the embryo reaches a certain size (approximately 50 μm in diameter), there is differential growth in various parts of the embryo. This process shapes the embryo into a *heart-shaped* body, which continues development into the *torpedo stage* embryo. Following this period, the embryo matures and undergoes a "dormancy" developmental program along with the desiccation of the seed. The faithful progression of developmental steps is apparently determined by the genetic program. It is not possible to skip steps to go from the zygote directly to a torpedo stage embryo, bypassing the globular stage. Embryogeny can, however, be interrupted or speeded up by changing the hormone concentrations in the medium. For example, 2,4-D can prevent the transition from the globular stage to the heart stage (Sung and Okimoto, 1981). Cytokinin can speed up embryo development (Sung *et al.*, 1979).

IV. Gene Expression during Somatic Embryogenesis

Differential gene expression has been observed during the transition stage from callus to globular stage embryo and to the heart stage embryo (Fig. 1). Two embryo-specific proteins have been identified by separating newly synthesized soluble proteins on native–SDS two-dimensional gels (Sung and Okimoto, 1981). These two proteins are also synthesized in the zygotic embryos of carrots. Carrot seeds were soaked for 3 h in water and cut in half. The half seeds containing the embryos were soaked in a medium containing [^{35}S]methionine for 4 h to label newly synthesized proteins. About 40 embryos were dissected, extracted, and electrophoretically separated on native–SDS 2-D PAGE. Figure 2 shows the *in vivo* translational profile of the mature embryos in the seed. Because of technical difficulties we have not studied these proteins in young embryos developing in the ovaries. However, the temporal period of synthesis of these proteins starts as soon as embryogenesis is initiated and continues throughout embryogeny in carrot somatic embryos (Sung and Okimoto, 1981). Therefore the fact that these proteins are synthesized in mature embryos strongly suggests that they are also synthesized in young embryos. Callus culture of a carrot cell line is characterized by a few biochemical markers, such as the accumulation of

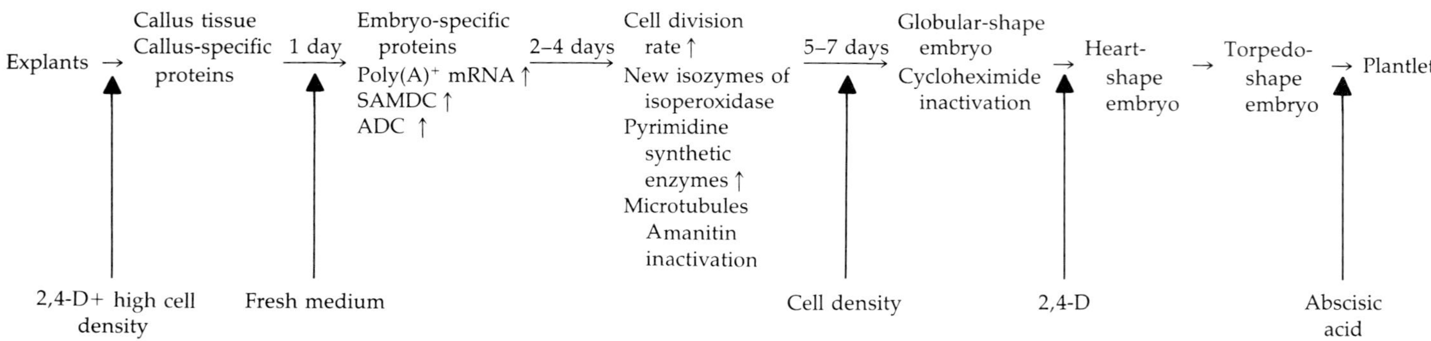

Figure 1. Temporal events during somatic embrogeny of a wild carrot culture. ADC, arginine decarboxylase activity; ↑, increase; ▲, time period when culture reacts to treatment; SAMDC, *S*-adenosylmethionine decarboxylase.

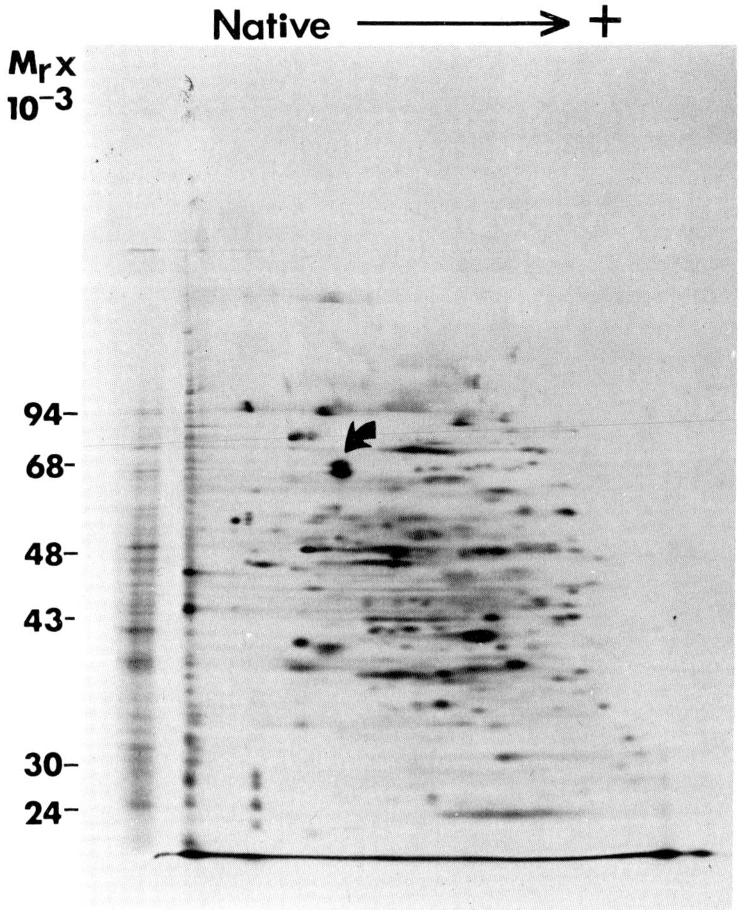

Figure 2. A fluorogram of a native–SDS 2-D PAGE gel showing *in vivo* translational profile of mature embryos from carrot seeds. Embryos were dissected from [^{35}S]methionine-labeled half seeds and extracted in 50 m*M* Tris-HCl, pH 6.8, 5% (v/v) 2-mercaptoethanol, 15% (v/v) glycerol. Native slab gels contained 6.5% (w/v) acrylamide and 0.133% (w/v) *N,N'*-methylenebisacrylamide and were run on an apparatus cooled to 0°C. Approximately 250,000 cpm (about 10 μl) were loaded. The arrows point at embryo-specific proteins. From Sung and Okimoto (1981).

about 10 m*M* ethanol in the growth medium, synthesis of callus-specific proteins (C_1 and C_2), and factors in the medium that promote callus growth as well as C_1 and C_2 synthesis (Sung and Okimoto, 1983). Embryogenesis in carrot cultures is usually initiated by transferring the callus culture into a fresh medium devoid of 2,4-D (Steward *et al.*, 1964; Halperin and Wetherell, 1965). We find that embryogeny can be externally manipulated by fresh medium, cell density, 2,4-D, and abscisic acid. Figure 1 shows that the fresh medium "turns on" embryogenesis as evidenced by the synthesis of embryo-specific proteins (Sung and Okimoto 1982), change in the profile of poly(A)$^+$mRNA synthesis (Sengupta and Raghavan, 1980a,b), elevated arginine decarboxylase and *S*-adenosylmethionine decarboxylase activity, and polyamine levels (Montague et al, 1978, 1979; Fienberg and Sung, 1982).

About three days later, the tissues show an increase in the rate of cell division and organized growth (Warren and Fowler, 1978; Fujimura and Komamine, 1980). Also, a new form of isoperoxidase isozyme appears (Wochok and Burleson, 1974), and there are elevated enzyme activities for pyrimidine biosynthesis, (Ashihara *et al.*, 1981). Further, the cells developed an ability to inactivate α-amanitin. Some of the biochemical events (e.g., the embryo-specific proteins and the elevated polyamine metabolism) are shown to continue if embryogeny is allowed to proceed and to disappear if cells are cultured in 2,4-D or at high cell density. Cell lines impaired in embryogenesis usually do not respond to the fresh medium (Sung and Okimoto, 1981). Although a high density condition can not prevent the onset of the fresh medium-induced events, it prevents the normal development of a globular-shape embryo. Under high-density conditions the semiorganized tissues do not form the globular-shape embryos in 7 days; they become more disorganized and eventually produce callus and some abnormal embryos.

At 0.1 mg/l 2,4-D does not interfere with the biochemical activities occurring during the first 7 days; however, it prevents embryogeny at the transition period from globular to heart-stage. A low-density culture containing 2,4-D enables the formation of globular stage embryos that will enlarge rather than transform into heart-shaped embryos. It is postulated that 2,4-D disturbs the auxin gradient inside the embryo; the gradient determines the position of shoot and root meristem and the differential growth within the embryo.

The cell density effect has been further investigated. The fact that high density prevents embryogeny and promotes callus growth suggests that the medium may be involved. Culture medium (conditioned medium) incubated for 6 days with a high-density culture (inoculum size, 8×10^5 cells/ml) was separated from the cells and incubated again with cells at low density (2×10^4 cells/ml). This medium will prevent embryogeny (Sung and Okimoto, 1983). We have found that the medium effect is not caused by the depletion of nutrients, change in osmotic pressure, or change in pH (R. Chorneau and Z. R. Sung, unpublished). These results indicate that a diffusable substance(s) in the medium is responsible for suppressing embryogeny and promoting callus growth. The medium has been fractionated and the inhibitory capability is found to be located in the fraction containing molecules smaller than 500 daltons (R. Chorneau and Z. R. Sung, unpublished). One of the substances is found to be ethanol (R. Chorneau and M. Christianson, unpublished).

The conditioned medium also contains some substance that will induce the synthesis of two series of polypeptides, C_1 and C_2, in the molecular weight range of 20,000 and 24,000 detected by native–SDS two-dimensional gel electrophoresis (Sung and Okimoto, 1983). The production of the substance that induces C_1 and C_2 synthesis is dependent not only on a high-density condition but also on the presence of 2,4-D in the medium because a high-density culture grown in the absence of 2,4-D does not make C_1 and C_2 (Sung and Okimoto, 1983).

V. Coordinate Regulation of Sets of Biochemical Phenotypes

Because both callus growth and C_1 and C_2 synthesis are dependent on substances in the conditioned medium, it is of interest to find out if the same substance is responsible for the morphological and biochemical effects. Furthermore, when callus is switched to embryonic growth, the switch is accompanied by the simultaneous appearance of a number of embryo-specific traits, including the polypeptides E_1 and E_2, the phenomenon of cycloheximide inhibition (CH^i), and the disappearance of C_1 and C_2. We are also interested in determining if a common mechanism exerts a positive regulation on one set of traits but a negative regulation on another.

To address these two questions we have made use of the phenotype cycloheximide inactivation (CH^i), which renders cells resistant to growth inhibition by cycloheximide (CH), and selected a cell line that expresses CH^i constitutively: WCH105. Like normal embryos, callus tissue of WCH105 can inactivate CH, thus growing in its presence. We examined WCH105 callus for its ability (1) to synthesize E_1 and E_2, (2) to condition the medium for callus growth, and (3) to condition the medium for C_1 and C_2 synthesis. Callus of WCH105 synthesizes E_1 and E_2 but not C_1 and C_2, thus suggesting that the turning on of CH^i also turns on E_1 and E_2 but turns off C_1 and C_2. Apparently a common regulatory mechanism *X* controls both the embryo-specific and the callus-specific traits but in opposite directions. In WCH105 *X* is stuck in a position that constitutively expresses the embryo-specific traits and will never express the callus-specific traits. Because the production of C_1 and C_2 is dependent on the conditioned medium, it is not clear if WCH105 has lost the ability to condition the medium or the ability to synthesize C_1 or C_2 or both. To investigate this question, we tested the conditioned medium of WCH105 on wild-type (*wt*) cells grown in low density, and the conditioned medium of *wt* cells on low-density culture of WCH105. WCH105 cannot condition *wt* cells for the synthesis of C_1 and C_2, nor can *wt* cells condition WCH105 to synthesize C_1 and C_2. However, it can condition *wt* cells to grow as callus. Therefore the diffusable substance responsible for callus growth is different from that responsible for C_1 and C_2 synthesis.

This study also reveals that although the synthesis of C_1 and C_2 correlates with the callus morphology, it is apparently coincidental. Without these polypeptides cells can still grow as callus as in WCH105. E_1 and E_2 are forced to turn on in WCH105 callus tissue because of the altered regulation. This result indicates that E_1 and E_2 alone cannot induce embryogenesis; they may or may not be necessary for embryogenesis. A cell line that accumulates higher levels of indoleacetic acid cannot undergo embryogenesis in the absence of 2,4-D (Sung, 1979); it cannot be induced to synthesize E_1 and E_2 (Sung and Okimoto, 1981). However, to prove that embryogenesis requires E_1 and E_2, it is necessary to isolate cell lines with primary defects in E_1 and E_2 synthesis.

In summary, we do not know the functions of these polypeptides detected by 2-D PAGE; however, they have been useful in studying differential gene expression and biochemical basis of morphogenesis. They also provide temporal markers for developmental processes and can be employed to study gene regulation in response to developmental signals.

Native–SDS gels were developed by Okimoto and Sachs in M. Freeling's lab to study protein synthesis in response to anaerobiosis (Sachs *et al.*, 1980). In general, 2-D gels require tremendous care by a meticulous worker to achieve good resolution and reproducibility. Native–SDS 2-D gels are more demanding than IEF–SDS 2-D gels. Although IEF gels can be stored at −70°C, proteins in the native gel have to be electrophoresced on the second-dimension gel immediately, before they become denatured. This means a 12-h continuous work schedule. Compared to IEF–SDS 2-D PAGE, more streaking occurs in native gels because of the nondenaturing condition. However, native–SDS 2-D PAGE is a simpler system than IEF–SDS 2-D PAGE. The fact that the first dimension is a slab gel and is never frozen, defrosted, and reequilibrated as in the case of IEF tube gels reduces variability. However, the disadvantage of the native–SDS gel is that the first dimension separates by charge and molecular weight and the latter overlaps with the principle of separation in the 2-D SDS gel; only about 200 radioactive spots are resolved compared to the 800 spots that can be resolved on IEF–SDS 2-D gel. Thus IEF–SDS 2-D gels are employed for the study described below.

VI. Gene Expression in Temperature-Sensitive Cell Lines Impaired in Embryogeny

In an attempt to study the genetics of early embryogenesis, we have isolated temperature-sensitive lines (Breton and Sung, 1982) impaired in embryo development (*ts-emb*$^-$) from a haploid cell line of *Daucus carota* L., that is, HA (Smith *et al.*, 1981). They grow and develop normally at 24°C; at 32°C they grow as callus but cannot develop into normal mature embryos. The conditional defect and the ability to separate unorganized growth from organized growth, which leads to differentiation, are the two criteria used to identify developmental mutants. These *ts-emb*$^-$ are not altered in processes necessary for growth only, nor are they *habituated*, a term used to describe long-term tissue culture cell lines that can proliferate autonomously as callus tissues without exogenously supplied growth regulators.

The *ts-emb*$^-$ lines are grouped into three categories. They separate the three developmental stages (callus → globular shape → heart-shape embryo) into four functional stages (unorganized growth with 2,4-D → unorganized growth without 2,4-D → organized growth → differential growth). The first functional stage involves the switch from growth in 2,4-D to growth without 2,4-D; variants altered at this step are unable to grow without 2,4-D. The second stage involves organization; variants unable to carry out organized growth proliferate as callus rather than forming embryos at the globular stage in embryogenic medium. The third stage involves the morphogenesis of embryos from the globular- to the heart-shape stage. This morphogenetic step results from differential growth, a process that shapes the embryos from the spherical to the heart shape. Variants that cannot undergo this differential growth form globular embryos of unusual size but fail to develop into heart-shape embryos. Eventually, the large spherical embryos are converted to callus (Breton and Sung, 1982).

To investigate the primary or pleiotrophic effect of the alteration on gene expression, we have examined the translational profiles of the embryo and callus tissues of a *ts-emb*$^-$ line, *ts-59*, and HA grown at both 24 and 32°C (M. Lempert and Z. R. Sung, unpublished). Proteins extracted and lysed in Tris buffer containing 0.1% NP-40 (detergent) and 9.5 *M* urea were separated by IEF–SDS 2-D PAGE according to O'Farrell and O'Farrell (1977). There are about five polypeptides that differ between the callus and embryo tissues grown at both temperatures. Preliminary results showed that one of the embryo-specific proteins in the molecular weight range of 70,000 is synthesized in *ts-59* grown in embryogenic condition but to a much greater extent at 24 than at 32°C (Fig. 3). The close association of this protein with embryogenesis is very intriguing. We are interested in studying its time course of appearance and spatial distribution within the embryo, whether it is the primary or secondary product of the *ts-emb*$^{-}$ phenotype.

VII. Gene Expression during Late Embryogeny

Late embryogeny can easily be studied by harvesting seeds or even dissecting embryos after anthesis because the embryos have reached a size that can be handled manually. In wheat, Quatrano and his associates have studied embryogeny under the influence of abscisic acid (ABA). Abscisic acid applied to the embryos cultured *in vitro* will stimulate embryogenesis and arrest germination (Triplett and Quatrano, 1982). Accompanying the developmental decision is the choice of a number of proteins whose synthesis is controlled by ABA. For example, in the absence of ABA, the embryos germinate precociously; ribulose bisphosphate carboxylase accumulates, but a group of basic proteins are not synthesized. Figure 4 shows the basic proteins separated by acid–urea–SDS 2-D PAGE; they are in the molecular weight range of 13,000 to 15,000. These proteins are only synthesized in wheat embryos treated with ABA. Quatrano *et al.* (1982) suggested that they are scutellum-specific storage proteins.

VIII. Gene Expression in Cotton Cotyledons

In some plant species such as legumes and cotton, the cotyledons rather than the endosperm serve as the storage organ for nutrient supply during germination of the seedling. In these plants there is further development of the embryo following the torpedo stage. The development usually involves cell division followed by cell expansion in the cotyledons. The last stage of seed formation involves the desiccation of the entire seed including the embryo and the storage tissue either in the form of cotyledon or endosperm; the seed enters the dormancy stage.

Dure *et al.* (1980) have compiled protein catalogs of cotyledon development at the cell division, cell expansion, and seed stage in cotton by two-dimensional gel electrophoresis. In general, there are several changes in about 200 Coomassie Blue–stained protein spots during cotyledon development. The most dominant proteins accumulating during the cell division stages, 70- and 40-kdalton proteins, gradually disappear as cotyledonous cells enlarge and finally vanish in the seed. Further analysis based on pulse-chase and rocket immunoelectrophoresis showed that the 70-kdalton protein is a precursor for a storage protein in the cotyledon and is slowly processed during the last stage of seed maturation.

The 40-kdalton protein was believed to be actin, based on its MW, pI, biochemical nature, and temporal correlation with the period of cell division. Accompanying seed dormancy, ABA accumulates. In the absence of ABA embryos can undergo precocious germination, bypassing the dormancy stage (Ihle and Dure, 1970). During precocious germination, some abundant proteins, including the 70-kdalton precursor and its 40-kdalton protein, all cease to accumulate. Abscisic acid–treated, precociously germinating seedlings will still accumulate the abundant proteins except for the 70-kdalton precursor and 40-kdalton proteins. In addition a new set of proteins found only in the mature seed is accumulated in the cotyledons in response to ABA treatment. Although wheat is a monocot and cotton a dicot, it would be interesting to compare the regulatory and structural similarity between the proteins induced by ABA during seed development.

IX. Gene Expression in Germinating Seeds

The hypocotyls have been favored experimental organs because of their relative uniformity in cell population. The seedlings can be germinated under controlled conditions. Within a few days the materials are ready for experiments. The hypocotyls grow rapidly, mainly by cell elongation, and this elongating process is stimulated by auxin. Excised hypocotyls are commonly employed to study auxin-induced elongation because of the ease of *in vitro* manipulation, for example, incorporation of

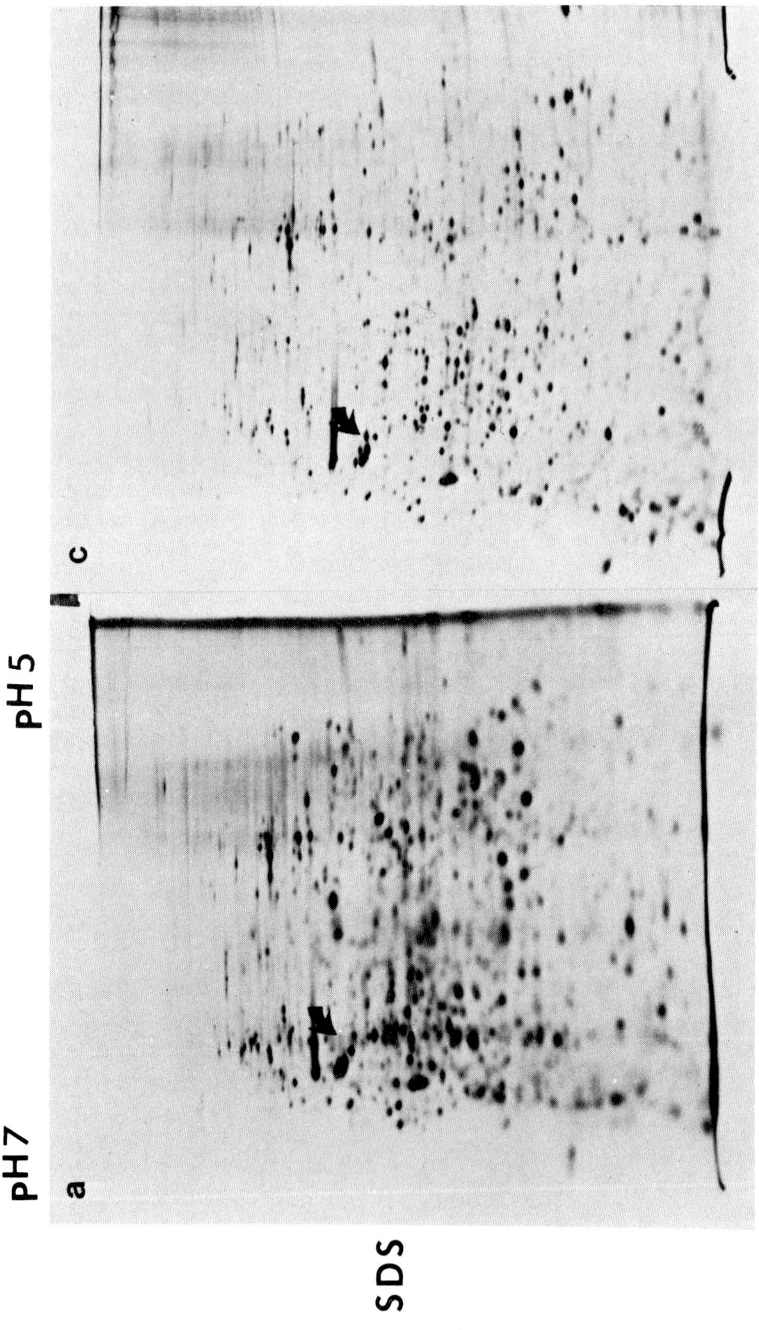
pH 5
pH 7
SDS
a
c

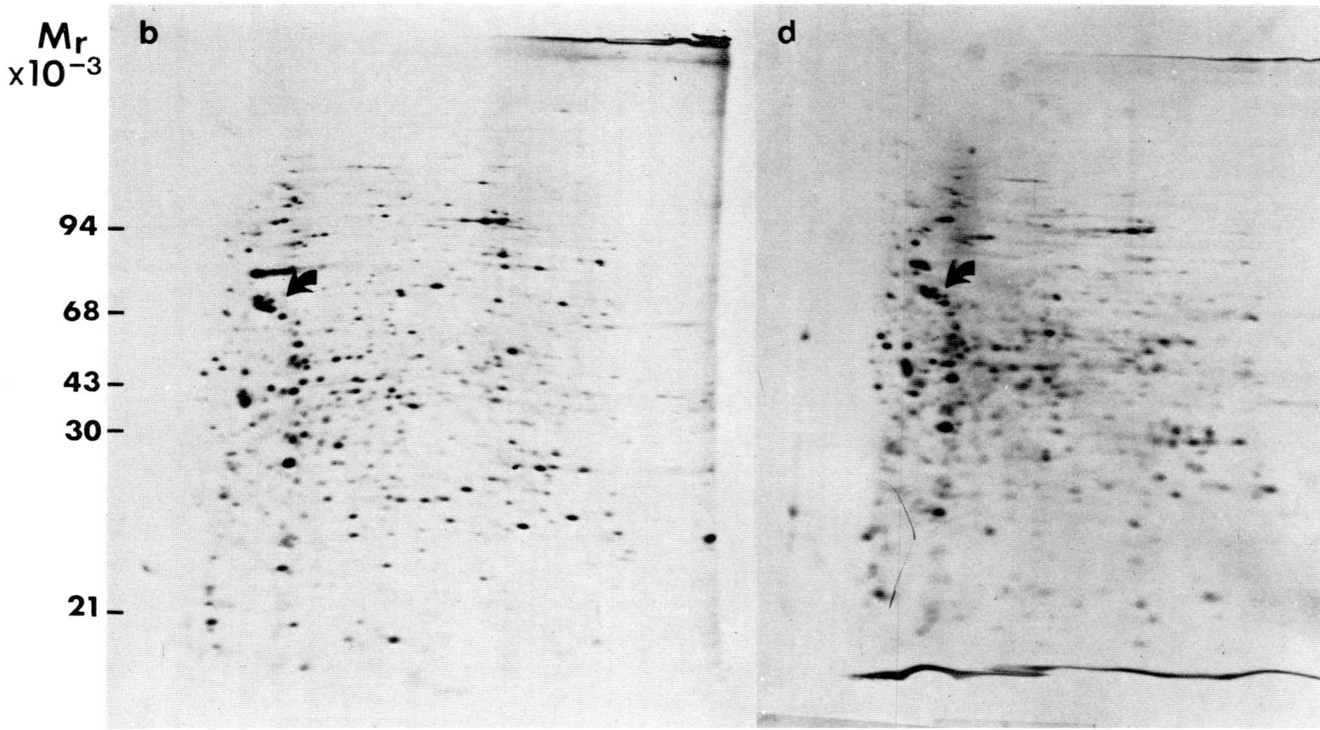

Figure 3. Autoradiograms of IEF–SDS 2-D PAGE gels showing *in vivo* translational profile of *ts-59* incubated with [^{35}S]methionine for 4 h. Tissues were washed and homogenized. Cell extracts containing about 100,000 cpm were loaded and 2-D gel electrophoresis performed according to O'Farrell and O'Farrell (1977). (a) callus tissue grown at 24°C; (b) callus tissue grown at 32°C; (c) embryo tissue grown at 24°C; (d) embryo tissue grown at 32°C. Arrow points at 70 kdalton protein not present in callus tissues and present in large quantity in 24°C embryo but only in small quantity in 32°C embryo.

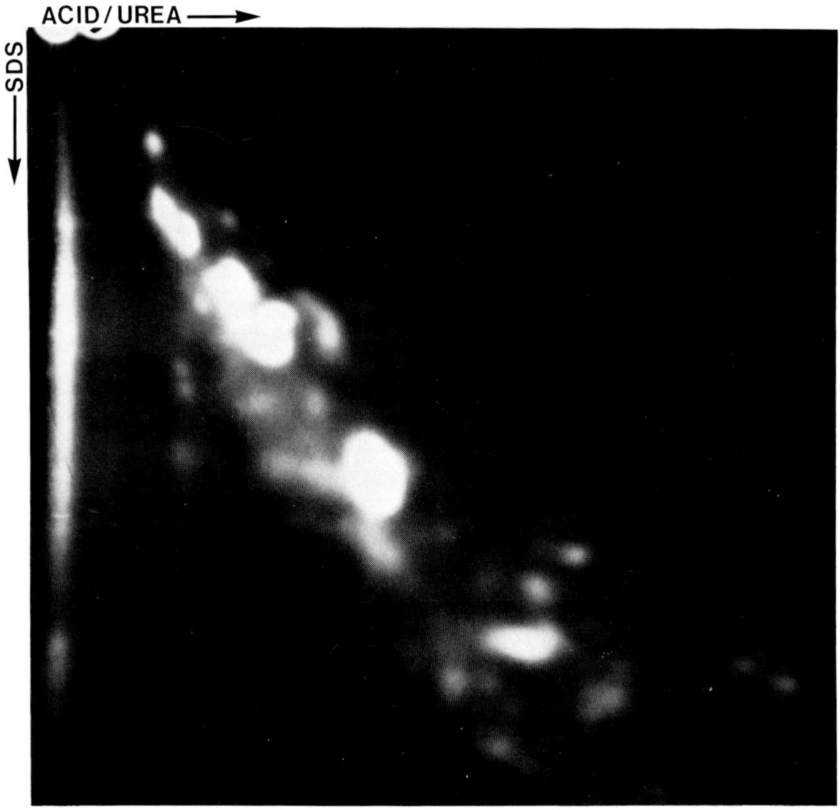

Figure 4. An autoradiogram of an acetic acid-urea–SDS 2-D PAGE gel showing abscisic acid-induced basic proteins of wheat embryo. Excised wheat embryos were homogenized and [^{35}S]methionine-labeled material extracted in 0.05 *M* HCl. The homogenate was centrifuged for 10 min at 10,000 rpm and the supernatant dialyzed against 10 m*M* ammonium carbonate. The first dimension gel contained 0.9 *N* acetic acid, 2.5 *M* urea, and 15% acrylamide. The second dimension contained 0.1% SDS with 6% acrylamide in the stacking and 18% acrylamide in the separating gel. Courtesy of Dr. Ralph Quatrano.

radioactive materials. Zurfluh and Guilfoyle (1980) have found that auxin treatment induces the appearance of one 100-kdalton and two 40-kdalton proteins and the disappearance of another 40-kdalton protein from the soybean hypocotyls. Poly(A)mRNA has been isolated from auxin-treated and untreated tissues and translated *in vitro*. The proteins synthesized *in vitro* can be separated on 2-D gels (Baulcombe *et al.*, 1980). While it is difficult to compare *in vivo* translational products with those of an *in vitro* experiment, mRNA from the auxin-treated tissues did produce a different translational profile from that of the control tissue. Some polypeptides are newly synthesized or elevated in quantities in response to auxin treatment (Baulcombe *et al.*, 1980; Zurfluh and Guilfoyle, 1981). Upon a longer period of incubation with auxin some polypeptides disappear in the products translated from auxin-treated mRNA (Baulcombe *et al.*, 1980).

These results suggest that auxin treatment alters gene expression at the message level. A new mRNA is induced as soon as 15 min after auxin treatment (Zurfluh and Guilfoyle, 1982a). Auxin-induced modulation of specific mRNAs occurs not only in excised but also in intact seedlings (Zurfluh and Guilfoyle, 1982b). The same authors have also studied the effect of ethephone, an ethylene producing compound, on protein synthesis because ethylene induces a similar kind of morphological change on soybean seedlings (Holm and Abeles, 1968). They found that a group of polypeptides in the molecular weight range of 40,000 to 48,000 are induced both by 2,4-D and ethylene. However, there are specific sets of proteins whose mRNA levels are elevated in response to one hormone only. For example, a group of proteins in the molecular weight range of 24,000 to 32,000 are auxin specific.

X. Conclusion

Two-dimensional gels provide a sensitive assay for studying the expression of proteins comprising as little as 10^{-3} to $10^{-4}\%$ of the total cellular protein. This technique has enabled developmental biologists to ask questions relating to the modulation of protein population in different tissues or organs. It has also been useful in detecting tissue-specific proteins whose appearance correlates with developmental changes. However, because the function of the proteins is often unknown, the utility of the technique ends unless it is coupled with genetic, immunological, or molecular studies. Because of the tremendous advancement in recombinant DNA technology, most investigators are undertaking the approach of isolating molecular probes for the stage-specific proteins. These probes are potentially useful in studying the mechanism of coordinate regulation. However, to understand the function of a specific protein, it is imperative that genetic analysis or *in vitro* expression system such as DNA transformation be employed.

Acknowledgments

The author thanks Drs. Ralph Quatrano and Linda Zurfluh for their helpful discussions and for providing unpublished results. This work is supported by NSF Grant No. PCM 8021209 and USDA Competitive Grant No. 79-59-2061-1-1-373–1.

References

Ashihara, R.,Fujimura, T., and Komamine, A. (1981). Pyrimidine nucleotide biosynthesis during somatic embryogenesis in a carrot cell suspension culture. *Z. Pflanzenphysiol.* **104**, 129–237.

Baulcombe, D., Giorgini, J., and Key, J. L. (1980). The effect of auxin on polyadenylated RNA of soybean hypocotyls. *In* "Genome Organization and Expression in Plants" (C. J. Leaver, ed.), pp. 175–185. Plenum, New York.

Breton, A., and Sung, Z. R. (1982). Temperature-sensitive carrot variants impaired in somatic embryogenesis. *Dev. Biol.* **89**, 58–66.

Cionini, P. G., Bennici, A., Alpi, A., and D'Amato, F. (1976). Suspensor, gibberellin and *in vitro* development of *Phaseolus coccineus* embryos. *Planta* **131**, 115–117.

Dure, L. S., III, Capdevila, A. M., and Greenway, S. C. (1980). Messenger RNA domains in the embryogenesis and germination of cotton cotyledons. *In* "Genome Organization and Expression in Plants" (J. C. Leaver, ed.), pp. 127–146. New York.

Fienberg, A., and Sung, Z. R. (1982). Role of arginine decarboxylase and polyamines during somatic embryogenesis. *Conf. Genet. Eng. Plants Univ. Calif. Davis,* Aug. 15–19.

Fujimura, T., and Komamine, A. (1979). Synchronization of somatic embryogenesis in a carrot suspension culture. *Plant Physiol.* **64**, 162–166.

Fujimura, T., and Komamine, A. (1980). The serial observation of embryogenesis in a carrot cell suspension culture. *New Phytol.* **86**, 213–218.

Halperin, W., and Wetherell, E. F. (1965). Ontogeny of adventive embryos of wild carrot. *Science (Washington, D.C.)* **147**, 756–758.

Holm, R. E., and Abeles, F. B. (1968). The role of ethylene in 2,4-D induced growth inhibition. *Planta* **78**, 293–304.

Ihle, J. N., and Dure, L., III (1970). Hormonal regulation of translation inhibition requiring RNA synthesis. *Biochem. Biophys. Res. Commun.* **38**, 995.

Montague, M. J., Koppenbring, J. W., and Jaworski, E. G. (1978). Polyamine metabolism in embryogenic cells of *Daucas carota.* I. Changes in intracellular content and rates of synthesis. *Plant Physiol.* **62**, 430–433.

Montague, M. J., Armstrong, T. Z., and Jaworski, E. G. (1979). Polyamine matabolism in embryogenic cells of *Daucus carota.* II. Changes in argining decarboxylase activity. *Plant Physiol.* **63**, 341–345.

O'Farrell, P., and O'Farrell, P. (1977). Two-dimensional polyacrylamide gel electrophoretic fractionation. *Methods Cell Biol.* **14,** 407–420.

Panyim, S., and Chalkley, R. (1969). High resolution acrylamide gel electrophoresis of histones. *Arch. Biochem. Biophys.* **130**, 337–346.

Przybyllok, T., and Nagl, W. (1977). Auxin concentration in embryo suspensors of *Tropaeolum majus,* as determined by mass fragmentation. *Z. Pflanzenphysiol.* **84**, 463–465.

Quatrano, R. S., Mansfield, M. A., and Hamblin, M. T. (1982). ABA controlled expression of embryo-specific genes during wheat grain development. *Conf. Genet. Eng. Plants Univ. Calif. Davis*, Aug. 15–19, p. 10.

Sachs, M. M., Freeling, M., and Okimoto, R. (1980). The anaerobic proteins of maize. *Cell* **20**, 761–767.

Sengupta, C., and Raghavan, V. (1980a). Somatic embryogenesis in carrot cell suspension. I. Pattern of protein and nucleic acid synthesis. *J. Exp. Bot.* **31**, 247–258.

Sengupta, C., and Raghaven, V. (1980b). Somatic embryogenesis in carrot cell suspension. II. Synthesis of ribosomal RNA and nucleic acid synthesis. *J. Exp. Bot.* **31** 259–268.

Smith, J., Furner, I., and Sung, Z. R. (1981). Nutritional and karyotypic characterization of a haploid cell culture of *Daucus catora. In Vitro* **17**, 315–321.

Steward, F. C., Mapes, M., Kent, A., and Holston, R. (1964). Growth and development of cultured plant cells. *Science (Washington, D.C.)* **143**, 20–27.

Sung, Z. R. (1979). Relationship of indole-3-acetic acid and tryptophan concentrations in normal and 5-methyltryptophan resistant cell lines of wild carrots. *Planta* **145**, 339–345.

Sung, Z. R., and Okimoto, R. (1981). Embryonic proteins in somatic embryos of carrot. *Proc. Natl. Acad. Sci. USA* **78**, 3682–3687.

Sung, Z. R., and Okimoto, R. (1983). Coordinate gene regulation during somatic embryogenesis in carrots. *Proc. Natl. Acad. Sci. USA* **80**, 2661–2665.

Sung, Z. R., Horowitz, J., and Smith, R. (1979). Quantitative studies in normal and 5-methyltryptophan resistant cell lines of wild carrot: the effects of growth regulators. *Planta* **147**, 236–240.

Sung, Z. R., Lazar, G. B., and Dudits, D. (1981). Cycloheximide resistance in carrot cultures: A differentiated function. *Plant Physiol.* **68**, 261–264.

Triplett, B. A., and Quatrano, R. S. (1982). Timing, localization and control of wheat germ agglutinin synthesis in developing wheat embryos. *Dev. Biol.* **91**, 941–946.

Warren, G. S., and Fowler, M. W. (1978). Cell number and cell doubling time during development of carrot embroids in suspension culture. *Experientia* **34**, 356–357.

Wochok, Z., and Burleson, B. (1974). Isoperoxidase activity and induction in cultured tissues of wild carrot: A comparison of proembryos and embryos. *Physiol. Plant.* **31**, 73–75.

Zurfluh, L. L., and Guilfoyle, T. J. (1980). Auxin-induced changes in the patterns of protein synthesis in soybean hypocotyl. *Proc. Natl. Acad. Sci. USA* **77**, 357–361.

Zurfluh, L. L., and Guilfoyle, T. J. (1981). Auxin-induced nucleic acid and protein synthesis in the soybean hypocotyl. *In* "Levels of Genetic Control in Development" (S. Subtelny and U. K. Abbott, eds.), pp. 99–118. Liss, New York.

Zurfluh, L. L., and Guilfoyle, T. J. (1982a). Auxin-induced changes in the population of translatable messenger RNA in elongating sections of soybean hypocotyl. *Plant Physiol.* **69**, 332–337.

Zurfluh, L. L., and Guilfoyle, T. J. (1982b). Auxin- and ethylene-induced changes in the population of translatable messenger RNA in basal sections and intact soybean hypocotyl. *Plant Physiol.* **69**, 338–340.

PART III

Protein Catalogs

CHAPTER 13

The Protein Catalog of *Escherichia coli*

FREDERICK C. NEIDHARDT
TERESA A. PHILLIPS
Department of Microbiology and Immunology
University of Michigan
Ann Arbor, Michigan

I.	Introduction	417
II.	Identification of Proteins	418
III.	Current *E. coli* Catalog	426
IV.	Applications in *E. coli* Research	439
	References	442

I. Introduction

For the past thirty years the bacterium *Escherichia coli* has served as the paradigm of a free-living single-celled organism. It is a small cell ($\sim 3 \times 10^{-13}$ g dry weight) with a relatively small genome encompassed in a single DNA molecule of 3800 kilobase pairs. Approximately 64 genes produce stable RNA as a final product (22 genes for rRNA; 42 known genes for tRNA); the remainder presumably code for no more than 2500–3000 proteins. This cell grows very rapidly in a variety of chemically defined media and is readily studied by a number of *in vivo* and *in vitro* biochemical techniques including isotopic labeling.

Many bacterial species share the structural and biochemical properties of *E. coli* and might be equally excellent choices for study, but in two respects *E. coli* is without peer. First, together with its close relative, *Salmonella typhimurium*, it offers an unequaled array of powerful *in vivo* and *in vitro* techniques for genetic analysis. Second, the body of informa-

TWO-DIMENSIONAL GEL
ELECTROPHORESIS
OF PROTEINS

ISBN 0-12-164720-X

tion accumulated about *E. coli* is by now so considerable as to be itself a significant reason for continued focus on this cell. Between one-third and one-half of its genes have been defined by the analysis of mutant alleles; over 500 of its proteins have been purified or characterized to some degree; probably over 90% of its catabolic, amphibolic, and biosynthetic pathways are known; its biochemical architecture is known to an impressive extent, as are the molecular details of how it makes macromolecules and regulates gene expression. We are far from understanding the growth of any cell, but the prospect of understanding *E. coli* appears to be an attainable goal.

In this context the introduction of two-dimensional gels to resolve complex mixtures of proteins (O'Farrell, 1975; O'Farrell *et al.*, 1977) has had special significance in *E. coli* biology. Two-dimensional gels virtually force a global attitude on us by visual display of the major proteins of a cell. They certainly make it feasible to commence work on a quantitative molecular catalog of this cell. In less expected ways, however, they also open the door to unique exploration of both the genetics and physiology of this organism.

A central repository of information about the proteins of *E. coli* revealed on gels was established in our laboratory late in the 1970s. In this chapter we present a status report on this catalog and an overview of its applications in *E. coli* biology.

II. Identification of Proteins

By 1978 many laboratories were producing high-quality two-dimensional gels of the proteins of total cell extracts and subcellular fractions, and it was evident that exchange of information between laboratories was both possible and desirable. The first step was to establish a system of nomenclature based on internal standards; certain easily recognized protein spots, identified as constitutive proteins, were chosen as boundaries separating nine (A–I) arbitrary ranges of pH in the isoelectric focusing (horizontal) direction. Proteins of known molecular weight were similarly chosen to establish a molecular weight scale in the vertical direction. Each protein spot could then be assigned an alphanumeric designation that was visually descriptive, easily communicated, and reasonably helpful even without reference to the standard gel (Pedersen *et al.*, 1978).

The alphanumeric nomenclature has been useful, but for a definitive catalog of the individual protein spots, precise reference to a standard gel is necessary. In Figs. 1a and 1b are reproduced the reference gels of *E. coli* strain NC3 (B/r) grown in glucose minimal medium that have been used to communicate the identities of individual protein spots. These autoradiograms were originally published (Bloch *et al.*, 1980) nearly "life-size" in our first identifications paper to aid other laboratories in matching the patterns of their gels with that of ours.

One unfortunate aspect of *E. coli* research is that it has been conducted not on a single strain and its mutant variants but on a variety of independent isolates from nature. Most genetic work has been done with so-called K-12 strains, many of which are laboratory variants more or less related to each other; however, a considerable amount of physiological and biochemical work has been done on the faster growing and somewhat hardier B strains (B, B/r, etc., including NC3). As a result our identification process has involved the use of purified proteins, which are typically from strain B, and mutants, which are usually of strain K-12. Given the continued use of both these strains and their many variants in *E. coli* research, it has become clear that two sets of *E. coli* reference gels are necessary. Accordingly new identifications will include a K12 set of reference gels cross-indexed with the set for strain B.

Figure 2a is a composite of equilibrium and nonequilibrium gels for a reference culture of a K-12 strain (W3110) grown in glucose minimal medium. Adjacent, in Figure 2b, is a tracing of the K-12 reference gels that shows only the spots of known biochemical or genetic identity. These 154 identifications were made by a variety of means. In Table I we have listed the 10 procedures that have proved useful so far in identifying or confirming the identity of individual protein spots.

Most of the *E. coli* protein identifications have been made by comigration of purified marker proteins and radioactive whole cell extracts. In addition to the 200 protein samples already sent to us, there are nearly an equal number of additional purified proteins that have been reported and should be available to us. Reports of new protein purifications appear in the literature at the rate of 30 to 50 per year. We conservatively expect to identify approximately 50 proteins per year for each of several years to come.

The other methods will be increasingly used, especially for proteins not yet obtainable in purified form. Mutant strains (regulatory, deletion, insertion, and missense ones) have until now been used chiefly for confirmation but can in the future be used to a greater extent for primary identification. Mutants produced by transposon insertions are particularly useful because the protein product is either missing or drastically altered in position on the gels; there is a growing catalog of genes with such insertions (Berg and Berg, 1981). Increasing use of the bacterio-

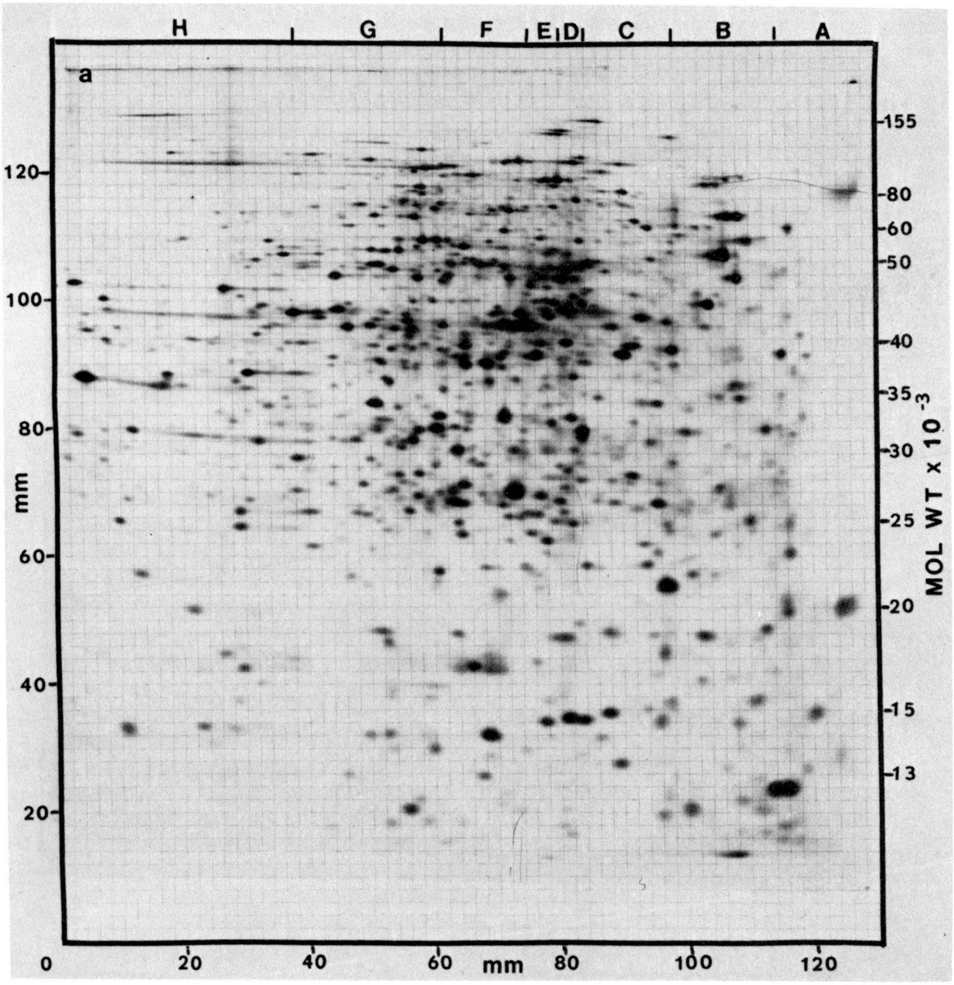

Figure 1. *E. coli* NC3 (B/r) reference two-dimensional gels. A grid overlay provides coordinates for locating individual spots. The letters A to I indicate zones of increasing isoelectric point (Pedersen *et al.*, 1978). The approximate molecular weight scale was determined from the migration distances of identified proteins of known molecular weight. Strain NC3 was labeled with $^{35}SO_4$ during growth in glucose minimal medium at 37°C. (a) Isoelectric focusing to equilibrium (7000 Vh); 1.6% pH 5–7, 0.4% pH 3.5–10 ampholine mixture; SDS-polyacrylamide gel, 11.75%; method according to O'Farrell (1975).

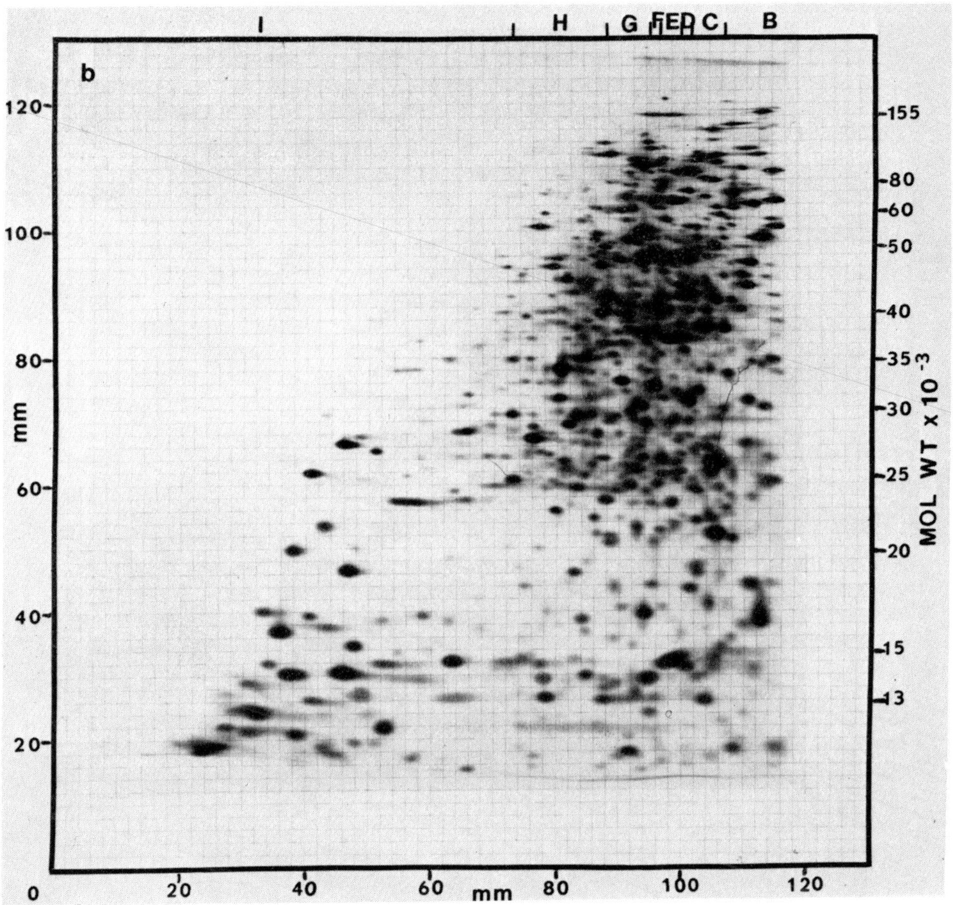

Figure 1 (*continued*) (b) Nonequilibrium pH gradient electrophoresis (1250 Vh); 2% pH 3.5–10 ampholine mixture; SDS-polyacrylamide gel, 11.5%; method according to O'Farrell *et al.* (1977). From Bloch *et al.* (1980).

phage transposon, Mu *d*1 (Ap^R, *lacZ*), to fuse *lacZ* (β-galactosidase) to promoters of specified properties is generating catalogues of mutants that will be useful for protein identifications and confirmations. A fusion mutation of this sort results in β-galactosidase being produced under the direction of the promoter to which *lacZ* is fused. Because β-galactosidase is readily recognized by its production of pink colonies on MacConkey plates, a collection of mutants can be produced and recognized that, for example, produce this enzyme only anaerobically or only at high temperatures. The protein normally produced from that promoter is not

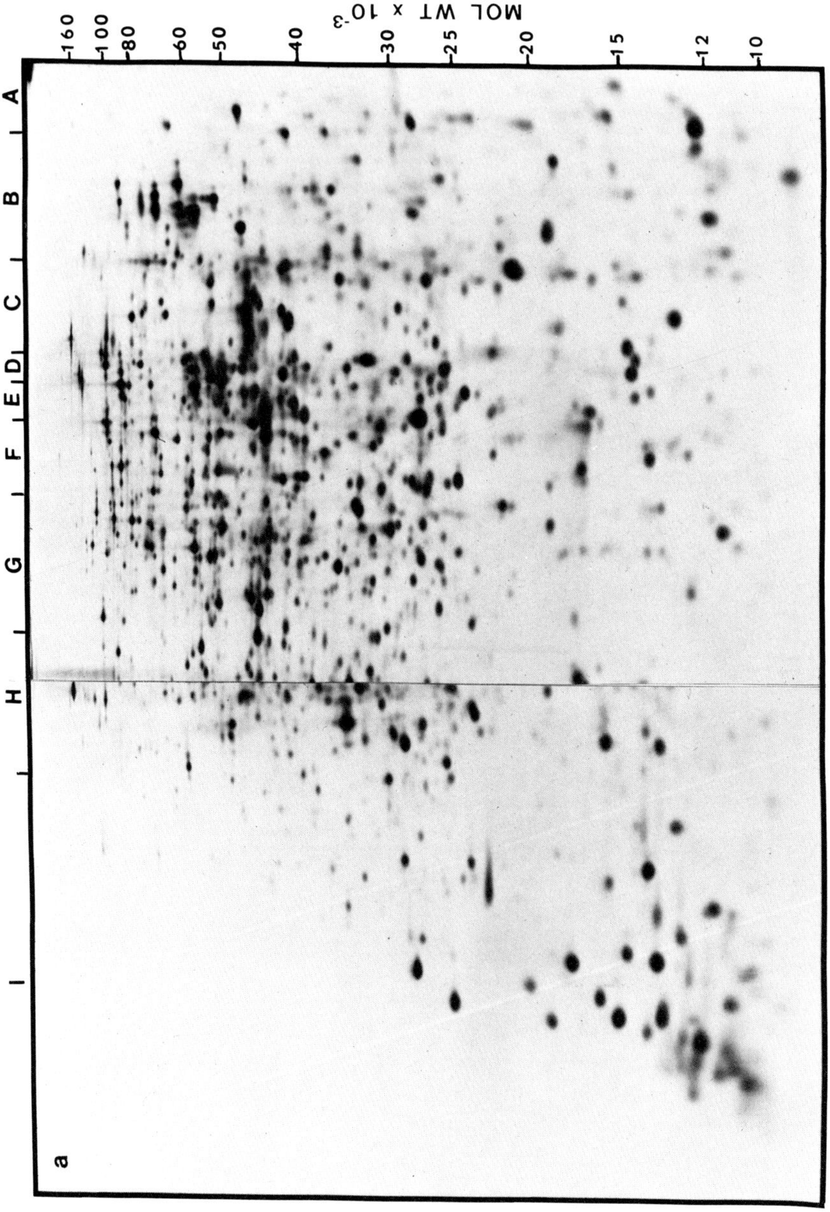
MOL WT x 10-3
160
100
80
60
50
40
30
25
20
15
12
10
A
B
C
D
E
F
G
H
I
a

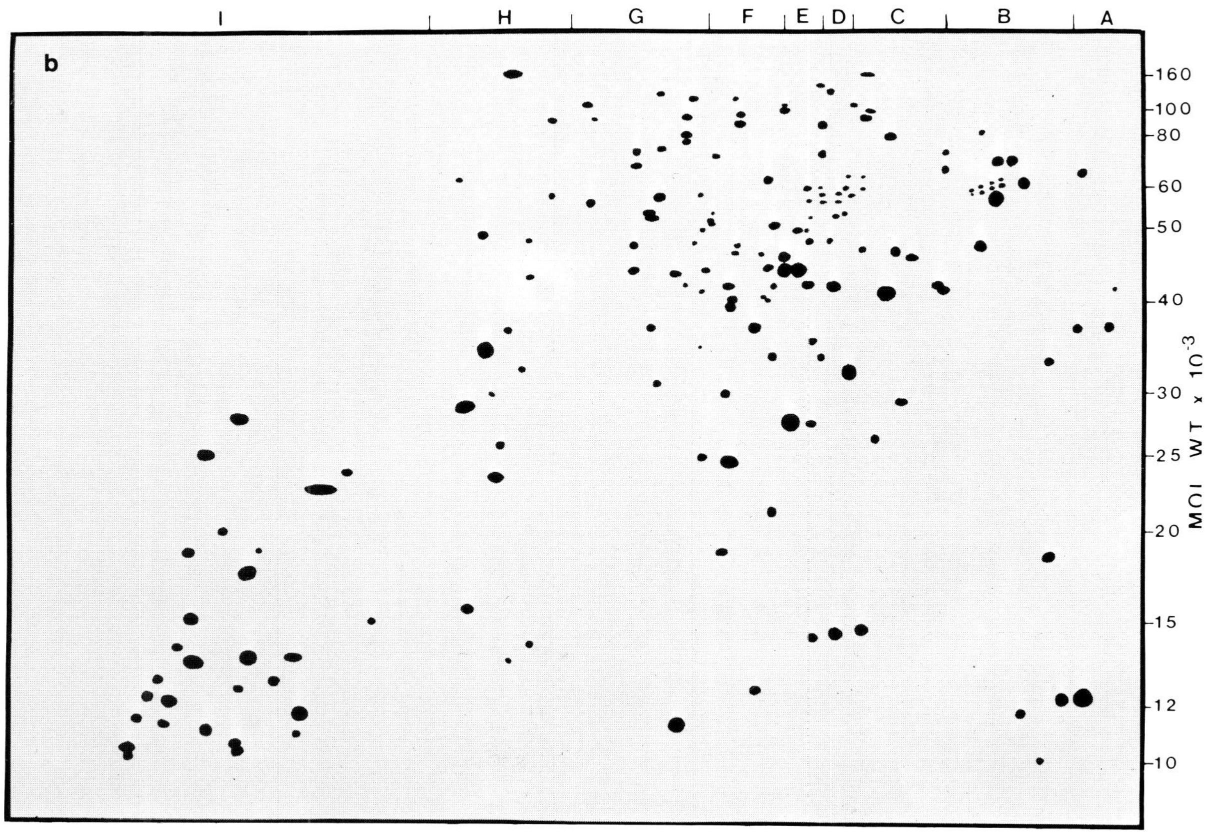

Figure 2. *E. coli* proteins resolved on two-dimensional gels. The letters designating isoelectric zones and the approximate molecular weight scale are as in Fig. 1. *E. coli* K12, strain W3110, was labeled with $^{35}SO_4$ during growth in glucose minimal medium at 37°C. (a) Composite autoradiogram made from nonequilibrium (left side) and pH 5–7 (right side) gels; (b) Proteins that have been identified (listed in Table II), traced from (a).

TABLE I

Procedures for Identifying Bacterial Proteins on Gels

Procedure[a]	Comments	References[b]
A. Chemical		
1. Comigration (C)	Single most useful procedure for purified markers; useful also for testing identity of plasmid coded proteins	Bloch *et al.* (1980); Drahos and Hendrix (1982); Hayward and Fyfe (1978); McConkey (1979); Neidhardt *et al.* (1980); Phillips *et al.* (1980)
2. Peptide mapping (M)	Useful either with radioautography or silver stain; sufficiently sensitive for analysis of individual spots; particularly helpful if a protein is suspected to have multiple isoelectric species	Cleveland *et al.* (1977); Hirshfield *et al.* (1981); Neidhardt *et al.* (1981); see also chapter by Fey *et al.*
3. Amino acid composition (A)	Most convenient procedure is double isotope labeling with different amino acids; helpful in confirming homogeneity of an individual spot; helpful in testing identity of closely migrating spots, particularly under different growth conditions, and in testing whether the protein in a spot has an amino acid composition consistent with the nucleotide sequence of its putative structural gene	Hayward and Fyfe (1978); Neidhardt *et al.* (1977); Pedersen *et al.* (1978)
4. Selective labeling (L)	Useful in special instances, as for methylated or phosphorylated proteins	Hazelbauer and Engstrom (1981)
5. Prosthetic group detection (R)	Useful in special instances, as for lipoic acid linked subunit of pyruvate dehydrogenase	M. Smith (unpublished)
6. Selective derivatization (Z)	Useful in special instances, as for proteins that covalently bind penicillin or TPCK[c]	Spratt (1977); Young and Neidhardt (1978)
7. Immunological reactivity (I)	Useful in searching for cross-reacting proteins as well as in primary identification; can be tested directly on gels or antibodies can be used to precipitate the cellular protein of interest	Drahos and Hendrix (1982); Hackett and MacGregor (1981); Kessler (1981)

TABLE I (*continued*)

Procedure[a]	Comments	References[b]
B. Biological		
1. Genetic (G)	Extremely helpful, particularly for *E. coli* with ~1200 known genes; one-third of missense mutations alter the isoelectric point; nonsense, frame-shift, insertion, and deletion mutations change the cellular level; maxicells, minicells, and chloramphenicol-treated cells permit selective expression of genes carried on hybrid plasmids and phage and therefore can be used to find the protein products of these genes; hybrid multicopy plasmids lead frequently to elevated cellular levels of their protein products; use of new *in vivo* techniques such as phage Mu-mediated *lac* fusions permits explicit searches for proteins of defined physiological properties	Bloch *et al.* (1980); Neidhardt *et al.* (1980, 1981); Pedersen *et al.* (1976); Phillips *et al.* (1980); Reeve (1979); Sancar *et al.* (1979)
2. Physiological (P)	Approximately 80% of *E. coli* proteins are inducible, repressible, or subject to growth rate-related control, and thus the conditions under which a protein is made can often be used to aid identification	Bloch *et al.* (1980); Pedersen *et al.* (1978); Phillips *et al.* (1980)
3. Cellular fraction (F)	Fractionation of the *E. coli* cell into envelope, periplasm, ribosomes, nucleoid, and cytosol, followed by gel analysis of the proteins in each structure provides helpful information	Ames and Nikaido (1976); Bloch *et al.* (1980); Chopra and Shales (1980); Pedersen *et al.* (1978); Phillips *et al.* (1980); Moriya and Hori (1981); Sato *et al.* (1977); Wilson (1982)

[a] Each of the major procedures is listed followed in parentheses by a letter used in Table II to indicate which procedures have been used for each identified protein.

[b] The references are not comprehensive. They have been selected to illustrate, wherever possible, the application of a given procedure and thereby guide the reader to further information and directions.

[c] TPCK, L-1-tosylamido-2-phenylethyl chloromethyl ketone.

produced, in most cases, as a result of the insertion mutation. Wanner and McSharry (1982) have defined 18 separate phosphate starvation–inducible (*psi*) genes that code for proteins (some known, some unknown) under phosphate regulation. Our laboratory is currently scanning gels of each of these mutants to locate the Psi proteins.

One other genetic procedure will be of increasing usefulness in the near future. Gene banks of *E. coli* produced as hybrid ColE1 plasmids (Clarke and Carbon, 1976) and as cosmids (Collins and Bruning, 1978) are being surveyed using special procedures that permit selective expression of plasmid-borne genes (minicells: Reeve, 1979; maxicells: Sancar *et al.*, 1979; and chloramphenicol-treated cells: Neidhardt *et al.*, 1980). Identification usually involves knowing that a given gene is on a plasmid, seeing the plasmid proteins expressed and resolved on two-dimensional gels, and having an idea of the size of the protein or its gene. Expression of plasmid genes can be the foundation of a new method for mapping the structural genes of proteins (see Section IV). As this method becomes more fully utilized protein identifications will more frequently result.

It should be emphasized that all identifications are considered tentative until independent methods (e.g., genetic and biochemical) give supporting evidence. Our published reports, which include tentative assignments to elicit confirming or conflicting information from others, always indicate the kind of information that has led to each assignment.

III. Current *E. coli* Catalog

Table II presents the locations of 160 *E. coli* proteins on the reference two-dimensional gel shown in Fig. 1, with the criteria used to establish or confirm the identifications.

Most of the major physiological classes of proteins are represented: catabolic, amphibolic, and anapleurotic enzymes; biosynthetic enzymes; enzymes and factors involved in the polymerization, repair, and modification of DNA, RNA, and protein; transport and binding proteins; proteins involved in motility and chemotaxis; regulatory proteins; and proteins necessary for morphogenesis. The list includes proteins from each of the subcellular compartments of *E. coli:* flagella, outer membrane, periplasm, cytoplasmic membrane, ribosomes, cytosol, and nucleoid.

The coordinates listed in Table II refer to Fig. 1, a reference gel of the B strain NC3 grown in glucose minimal medium. For the few proteins that

TABLE II.

***E. coli* Proteins Identified on Two-Dimensional Gels**

Protein[a]	Location[b] pH 5–7	Location[b] Nonequilibrium	Alphanumeric designation[c]	Basis for identification[d]
1. Acetate CoA-transferase (EC 2.8.3.8)[e]	87 × 70	—	C26.8	CP
	61 × 64	—	G24.2	CP
2. Adenosinetriphosphatase, membrane-bound [Mg^{2+}–Ca^{2+}](EC 3.6.1.3.)				
α-subunit	51 × 106	87 × 94	G51.0	C
β-subunit	103 × 99	110 × 91	B46.7	C
3. Alanyl-tRNA synthetase (EC 6.1.1.7)	61 × 121	97 × 111	F93.0	C
4. Alkaline phosphatase (EC 3.1.3.1)	70 × 99	—	F46.6	CP
5. Anthranilate phosphoribosyltransferase (EC 2.4.2.18)	36 × 108	—	H54.5	CGP
6. Anthranilate synthase (EC 4.1.3.27)	77 × 105	—	E50.0	CGP
7. *ara* Operon activator and repressor protein (*araC* gene product)	—	80 × 69	H30.2	CGP
8. L-Arabinose isomerase (EC 5.3.1.4)[e]	42 × 108	—	G54.1	GP
9. Arginyl-tRNA synthetase (EC 6.1.1.19)	77 × 110	—	E58.0	CP
10. Asparaginyl-tRNA synthetase (EC 6.1.1.22)[f]	59 × 121	—	G99.0	C
11. Aspartate aminotransferase (EC 2.6.1.1)	65 × 93	—	F39.7	CM
12. Aspartate carbamoyltransferase (EC 2.1.3.2)				
α-subunit	—	67 × 37	I17.5	C
β-subunit	—	84 × 76	H31.3	C
13. Aspartyl-tRNA synthetase (EC 6.1.1.12)	71 × 111	—	F58.5	CP
14. Binding protein, high-affinity branched-chain amino acid transport system (*livJ* gene product)	81 × 93	—	D40.7	CGP
15. Binding protein, high-affinity branched-chain amino acid transport system (*livK* gene product)[g]	97 × 93	—	B40.8	CGP

(*continued*)

TABLE II (*continued*)

Protein[a]	Location[b]: pH 5–7	Location[b]: Nonequilibrium	Alphanumeric designation[c]	Basis for identification[d]
16. Branched-chain amino acid aminotransferase (transaminase B) (EC 2.6.1.42)	71 × 82	95 × 75	F32.5	C
17. Carbamoyl-phosphate synthetase (glutamine-hydrolysing) (EC 6.3.5.5)				
α-subunit	50 × 96	—	G41.4	C
β-subunit	80 × 126	104 × 116	E133.0	C
18. Chorismate mutase T-prephenate dehydrogenase (EC 5.4.99.5/1.3.1.12)	61 × 92	—	F37.9	C
19. Citrate (*si*)-synthase (EC 4.1.3.7)	27 × 102	—	H47.4	MP
	32 × 102	—	H47.5	CM
20. Cyclic AMP receptor protein	—	65 × 57	I21.4	CG
21. DNA binding protein NS1	—	42 × 19	I11.5	C
22. DNA binding protein NS2	—	44 × 18	I11.3	C
23. *dnaK* Gene product	108 × 113	114 × 104	B66.0	G
24. DNA nucleotidyltransferase (DNA polymerase I) (EC 2.7.7.7)	74 × 123	—	F113.0	C
25. Dihydrolipoamide acetyltransferase (EC 2.3.1.12)	99 × 112	108 × 103	C62.7	CGMR
	98 × 114	108 × 106	C70.0	CGMR
26. Dihydrolipoamide reductase (lipoamide dehydrogenase) (EC 1.6.4.3)	53 × 105	—	G50.0	C
27. Dihydrolipoamide succinyltransferase (EC 2.3.1.61)	72 × 104	—	F50.3	CMR
28. Enolase (EC 4.2.1.11)	74 × 98	96 × 89	F43.8	C
29. Exodeoxyribonuclease III (endonuclease II) (EC 3.1.11.2)	52 × 78	—	G28.2	C
30. Formate acetyltransferase (pyruvate-formate lyase) (EC 2.3.1.54)				
α-subunit	58 × 118	96 × 108	G74.0	CP
β-subunit	58 × 117	96 × 107	G70.0	CP
31. β-D-Galactosidase (EC 3.2.1.23)[e]	82 × 124	—	E123.0	CGP
32. Glucose phosphotransferase system: factor III	113 × 49	—	B18.7	C
33. Glutamate decarboxylase (EC 4.1.1.15)[e,g]	81 × 101	—	D46.5	CP
	77 × 101	—	E46.5	CP

TABLE II (*continued*)

Protein[a]	Location[b] pH 5–7	Location[b] Nonequilibrium	Alphanumeric designation[c]	Basis for identification[d]
34. Glutamine synthetase (EC 6.3.1.2)				
Unadenylylated	81 × 105	—	D49.9	C
Adenylylated	83 × 105	—	D50.0	C
35. Glutaminyl-tRNA synthetase (EC 6.1.1.18)	51 × 113	—	G61.0	CP
36. Glutamyl-tRNA synthetase (EC 6.1.1.17)	62 × 104	—	F47.8	C
37. Glutaredoxin	109 × 16	—	B11.0	C
38. Glutathione reductase (NAD(P)H) (EC 1.6.4.2)	60 × 100	—	F45.6	C
39. Glyceraldehyde-phosphate dehydrogenase (EC 1.2.1.12)	—	83 × 78	H34.3	C
	—	80 × 78	I33.5	C
40. Glycerol kinase (EC 2.7.1.30)[e]	77 × 102	—	E48.7	CP
41. Glycerol-3-phosphate dehydrogenase (NAD^+) (aerobic) (EC 1.1.1.8)[e]	—	76 × 96	H47.3	C
42. Glycyl-tRNA synthetase β-subunit (EC 6.1.1.14)	79 × 115	101 × 106	E77.5	CGP
43. GP2 (a glycerol-3-phosphate binding protein)	32 × 95	—	H41.0	CP
44. *groEL* Gene product (A-protein)	106 × 107	112 × 99	B56.5	CG
45. *hisP* Gene product	—	69 × 64	I26.2	G
46. Histidine binding protein (*hisJ* gene product)	91 × 73	—	C28.5	G
47. Histidyl-tRNA synthetase (EC 6.1.1.21)	61 × 103	—	F48.1	CP
48. HTP regulator[h]	60 × 83	—	F33.4	G
49. 4-hydroxy-2-oxoglutarate aldolase (EC 4.1.3.16)[i]	64 × 63	—	F22.5	CM
50. IHF (integration host factor)[j]	—	50 × 19	I10.5	C
51. Indole-3-glycerol-phosphate synthase (EC 4.1.1.48)	71 × 98	—	F45.5	CGMP
52. Isocitrate dehydrogenase ($NADP^+$) (EC 1.1.1.42)	85 × 98	—	C43.8	CP
53. Isoleucyl-tRNA synthetase (EC 6.1.1.5)	58 × 124	—	F107.0	CG
54. *lac* Operon repressor protein	—	84 × 86	H39.0	C
55. Leucyl-tRNA synthetase (EC 6.1.1.4)	90 × 121	—	D100.0	CG
56. *lon* (*capR*) Gene product[g]	35 × 121	—	H94.0	CG
	41 × 121	—	H94.1	C

(*continued*)

TABLE II (*continued*)

Protein[a]	Location[b] pH 5–7	Location[b] Nonequilibrium	Alphanumeric designation[c]	Basis for identification[d]
57. Lysine decarboxylase (EC 4.1.1.18)[e]	50 × 116	—	G75.0	CP
58. Lysyl-tRNA synthetase (EC 6.1.1.6)				
Form I	83 × 109	—	D58.5	CG
Form II	83 × 112	—	D60.5	CG
Form III	86 × 109	—	C58.4	CG
Form IV	86 × 112	—	C60.5	CG
59. Methionine adenosyltransferase (EC 2.5.1.6)	93 × 97	105 × 89	C44.6	C
60. Methionyl-tRNA formyltransferase (EC 2.1.2.9)	64 × 83	—	F33.6	C
61. Methionyl-tRNA synthetase (EC 6.1.1.10)	60 × 114	—	G72.0	C
62. Methyl-accepting chemotaxis protein I[g]	103 × 107	—	B60.1	L
	103 × 108	—	B60.2	L
	104 × 108	—	B60.3	L
	104 × 109	—	B60.4	L
	105 × 108	—	B60.5	L
	105 × 109	—	B60.6	L
	106 × 109	—	B60.7	L
	106 × 110	—	B60.8	L
63. Methyl-accepting chemotaxis protein II[g]	82 × 108	—	D58.1	L
	82 × 109	—	D58.2	L
	84 × 109	—	D58.3	L
	76 × 108	—	E58.1	L
	78 × 107	—	E58.2	L
	78 × 108	—	E58.3	L
64. Murein transglycosylase, soluble form	—	83 × 101	H63.0	C
65. *nusA* Gene product	116 × 111	—	B61.0	C
66. *nusB* Gene product	—	81 × 29	H13.8	G
67. Origin binding protein a	61 × 108	—	G55.3	C
68. Outer membrane protein b (Ia); matrix protein (*ompF* gene product)[k]	115 × 85	—	B36.0	G
69. Outer membrane protein c (Ib); (*ompC* gene product)[g,k]	119 × 85	—	A35.5	G
70. Outer membrane protein d (II*); SSP (*ompA* gene product)	73 × 70	—	F24.5	C

TABLE II (*continued*)

Protein[a]	Location[b] pH 5–7	Location[b] Nonequilibrium	Alphanumeric designation[c]	Basis for identification[d]
71. 3-oxoacyl-[acyl-carrier-protein] synthase I (KASI) (EC 2.3.1.41)	71 × 97	—	F42.2	C
72. Oxoglutarate dehydrogenase (EC 1.2.4.2)	41 × 123	—	G97.0	CGM
73. PII_D (a regulatory protein, catalyzes deadenylylation of glutamine synthetase)	67 × 25	94 × 24	F12.2	C
74. Penicillinase (chromosomal β-lactamase) (EC 3.5.2.6)	—	70 × 81	I35.7	CGP
75. Phenylalanyl-tRNA synthetase (EC 6.1.1.20)				
α-subunit	53 × 87	—	G36.0	CP
β-subunit	87 × 120	—	D94.0	CP
76. Phosphoenolpyruvate carboxykinase (GTP) (EC 4.1.1.32)[e]	79 × 109	—	E56.0	CP
77. Phosphoenolpyruvate carboxylase (EC 4.1.1.31)	66 × 120	—	F84.0	C
78. 6-Phosphofructokinase (EC 2.7.1.11)				
Activity I	69 × 87	—	F35.8	C
Activity II	77 × 88	—	E36.6	C
79. Phospho-2-keto-3-deoxygluconate aldolase (EC 4.1.2.14)[i]	64 × 63	—	F22.5	G
80. Phospho-2-keto-3-deoxyheptonate aldolase (DAHP synthase, tyr-sensitive) (EC 4.1.2.15)	70 × 91	—	F38.1	C
81. Phosphohistidinoprotein-hexose phosphotransferase (EC 2.7.1.69)	56 × 21	91 × 18	F7.0	C
82. Phosphotransferase system enzyme I	110 × 109	—	B58.3	CM
83. Polyhook[g]	124 × 91	—	A39.8	C
84. Polyribonucleotide nucleotidyltransferase (polynucleotide phosphorylase) (EC 2.7.7.8)	90 × 117	—	C78.0	C
85. Protein chain elongation factor G	80 × 119	99 × 109	D84.0	CG
86. Protein chain elongation factor Ts	83 × 79	101 × 73	C30.7	CG
	83 × 80	101 × 74	C31.6	

(*continued*)

TABLE II (*continued*)

Protein[a]	Location[b]: pH 5–7	Location[b]: Nonequilibrium	Alphanumeric designation[c]	Basis for identification[d]
87. Protein chain elongation factor Tu	74 × 96	96 × 88	E42.0	CG
88. Protein chain initiation factor				
2A	52 × 125	—	G117.0	C
2B	67 × 124	—	F119.0	C
89. Protein chain initiation factor				
3L	—	42 × 53	I19.5	CG
3S	—	47 × 50	I20.1	CG
90. Pyruvate dehydrogenase (lipoamide) (EC 1.2.4.1)	74 × 122	—	F99.0	CGM
91. Pyruvate kinase I (EC 2.7.1.40)	54 × 108	—	G54.7	C
92. Pyruvate, water dikinase (phosphoenolpyruvate synthase) (EC 2.7.9.2)	104 × 118	—	B83.0	CP
93. *recA* Gene product (protein X)	90 × 91	103 × 84	C39.3	CGP
94. Rhamnulosephosphate aldolase (EC 4.1.2.19)[e]	60 × 76	—	G30.0	CMP
95. 50S Ribosomal subunit protein L1	—	46 × 66	I26.0	CF
96. 50S Ribosomal subunit protein L2	—	64 × 72	I30.2	CM
97. 50S Ribosomal subunit protein L3	—	40 × 62	I23.0	CF
98. 50S Ribosomal subunit protein L4	—	55 × 57	I21.1	CF
99. 50S Ribosomal subunit protein L5	—	46 × 46	I16.4	CF
100. 50S Ribosomal subunit protein L6	—	37 × 50	I17.2	CF
101. 50S Ribosomal subunit protein L7	116 × 24	—	A13.0	CF
102. 50S Ribosomal subunit protein L9	29 × 33	84 × 30	H14.0	CF
103. 50S Ribosomal subunit protein L11	—	37 × 30	I14.1	CF
104. 50S Ribosomal subunit protein L12	114 × 24	—	B13.0	CF
105. 50S Ribosomal subunit protein L17[l]	—	24 × 33	I14.9	C
106. 50S Ribosomal subunit protein L21	—	40 × 26	I13.2	CF
107. 50S Ribosomal subunit protein L23	—	30 × 25	I13.0	C

TABLE II (*continued*)

Protein[a]	Location[b] pH 5–7	Nonequilibrium	Alphanumeric designation[c]	Basis for identification[d]
108. 50S Ribisomal subunit protein L25	—	38 × 21	I11.8	CF
109. 50S Ribosomal subunit protein L27[l]	—	15 × 25	I12.8	C
110. 50S Ribosomal subunit protein L28[l]	—	14 × 22	I12.2	C
111. 50S Ribosomal subunit protein L29	—	38 × 17	I11.1	C
112. 50S Ribosomal subunit protein L30	—	23 × 18	I11.4	CF
113. 50S Ribosomal subunit protein L32[l]	—	12 × 20	I11.7	C
114. 30S Ribosomal subunit protein S1	106 × 113	111 × 104	B65.0	CF
115. 30S Ribosomal subunit protein S2	—	76 × 67	H27.0	F
116. 30S Ribosomal subunit protein S5	—	35 × 37	I14.4	CF
117. 30S Ribosomal subunit protein S6				
A	78 × 34	97 × 31	D14.7	CF
B	81 × 34	98 × 32	C14.8	CF
C	87 × 35	99 × 32	C15.3	F
118. 30S Ribosomal subunit protein S8	—	48 × 27	I13.5	CF
119. 30S Ribosomal subunit protein S9	—	52 × 32	I14.6	CM
120. 30S Ribosomal subunit protein S10	—	52 × 21	I12.1	CF
121. 30S Ribosomal subunit protein S15	—	30 × 21	I12.0	CF
122. 30S Ribosomal subunit protein S16	—	30 × 20	I11.9	C
123. 30S Ribosomal subunit protein S17	—	32 × 24	I12.6	CF
124. 30S Ribosomal subunit protein S20[l]	—	13 × 25	I12.9	C
125. Ribosomal subunit protein	—	34 × 32	I14.5	F
126. Ribosomal subunit protein	—	45 × 30	I14.3	F
127. Ribosomal subunit protein	—	30 × 29	I13.7	F
128. Ribosomal subunit protein	—	26 × 27	I13.3	F
129. Ribosomal subunit protein	—	27 × 22	I12.3	F
130. Ribosomal subunit protein	—	47 × 19	I11.6	F
131. Ribosomal subunit protein L7/L12 acetyltransferase	71 × 92	—	F38.5	C
	65 × 91	—	F37.8	C

(*continued*)

TABLE II (*continued*)

Protein[a]	Location[b] pH 5–7	Nonequilibrium	Alphanumeric designation[c]	Basis for identification[d]
132. Ribosome release factor	13 × 57	88 × 51	H21.6	C
	—	88 × 52	H21.7	C
133. Ribulokinase (EC 2.7.1.16)[e]	79 × 107	—	D55.0	CP
134. L-Ribulosephosphate 4-epimerase (EC 5.1.3.4)[e]	53 × 77	—	G28.1	CP
135. RNA nucleotidyltransferase (RNA polymerase) (EC 2.7.7.6)				
α-subunit	98 × 92	—	B40.7	CG
β-subunit	85 × 128	112 × 118	D157.0	CG
β′-subunit	—	95 × 118	I160.0	CG
136. Seryl-tRNA synthetase (EC 6.1.1.11)	75 × 102	—	E48.8	C
137. Single-stranded DNA binding protein (*ssb* gene product)	63 × 48	93 × 44	F18.8	C
138. Succinate dehydrogenase (fumarate reductase) (EC 1.3.99.1)[e]				
I	—	92 × 104	G63.2	CP
II	—	83 × 61	H24.7	CP
139. Succinyl-CoA synthetase (ADP-forming) (EC 6.2.1.5)				
α-subunit	25 × 78	—	H28.0	CP
β-subunit	77 × 93	—	E39.8	CP
140. Superoxide dismutase (EC 1.15.1.1)				
iron protein	70 × 54	—	F21.3	C
manganese protein	—	79 × 56	I21.3	C
141. Tetrahydrofolate dehydrogenase (dihydrofolate reductase) (EC 1.5.1.3)	100 × 52	—	B20.0	C
142. Tetrahydropteroyltriglutamate methyltransferase (EC 2.1.1.14)[f]	59 × 121	—	F88.0	GP
143. Thioredoxin	109 × 22	105 × 17	B12.0	C
144. Thioredoxin reductase (NADPH) (EC 1.6.4.5)	79 × 81	—	D31.5	C
145. Threonine dehydratase (biosynthetic) (EC 4.2.1.16)	63 × 106	—	F50.1	CM
146. Threonyl-tRNA synthetase (EC 6.1.1.3)	55 × 116	—	G65.0	C
147. Topoisomerase II′, subunit ν	67 × 100	—	F44.6	C
148. Transcription termination factor rho	—	79 × 94	I48.8	G
149. tRNA (Uracil-5)-methyltransferase (EC 2.1.1.35)	58 × 93	—	G39.6	C

TABLE II (*continued*)

Protein[a]	Location[b] pH 5–7	Location[b] Nonequilibrium	Alphanumeric designation[c]	Basis for identification[d]
150. Tryptophan synthase (EC 4.2.1.20)				
α-subunit (A protein)	77 × 69	—	E25.5	CGP
β-subunit (B protein)	56 × 95	—	G40.0	CP
151. Tryptophanase (EC 4.1.99.1)[e]	50 × 101	—	G46.5	CP
152. Tryptophanyl-tRNA synthetase (EC 6.1.1.2)	28 × 87	83 × 79	H37.0	CG
153. Tyrosyl-tRNA synthetase (EC 6.1.1.1)	61 × 96	—	G44.0	C
154. Valyl-tRNA synthetase (EC 6.1.1.9)	84 × 123	—	E106.0	CG

[a] Protein names are from Nomenclature Committee (1979) and Bachmann and Low (1980). Common names by which some proteins are more generally known are in parentheses. EC numbers are from Nomenclature Committee (1979).

[b] Each protein spot is located on the reference autoradiograms in Fig. 1a (pH 5–7) and 1b (nonequilibrium) by its coordinates (abscissa × ordinate). When necessary, coordinate numbers are rounded down.

[c] Alphanumeric characters are as described in Pedersen *et al.* (1978). Letter prefixes refer to the position of the protein in the axis parallel to the isoelectric focusing dimension (A to I, acid to base), Numbers refer to the apparent molecular weight of the protein, determined by its mobility in the SDS electrophoresis dimension (26.8 indicates molecular weight of 26,800). This nomenclature is approximate.

[d] The letters designating the methods used to identify proteins are as described in Table I.

[e] Proteins repressed by growth in glucose-containing media or induced by specific growth conditions do not appear on the autoradiograms in Fig. 1. Coordinates given refer to their locations when cells are grown under appropriate conditions.

[f] The large, dark spot at these coordinates is F88.0, which obscures G99.0 when cells are grown in methionine-containing medium, synthesis of F88.0 is repressed and G99.0 can be clearly seen. Similarly, gels of extract of a mutant, CBK040 (*metE*::Tn5), which does not produce F88.0, allow the clear identification of G99.0 as asparaginyl-tRNA synthetase.

[g] Not present in extracts of *E. coli* B/r strain NC3. Coordinates given were obtained by comparison with gels of *E. coli* K12 strain W3110.

[h] This small protein spot is obscured by the larger spot located at 61 × 82. It is discernible on gels of extracts from other strains of *E. coli.*

[i] The unexpected finding that this spot corresponds to two different enzymes (proteins number 49 and 79) has prompted further studies, currently underway.

[j] IHF subunits α and β have previously been resolved only on urea–SDS gels (H. Nash, personal communication).

[k] These proteins are solubilized when the cell extract is prepared by boiling in the presence of SDS and, therefore appear faintly or not at all on the autoradiograms of Fig. 1.

[l] These proteins do not appear on the autoradiogram in Fig. 1B. Some nonequilibrium gels have more proteins to the basic side of those shown in Fig. 1b, with which these ribosomal subunit proteins comigrate. Coordinates given are the approximate locations of the proteins if they were present on the reference gel.

TABLE IIIA

Protein Spots Identified on pH 5–7 Isoelectric Focusing Two-Dimensional Gels of *E. coli* Extracts[a]

Location	Protein Number	Location	Protein Number
13 × 57	132	63 × 48	137
25 × 78	139	63 × 106	145
27 × 102	19	64 × 63	49,79
28 × 87	152	64 × 83	60
29 × 33	102	65 × 91	131
32 × 95	43	65 × 93	11
32 × 102	19	66 × 120	77
35 × 121	56	67 × 25	73
36 × 108	5	67 × 100	147
41 × 121	56	67 × 124	88
41 × 123	72	69 × 87	78
42 × 108	8	70 × 54	140
50 × 96	17	70 × 91	80
50 × 101	151	70 × 99	4
50 × 116	57	71 × 82	16
51 × 106	2	71 × 92	131
51 × 113	35	71 × 97	71
52 × 78	29	71 × 98	51
52 × 125	88	71 × 111	13
53 × 77	134	72 × 93	11
53 × 87	75	72 × 104	27
53 × 105	26	73 × 70	70
54 × 108	91	74 × 96	87
55 × 116	146	74 × 98	28
56 × 21	81	74 × 122	90
56 × 95	150	74 × 123	24
58 × 93	149	75 × 102	136
58 × 117	30	76 × 108	63
58 × 118	30	77 × 69	150
58 × 124	53	77 × 88	78
59 × 121	10,142	77 × 93	139
60 × 76	94	77 × 101	33
60 × 83	48	77 × 102	40
60 × 100	38	77 × 105	6
60 × 114	61	77 × 110	9
61 × 64	1	78 × 34	117
61 × 92	18	78 × 107	63
61 × 96	153	78 × 108	63
61 × 103	47	79 × 81	144
61 × 108	67	79 × 107	133
61 × 121	3	79 × 109	76
62 × 104	36	79 × 115	42

TABLE IIIA (*continued*)

Location	Protein Number	Location	Protein Number
80 × 119	85	97 × 93	15
80 × 126	17	98 × 92	135
81 × 34	117	98 × 114	25
81 × 93	14	99 × 112	25
81 × 101	33	100 × 52	141
81 × 105	34	103 × 99	2
82 × 108	63	103 × 107	62
82 × 109	63	103 × 108	62
82 × 124	31	104 × 108	62
83 × 79	86	104 × 109	62
83 × 80	86	104 × 118	92
83 × 105	34	105 × 108	62
83 × 109	58	105 × 109	62
83 × 112	58	106 × 107	44
84 × 109	63	106 × 109	62
84 × 123	154	106 × 110	62
85 × 98	52	106 × 113	114
85 × 128	135	108 × 113	23
86 × 109	58	109 × 16	37
86 × 112	58	109 × 22	143
87 × 35	117	110 × 109	82
87 × 70	1	113 × 49	32
87 × 120	75	114 × 24	104
90 × 91	93	115 × 85	68
90 × 117	84	116 × 24	101
90 × 121	55	116 × 111	65
91 × 73	46	119 × 85	69
93 × 97	59	124 × 91	83

[a] Locations of identified protein spots in Fig. 1a are listed in order of increasing abscissa coordinate. Protein number refers to the rank order of identified proteins in Table II.

exhibit strain differences only the B strain coordinates are given. In many cases no protein spot is visible in Fig. 1 at a set of coordinates for a particular protein. These are usually instances of proteins that are induced only under growth conditions other than glucose minimal medium at 37°C.

Original autoradiograms of the reference gels display approximately 1100–1200 protein spots, which represent perhaps 30–40% of the proteins that could maximally be encoded by the *E. coli* chromosome. The

TABLE IIIB

Protein Spots Identified on Nonequilibrium pH Gradient Two-Dimensional Gels of *E. Coli* Extracts[a]

Location	Protein Number	Location	Protein Number
12 × 20	113	80 × 69	7
13 × 25	124	80 × 78	39
14 × 22	110	81 × 29	66
15 × 25	109	83 × 61	138
23 × 18	112	83 × 78	39
24 × 33	105	83 × 79	152
26 × 27	128	83 × 101	64
27 × 22	129	84 × 30	102
30 × 20	122	84 × 76	12
30 × 21	121	84 × 86	54
30 × 25	107	87 × 94	2
30 × 29	127	88 × 51	132
32 × 24	123	88 × 52	132
34 × 32	125	91 × 18	81
35 × 37	116	92 × 104	138
37 × 30	103	93 × 44	137
37 × 50	100	94 × 24	73
38 × 17	111	95 × 75	16
38 × 21	108	95 × 118	135
40 × 26	106	96 × 88	87
40 × 62	97	96 × 89	28
42 × 19	21	96 × 107	30
42 × 53	89	96 × 108	30
44 × 18	22	97 × 31	117
45 × 30	126	97 × 111	3
46 × 46	99	98 × 32	117
46 × 66	95	99 × 32	117
47 × 19	130	99 × 109	85
47 × 50	89	101 × 73	86
48 × 27	118	101 × 74	86
50 × 19	50	101 × 106	42
52 × 21	120	103 × 84	93
52 × 32	119	104 × 116	17
55 × 57	98	105 × 17	143
64 × 72	96	105 × 89	59
65 × 57	20	108 × 103	25
67 × 37	12	108 × 106	25
69 × 64	45	110 × 91	2
70 × 81	74	111 × 104	114
76 × 67	115	112 × 99	44
76 × 96	41	112 × 118	135
79 × 56	140	114 × 104	23
79 × 94	148		

[a] Locations of identified protein spots in Fig. 1b are listed in order of increasing abscissa coordinate. Protein number refers to the rank order of identified proteins in Table II.

discrepancy is unlikely to be caused by saturation of the resolving power of the gels (see O'Farrell, 1975). More likely explanations are (1) failure to detect polypeptides present in only a few copies per cell, especially small ones; (2) absence of appropriate conditions for inducing some proteins; (3) loss of some hydrophobic or unstable proteins during the extraction procedure; and (4) the existence of silent (mutationally inactivated) genes or noncoding stretches of DNA. Instances are known for each of these situations. Multicopy hybrid plasmids have allowed us to detect many more polypeptides than are present on the reference gel (P. L. Bloch and V. Vaughn, unpublished). Our study of other growth conditions (anaerobiosis, high and low temperature, etc) has revealed many additional proteins (see Herendeen *et al.*, 1979). Different extraction procedures have different effectiveness in the recovery of ribosomal and membrane proteins. Several instances of silent genes are known. What we do not know is the relative contribution of each of these four factors in accounting for the discrepancy.

IV. Applications in *E. coli* Research

The gel electrophoresis technique has proved useful in a dozen or more different types of applications in *E. coli* studies. Because polypeptides can be sampled from the gels and their content of different radioactive isotopes readily measured, the use of appropriate labeling protocols makes it possible for one to determine for a protein of interest: (1) *in vivo* amount, (2) rate of synthesis, (3) rate of degradation, (4) processing rate, (5) mRNA life time, (6) distribution among subcellular compartments, and (7) regulation under different environmental conditions. Lengthy and tedious purifications are unnecessary, and the protein need not have an assayable enzyme activity (some examples of these applications are referenced in Bloch *et al.*, 1980).

These applications are of primary interest in more or less classical studies of individual protein structure, function, and regulation. Were these the sole uses of two-dimensional gels the expense and effort of constructing a catalog of protein spots would still be worthwhile and would be necessary to complete a molecular accounting of the cell. The value of two-dimensional gel analysis of cellular proteins for these studies is far exceeded, however, by unique applications in the areas of physiology and genetics. Two of these applications are discussed briefly.

Increasingly it is being recognized that there are large sets of genes controlled *en bloc* by various cellular regulatory signals. In higher plants and animals coordinate activation of genes is a common feature in cellular differentiation, hormonal regulation of metabolism, and the response to physical, chemical, or nutritional stress. One of the most dramatic and ubiquitous examples of a coordinate multigene response to stress is the heat shock response exhibited by many if not all living cells. Dozens of similar multigene control systems are known.

Bacteria share with eucaryotic cells the property of coordinately regulating genes that are functionally related. In many instances genes that are exclusively associated with a single function (e.g., biosynthesis of histidine or of tryptophan) are structurally linked into a single transcriptional unit, an operon. This arrangement is efficient and effective but is appropriate chiefly for genes that should nearly always be coregulated. This limitation, and perhaps other factors, has led to a second, very common situation: some operons, which are scattered on the genetic map and are usually adjusted to different individual levels of activity, can in some circumstances come under a common global control that coordinates their expression. Examples of this mode of multigene control include the guanosine tetraphosphate (ppGpp)-mediated response (stringency) to amino acid limitation (Gallant, 1979), the cyclic AMP (cAMP)-mediated response (release of catabolite repression) to carbon and energy source limitation (Adhya and Garges, 1982), and the SOS response to DNA damage (Little and Mount, 1982). The large number of genes that are controlled by ppGpp and cAMP has led to these signal molecules being designated *alarmones* (Stephens *et al.*, 1975). These three examples illustrate the variety of molecular devices that bacteria such as *E. coli* have evolved to override individual controls on single operons.

These control systems were in large part discovered almost by accident in studies of individual genes and their regulation. The resolution of individual cellular proteins on two-dimensional polyacrylamide gels is uniquely suited for discovering sets of coregulated gene products and makes it possible to launch studies aimed explicitly at discovering and analyzing global (multioperon) control systems. Recognizing coregulated gene products is done by direct observation of extracts of normal cells—mutants need not be isolated—and therefore even proteins essential to growth can be studied. Furthermore, the function of the gene products need not be known initially, because activity assays are unnecessary.

An independent approach was made possible by the development of techniques for fusing genes (such as *lacZ*) randomly to chromosomal promoters (Casadaban and Cohen, 1979; Beckwith, 1981). After being treated with Mu phage engineered to produce fusions, cells can be plated and large numbers of colonies screened for production of the fusion gene product under whatever environmental condition is being studied. In this way promoters can be identified that respond to the stimulus and they can then be mapped and their regulation analyzed. This approach is limited to discovering promoters of nonessential genes, but it is reasonable to assume that many genes that are strongly induced by a given stimulus are not essential under other conditions.

Both approaches have led to recent advances in the study of global control systems in *E. coli*. The use of two-dimensional gels to characterize the effect of temperature on the regulation of synthesis of individual proteins (Herendeen *et al.*, 1979) led to the discovery of the high-temperature production (HTP) regulon of *E. coli*, a set of 13 genes that are transiently and strongly induced by a shift up in temperature (Neidhardt and VanBogelen, 1981; Yamamori and Yura, 1982; Neidhardt *et al.*, 1982). (Coregulated genes that share a common regulator molecule are called a regulon.) The genetic approach using Mu *d*1 fusions was used by Wanner and McSharry (1982) to reveal twenty promoters that are phosphate starvation–induced (*psi*). Work on the HTP and *psi* control systems illustrates the usefulness of these individual biochemical and genetic approaches but shows also the complementary nature of the techniques. The *psi* system can now be usefully investigated with two-dimensional gels to identify the protein products of the genes governed by the *psi*-promoters, as well as to discover whether other *psi* genes have escaped detection (because, for example, their products are always essential). Because only four of the thirteen HTP proteins have mapped structural genes, Mu *d*1 approach is one of several techniques that should be used to complete the task of finding the HTP genes.

The second unique application of two-dimensional gels is in the area of genetics, and it provides, in principle, a means to genetically map the structural genes for all the proteins made by a bacterial cell. Recombinant DNA techniques make it easy to construct plasmids containing short segments of the *E. coli* (or other) chromosome. Banks of colonies have been constructed that contain nearly all of the *E. coli* genome on overlapping segments. Techniques exist that allow selective expression of genes carried on recombinant plasmids. By two-dimensional gel analysis one can therefore learn what proteins are encoded by genes on each segment. Because the segments are overlapping relative ordering of the genes is possible. If a plasmid produces a known gene product or complements a particular mutant gene, the ordered array can be related to the standard genetic map.

In practice of course there are difficulties. The procedures for selective expression are not perfect, hundreds of plasmid clones must be examined, the tests for protein matches are extremely tedious and must be confirmed by comigration, and some proteins are made in multiple isoelectric species under the conditions required for selective plasmid gene expression. Nonetheless, using minicells (Reeve, 1979), maxicells (Sancar *et al.*, 1979), and cells recovering from chloramphenicol treatment (Neidhardt *et al.*, 1980), our laboratory has examined over 200 plasmids of the Clarke–Carbon colony bank (Neidhardt *et al.*, 1983). A total of 1234 individual protein products have been detected. Not all of these proteins can be assigned chromosomal locations, but a large number can. Approximately 150 of the 200 tested plasmids can be placed on the genetic map, some because one or more of the proteins they produce is the product of a known gene and some because they produce proteins that are also produced by plasmids already located on the map. In all, we have placed approximately 1000 different proteins on the genetic map with a precision of ~0.3 units (the average length of DNA in the plasmids is 14 kilobase pairs or 0.3 units).

Work now in progress is hoped to lead to the construction of a combined genetic and biochemical catalog of the proteins of *E. coli*. Examples of the usefulness of such a catalog include (1) defining and locating genes for newly recognized proteins, (2) identifying the protein products of newly discovered genes, (3) identifying proteins on gels, (4) discovering the protein components of regulons and other "global" control systems, (5) aiding the preparation of clones of particular genes, and (6) describing the functional organization of the *E. coli* chromosome.

References

Adhya, S., and Garges, S. (1982). How cyclic AMP and its receptor protein act in *Escherichia coli*. *Cell* **29**, 287–289.

Ames, G. F. L., and Nikaido, K. (1976). Two-dimensional gel electrophoresis of membrane proteins. *Biochemistry* **15**, 616–623.

Bachman, B. J., and Low, K. B. (1980). Linkage map of *Escherichia coli* K-12, edition 6. *Microbiol. Rev.* **44**, 1–56.

Beckwith, J. (1981). A genetic approach to characterizing complex promoters in *E. coli*. *Cell* **23**, 307–308.

Berg, C. M., and Berg, D. E. (1981). Bacterial transposons. *In* "Microbiology 1981" (D. Schlessinger, ed.), pp. 107–116. Amer. Soc. for Microbiol., Washington, D.C.

Bloch, P. L., Phillips, T. A., and Neidhardt, F. C. (1980). Protein identifications on O'Farrell two-dimensional gels: Locations of 81 *Escherichia coli* proteins. *J. Bacteriol.* **141**, 1409–1420.

Casadaban, M. J., and Cohen, S. (1979). Lactose genes fused to exogenous promoters in one step using a Mu-*lac* bacteriophage: *In vivo* probe for transcriptional control sequences. *Proc. Natl. Acad. Sci. USA* **76**, 4530–4533.

Chopra, I., and Shales, S. W. (1980). Comparison of the polypeptide composition of *Escherichia coli* outer membranes prepared by two methods. *J. Bacteriol.* **144**, 425–427.

Clarke, L., and Carbon, J. (1976). A colony bank containing synthetic ColE1 hybrid plasmids representative of the entire *E. coli* genome. *Cell* **9**, 91–99.

Cleveland, D. W., Fischer, S. G., Kirschner, M. W., and Laemmli, U. K. (1977). Peptide mapping by limited proteolysis in sodium dodecyl sulfate and analysis by gel electrophoresis. *J. Biol. Chem.* **252**, 1102–1106.

Collins, J., and Bruning, H. J. (1978). Plasmids useable as gene-cloning vectors in an *in vitro* packaging by coliphage λ: "cosmids." *Gene* **4**, 85–107.

Drahos, D. J., and Hendrix, R. W. (1982). Effect of bacteriophage lambda infection on synthesis of *groE* protein and other *Escherichia coli* proteins. *J. Bacteriol.* **149**, 1050–1063.

Gallant, J. (1979). Stringent control in *E. coli*. *Annu. Rev. Genet.* **13**, 393–415.

Hackett, C. S., and MacGregor, C. H. (1981). Synthesis and degradation of nitrate reductase in *Escherichia coli*. *J. Bacteriol.* **146**, 352–359.

Hayward, R. S., and Fyfe, S. (1978). Over-synthesis and instability of sigma protein in a merodiploid strain of *Escherichia coli*. *Mol. Gen. Genet.* **159**, 89–99.

Hazelbauer, G. L., and Engstrom, P. (1981). Multiple forms of methyl-accepting chemotaxis proteins distinguished by a factor in addition to multiple methylation. *J. Bacteriol.* **145**, 35–42.

Herendeen, S. L., VanBogelen, R. A., and Neidhardt, F. C. (1979). Levels of major proteins of *Escherichia coli* during growth at different temperatures. *J. Bacteriol.* **139**, 185–194.

Hirshfield, I. N., Bloch, P. L., VanBogelen, R. A., and Neidhardt, F. C. (1981). Multiple forms of lysyl-transfer ribonucleic acid synthetase in *Escherichia coli*. *J. Bacteriol.* **146**, 345–351.

Kessler, R. E. (1981). Crossed immunoelectrophoresis from sodium dodecyl sulfate–polyacrylamide gels after fixation and staining and without nonionic detergent intermediate layers. *Anal. Biochem.* **116**, 129–132.

Little, J. W., and Mount, D. W. (1982). The SOS regulatory system of *Escherichia coli*. *Cell* **29**, 11–22.

McConkey, E. H. (1979). Double-label autoradiography for comparison of complex protein mixtures after gel electrophoresis. *Anal. Biochem.* **96**, 39–44.

Moriya, T., and Hori, K. (1981). Studies on *Escherichia coli* chromosome proteins. I. Analysis of the proteins by two-dimensional gel electrophoresis. *Biochim. Biophys. Acta* **653**, 169–184.

Neidhardt, F. C., and VanBogelen, R. (1981). Positive regulatory gene for temperature-controlled proteins in *Escherichia coli*. *Biochem. Biophys. Res. Commun.* **100**, 894–900.

Neidhardt, F. C., Bloch, P. L., Pedersen, S., and Reeh, S. (1977). Chemical measurement of steady-state levels of ten aminoacyl-transfer ribonucleic acid synthetases in *Escherichia coli*. *J. Bacteriol.* **129**, 378–387.

Neidhardt, F. C., Wirth, R., Smith, M. W., and VanBogelen, R. A. (1980). Selective synthesis of plasmid-coded proteins by *Escherichia coli* during recovery from chloramphenicol treatment. *J. Bacteriol.* **143**, 535–537.

Neidhardt, F. C., Phillips, T. A., VanBogelen, R. A., Smith, M. W., Georgalis, Y., and Subramanian, A. R. (1981). Identity of the B56.5 protein, the A-protein, and the *groE* gene product of *Escherichia coli*. *J. Bacteriol.* **145**, 513–520.

Neidhardt, F. C., VanBogelen, R. A., and Lau, E. T. (1982). The high temperature regulon of *Escherichia coli*. *In* "Heat Shock from Bacteria to Man" (M. Schlesinger, M. Ashburner, and A. Tissières, eds.), pp. 139–145. Cold Spring Harbor Lab., Cold Spring Harbor, New York.

Neidhardt, F. C., Vaughn, V., Phillips, T. A., and Bloch, P. L. (1983). Gene-protein index of *Escherichia coli* K-12. *Microbiol. Revs.* **47**, 231–284.

Nomenclature Committee of the International Union of Biochemistry. (1979). "Enzyme Nomenclature." Academic Press, New York.

O'Farrell, P. H. (1975). High resolution two-dimensional electrophoresis of proteins. *J. Biol. Chem.* **250**, 4007–4021.

O'Farrell, P. Z., Goodman, H. M., and O'Farrell, P. H. (1977). High resolution two-dimensional electrophoresis of basic as well as acidic proteins. *Cell* **12**, 1133–1142.

Pedersen, S., Reeh, S., Parker, J., Watson, R. J., Friesen, J. D., and Fiil, N. P. (1976). Analysis of the proteins synthesized in ultraviolet light-irradiated *Escherichia coli* following infection with the bacteriophages λ *drif-18* and λ *dfus-3*. *Mol. Gen. Genet.* **144**, 339–344.

Pedersen, S., Bloch, P. L., Reeh, S., and Neidhardt, F. C. (1978). Patterns of protein synthesis in *E. coli:* A catalog of the amount of 140 individual proteins at different growth rates. *Cell* **14**, 179–190.

Phillips, T. A., Bloch, P. L., and Neidhardt, F. C. (1980). Protein identifications on O'Farrell two-dimensional gels: locations of 55 additional *Escherichia coli* proteins. *J. Bacteriol.* **144**, 1024–1033.

Reeve, J. (1979). Use of minicells for bacteriophage directed polypeptide synthesis. *Methods Enzymol.* **68**, 493–503.

Sancar, A., Hack, A. M., and Rupp, W. D. (1979). Simple method for identification of plasmid-coded protein. *J. Bacteriol.* **137**, 692–693.

Sato, T., Ito, K., and Yura, T. (1977). Membrane proteins of *Escherichia coli* K-12: two-dimensional polyacrylamide gel electrophoresis of inner and outer membranes. *Eur. J. Biochem.* **78**, 557–567.

Spratt, B. G. (1977). Properties of the penicillin-binding proteins of *Escherichia coli* K12. *Eur. J. Biochem.* **72**, 341–352.

Stephens, J. C., Artz, S. W., and Ames, B. N. (1975). Guanosine 5'-diphosphate 3'-diphosphate (ppGpp): Positive effector for histidine operon transcription and general signal for amino acid deficiency. *Proc. Natl. Acad. Sci. USA* **72**, 4389–4393.

Wanner, B. L., and McSharry, R. (1982). Phosphate-controlled gene expression in *Escherichia coli* K12 using Mu*dl*-directed *lacZ* fusions. *J. Mol. Biol.* **158**, 347–363.

Wilson, D. B. (1982). Effect of the lambda *S* gene product on properties of the *Escherichia coli* inner membrane. *J. Bacteriol.* **151**, 1403–1410.

Yamamori, T., and Yura, T. (1982). Genetic control of heat-shock protein synthesis and its bearing on growth and thermal resistance in *Escherichia coli* K-12. *Proc. Natl. Acad. Sci. USA* **79**, 860–864.

Young, F. S., and Neidhardt, F. C. (1978). Effect of inhibitors of elongation factor Tu on the metabolic regulation of protein synthesis in *Escherichia coli*. *J. Bacteriol.* **135**, 675–686.

CHAPTER 14

Catalog of HeLa Cell Proteins

RODRIGO BRAVO*
JULIO E. CELIS
Division of Biostructural Chemistry
Department of Chemistry
Aarhus University
Aarhus, Denmark

I. Introduction 445
II. Two-Dimensional Gel Electrophoretic Separations of HeLa Cell Proteins 446
III. Some Characteristics of HeLa Cell Proteins 447
A. Variation during the Cell Cycle 447
B. Distribution in Cytoplasts and Karyoplasts 466
C. Distribution in Triton Cytoskeletons and Triton Supernatants 469
D. Presence in High Salt-Extracted Cytoplast Skeletons 471
E. Phosphorylation 473
IV. Transformation-Sensitive Polypeptides 474
V. Conclusions 474
References 475

I. Introduction

The advantages of preparing a catalog of human proteins resolved by two-dimensional gel electrophoresis is obvious. It would greatly facilitate the flow of information among laboratories using this technology to study a variety of biological problems. (For references see special issue

* Present affiliation: European Molecular Biology Laboratory, Heidelberg 6900, Federal Republic of Germany.

TWO-DIMENSIONAL GEL
ELECTROPHORESIS
OF PROTEINS

ISBN 0-12-164720-X

of *Clinical Chemistry* **28**, 1982.) Fortunately, the two-dimensional protein patterns of cultured human cells are qualitatively similar and a large number of proteins seem to be expressed, albeit at different levels, in various human cultured cells so far studied (Bravo and Celis, 1980, 1982b; Celis and Bravo, 1981; Bravo *et al.*, 1983). Because our original work was devoted mainly to HeLa cells and given the significant amount of information already gathered (Bravo and Celis, 1980; Bravo *et al.*, 1981a, 1982a,b; Bellatin *et al.*, 1982), we have chosen this cell line to prepare such a catalog (Bravo *et al.*, 1981a; Bravo and Celis, 1982a). Improvement in the gel separation as well as in labeling techniques have so far resulted in the separation of nearly 1400 polypeptides. Given the reproducibility of the gel patterns and the fact that polypeptides can be assigned standard coordinates, it now seems possible to use computer analysis to detect, quantitate, and store the information contained in these gels. (Lutin *et al.*, 1978; Garrels, 1979; Bossinger *et al.*, 1979; Anderson and Anderson, 1979; Likpin and Lemkin, 1980; Anderson *et al.*, 1981; Lester *et al.*, 1982; Taylor *et al.*, 1982; Miller *et al.*, 1982; Skolnick *et al.*, 1982; Merril and Goldman, 1982; see also chapter by Garrels).

II. Two-Dimensional Gel Electrophoretic Separation of HeLa Cell Proteins

A total of 1357 polypeptides, 946 acidic (IEF; O'Farrell, 1975) and 411 basic (NEPHGE; O'Farrell *et al.*, 1977), from whole asynchronous HeLa cells have been separated by two-dimensional gel electrophoresis and cataloged (Bravo *et al.*, 1981a; Bravo and Celis, 1982a). Of these 1266 were detected by using [^{35}S]methionine labeling and the rest were re-

Figure 1. Two-dimensional gel electrophoresis map (IEF, NEPHGE) of [^{35}S]methionine-labeled polypeptides from 100 asynchronous HeLa cells labeled for 24 h. In IEF the pH ranges from 7.5 (left) to 4.5 (right). In NEPHGE the pH varies from 7.5 (right) to 10.5 (left). To number the spots, we selected a total of 101 original numbers, 34 basic (0–34, NEPHGE) and 67 acidic (0–66, IEF). In order to extend the numbering system to all 1357 polypeptides detected, we have indicated the extra spots between two original numbers with letters or with letters followed by a number. In cases where there were more than 26 spots between two of the original numbers (for example, between IEF 1 and 2, Table I) we have indicated those spots with the letter *z*, followed by small numbers. Polypeptides have been numbered from higher to lower molecular weight and from the origin of the first dimension. All polypeptides are numbered in this figure and, because of space problems, in many cases we have placed the original number above the arrow. From Bravo *et al.* (1981a) and Bravo and Celis (1982a).

vealed by silver staining or by labeling with a mixture of 16 ^{14}C-labeled amino acids. All 1357 polypeptides have been numbered and are indicated in Fig. 1. This figure shows fluorograms of two-dimensional gels (IEF, NEPHGE) of [^{35}S]methionine-labeled polypeptides from 100 asynchronous HeLa cells labeled for 24 h. The positions of those polypeptides not detected by [^{35}S]methionine labeling were determined by superimposing negatives of ^{14}C-labeled or silver-stained gels onto Fig. 1. Examples of basic polypeptides revealed only or mainly by silver staining or by ^{14}C-labeling are shown in Fig. 2 (arrows) and 3 (arrowheads). Similar examples of acidic polypeptides are indicated with arrowheads in Fig. 4.

The percentages of some of the major polypeptides labeled with a mixture of 16 ^{14}C-labeled amino acids (see Figs. 3 and 4) have been determined, and the data are presented in Tables I and II. Total actin (β and γ) is the major protein synthesized by HeLa cells (3.32% of total protein label). Other quantitated proteins of known identity correspond to α-actinin (0.55%; IEF 8), α- (0.69%; IEF 19) and β-tubulin (0.77%; IEF 25); IEF 24 (0.59%; mitochondrial) (Anderson, 1981; Mose Larsen *et al.*, 1982, 1983); vimentin (0.55%; IEF 26); keratin IEF 31 (0.08%) (Franke *et al.*, 1981; Bravo *et al.*, 1982a,b, 1983; Fey *et al.*, 1983); keratin IEF 36 (0.09%) (Franke *et al.*, 1981; Bravo *et al.*, 1982a,b, 1983; Fey *et al.*, 1983); keratin IEF 44 (0.07%) (Franke *et al.*, 1981; Bravo *et al.*, 1982a,b, 1983; Fey *et al.*, 1983) keratin IEF 46 (0.15%) (Franke *et al.*, 1981; Bravo *et al.*, 1982a,b, 1983; Fey *et al.*, 1983) and EF-1α (0.63% NEPGHE 12) (Coppard *et al.*, 1982).

III. Some Characteristics of HeLa Cell Proteins

A. Variation during the Cell Cycle

Tables I and II contain the data for 99 major polypeptides (IEF, NEPHGE) quantitated throughout the cell cycle (Bravo and Celis, 1980). Mitotic cells obtained by mechanical detachment as well as synchronized cells in the other stages of the cycle were labeled with [^{35}S]methionine for 30 min pulses or for long terms starting at the beginning of each phase. Given the reproducibility of the gels and the constancy of many polypeptides, we arbitrarily defined as variable only those polypeptides whose relative proportions varied by twofold or

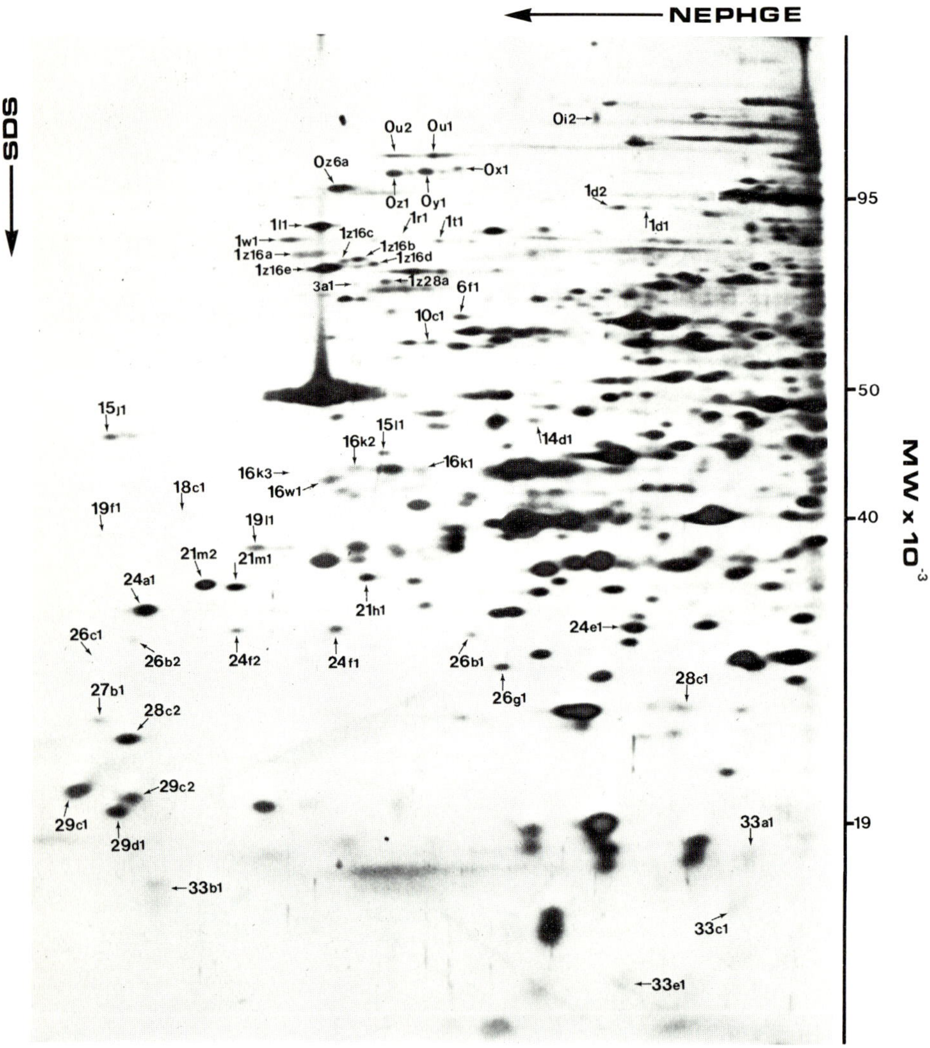

Figure 2. Two-dimensional map (NEPHGE) of HeLa basic proteins stained with $AgNO_3$. Proteins made visible with silver staining that are not detected with [^{35}S]methionine or a mixture of 16 ^{14}C-labeled amino acids are indicated in this figure by arrows. From Bravo and Celis (1982a).

more (Tables I and II). Polypeptides whose relative proportion is constant through the cell cycle are indicated with the letter *g* in Tables I and II (other characteristics). The letters *q* (S phase), *h* (mitosis), and *m* (mitosis and G2) indicate polypeptides whose relative proportion changed

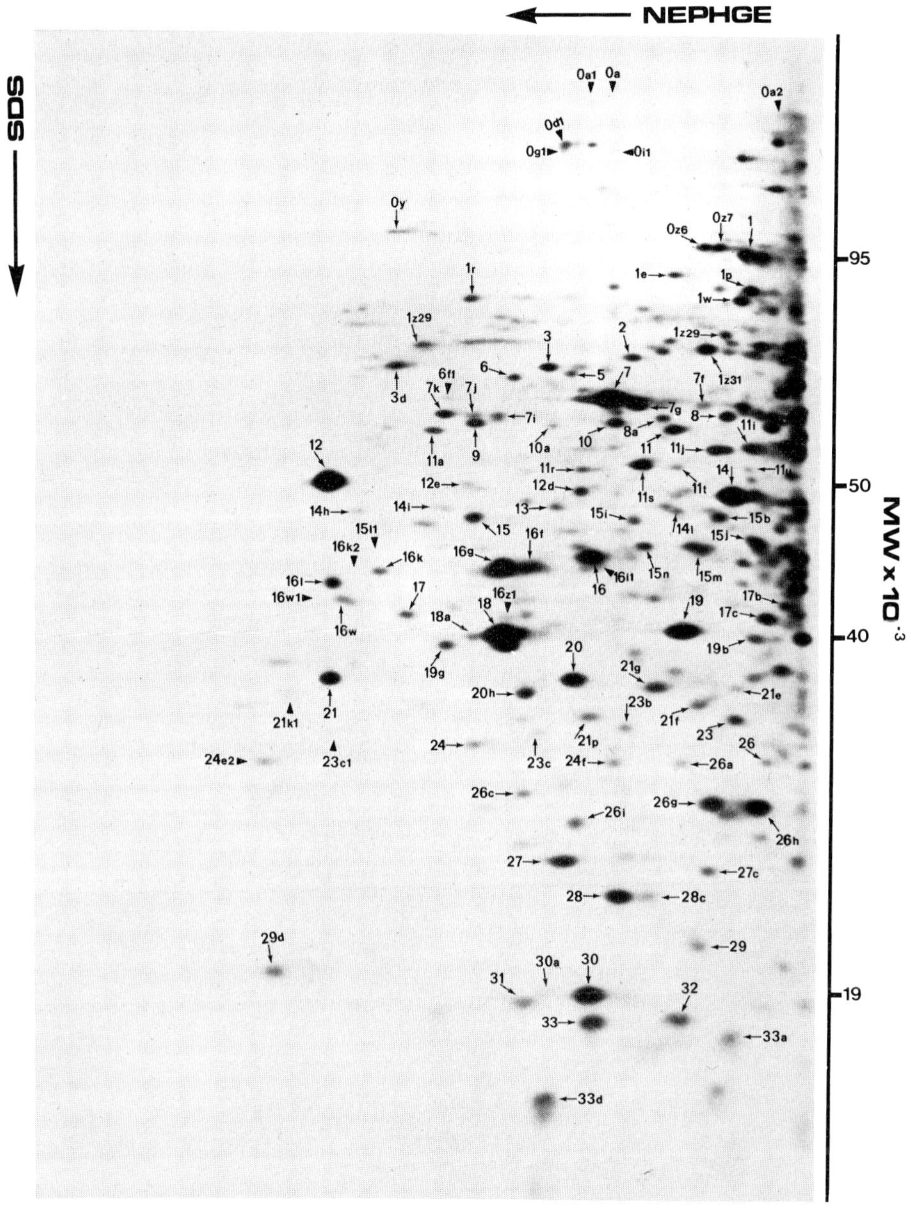

Figure 3. Two-dimensional map (NEPHGE) of HeLa basic proteins labeled for 16 h with a mixture of 16 ^{14}C-labeled amino acids. Cells grown directly in microtiter plates were labeled with 0.1 ml of homemade DMEM (1 g/liter, $NaHCO_3$) lacking amino acids and containing 10% dialyzed fetal calf serum and 50 μCi of a mixture of 16 ^{14}C-labeled amino acids (CFB 104, Amersham). The amino acid mixture was lyophilized before it was mixed with the medium. Approximately 1500 trichloroacetic acid–precipitable counts can be incorporated per cell with this labeling procedure (Bravo *et al.*, 1981c). Proteins indicated with arrowheads correspond to those not detected by [^{35}S]methionine labeling or that are weakly stained by silver. Proteins indicated with arrows have been quantitated (see Table II). From Bravo and Celis (1982a).

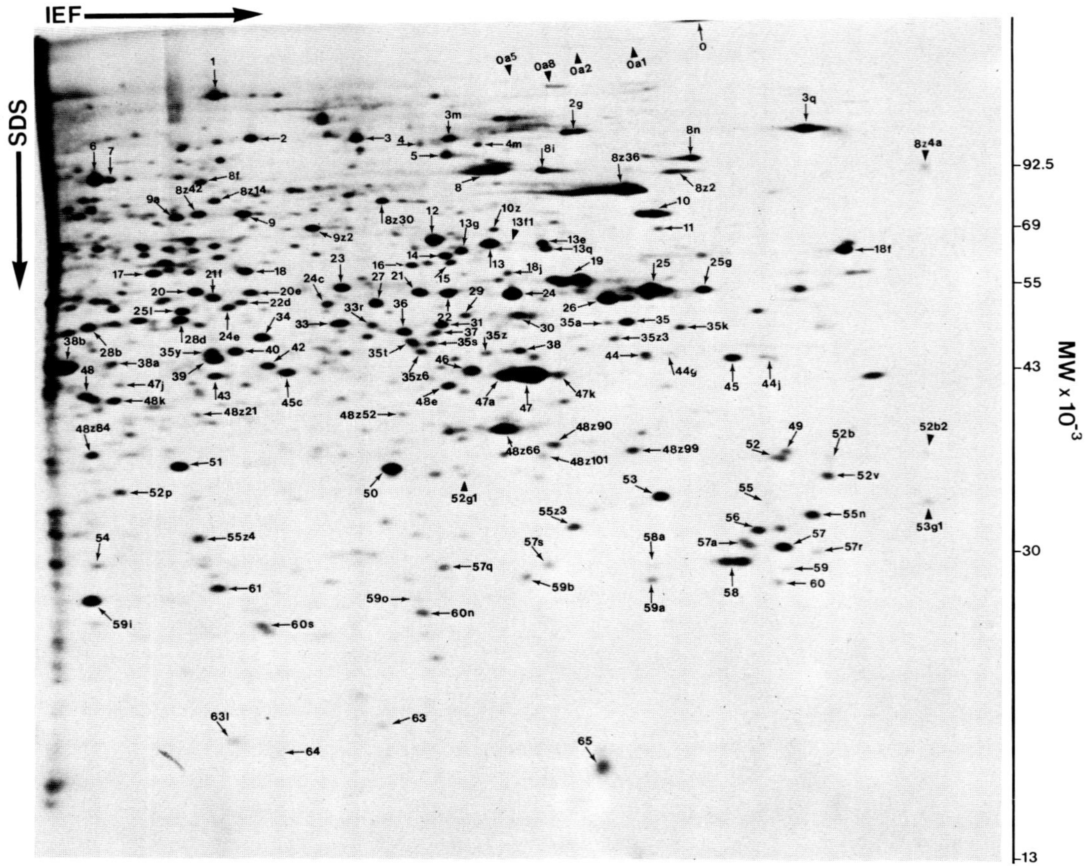

Figure 4. Two-dimensional map (IEF) of HeLa acidic proteins labeled for 16 h with mixture of 16 ^{14}C-labeled amino acids. Proteins indicated with arrowheads correspond to some of the polypeptides not detected by [^{35}S]methionine labeling. These polypeptides can also be detected with silver staining. Proteins indicated with arrows have been quantitated (see Table I). From Bravo and Celis (1982a).

TABLE I

Percentages and Characteristics of Some Major Acidic Proteins from HeLa Cells

Polypeptide	Coordinates molecular weight/mobility relative to β-actin[a]	Total label (%)	Number of spots with same original number	Amount found in (%)[c]		Amount found in (%)[d]		Other characteristics
				Cyto-plast	Karyo-plast	Triton cyto-skeleton	Triton super-natant	
0	nd/1.35	0.11	13 (a–c)					*e*
0a9								*e*
1	132/0.33	0.15	43 (a–z16)	+	+	48	52	
1s				−	+	+	−	*e*
1z4								*e*
1z8						+	−	
2	110/0.44	0.11	9 (a–h)	52	48	17	83	*f,g*
2b						+	−	
2g		0.12						
3	108/0.68	0.12	18 (a–q)	+	+	13	87	*g*
3K						−	+	
3m		0.12						
3q		0.26				−	+	*e*
4	100/0.83	0.04	20 (a–r)	+	+	17	83	*g*
4f						+	−	
4m		0.05						
5	96/0.89	0.04	19 (a–r)	+	+	12	88	*g*
5l						−	+	
5n						−	+	
5r						−	+	
6	95/0.09	0.48	2 (a)	+	+	8	92	*h*

(continued)

TABLE I (*continued*)

Polypeptide	Coordinates molecular weight/ mobility relative to β-actin[a]	Total label (%)[b]	Number of spots with same original number	Amount found in (%)[c]		Amount found in (%)[d]		Other characteristics
				Cyto-plast	Karyo-plast	Triton cyto-skeleton	Triton super-natant	
7	95/0.12	0.06	4 (a–c)	+	+	16	84	*h*
8(α-actinin)	94/0.96	0.55	73 (a–z45)	65	35	41	59	*g*
8g				–	+			
8f		0.06						
8i		0.17				–	+	
8n		0.16						
8v								*f*
8z2		0.11						
8z6						–	+	
8z9				+	–			
8z14		0.05		–	+	+	–	*f*
8z30		0.09		–	+	+	–	*f*
8z31						+	–	
8z36		1.40						*e,f,i*
8z42		0.09						
9	74/0.40	0.14	36 (a–z8)	+	+	17	83	*g*
9a		0.09						
9g				+	–	+	–	
9v				+	–			
9x						–	+	
9z2		0.08		+	+	–	+	
9z7						–	+	

10	72/1.28	0.29	29 (a–z1)	45	55	7	93	*e,f*
10g						–	+	
10q						–	+	
11	69/1.33	0.03	4 (a–c)	50	50	8	92	*g*
12	68/0.85	0.33	19 (a–r)	35	65	43	57	*j,f,g*
12g								*f*
13	67/0.93	0.31	23 (a–u)	32	68	44	56	*g*
13e		0.08		–	+	+	–	*f*
13g		0.07		–	+			*f*
13m				–	+			
13p				–	+			
13q		0.04		–	+	+	–	*f*
13t				–	+			
14	66/0.86	0.09	10 (a–i)	+	+	21	79	*f,g,i*
14h						–	+	*e*
15	65/0.87	0.04		58	42	29	71	*g*
16	65/0.78	0.04	3 (a–b)	+	+	23	77	*g*
17	64.5/0.70	0.10	14 (a–m)	+	+	24	76	*g*
17g				–	+			
18	64/0.39	0.11	24 (a–w)	+	+	23	77	*g*
18a								*e*
18f		0.19				–	+	
18j		0.04		–	+			
19, 19b (α-tubulin)	58/1.10; 1.14	0.69	6 (a–e)	+	+	30	70	*h*
20	61/0.27	0.14	9 (a–h)	+	+	28	72	*g*
20e		0.07						*e*
21	58/0.78	0.14	9 (a–h)	+	+	22	78	
21d				+	–			
21f		0.09						
22	57.5/0.83	0.18	6 (a–e)	+	+	10	90	*g*
22d		0.05						
23	60/0.64	0.12		+	+	26	74	*g*

(continued)

TABLE I (*continued*)

Polypeptide	Coordinates molecular weight/ mobility relative to β-actin[a]	Total label (%)[b]	Number of spots with same original number	Amount found in (%)[c]		Amount found in (%)[d]		Other characteristics
				Cyto-plast	Karyo-plast	Triton cyto-skeleton	Triton super-natant	
24 (mitochondrial)	56/0.97	0.59	8 (a–g)	46	54	66	34	*j,g,k*
24c		0.06						
24e		0.06						
24g								*h* (corresponds to m_1^l)
25 (β-tubulin)	56/1.28	0.77	13 (a–l)	60	40	30	70	*h*
25b				+	–			
25g		0.13				–	+	
25l		0.09				–	+	*f*
26 (vimentin)	54/1.19	0.66	9 (a–g)	75	25	99	1	*j,f,h*
26e (phospho-vimentin)				+	–			*j,e,h* (corresponds to m_2^l)
27	55/0.69	0.12		41	59	11	89	*h*
28	53.5/0.15	0.09	17 (a–o)	+	+	26	74	*g*
28b		0.13						
28d		0.12				–	+	
28f				+	+			
28o				–	+			
29	52.5/0.86	0.05	4 (a–c)	+	+	33	67	*g*
30	52.5/0.99	0.06	4 (a–b)	+	+	18	82	*h,m*
31[n] keratin	50/0.84	0.08		70	30	90	10	*j,g*

32 (vd1)[o]	52.5/1.24	0.01				87	13	*m*
33	49.5/0.64	0.12	23 (a–v)	–	100	88	12	*j,g*
33l				+	+			
33p								*e,h* (corresponds to m^{l}_{8})
33q								*e,h* (corresponds to m^{l}_{6})
33r		0.46						*f*
34	49/0.43	0.12	3 (a–b)	+	+	20	80	*g*
35	49/1.18	0.12	37 (a–z8)	60	40	90	10	*j,g*
35a		0.03						
35a1								*e*
35b								*e,h* (corresponds to m^{l}_{7})
35l								*e*
35m								*e*
35s		0.04				–	+	
35t		0.07				+	–	*e*
35y		0.10						
35z3		0.04				–	+	
35z5		0.03		–	+	–	+	
35z6		0.03		–	+	+	–	
36[n] keratin	48.5/0.74	0.09		60	40	89	11	*j,p,g*
37	48.5/0.83	0.08		60	40	83	17	*j,e,p,h*
38	46/0.98	0.06	3 (a–b)	–	100	90	10	*g*
38a		0.08		+	–			
38b		0.69				–	+	
39	45.5/0.31	0.18	2 (a)	+	+	22	78	*f,g*
40	47/0.36	0.10		+	+	31	64	*g*
41 (vd3)[o]	47/1.32					57	43	*m*
42	45/0.44	0.08	13 (a–k)	+	+	14	86	*g*
42d				–	+			
42f								*e*
42j				+	–			

(continued)

TABLE I (*continued*)

Polypeptide	Coordinates molecular weight/ mobility relative to β-actin[a]	Total label (%)[b]	Number of spots with same original number	Amount found in (%)[c]		Amount found in (%)[d]		Other characteristics
				Cyto- plast	Karyo- plast	Triton cyto- skeleton	Triton super- natant	
57a		0.08						
57q		0.04		–	+			*f*
57r		0.04						
57s		0.04				–	+	f
58	29/1.38	0.22	3 (a–b)	+	+	5	95	*g*
58a		0.02						*f*
59	28/1.38	0.02	17 (a–p)	+	+	–	100	*p,g*
59a		0.04				–	+	
59b		0.05				–	+	
59i		0.17				–	+	
59o		0.02				–	+	
60	26/1.43	0.03	27 (a–y)	+	+	40	60	*g*
60d1								*f*
60f				+	–			
60u		0.05				–	+	
60s		0.10				–	+	
61	27/0.29	0.08		+	+	46	54	*g*
62	23/1.34		20 (a–s)			4	96	*g*
63	20/0.64	0.04	13 (a–l)			–	100	*f,g*
63l		0.05						

64	18/0.45	0.04	9 (a–h)			8	92	*g*
64a				+	–			
64b				+	–			
65	16/1.15	0.14		+	+	12	88	*f,g*
66	16/1.24	0.01	17 (a–p)	+	+	–	100	*g*
66g				+	–			

[a] From Bravo *et al.* (1981a); nd, not determined.
[b] Average of at least two determinations.
[c] From Bravo *et al.* (1981b).
[d] From Bravo *et al.* (1982b).
[e] Phosphoprotein.
[f] Neoplastic transformation-sensitive polypeptides common to both epithelial and fibroblast human cultured cells (Bravo and Celis, 1982b).
[g] The relative proportion of these polypeptides is constant throughout the cell cycle (Bravo and Celis, 1980). Only polypeptides whose relative proportion changed by twofold or more were considered as variable in that study.
[h] The relative proportion of these polypeptides changes in mitosis (Bravo and Celis, 1980).
[i] Stress protein.
[j] Present in low and high salt (*LHL*)-extracted cytoplast cytoskeletons enriched in intermediate filaments (Bravo *et al.*, 1982b).
[k] Mitochondrial protein (Mose Larsen *et al.*, 1982, 1983)
[l] These polypeptides have been shown to correspond to modifications of cytoskeletal polypeptides that increase significantly in mitosis (Bravo and Celis, 1980; Bravo *et al.*, 1982a,b).
[m] The relative proportion of these polypeptides changes in G_2 and mitosis (Bravo and Celis, 1980).
[n] From Franke *et al.* (1981); Bravo *et al.* (1982a,b, 1983); Fey *et al.* (1983). A catalog of human keratins has been recently published by Moll *et al.* (1982).
[o] Vimentin degradation products (Bravo and Celis, 1980; Bravo *et al.*, 1982a,b).
[p] These polypeptides cannot be detected in human skin or lung fibroblasts.
[q] The relative proportion of this polypeptide changes in S phase (Bravo and Celis, 1980).

TABLE II

Percentages and Characteristics of Some Major Basic Polypeptides from HeLa Cells

Polypeptide	Coordinates[a] molecular weight/ mobility relative to NEPHGE 9	Protein label[b]	Number of spots with same original number	Amount found in (%)[c]		Amount found in (%)[d]		Other characteristics
				Cytoplast	Karyoplast	Triton cytoskeleton	Triton supernatant	
0	nd/nd		47 (a–z7)					
0d				–	+			
0e				–	+			
0g				–	+			
0i				–	+			
0k				–	+	+	–	
0y		0.07						*h*
0x								*h*
0z1				–	+			
0z2						+	–	
0z3				–	+			
0z6		0.04						
0z7		0.02						
1	95/0.18	0.23	73 (a–z35)	58	42	4	96	*g*
1e		0.02						
1f								*f*
1g								*f*
1p		0.04				+	–	
1q				–	+			
1r		0.03		–	+	+	–	
1w		0.06				+	–	
1t				–	+	+	–	

1y				–	+			
1z17				–	+			
1z18				–	+			
1z19				–	+			
1z24						+	–	*f*
1z28		0.05		–	+	+	–	
1z29		0.05						*h*
1z31		0.04				+	–	*f*
2	67/0.51	0.05		12	88	16	84	*g*
3	65/0.75	0.04	11 (a–i)	–	100	92	8	*g*
3b				–	+			
3c				–	+			
3d		0.08						
3e				–	+			
3f				–	+	+	–	
3h				–	+			
4	64/0.28							*g*
5	64/0.70	0.02	2 (a)	5	95	80	20	*f,g*
5a						+	–	
6	62/0.88	0.03	10 (a–h)	–	100	89	11	*e,g*
6a				–	+	+	–	
6c				–	+	+	–	*f*
6g				–	+	+	–	
7	60/0.59	0.60	14 (a–m)	52	48	–	100	*g*
7f		0.03						
7g		0.03						*f*
7i		0.04		–	+	+	–	
7j		0.05		–	+			
7k		0.05		–	+	+	–	
8	57/0.24	0.05	5 (a–d)	90	10			*f,g*
8a		0.03						
9	55/1.00	0.04		5	95	90	10	*g*

(*continued*)

TABLE II (*continued*)

Polypeptide	Coordinates[a] molecular weight/ mobility relative to NEPHGE 9	Protein label[b]	Number of spots with same original number	Amount found in (%)[c]		Amount found in (%)[d]		Other characteristics
				Cytoplast	Karyoplast	Triton cytoskeleton	Triton supernatant	
10	54.5/0.57	0.09	6 (a–d)	58	42	19	81	*g*
10a		0.02				+	–	*e,f*
10b				–	+	+	–	*f*
10c				–	+	+	–	
11	54/0.41	0.09	23 (a–v)	+	+	7	93	*g*
11a		0.03		–	+	+	–	
11c				–	+	+	–	
11e				–	+			
11i		0.12				+	–	
11j		0.06		+	–	+	–	
11r		0.08						
11s		0.08		+	–	+	–	*j*
11t		0.03						
11u		0.02						
12 (EF-1α)[r]	50/1.44	0.63	8 (a–g)	60	40	7	93	*g*
12d		0.03						
12e		0.06						
13	49/0.67	0.03		15	85	85	15	*f,g*
14	48.5/0.24	0.74	15 (a–m)	65	35	1	99	*f,g*
14b				80	20			
14e				+	–			
14f				+	–			
14i		0.02						
14l		0.04						

15	47/1.00	0.03	12 (a–q)	98	2	100	–	*f,s*
15a				+	–			
15h		0.11						
15i		0.04						
15j		0.10		+	–	–	+	
15m		0.08						
15n		0.06						
16	43/0.66	0.66	33 (a–z)	49	51	–	100	*g*
16f		0.14						
16g		0.27		+	–			
16k		0.03		+	+			
16l		0.03		–	+	+	–	
16m				+	–			
16w		0.03		–	+	+	–	
17	41/1.22	0.07	6 (a–e)	20	80	72	28	*f,g*
17b		0.04						
17e		0.06				+	–	
18	40/0.92	1.35	7 (a–e)	60	40	10	90	*g*
18a		0.06						
18b				+	–			
19	39.5/0.40	0.37	16 (a–m)	68	32	–	100	*f,g*
19b		0.05						
19e				+	–			
19f				+	–			
19g		0.04		+	–			
19h				–	+			
19i				+	–			*f*
19m				+	–			
20	36/0.71	0.21	7 (a–f)	57	43	9	91	*g*
20b				+	–			
20d								*e*
20h		0.09						

(*continued*)

TABLE II (*continued*)

Polypeptide	Coordinates[a] molecular weight/ mobility relative to NEPHGE 9	Protein label[b]	Number of spots with same original number	Amount found in (%)[c]		Amount found in (%)[d]		Other characteristics
				Cytoplast	Karyoplast	Triton cytoskeleton	Triton supernatant	
21	35/1.45	0.14	21 (a–p)	20	80	81	19	*f,t*
21e		0.02						
21f		0.02						
21g		0.10				+	–	
21h				+	–			*j*
21p		0.03						
22	35/0.21							*g*
23	33/0.23	0.03	8 (a–f)					*g*
23a				+	–			
23c		0.02		+	–	+	–	*j*
23d		0.02						*f*
24	32/0.98	0.05	12 (a–f)			–	+	*g*
24f		0.02		+	–			
25	32/0.04							*g*
26	31/0.15	0.02	19 (a–n)			–	+	*g*
26a		0.02						
26c		0.02						
26d				+	–			
26g		0.08		+	–			
26h		0.17				–	+	
26i		0.03						*f*
27	27/0.77	0.14	5 (a–c)	66	34	81	19	*g*
27b		0.02						*f*

27c		0.01						
28	25/0.60	0.06	7 (a–d)	67	33	–	100	*g*
28a				+	–			
28c		0.04						
28d				–	+			
29	23/0.35	0.04	9 (a–e)					*g*
29a				+	–			
29d		0.04						
30	20.5/0.65	0.15	3 (a–b)	55	45	–	100	*g*
30a		0.03						
31	19/0.89	0.07	3 (a–b)	65	35	24	76	*g*
32	18.5/0.42	0.11		+	+	–	100	*g*
33	18/0.68	0.13	14 (a–i)	31	69	–	100	*f,g*
33a		0.04						
33b				+	–			
33d		0.13				–	+	

[a] From Bravo *et al.* (1981a). nd, not determined.
[b] Average of at least two determinations.
[c] From Bravo *et al.* (1981b).
[d] From Bravo *et al.* (1982b).
[e] Phosphoprotein.
[f] Neoplastic transformation-sensitive polypeptides common to both epithelial and fibroblast human cultured cells (Bravo and Celis, 1982a).
[g] The relative proportion of these polypeptides is constant throughout the cell cycle (Bravo and Celis, 1980). Only polypeptides whose relative proportion changed by twofold or more were considered as variable in that study.
[h] The relative proportion of these polypeptides changes in mitosis (Bravo and Celis, 1980).
[j] Present in low and high salt (*LHL*)-extracted cytoplast cytoskeletons enriched in intermediate filaments (Bravo *et al.*, 1982b).
[r] Eukaryotic elongation factor Tu (EF-1α) (Coppard *et al.*, 1982).
[s] The relative proportion of this polypeptide changes in S phase as determined by short term labelings (Bravo and Celis, 1980).
[t] The relative proportion of this polypeptide changes in G_2 as determined by short term labeling (Bravo and Celis, 1980).

during the cell cycle (Tables I and II; other characteristics) (Bravo and Celis, 1980). Similar observations have been reported by Milcarek and Zahn (1978).

B. Distribution in Cytoplasts and Karyoplasts

Figures 5 and 6 show fluorograms of two-dimensional gels (IEF and NEPHGE) of [^{35}S]methionine-labeled polypeptides from cytoplasts and karyoplasts prepared from [^{35}S]methionine-labeled HeLa cells treated with Cytochalasin B and enucleated by centrifugal force (Small and Celis, 1978b; Bravo *et al.*, 1981b). The purity of the cytoplasts was higher than 90%, as determined by microscopic inspection of many cells (Fig. 7). The karyoplasts, however, were slightly contaminated with small cytoplasmic vesicles but this contamination did not interfere with the polypeptide analysis as the fluorograms allowed the assignment of many polypeptides to either the karyoplasts or the cytoplasts, showing that negligible cross examination has occurred (Figs. 5 and 6) (Bravo *et al.*, 1981b). For quantitating the distribution of acidic polypeptides both the cytoplasmic keratin IEF 46 (Figs. 5 and 6) and the karyoplastic polypeptide IEF 33 (Figs. 5 and 6) were used as references (Bravo *et al.*, 1981b). For example, the ratio total actin/IEF 46 in whole cells is 61.5 (Bravo *et al.*, 1981b), whereas that of the cytoplasts is 39.6. Because there is no detectable IEF 46 in the karyoplasts it is possible to calculate that 60% of the actin remained in the cytoplasts ($36.9/6.15 \times 100 = 60\%$). The same approach was used to determine the distribution of basic polypeptides. In this case polypeptides NEPHGE 9 (mainly present in karyoplasts; Figs. 5 and 6) and NEPHGE 15 (mainly present in cytoplasts; Figs. 5. and 6) were used as references. Polypeptides that are present in 80% or more in one cellular fragment are indicated with a *c* or a *k* in Figs. 5 and 6 and the quantitations are given in Tables I and II. Polypeptides showing less than 80% distribution in any particular cell fragment are indicated with plain arrows in Figs. 5 and 6 (see also Tables I and II). The distribution of many basic polypeptides was determined by visual inspection of many independent fluorograms exposed for different periods of time. In Tables I and II a plus sign in only one cellular fraction indicates that the polypeptide is found mainly or only there. A plus sign in both cellular fractions indicates that the polypeptide is about equally distributed in both fractions. Not all polypeptides listed in Tables I and II are shown in Figs. 5 and 6.

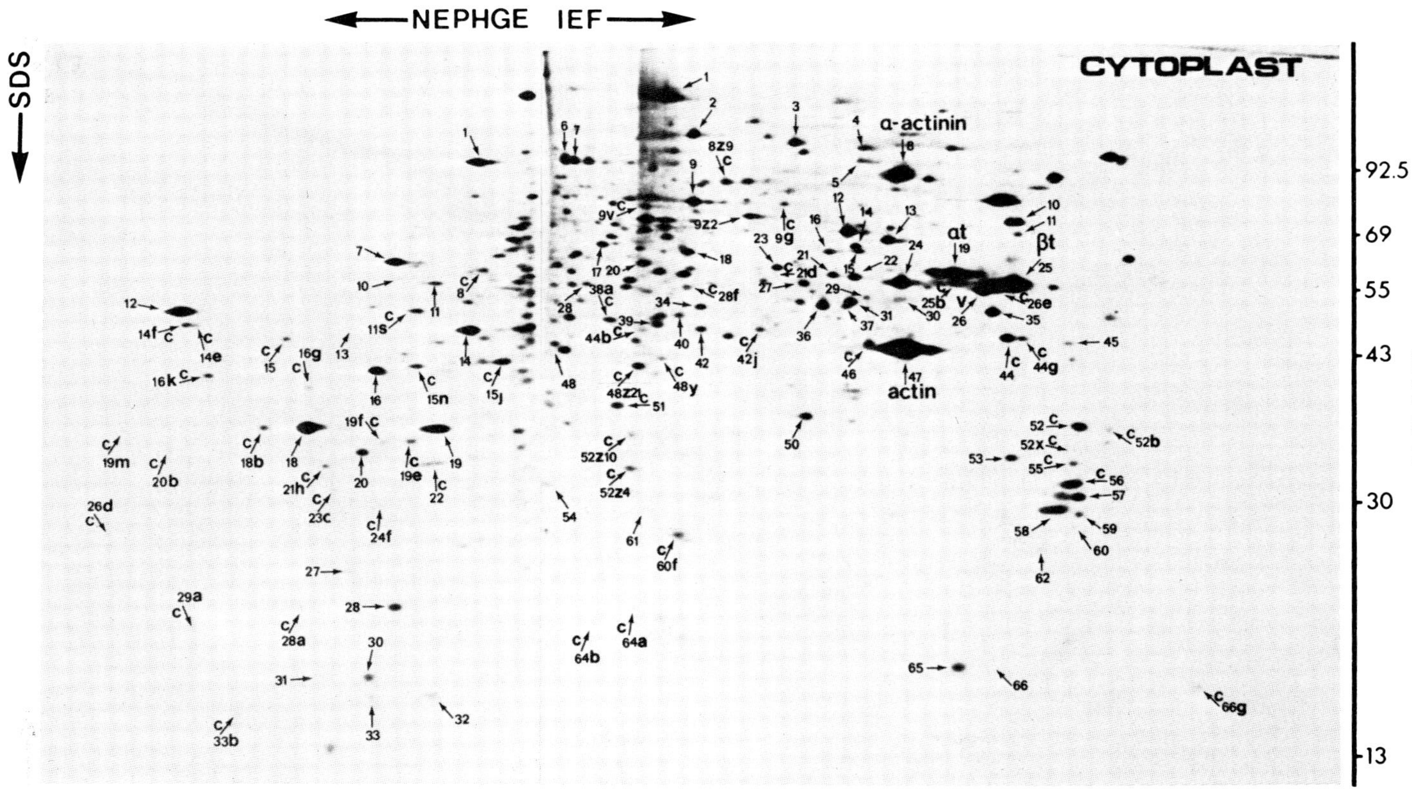

Figure 5. Two-dimensional gel electrophoresis maps of [^{35}S]methionine-labeled polypeptides from HeLa cytoplasts. Asynchronous cells grown on 2.5-cm plastic discs and labeled with [^{35}S]methionine were enucleated with cytochalasin B as previously described (Small and Celis, 1978b; Bravo *et al.*, 1981b). The gels have been exposed so as to keep the intensity of the spots within the linear range of the film response. Polypeptides indicated with a *c* above the arrow are present mainly in cytoplasts. Polypeptides present in both cellular fragments are indicated with plain arrows. From Bravo *et al.* (1981b).

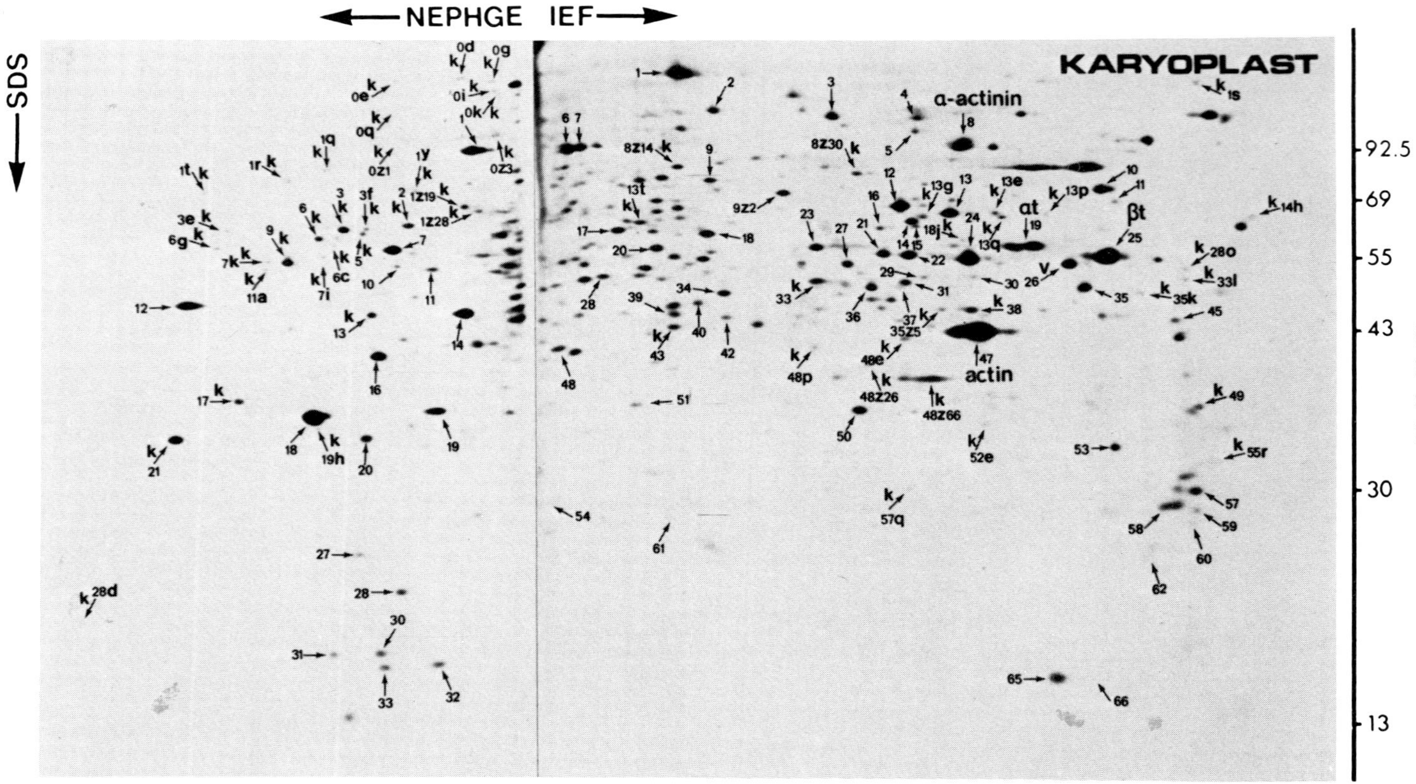

Figure 6. Two-dimensional gel electrophoresis maps of [^{35}S]methionine-labeled polypeptides from HeLa karyoplasts. The gels have been exposed so as to keep the intensity of the spots within the linear range of the film response. Polypeptides indicated with a *k* are present mainly in karyoplasts. Polypeptides present in both cellular fragments are indicated with plain arrows. From Bravo *et al.* (1981b).

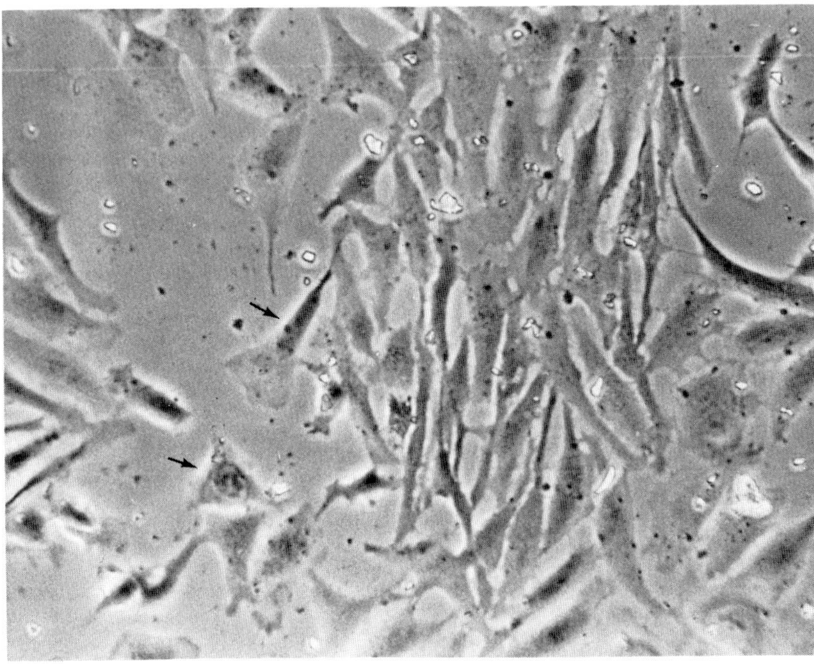

Figure 7. HeLa cell cytoplasts. Cytoplasts were prepared as described in Bravo *et al.* (1981b). The cells indicated with an arrow are the only nucleated cells in the population.

C. Distribution in Triton Cytoskeletons and Triton Supernatants

Figure 8 illustrates the polypeptide composition of the Triton cytoskeletons and Triton supernatants from [^{35}S]methionine-labeled asynchronous HeLa cell monolayers treated for 90 sec with 0.1% Triton *X*-100 in Pipes cytoskeleton buffer (Small and Celis, 1978a,b; Bravo *et al.*, 1982b). The *Triton cytoskeleton* contains about 90 major polypeptides (48 basic and 42 acidic) and about 30% of the total counts present in whole cells. Spots indicated with large arrows in Fig. 8a correspond to polypeptides that are less than 40% extractable. Polypeptides indicated with small arrows are about 50% extractable. Even though 70% of α- and β-tubulin can be extracted with 0.1% Triton X-100, they are indicated with large arrows in Fig. 8a as they are part of known cytoarchitectural structures. Triton extractable polypeptides (60% or more) are indicated with arrowheads in Fig. 8b. The Triton extractability of 91 polypeptides, indicated

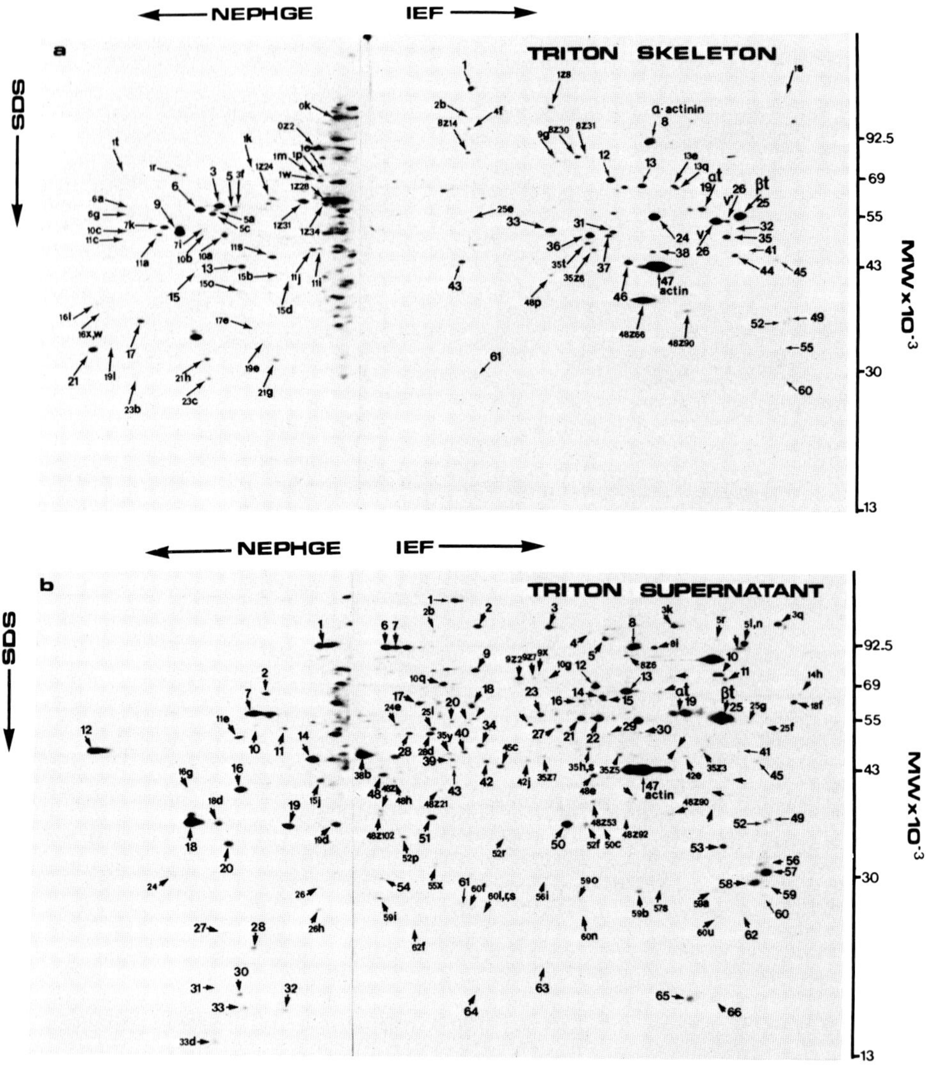

Figure 8. Two-dimensional gel electrophoresis (IEF, NEPHGE) of [^{35}S]methionine-labeled polypeptides from HeLa Triton cytoskeletons (a) and Triton supernatants (b). Cells to be treated with Triton alone or to be further extracted with buffers of low and high ionic strength were grown attached to 9 mm^2 coverslips (Belco) placed in 0.25 ml flat-bottomed microtiter plates (NUNC) containing 0.2 ml of DMEM. After the cells were labeled with [^{35}S]methionine the coverslips were removed, rinsed in Hank's solution and transferred to 0.1% Triton X-100 in a Ca^{2+}-free Hank's solution containing 2 m*M* $MgCl_2$, 2 m*M* EGTA, 5 m*M* PIPES (pH 6.1) (PIPES cytoskeleton buffer; Small and Celis, 1978a). Treatment with Triton X-100 was for 60–90 sec at room temperature. After being washed in Hank's solution, the coverslips were placed in microtiter wells containing 20 μl of lysis buffer. The Triton supernatant was resuspended in lysis buffer after lyophylization.

with large numbers, was quantitated and the results are presented in Tables I and II. (Bravo *et al.*, 1982b).

D. Presence in High Salt-Extracted Cytoplast Skeletons

Figure 9 shows two-dimensional gel separations (IEF, NEPHGE) of [^{35}S]methionine-labeled polypeptides from cytoplast Triton skeletons extracted with buffers of low and high salt concentrations (Bravo *et al.*, 1982b). The cytoskeletons, which are enriched in intermediate filaments (Small and Celis, 1978a,b; Bravo *et al.*, 1982b) and exhibit mitochondrial remnants (Mose Larsen *et al.*, 1982), contain in addition to vimentin (IEF 26) and keratins (IEF 31, 36, 44 and 46) (Franke *et al.*, 1981; Bravo *et al.*, 1982a,b, 1983; Fey *et al.*, 1983) a few other abundant polypeptides (IEF

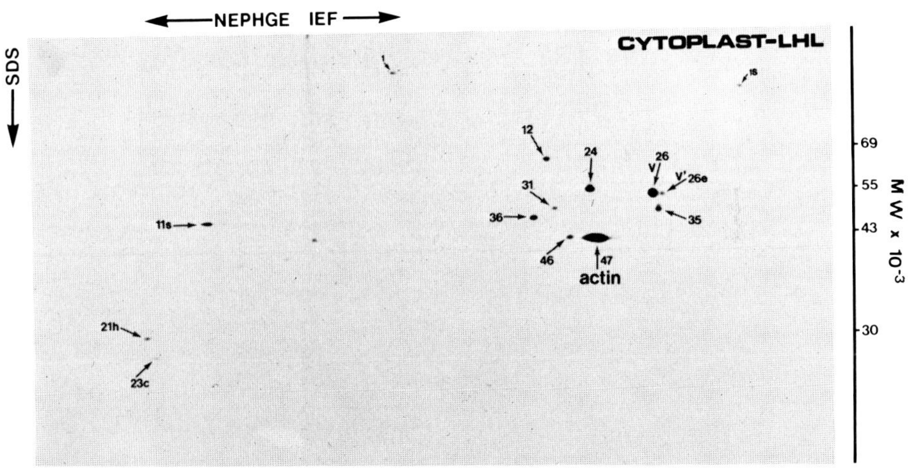

Figure 9. Acidic (IEF) and basic (NEPHGE) [^{35}S]methionine-labeled polypeptides present in high salt-extracted HeLa cytoplasts. HeLa cells grown in plastic discs (2.5-cm diameter) were labeled with [^{35}S]methionine and enucleated as described in the legend of Fig. 5. After appropriate times in culture (recovery) the discs were washed in Hank's and extracted with 0.1% Triton X-100 as described in the legend of Fig. 8. After Triton treatment, the discs were washed twice in PIPES buffer and dipped for 90 sec in each of the following ice-cold extraction solutions (Small and Sobieszek, 1977; Small and Celis, 1978b): (L): 60 m*M* KCl, 1 m*M* EDTA, 2 m*M* EGTA, 1 m*M* cysteine, 10 m*M* ATP, 40 m*M* imidazole (pH 7.1); (H): 0.6 *M* KCl, 1 m*M* EDTA, 2 m*M* ATP, 1 m*M* cysteine, 40 m*M* imidazole (pH 7.1); (L) solution again. The coverslips were then rinsed in the PIPES buffer and transferred to lysis buffer (O'Farrell, 1975). These skeletons extracted in high and low salt are referred to as *LHL* cytoskeletons.

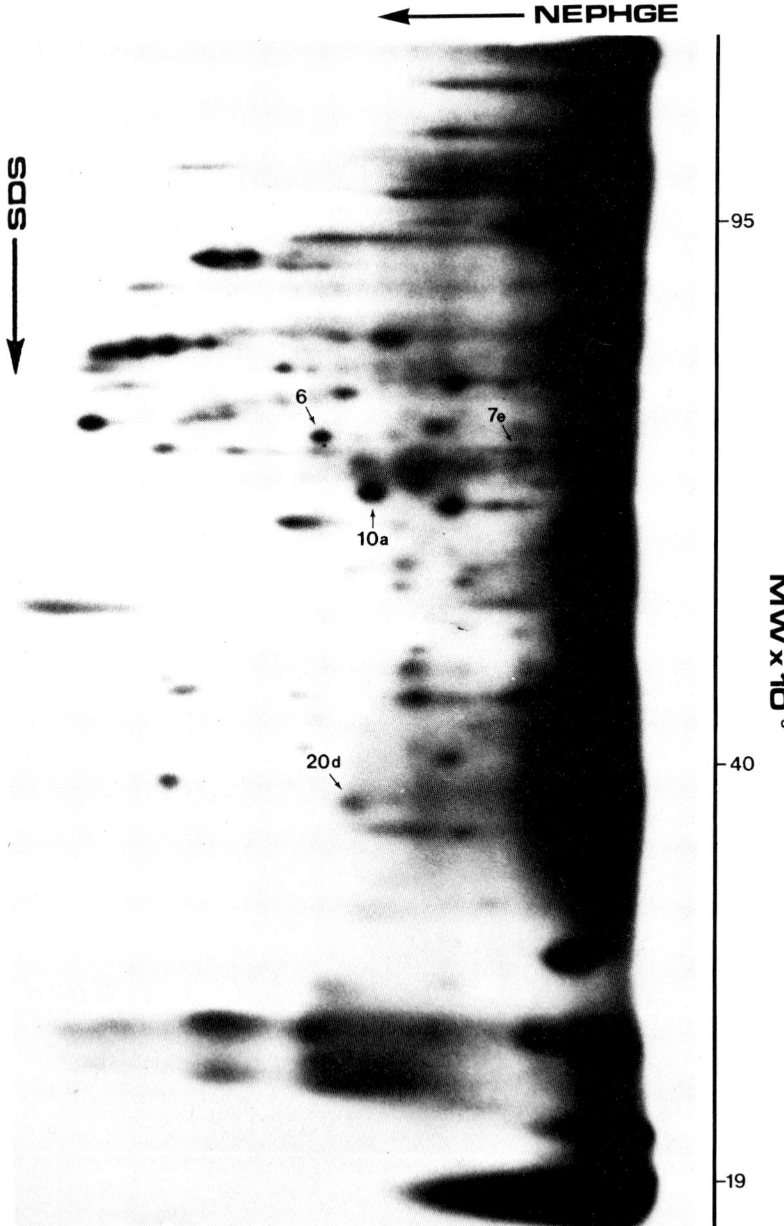

Figure 10. Two-dimensional map (NEPHGE) of HeLa basic proteins labeled for 8 h with [^{32}P]orthophosphate. Only the phosphoproteins that have been shown to comigrate with cataloged [^{35}S]methionine-labeled proteins are indicated. From Bravo and Celis (1982a).

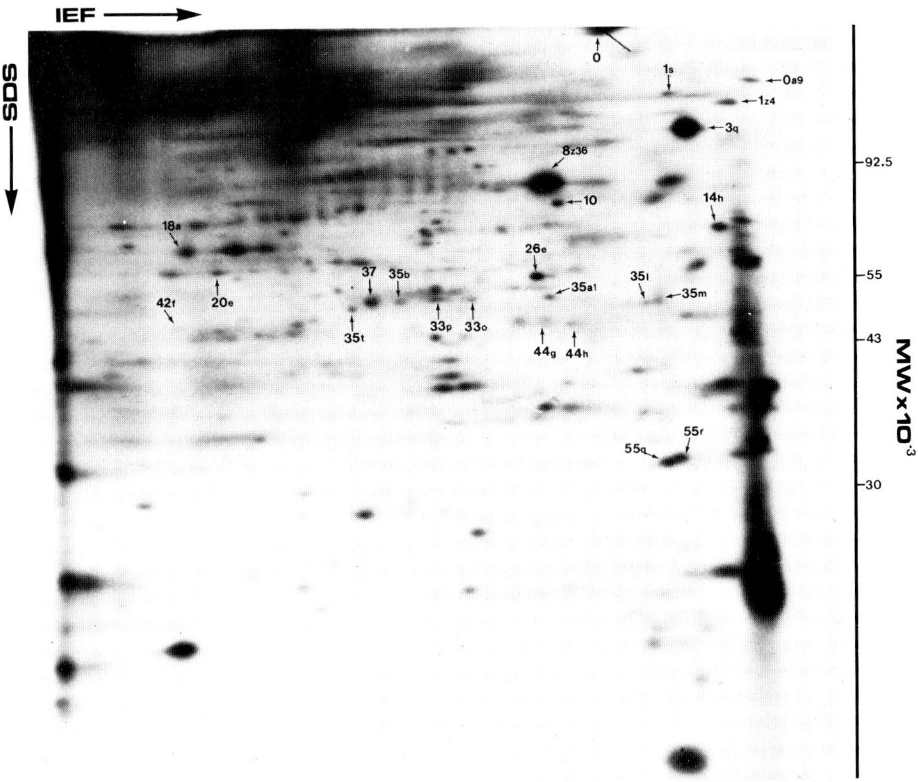

Figure 11. Two-dimensional map (IEF) of HeLa acidic proteins labeled for 8 h with [^{32}P]orthophosphate. Only those phosphoproteins that have been shown to comigrate with cataloged [^{35}S]methionine-labeled proteins are indicated. From Bravo and Celis (1982a).

12, 24 (mitochondrial), 35), which are tenaciously bound to these cytoskeletons. Some of these polypeptides may correspond to intermediate filament associated proteins (Bravo *et al.*, 1982b). The *LHL polypeptides* are indicated with the letter *j* (other characteristics) in Tables I and II.

E. Phosphorylation

Figures 10 and 11 show two-dimensional gel separations of basic (NEPHGE; Fig. 10) and acidic (IEF; Fig. 11) ^{32}P-labeled polypeptides from asynchronous HeLa cells labeled for 8 h with 2mCi per ml of [^{32}P]orthophosphate (Bravo *et al.*, 1982a,b; Bravo and Celis, 1982a,b).

About 250 phosphoproteins could be clearly resolved (Fig. 10 and 11) and a few are indicated with the letter *e* (other characteristics) in Tables I and II. The phosphoprotein 8z36 corresponds to a heat shock protein.

IV. Transformation-Sensitive Polypeptides

Given the qualitative similarity between the two-dimensional protein patterns of HeLa cells and those of other epithelial cells and fibroblasts, we have included in this catalog the human polypeptides whose rates of synthesis are sensitive to transformation in cultured epithelial and fibroblast cells (Bravo and Celis, 1982b). These polypeptides are indicated with the letter *f* in Tables I and II (other characteristics).

V. Conclusions

A total of 1357 polypeptides from HeLa cells (956 acidic and 411 basic) have so far been separated by two-dimensional gel electrophoresis and cataloged. It should be stressed that we have been unable to increase significantly the number of polypeptides detected by increasing the amount of radioactivity or protein (silver staining) applied to the gels (as much as tenfold). Similar observations have been reported by Duncan and McConkey (1982). Clearly under the current experimental conditions used, new, less abundant proteins, if present, may only be revealed by the analysis of purified subcellular fractions.

Acknowledgements

We thank our colleagues J. Bellatin, S. J. Fey, P. Mose Larsen, H. Macdonald-Bravo, and A. Celis for helpful discussion. We also thank O. Jensen for excellent photographic assistance. R. B. was a recipient of a fellowship from the Danish Medical and Natural Science Research Councils. This work was supported by grants from Euratom, the Danish Medical and Natural Science Research Councils, the Danish Cancer Society, Carlsberg, and Novo.

References

Anderson, N. G., and Anderson, N. L. (1979). Molecular anatomy. *Behring Inst. Mitt.* **63**, 169–210.

Anderson, N. L. (1981). Identification of mitochondrial proteins and some of their precursors in two-dimensional electrophoretic maps of human cells. *Proc. Natl. Acad. Sci. USA* **78**, 2407–2411.

Anderson, N. L., Taylor, J., and Scandora, A. E. (1981). The TYCHO system for computerized analysis of two-dimensional gel protein mapping data. *Clin. Chem.* **27**, 1807–1820.

Bellatin, J., Bravo, R., and Celis, J. E. (1982). Changes in the relative proportion of transformation sensitive polypeptides in giant HeLa cells produced by irradiation with lethal doses of X-rays. *Proc. Natl. Acad. Sci. USA* **79**, 4367–4370.

Bossinger, J., Miller, M. J., Vo, K. P., Geiduschek, E. P., and Xuong, N. H. (1979). Quantitative analysis of two-dimensional electropherograms. *J. Biol. Chem.* **254**, 7986–7998.

Bravo, R., and Celis, J. E. (1980). A search for differential polypeptide synthesis throughout the cell cycle of HeLa cells. *J. Cell Biol.* **84**, 795–802.

Bravo, R., and Celis, J. E. (1982a). Updated catalogue of HeLa cell proteins: percentages and characteristics of the major cell polypeptides labeled with a mixture of 16 ^{14}C-labeled amino acids. *Clin. Chem.* **28**, 766–781.

Bravo, R., and Celis, J. E. (1982b). Human proteins sensitive to neoplastic transformation in cultured epithelial and fibroblast cells. *Clin. Chem.* **28**, 949–954.

Bravo, R., Bellatin, J., and Celis, J. E. (1981a). [^{35}S]-Methionine labeled polypeptides from HeLa cells. Coordinates and percentages of some major polypeptides. *Cell Biol. Int. Rep.* **5**, 93–95.

Bravo, R., Celis, A., Mosses, D., and Celis, J. E. (1981b). Distribution of HeLa cells polypeptides in cytoplasts and karyoplasts. *Cell Biol. Int. Rep.* **5**, 479–489.

Bravo, R., Fey, S. J., Small, J. V., Mose-Larsen, P., and Celis, J. E. (1981c). Coexistence of three major isoactins in a single Sarcoma 180 cell. *Cell* **25**, 195–202.

Bravo, R., Fey, S. J., Mose Larsen, P., and Celis, J. E. (1982a). Modification of vimentin polypeptides during mitosis. *Cold Spring Harbor Symp. Quant. Biol.* **154**, 121–143.

Bravo, R., Small, J. V., Fey, S. J., Mose Larsen, P., and Celis, J. E. (1982b). Architecture and polypeptide composition of HeLa cytoskeletons. Modification of cytoarchitectural polypeptides during mitosis. *J. Mol. Biol.* **154**, 121–143.

Bravo, R., Fey, S. J., Mose Larsen, P., Coppard, N., and Celis, J. E. (1983). Proteins IEF 31 and IEF 46 are keratin type components of the intermediate-sized filaments: Keratins of various human cultured epithelial cells. *J. Cell Biol.* **96**, 416–423.

Celis, J. E., and Bravo, R. (1981). Towards cataloguing human and mouse proteins. *Trends Biochem. Sci.* **5**, 281–283.

Coppard, N. J., Cramer, F., and Clark, B. F. C. (1982). Identification of elongation factor 1α from mouse liver. *FEBS Lett.* **145**, 332–336.

Duncan, R., and McConkey, E. H. (1982). How many proteins are there in a typical mammalian cell? *Clin. Chem.* **28**, 749–755.

Fey, S. J., Mose Larsen, P., Bravo, R., Celis, A., and Celis, J. E. (1983). Differential immunological cross-reactivity of HeLa keratin antibodies with human epidermal keratins. *Proc. Natl. Acad. Sci. USA* **80**, 1905–1909.

Franke, W. W., Schiller, D. L., Noll, R., Winter, S., Schmid, E., Engelbrecht, I., Denk, H., Knepler, R., and Platzer, B. (1981). Diversity of cytokeratins. Differentiation specific expression of cytokeratin polypeptides in epithelial cells and tissues. *J. Mol. Biol.* **153**, 933–959.

Garrels, J. I. (1979). Two-dimensional gel electrophoresis and computer analysis of proteins synthesized by cloned cell lines. *J. Biol. Chem.* **254**, 1971–1977.

Lester, E. P., Lemkin, P., and Lipkin, L. (1982). A two-dimensional gel analysis of autologous T and B lymphoblastoid cell lines. *Clin. Chem.* **28**, 828–839.

Lipkin, L. E., and Lemkin, P. F. (1980). Data base techniques for multiple two-dimensional polyacrylamide gel electrophoresis analyses. *Clin. Chem.* **26**, 1403–1412.

Lutin, W. A., Kyle, C. F., and Freeman, J. A. (1978). Quantitation of brain proteins by computer analyzed two-dimensional electrophoresis. *In* "Electrophoresis '78" (N. Catsimpooles, ed.), pp. 93–106. Elsevier-North Holland, Amsterdam.

Merril, C. R., and Goldman, D. (1982). Quantitative two-dimensional protein electrophoresis for studies of inborn errors of metabolism. *Clin. Chem.* **28**, 1015–1020.

Milcarek, C., and Zahn, K. (1978). The synthesis of ninety proteins including actin throughout the HeLa cell cycle. *J. Cell Biol.* **79**, 833–838.

Miller, M. J., Vo, K. P., Nielsen, V., Geiduschek, E. P., and Xoung, N. H. (1982). Computer analysis of two dimensional gels: Semi-automatic matching. *Clin. Chem.* **28**, 867–875.

Moll, R., Franke, W. W., Schiller, D., Geiger, B., and Krepler, R. (1982). The catalogue of human cytokeratins: Patterns of expression in normal epithelia, tumours and cultured cells. *Cell* **31**, 11–24.

Mose Larsen, P., Bravo, R., Fey, S. J., Small, J. V., and Celis, J. E. (1982). Putative association of mitochondria with a sub-population of intemediate-sized filaments in cultured human skin fibroblasts. *Cell* **31**, 681–692.

Mose-Larsen, P., Fey, S. J., Bravo, R., and Celis, J. E. (1983). Mouse mitochondrial protein IEF 24: Identification and immunohistochemical localization of mitochondria in various tissue. *Electrophoresis*, **4**, 247–256.

O'Farrell, P. H. (1975). High resolution two-dimensional electrophoresis of proteins. *J. Biol. Chem.* **250**, 4007–4021.

O'Farrell, P. Z., Goodman, H. M., and O'Farrell, P. H. (1977). High resolution two-dimensional electrophoresis of basic as well as acidic proteins. *Cell* **12**, 1133–1142.

Skolnick, M. M., Stirnberg, S. R., and Neel, J. V. (1982). Computer programs for adapting two-dimensional gels to the study of mutation. *Clin. Chem.* **28**, 969–978.

Small, J. V., and Celis, J. E. (1978a). Filament arrangements in negatively stained cultured cells. *Cytobiologie* **16**, 308–325.

Small, J. V., and Celis, J. E. (1978b). Direct visualization of the 10 nm (100 Å) filament network in whole and enucleated cultured cells. *J. Cell Sci.* **31**, 393–409.

Small, J. V., and Sobieszek, A. (1977). Studies on the function and composition of the 10 m, (100 Å) filaments of vertebrate smooth muscle. *J. Cell. Sci.* **23**, 243–268.

Taylor, J., Anderson, N. L., Scandora, A. E., Willard, K. E., and Anderson, N. G. (1982). Design and implementation of a prototype human protein index. *Clin. Chem.* **28**, 861–866.

Index

A

Abscisic acid, late embryogeny and, 406, 407
Acrylamide, concentration, gel analysis and, 41–44
Actin(s)
  estimation of degree of mistranslation, 374–375
  mutant, variations in expression in chemically transformed human diploid fibroblasts, 334–337
  satellites, DNase I and, 371–373
Albumin
  pH gradient in gels and, 270–271
  removal from serum samples, 212
Amino acid(s)
  analogs, incorporation and cell longevity, 383
  starvation, stringent response to, 386
Amniotic fluid, 2-DGE of, 234
Analysis
  of clinical gels, 206–207
  of reduced data
    QUEST system and, 67–68
      measurement of data quality, 68–69
      plotted spot maps, 74–75
      sample comparisons, 69–72
      spot sets, 72–73
      tabular and graphic displays, 73–74
      textual data base, 74
Antibodies
  to protein eluted from SDS gels, 146–147
  specific, for identification of proteins in gels, 347
Arginine, starvation, mistranslation and, 391
Asparagine, starvation, error-containing proteins and, 369, 371, 391
Autoradiography, drawbacks of, 94

B

Bacteria
  frequency of mistranslation in, viability and, 383
  stringent control system in, effect on protein synthetic accuracy, 386–388
Blocking agents, probing electrophoretic protein transfer and, 147
Blood
  fractionation for study of mutation, 269
  study of mutations and, 265–267
Blood cells
  storage of, 197
  use for 2-DPGE
    erythrocytes, 226–227
    lymphocytes, 227–229
    platelets, 229–230
Brain, 2-DGE and, 232–233

C

Calibration gels, QUEST system and, 58–59
Cancer, serum proteins and, 216
Carcinogenesis, unifying theory of, 310
Carrot cells, in culture, production of embryos by, 398
Cell cycle, variations in HeLa cell proteins during, 447–448, 466
Cell density, plant embryogeny and, 402–403

Cell lines, impaired in embryogeny, gene expression in, 405–406
Cellular aging, error catastrophy theory of, 379–386
Charge effects, silver staining and, 105
Charge heterogeneity, aging and, 383
Charge trains, serum proteins and, 213
Chemical effects, silver staining and, 105–107
Chromogenic system, for immunoprobing
 detection by color reaction, 150
 options, 150
 reagents and materials, 149
 treatment with antibody, 149–150
Chromosomes, human, gene mapping of, 120–121
Clinical application, of electrophoresis, meaning of, 195
Clinical laboratory, electrophoresis in, 194
 team approach, 196
Codon–anticodon discrimination, ribosomal function and, 390
Color, use to identify proteins in different species, 120–121
Color silver stain (GELCODE system)
 application to 2-D mapping of proteins
  characterizing proteins in posttranslationally modified charge trains, 118–119
  distinguishing overlapping spots, 117–118
  modulation of proteins by tunicamycin, 120
  subcellular fractionation, 122
  tracking proteins in urine, 119–120
  use of animal models to find marker proteins for disease, 121–122
  use of color to identify proteins in different species: application to gene mapping of human chromosomes, 120–121
 computerized data acquisition of color-stained images: test for stoichiometric staining
  BioImage system, 122–123
  stoichiometry of GELCODE stain: integrated intensity versus concentration, 123–125
 variables affecting sensitivity and reproducibility of, 113–114
  fixation acid, 116
  fixation alcohol and time of fixation, 116
  reagents, 116–117
  thickness of gel, 114–115
  time and temperature in reducing solution, 115
Colorectal cancer, protein changes and, 231
Coomassie blue staining, 202–203
 disadvantage of, 93
Coordinates, use in 2-D gel electrophoresis, 24–28
Cotton cotyledons, gene expression in, 406–407
Creatine phosphokinase, muscle damage and, 214
Cyclic adenosine monophosphate, control of gene expression and, 440
Cyclin
 occurrence of, 326–328
 synthesis, state of cell proliferation and, 328–332
Cycloheximide, coordinate regulation of sets of biochemical phenotypes, 403–404
Cytocomputer, gel analysis and, 289–292
Cytokinin, somatic embryogenesis and, 399
Cytoplast skeletons
 high salt-extracted, HeLa cell proteins in, 471–473
Cytoskeletons
 Triton, HeLa cell proteins and, 469–471

D

Data
 measures of quality, QUEST system and, 68–69
 programs for final reduction, QUEST system and, 84–86
Data analysis, 2-D gel electrophoresis and, 39–40
Data base, textual, QUEST system and, 74
Data base form, standard, specimen organization and, 198

Data management, in mutation monitoring, 296–297
friendly computer interface, 297
interactive data acquisition, 297–298
networking, 298
Densitometer, analysis of clinical gels and, 206
Deoxyribonuclease I, reaction with actin satellites, 371, 373
Dermatomyositis, protein changes in, 233
Destaining, silver staining and, 100
Detection, silver staining and, 100–103
Detergents, electrophoretic transfer of proteins and, 147
Diabetes, gene products and, 121–122
Diazobenzyloxymethyl paper, protein detection and, 129–130, 134–135
Diazo paper, electrophoretic transfer of proteins to, 131
materials and reagents, 142
options, 145
posttransfer treatment, 143
removal of probes from, 154
transfer conditions, 143
Diazophenyl thioether paper, electrophoretic transfer of proteins to, 130–131, 134–135
2, 4-Dichlorophenoxyacetic acid
seed germination and, 407–411
somatic embryogenesis and, 399, 402
Disease, marker proteins, use of animal models to find, 121–122
Displays, tabular and graphic, QUEST system and, 73–74
Dodecyl sulfate dimension, for clinical samples, composition of gels, 201–202
Drosophila, genetic map of, 243
Duchenne's muscular dystrophy, protein deficiency in, 254

E

Electroblotting, use in clinical laboratory, 210
Electrophoresis, in clinical laboratory, 194
team approach in, 196
Electrophoresis system, in clinical laboratory
additional techniques, 209–210
analysis, 206–207
dodecyl sulfate dimension, 201–202
isoelectric focusing dimension, 198–200
need for high throughput, 196–197
reproducibility and quality assurance, 207–209
specimen storage and management, 197–198
visualization, 202–205
Embryogenesis
gene expression in plants during
early, 398
late, 406
somatic, 399–403
somatic, in plants, 399
Epidermal growth factor, tyrosine phosphorylation in proteins and, 341–342
Equipment, electrophoretic, for clinical use, 196–197
Error catastrophy theory, of cellular aging, 379–386
Erythrocytes, 2-DGE and, 226–227
Escherichia coli
current protein catalog, 426–439
applications in research, 439–442
EF-G and EF-Tu, mistranslation of, 376
genome size, 417
identification of proteins of, 418–426
procedures, 424–425
strains used for protein identification, 419
Ethephone, seed germination and, 441
Exposure value, QUEST system and, 49

F

Fibroblasts, mouse kidney, updated protein catalog of, 177, 179–181
FILMSPOTS
files and standards, matching procedures and, 60–61
strategies and procedures for matching a file to a standard, 63–64

First dimension, procedures for gel electrophoresis
  common problems, 28–32
  equilibration of, 18
  preparation and running of, 9–11
  preparation of tubes, 8–9
  simple device for casting gels, 17–18
Fixation, GELCODE system and
  acid, 116
  alcohol, 116
Fluorography, 2-D gel electrophoresis and, 13, 15, 19–21
Forms, QST data management system and, 46–49, 51–57
Future program enhancements, of QUEST system, 87–88

G

Gel(s)
  analysis for mutants, 288–292
  multiple
    multiple exposures and, 44
    stratagems for combining data from, 44–45
  subdivision into constellations, 283
  thickness, GELCODE system and, 114–115
GELCODE system, *see* Color silver stain
Gel permeation, preparation of urine for 2-DPGE, 221–222
Gene(s)
  coordinately regulated, protein catalogs and, 440
  structural, number of, 250
Gene expression
  in plants
    cotton cotyledons, 406–407
    early embryogenesis, 398
    germinating seeds, 407–411
    late embryogeny, 406
    somatic embryogenesis, 399–403
  in temperature-sensitive cell lines impaired in embryogeny, 405–406
Gene fusion, identification of promoters and, 440–441
Gene mapping, protein catalog and, 441–442
Genetic code, misreading of, 367, 369, 390
Genetic damage, aging and, 384
Genetic variants
  ability of 2-D system to defect, 274–282
  serum proteins and, 217
Germinating seeds, gene expression in, 407–411
Glanzmann's thrombasthenia, platelet proteins and, 229–230
Glycoproteins, tunicamycin and, 120
Guanosine nucleotides, stringent response and, 386, 388, 440

H

Heating, of 2-D gel electrophoresis samples, 15–16
HeLa cell
  catalog of
    major acidic proteins from, 451–459
    major basic proteins from, 460–465
  characteristics of proteins
    distribution in cytoplasts and karyoplasts, 466–469
    distribution in Triton cytoskeletons and Triton supernatants, 469–471
    phosphorylation, 473–474
    presence in high salt-extracted cytoplast skeletons, 471–473
    variation during cell cycle, 447–448, 466
  proteins, separation by 2-DGE, 446–447
  synthesis of transformation-sensitive polypeptides in, 320–321
Hemoglobin, carbamylated, as charge standard, 200
Heterozygosity, average, of human population, 247
Heterozygosity index, estimation of, 281–282
Heterozygotes, detection of, 288
Histidine
  starvation
    error-containing proteins and, 367–369, 371, 378
    as test of error catastrophy theory of aging, 381–382
Histological analysis, of solid tissues used for 2-DGE, 230

Human
homologous peptides in other species, 177–178, 182–184
protein polymorphism in, 243–245
sarcomas and normal tissue, polypeptide synthesis in, 343–346
spontaneous mutation rate in, 261–262
two-dimensional gel patterns, correlation with mouse, 177–184
Human fibroblasts
chemically transformed, variation in expression of mutant actins in, 334–337
growing and senescent, synthesis of transformation-sensitive polypeptides in, 321–323
Huntington's disease, brain protein alterations in, 253

I

Idiopathic scoliosis
muscle protein variations in, 253–254
protein changes in, 233
Image(s)
colored, silver staining and, 104
intensification in silver staining, other procedures for, 99–100
Image analysis facility, of QUEST system, 50, 58
Immunoglobulins
effect on gel patterns, 271
removal from serum, 212
Immunoprobing, by protein transfer analysis
chromogenic system for visualization, 149–150
radiolabeled components for visualization, 150–153
Infectious mononucleosis, lymphocyte proteins and, 227
Inflammation, changes in serum proteins and, 214
IPMANAGER, QUEST system and, 67
Isoelectric focusing dimension, for clinical samples, equipment and gel composition, 198–200
Isoelectrofocusing gels, preparation and running of, 10–11

K

Karyoplasts, distribution of HeLa cell proteins in, 466–469
Kidney, carcinoma, protein changes and, 231–232
Kinetic amplification, specificity of substrate selection, stringent response and, 386, 388

L

Lesch–Nyhan syndrome
2-DGE and, 249–250
lymphocytes, trait-associated protein alterations in, 251, 252
Loading capacity, in 2-D gel electrophoresis, 17
Lymphocytes, 2-DGE of, 227
Lysine, starvation, error-containing proteins and, 369, 391

M

Markers, use in 2-D gel electrophoresis, 24–28
Matching
automatic
program for, 80–82
QUEST system and, 62–63, 64
interactive, QUEST system and, 61–62, 63
one standard to another, QUEST system and, 64–65
Matching procedures, QUEST system and
automatic matching, 62–63
data reduction to GELSPOTS and SMPSPOTS, 65–67
FILMSPOTS files and standards, 60–61
interactive matching, 61–62
matching one standard to another, 64–65
strategy and procedures for matching a FILMSPOTS file to a standard, 63–64
Materials, for 2-D gel electrophoresis, 5
Methionine, isotopic, labeling cells with, 7
Microsurgery, for isolation of tumor cells, 346–347

Mitosis, vimentin and, 326
Mouse
homologous peptides in other species, 177–178, 182–184
kidney fibroblast, updated protein catalog of, 177–181
two-dimensional gel patterns, correlation with human, 177–184
Multiple myeloma
serum proteins and, 216–217
urinary proteins and, 225
Multiple sclerosis, protein changes and, 234
Muscle, 2-DGE and, 233
Muscle damage, serum proteins and, 214
Muscular dystrophy, urinary proteins and, 225
Mutants
identification of *E. coli* proteins and, 419
monitoring, data management and, 296–298
putative, steps necessary for identification, 292–296
selection of populations to study, 299–302
Mutation
basic facts concerning, 260–262
design of a program for detection of, general considerations, 263–265
reasons to use 2-D gels for study of, 262–263
selection of most appropriate material for study, 265–269
study, technical considerations, 269–272

N

Negative image, silver staining procedure, 95–96
Neuraminidase, charge train proteins and, 118–119
Nitrocellulose membrane, electrophoretic transfer of proteins to, 130, 132, 135–136, 210
materials and reagents, 139–140
options, 141–142
posttransfer treatment, 141
removal of probes from, 153–154
transfer conditions, 140
Nomenclature, for proteins on 2-D gels
basic considerations, 282–283
designation of variants, 283–288
subdivision of gel into constellations, 283
Nonequilibrium gels, preparation and running of, 11
Nylon membrane, positively charged
electrophoretic transfer of proteins to, 132, 136–137
materials and reagents, 145
options, 146
posttransfer treatment, 146
removal of probes from, 154–155
transfer conditions, 145

O

OBSERVTION form, QUEST system and, 74
Operons, multigene, control of, 440

P

Pancreatic fluid, 2-DGE of, 233–234
Parentage, determination of, 292
Peptide mapping, methods, 170–171
one-dimensional, 171–173
two-dimensional, 174–177
Perspectives, for QUEST system, 88–90
pH, gel analysis and, 41–44
Phenotypes, biochemical, coordinate regulation in plants, 403–405
Phosphoproteins, transformation-sensitive, in cultured cells, 337–341
Phosphorylation, of HeLa cell proteins, 473–474
Photography, of gels, 204–205
Plants
biochemical phenotypes, coordinate regulation of, 403–405
gene expression in
cotton cotyledons, 406–407
early embryogenesis, 398
germinating seeds, 407–411
late embryogeny, 406
somatic embryogenesis, 399–403
somatic embryogenesis in, 399

Plasma
  alterations in reference pattern, 214, 216–220
  reference serum protein map, 212–216
  sample preparation, 211–212
  specimen procurement, 211
  storage of, 197
Plasma proteins, posttranslational modification of, 266
Plasmids, identification of *E. coli* proteins and, 426, 441
Platelets
  2-DGE and, 229–230
  study of mutations and, 266
Platelet disorders, protein deficiencies and, 254
Polymorphism, of proteins, detection of, 242–248
Polypeptides
  silver staining procedures
    colored protein images and, 104
    modified positive image stain, 98
    negative image stain, 95–96
    other image intensification and destaining procedures, 99–100
    positive image stain, 96–98
    quenching of radioactivity and, 103–104
    recycling for increased sensitivity, 99
    sensitivity, quantitation and detection, 100–103
  synthesis in human sarcomas and normal tissues, 343–346
  transformation-sensitive, of HeLa cells, 474
Polypeptide modulation, characteristic pattern of, 251–254
Positive image
  modified silver staining procedure, 98
  silver staining procedure and, 96–98
Premature termination, mistranslation and, 378–379
Preservatives, urine specimens and, 221
Processing times, QUEST system and, 86–87
Program descriptions, for QUEST system, 75–78
  assignment of species spot numbers, 83–84
  automatic matching program, 80–82
  future program enhancements, 87–88
  processing time and storage requirements, 86–87
  programs for final data reduction, 84–86
  programs for spot detection and quantitation, 78–80
  programs to create and update standards, 82–83
Prostatic carcinoma, urinary proteins and, 225
Protein(s)
  assay of, 209–210
  complex mixtures, comparison of, 24
  concentration by lyophilization, 209
  electrophoretic transfer of
    to diazo paper, 142–145
    to nitrocellulose membrane, 139–142
    to positively charged nylon membrane, 145–146
    preparation of 2-D polyacrylamide gels for transfer, 132–134
    principle and relevant factors for optimizing transfer, 131
    transfer sheets and apparatus, 134–139
  embryo specific, synthesis of, 402
  error-containing, detection of, 364–373
  loss, in 2-DGE, 207–209
  mistranslated, rapid degradation of, 389
  modulation by tunicamycin, GELCODE system and, 120
  in posttranslationally modified charge trains, GELCODE system and, 118–119
  serum, reference map, 212–214
  specific, detection by protein transfer analysis, 155–158
  urinary, origin of, 223–225
  variant
    qualitative, 286–287
    quantitative, 287–288
Protein data bases, computerized, problems to be faced in development of, 39–40
Protein kinase, activity, of oncogenes, 309
Protein mutations, 2-DGE and, 248–251
Protein polymorphism
  detection of, 242–248

Protein synthesis
accuracy, effect of stringent control system of bacteria on, 386–388
quantification of error frequency in, 373–379
Protein transfer analysis
examples and applications of
detection of specific proteins, 155–158
potential of method, 158–162
probing the transfer
alternative probes, 153
chromogenic system for visualization, 149–150
principle and relevant factors for optimization of detection, 146–148
radiolabeled components for visualization, 150–153
procedure
development of, 129–131
general description of, 128–129
removal of probes
from diazo paper, 154
from nitrocellulose membrane, 153–154
from positively charged nylon membrane, 154–155
Pseudouridylate synthetase, mutations, mistranslation and, 389

Q

QSTACP
function of, 47
programs, function of, 78
QST data management system, QUEST system user and, 46–49
Quantitation, silver staining and, 100–103
QUEST system
data reduction to GELSPOTS and SMPSPOTS, 65–67
program descriptions, 75–78
assignment of species spot numbers, 83–84
automatic matching program, 80–82
future program enhancements, 87–88
processing time and storage requirements, 86–87
programs for final data reduction, 84–86
programs for spot detection and quantitation, 78–80
programs to create and update standards, 82–83
as seen by user, 45–46
analysis of reduced data, 67–75
data reduction to GELSPOTS and SMPSPOTS, 65–67
image analysis facility, 50–58
IPMANAGER, 67
matching procedures, 60–65
QST data management system, 46–49
QUEST-2D gel laboratory, 49–50
scanning and spot quantitation, 58–60
strategy for, 40–45
QUEST-2D gel laboratory, procedures, 49–50

R

Radioactivity
percentage of that initially applied recovered in second dimension, 22–23
quenching, silver staining and, 103–104
TCA-precipitable, determination of, 8
Radioiodination, of proteins, 177, 204, 210
Radiolabeled components, immunoprobing and
detection by autoradiography, 152
options; 152–153
reagents and materials, 150–151
treatment with antibody, 151–152
Reagents, GELCODE system and, 116–117
RECONSILE process, QUEST system and, 65–66
Recycling, for increased sensitivity, silver staining and, 99
Reducing solution, GELCODE system and, 115
Renal disease, urinary proteins and, 225
Retroviruses, oncogenes and, 308–309
Rheumatoid arthritis, urinary proteins and, 225
Ribosomes, mistranslations and, 378–379

S

Saliva, 2-DGE of, 234
Sample(s), preparation for a 2-D gel electrophoresis, 7–8
Sample comparisons, QUEST system and, 69–72
Scanning and spot quantitation, QUEST system and, 58–60
Second dimensions
  common problems, 32–34
  procedures for gel electrophoresis
    application of first dimension on top of second dimension and running of slab gel, 13
    preparation of slab gel, 11–12
Semen, 2-DGE of, 233
Sensitivity
  of detection, 2-D gel electrophoresis, 21–22
  silver staining and, 100–103
Serine, starvation, mistranslation and, 391
Serum
  alterations in reference pattern, 214, 216–220
  reference serum protein map, 212–216
  sample preparation, 211–212
  specimen procurement, 211
  storage of, 197
Silver staining
  historical background, 94–95
  mechanisms of, 104–107
  potential and real problems in, 107
  procedures
    colored protein images and, 104
    modified positive image stain, 98
    negative image stain, 95–96
    other image intensification and destaining procedures, 99–100
    positive image stain, 96–98
    quenching of radioactivity and, 103–104
    recycling for increased sensitivity, 99
    sensitivity, quantitation and detection, 100–103
  use in clinical laboratory, 203–204
Slab gels, simplifying assembly and casting of, 18–19
Software, for 2-D gel electrophoresis data analysis, 40, 41
Solid tissue, 2-DGE and
  brain, 232–233
  colon and colorectal cancer, 231
  kidney, 231–232
  muscle, 233
  need for histological analysis, 230
  sample preparation, 230
Solubilization, of blood fractions, 269–270, 271
Solutions and buffers, for 2-D gel electrophoresis, 6–7
Specimens, clinical, storage and management of, 197–198
Species spot numbers, assignment of, QUEST system and, 83–84
Spinal fluid, 2-DGE of, 234
Spots
  detection and quantitation, programs for, 78–80
  location, statistical analysis of factors influencing, 272–274
  overlapping, GELCODE system and, 117–118
  quantitation, 2-D gel electrophoresis and, 15
  selection for study of mutation, 268
Spot maps, plotted, QUEST system and, 74–75
Spot sets, QUEST system and, 72–73
Staining, of gels for study of mutation, 270, 271–272
Standards, programs to create and update, QUEST system and, 82–83
Starvation, for certain amino acids, aberrant protein synthesis and, 364–371
Stoichiometric staining, GELCODE system and
  BioImage system, 122–123
  integrated intensity versus concentration, 123–125
Storage, of 2-D gel electrophoresis samples, 15–16
Storage requirements, QUEST system and, 86–87
Streaking, GELCODE system and, 117

Streptomycin, ribosomal fidelity and, 371, 389, 391
Subcellular fractionation, GELCODE system and, 122
SV40 virus, transformation by, mistranslation and, 384

T

Thin layer plates
 ascending chromatography in second dimension, 176–177
 electrophoresis in first dimension, 174–175
Tissue specimens, storage of, 197
Transfer sheets and apparatus, for electrophoretic transfer of proteins
 assembling the sandwich, 138–139
 selection of sheets, 134–137
 transfer chamber and power source, 134–137
Transformation, mistranslation and, 384
Transformation-sensitive polypeptides
 in cultured human cells, working with small numbers of cells, 311–320
 synthesis
  in giant HeLa cells, 320–321
  in growing and senescent human fibroblasts, 321–323
Transposons, identification of *E. coli* proteins and, 419, 421
Trisomy 21, secondary protein modulations in, 251–253
Triton X-100, cytoskeletons and supernatants, distribution of HeLa cell proteins in, 469–471
Tropomyosin, changes in synthesis in transformed cells, 332–333
Tryptophan, starvation, premature termination and, 378
Tumor cells, human, tyrosine phosphorylation of proteins in, 341–342
Tunicamycin, modulation of proteins by, GELCODE system and, 120
Two-dimensional gel electrophoresis
 experimental procedures
  determination of TCA-precipitable radioactivity in samples, 8
  for first dimension, 8–11
  fluorography, 13–15
  labeling of cells with [^{35}S]methionine, 7
  quantitation of spots, 15
  sample preparation, 7–8
  for second dimension, 11–13
 hints and comments
  common problems, 28–34
  comparison of complex protein mixtures, 24
  effects of storage or heating, 15–16
  equilibration time of first dimension, 18
  fluorography, 19–21
  loading capacity, 17
  percentage of applied radioactivity recovered in second dimension, 22–23
  sensitivity of detection, 21–22
  simple device for casting first dimension gels, 17
  simplifying assembly and casting of slab gels, 18–19
  use of coordinates and markers, 24–28
 materials, 5
 optimized procedure, 41
 stock solutions and buffers
  for first dimension, 5–6
  for second dimension, 6–7
Two-dimensional polyacrylamide gels, preparation for transfer
 first dimension, 132–133
 preequilibration, 133–134
 second dimension, 133
Two out of three reading, of genetic code, 390–391
Tyrosine phosphorylation, of proteins in A431 human tumor cells, 341–342

U

Urine
 alterations in reference pattern, 225–226
 reference urine map, 222–225
 sample preparation, 221–222
 specimen procurement, 220–221
 storage of, 197
 tracking proteins in, GELCODE system and, 119–120

V

Variants, of human plasma proteins, 275–282
  designation of, 283–288
Vimentin, changes in synthesis in transformed cells, 323–326
Visualization, in clinical laboratory
  Coomassie blue stain, 202–203
  photography, 204–205
  radiolabeling, autoradiography and fluorography, 204
  silver stain, 203–204